Animal behaviour

Psychobiology, ethology and evolution

DAVID McFARLAND

Third Edition

PEARSON
Prentice
Hall

Harlow, England • London • New York • Boston • San Francisco • Toronto
Sydney • Tokyo • Singapore • Hong Kong • Seoul • Taipei • New Delhi
Cape Town • Madrid • Mexico City • Amsterdam • Munich • Paris • Milan

Pearson Education Limited
Edinburgh Gate
Harlow
Essex CM20 2JE
England

and Associated Companies throughout the world

Visit us on the World Wide Web at:
www.pearsoned.co.uk

First Published 1985
Second Edition Published 1993
This Edition Published 1999

ISBN 978-0-582-32732-0

British Library Cataloguing-in-Publication Data
A catalogue record for this book is available from the British Library

Library of Congress Cataloging-in-Publication Data
A catalog entry for this title is available from the Library of Congress.

10   9   8   7
11   10   09   08   07

Set by 3 in Linotype Meridien 9.5/12
Printed in Malaysia, VVP

Contents

Preface to the third edition

The modern study of animal behaviour has its roots in three different lines of scientific thought: the psychological and physiological and zoological. These three traditions have given rise to comparative psychology, physiological psychology, ethology and sociobiology. This lattice of intersecting fields can bewilder the new student, and may lead to the temptation to study only one of these aspects of animal behaviour. Such early specialization is unfortunate. The biology student who is ignorant of the work of comparative psychologists loses as much as the psychology student who fails to learn about the evolutionary approach.

This book provides an integrated introductory treatment of the entire domain of animal behaviour. It is intended for both biology and psychology students enrolled in their first course in animal behaviour. As in the first edition, I have been careful to define and develop each new concept without reliance on previous coursework, in recognition of the diversity of the students' backgrounds in the natural sciences and psychology.

In this third edition, some chapters remain largely unaltered. These are mainly those providing background material in the fields of genetics, psychology and physiology. Those chapters relating directly to animal behaviour have been considerably updated and reorganized. New chapters on primate social behaviour, communication and animal economics were introduced in the second edition. In the third edition, there are new chapters on behavioural ecology, human ethology and animal robotics.

Many sections have been updated, including those on decision-making, evolutionary optimality, foraging, hormones, imprinting, navigation, visual recognition, and animal welfare.

As in the first edition, each of the three parts of the book, and each section, is prefaced by an introductory overview. Each section is headed by a profile of the life of a scientist whose work is central to the subject treated within the section. In addition, a set of points to remember and a list of suggested further reading appears at the end of each chapter. Original work has been cited directly throughout the text, and is fully referenced in a list at the back of the book.

DAVID McFARLAND
Oxford, January 1998

Acknowledgements

We are grateful to the following for permission to reproduce copyright material:

Academic Press Inc. for Figs 16.5 (After Adolph, 1972), 16.10 (Berthold, 1974) & 19.8 (Rescorla, 1971); Academic Press Limited for Figs 12.21 & 12.22 (Dawkins, 1971), 12.24 & 12.25 (Baerends & Kruijt, 1973), 14.14 & 16.11 (Schmidt-Koenig, 1979), 14.12 (von Holst, 1954), 20.9 (Bateson, 1979), 21.11 (Moynihan, 1955), 21.12 (Blurton & Jones, 1979) & Box 9.4 (Kacelnik, 1977); Addison Wesley Longman and the author, Dr. M. Edmunds for Figs 7.1–7.3, 7.5, 21.16 & 21.17 (Edmunds, 1974); Allyn & Bacon for Fig. 26.1 (Lieberman, 1975); American Association for the Advancement of Science for Figs 4.2 & 4.3 (Hirsch, 1963) 17.2 & 17.4 (Schneiderman *et al.*, 1962); American Psychological Association for Fig. 17.3 (Braun & Geiselhart, 1959); Dr. G. P. Baerends for Fig. 12.23; Prof. P. Bateson for Fig. 20.10 (Bateson *et al.*, 1980); Prof. D. Bentley for Figs 4.5 & 4.6 (Bentley & Hoy, 1972); Blackwell (Berlin) for Fig. 12.30 (von Ferson & Delius, 1989); Blackwell Science Ltd. for Figs 6.10, 7.6 & 11.4 (Perins, 1965), 9.3 & 25.2 (Wehner, 1997), 16.7 & 24.1 (After Goss-Custard, 1977), Boxes 9.3 and 21.1 (Krebs and Davies, 1993), Box 16.1 (Bligh, 1976); Prof. B. J. Le Boeuf for Fig. 8.4; E. J. Brill (Leiden) for Figs 6.2–6.5 (Hoogland *et al.*, 1957) 15.16 (Baerends, 1955), 21.2 (Tinbergen & Perdeck, 1950) & Box 9.1 (Zach, 1979); Prof. H. J. Brockmann for Fig. 7.8; Prof. L. P. Brower for Fig. 7.4; Bruce Coleman Collection for Section Figs 1.1 & 1.3 & Fig. 27.6; Cambridge University Press for Figs 19.5 & 19.6 (Dickenson, 1980) & 21.13 (Catchpole & Slater, 1995); The Company of Biologists Ltd. for Figs 16.8 (Pengelley & Asmundsen, 1974) & 17.5 (Rowell & Horn, 1968); Croom Helm for Fig. 10.5 (Dunbar, 1988); N. Davies for Figs 6.11 & 6.12 (Davies & Halliday, 1977) & 21.19 (Davies & Brook, 1989); Prof. Dr. J. Delius for Figs 12.28 (Lombardi & Delius, 1989) & 12.29 (After Delius, 1986); Elsevier Science for Fig. 22.4 (Kendon, 1967); Emmet Spier for Figs 25.13, 25.14 & Box 25.4; Getty Images Ltd. for portrait photographs of Claude Bernard, Gregor Mendel, Johannes Muller & Ivan Pavlov; Harvard University Press for Figs 23.4 & 23.5 (Lindauer, 1961), 24.9 & 24.12 (Heinrich, 1970). Harvard University Press, Copyright © 1961 & 1979 by the President and Fellows of Harvard College; Heather Angel for Fig. 27.4; John Murray Ltd. for Fig. 21.3; John Wiley & Sons for Figs 15.9 (Toates, 1980), 15.11, 15.12 & Box 15.1 (Toates, 1995); Prof. A. Kacelnik for Box 9.2 (Kacelnik, 1979); Kluwer Academic Publishers incorporating Chapman & Hall and Rapid Science for Fig. 13.1 (Tansley, 1965); Dr. H. Kruuk for Fig. 8.9; Mary Evan's Picture Library for the portrait photograph of Charles Darwin; The MIT Press for Box 23.1 (Maes); Dr. B. Moore for Fig. 18.2; National

Portrait Gallery for the portrait photograph of Sir Ronald Fisher; Oxford Scientific Films for Fig. 27.5; Oxford University Press for Figs 6.7 (Fisher, 1930), 15.2 (Schmidt-Nielsen, 1964), 20.2 & 21.8 (Tinbergen, 1951). Reproduced by the permission of Oxford University Press; Dr. I. Patterson for Fig. 6.6 (Patterson, 1965); Pergamon Press Ltd. for Fig. 15.10 (Wiepkema *et al.*, 1972) © Pergamon Press Ltd; Physiological Society for Fig. 11.2 (Hodgkin & Huxley, 1945); Plenum Publishing Corporation and the author, Prof. S. Daan for Figs 16.16 & 16.17 (Daan, 1981); Prentice-Hall, Inc. for Fig. 15.6 (Wilson, © 1979); Princeton University Press for Figs 9.2 & 14.7 (Stephens & Krebs, 1986); Princeton University Press and the author, N. Collias for Fig. 9.4; F. Schutz for Fig. 20.7; Scientific American and the author, I. T. Garber for Fig. 3.1 (Adapted from Scheller & Axel, 1984); Scientific American and the author, N. Prentis for Fig. 13.3 (Knudsen, 1981); Sinaur Associates, Inc. for Fig. 5.4; Prof. D. Singh for Fig. 22.5 (Singh, 1994); S. Karger AG, Basel for Figs 12.16 (Kaas *et al.*, 1972) & 22.6 (Passingham, 1975); J. Sparks for Fig. 20.1 (Sparks, 1982); E. Spier for Figs 25.13 & 25.14 & Box 25.4 (Spier, 1997); Springer-Verlag GmbH & Co. KG and the author, Prof. J. P. Ewert for Figs 13.6–13.8 & 13.10–13.12 (Ewert, 1980); R. Squibb for Fig. 5.12; University of Chicago Press for Figs 3.3 (Gottlieb, 1971), 4.8 (Blome, 1966), 4.9 & Box 4.1 (Photographs by J. Scott), 10.7 (Hinde, 1987); F. Webster for Section 2.1 photograph and Fig. 12.5; Weidenfield & Nicolson Ltd. for Fig. 16.14 (Cloudsley-Thompson, 1980); W. H. Freeman for Figs 2.3 (McClean & DeFries, 1973), 2.7 (Stern, 1973), 5.6 (Cavalli-Sforza, 1976), 10.1 & 11.21 (Passingham, 1982) & 11.15 (Bullock, 1977); W. W. Norton & Company, Inc. for Figs 12.6, 23.1 & 23.11 (Gould, 1982); The Zoological Society of San Diego for Figs 10.3 & 10.4.

Whilst every effort has been made to trace the owners of copyright material, in a few cases this has proved impossible and we take this opportunity to offer our apologies to any copyright holders whose rights we may have unwittingly infringed.

CHAPTER

Introduction to the study of animal behaviour

1.1 Interpreting animal behaviour

1.2 Historical outline

1.3 How this book is organized

The scientific study of animal behaviour involves a variety of approaches. Behaviour can be explained in terms of its evolutionary history, in terms of the benefits that it brings to the animal, in terms of psychological mechanisms, and in terms of physiological mechanisms. Which approach you take depends upon what you want to know about animal behaviour. In this book we explore the numerous ways of investigating animal behaviour as well as many different aspects of behaviour, from the simple responses of primitive animals to the mental life of the great apes.

1.1 Interpreting animal behaviour

Niko Tinbergen, a pioneer ethologist, distinguished four types of answer to 'why' questions about animal behaviour.

1. Why do animals respond to environmental stimuli in a particular way?
2. Why do animals respond to internal stimuli in a particular way?
3. Why do some animals respond in one way and others in another way to the same situation?
4. Why do animals of a particular species, or group, characteristically behave in particular ways in particular situations?

The answers to these questions place the emphasis on different aspects of the biological context within which animal behaviour occurs. Let us illustrate this approach by looking at a particular example.

Acorn woodpeckers (*Melanerpes formicivorus*) have been studied at the Hastings Natural History Reservation, located some 40 km southeast of Monterey, California, since 1971 (MacRoberts and MacRoberts, 1976). They are distinctive birds, which nest in holes in trees. They often breed in groups containing mate-sharing males and joint-nesting females (cobreeders). Cobreeders are usually closely related and thus, in addition to raising their own young, they sometimes raise the young of relatives. Walter Koenig of the University of California (Koenig, 1990) carried out a study of these interesting birds, which we will refer to a number of times in this book. He performed some experiments, which involved removing male birds from their natural habitat, prior to nesting, and keeping them in captivity until the group produced a nest and clutch of eggs, and then releasing them into their home region. He discovered that, when they were returned, dominant males destroyed the eggs in the nest (see Figure 1.1), but subordinate males did not do this. This study is interesting because it raises various questions that are important for animal behaviour studies in general.

A number of questions immediately spring to mind. What makes the male birds destroy the eggs? Do they think that the chicks in the eggs have been fathered by some other male? Why do subordinate males not behave in the same way as dominant males? What does it mean to say: that an animal knows something? These are all questions of the type that we will be discussing in this book.

In answering the first of Tinbergen's questions, the emphasis is on the environmental stimuli that the animal responds to in a given situation. In the case of the acorn woodpecker, we might want to know what it is about the eggs that induces the male to remove them from his nest. To answer this aspect of the question it would be necessary to show what stimulus characteristics of the egg, such as its shape, colour and markings, are important in eliciting the removal response. Many experiments have been done on the mechanisms of egg recognition in birds, and shape, size and coloration have been shown to be important, as we see in Chapter 12.

The second question emphasizes what the birds do. They remove, rather than incubate, or eat, the eggs. Here the answer must be given in terms of the animals' motivation. Thus, birds sit on eggs when they are broody, but they may eat them when they are hungry. The scientist must define what he/she means by broody and by hungry, and this task will require knowledge of the physiology of the animal. In the case of the acorn woodpecker we would want to know what motivates some males to eject the eggs.

Whereas questions 1 and 2 require answers in terms of immediate cause or mechanism, question 3 implies causality of a different order. Why do dominant males rather than subordinate males reject the eggs? Why do acorn woodpeckers show this type of behaviour, whereas some other species do

Fig. 1.1 Acorn woodpecker ejecting an egg from its nest (Nick Davies).

not? This is basically a question of how a particular activity gets into the repertoire of an individual. Part of the answer is that birds have a hereditary predisposition to produce eggs and to develop the behavioural skills to respond to them appropriately. The genetic factors involved are ultimately necessary for the development of the behaviour towards eggs that is characteristic of each species. The manner in which such traits are handed from one generation to the next is the subject of the study of behaviour genetics, an important part of the study of behaviour.

Another part of the answer to questions about the behavioural repertoire of an individual animal has to do with experience. Some individuals may experience circumstances in early life, or in adulthood, which cause them to learn certain types of behaviour. In the case of the acorn woodpecker, we may want to know whether individuals learn to eject eggs, whether they know the eggs are not theirs, etc.

In answering the fourth type of question, we might say that a male acorn woodpecker ejects the eggs in order to induce the female to lay more eggs which he has fertilized. This need not mean that such birds anticipate the consequences of their behaviour, but simply that birds that are programmed to behave in this way leave more offspring than those that are not so programmed. This type of argument, first put forward by Charles Darwin, implies that the survival value of any inherited trait is determined by natural selection; that is, the extent to which a trait is passed from one generation to the next in a wild population is determined by the breeding success of the parent generation and the value of the trait in enabling the animals to survive natural hazards such as food shortage, predators and sexual rivals.

Thus, in answering questions about behaviour, the scientist can take several points of view. In general, psychologists are interested in mechanisms that control behaviour, and evolutionary biologists are interested in why the mechanisms came to be as they are. Ethologists believe that the distinction between mechanism and design is fundamental to the study of animal behaviour. Birds behave towards eggs in certain ways because certain mechanisms cause them to do so. They are designed (by natural selection) to behave in those ways, because the behaviour fulfils a certain function which is important to their survival and reproduction. Questions about both design and mechanism are essential for a full understanding of animal behaviour.

In this book we follow the ethological viewpoint in emphasizing the importance of both mechanistic and evolutionary explanations of behaviour. Another way of appreciating the distinction is to consider how animals adapt to changes in environmental conditions. Animals can adapt in a variety of ways. First, we should distinguish between genotypic adaptation, in which the adjustment is genetic and takes place through evolution by natural selection, and phenotypic adaptation, which takes place within the individual animal on a non-genetic basis.

Evolutionary adaptations among birds sitting upon eggs can be seen by comparing the nest behaviour of birds that nest on cliff ledges with that of their close relatives that nest on relatively flat ground, as we do in Chapter 5. When we say that kittiwakes adapt to cliff nesting by adopting various behaviour patterns, we mean that those individuals which employ the behaviour patterns in a cliff-nesting situation have more surviving offspring than those that do not show the appropriate behaviour to the same extent. Phenotypic adaptation involves processes such as learning, maturation and temporary physiological adjustment. For example, a gull incubating its eggs on a cold windy day will face into the wind, sit tight on its eggs, and increase its heat production. On a hot sunny day it will cover the eggs just sufficiently to keep them shaded, and it will pant, spread its wings and show other forms of cooling behaviour. These are all short-term behavioural and physiological adaptations to the prevailing weather.

Some birds are able to learn to recognize their own eggs by their colour patterning (Baerends and Drent, 1982). This form of adaptation by learning may enable them to adjust to changes in the nest situation due to predation, or other forms of disturbance. The ability of individuals to respond appropriately to changing circumstances by learning or physiological adaptation is an important feature of animals that live in variable environments.

Although biologists use the term adaptation in many different ways, they are usually careful to distinguish between evolutionary or genotypic adaptation and individual or phenotypic adaptation. During his voyage on the *Beagle*, Darwin noticed many of the beautiful adaptations that are typical of particular environments, and he came to regard these as evidence for evolution, as we see in Box 1.1.

The Galapagos finches not only illustrate adaptive radiation but also show evolutionary convergence with species in other parts of the world. Thus, we speak of warbler finches and woodpecker finches because we recognize the similarities between these and true warblers and woodpeckers. Convergent evolution occurs when different species inhabit similar environments. It sometimes results in remarkable similarities in the appearance and behaviour of unrelated species, as illustrated in Figure 1.2.

Fig. 1.2 Convergent evolution leads to similarities in appearance and behaviour of unrelated species. The examples shown here are (a) the eastern meadow-lark (*Sturnella magna*) from America and (b) the yellow-throated longclaw (*Macronyx croceus*) from Africa; (c) an Australian wombat (*Phascolonus ursinus*), and (d) an American woodchuck (*Marmota monax*) (From *The Oxford Companion to Animal Behaviour*, Oxford University Press, 1981).

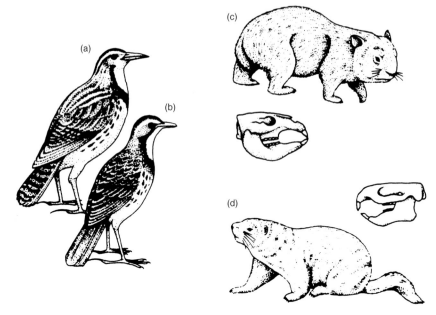

Box 1.1 Adaptive radiation of the Galapagos finches

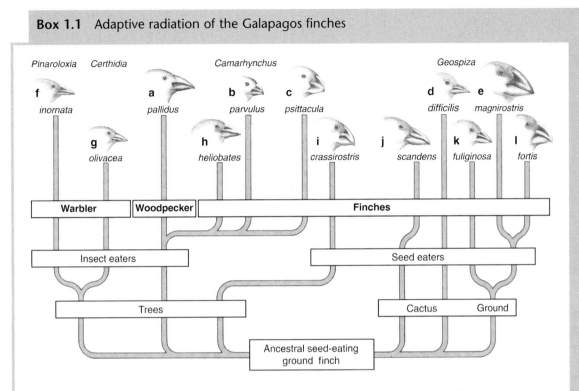

a. *Camarhynchus pallidus*
b. *Camarhynchus parvulus*
c. *Camarhynchus psittacula*
d. *Geospiza difficilis*
e. *Geospiza magnirostris*
f. *Pinarolaxia inornata*
g. *Certhidia olivacea*
h. *Camarhynchus heliobates*
i. *Camarhynchus crassirostris*
j. *Geospiza scandens*
k. *Geospiza fuliginosa*
l. *Geospiza fortis*
m. *Camarhynchus pauper*
n. *Geospiza coniostris*

Adaptive radiation of Darwin's finches from an ancestral seed-eating ground finch. Isolation on the Galapagos Islands led to intense competition and to subsequent dietary specialization. This has given rise to the different types of beak found among the present-day species.

The Galapagos finches have diversified considerably during their evolution from a common ancestor. The differences among the various types are undoubtedly adaptive, as can be seen by comparing them with other birds. This applies particularly to their beaks, but it is also true of many related aspects of their behaviour. Thus, the insectivorous tree finches move with agility among tree branches like titmice, the woodpecker finch climbs vertical trunks and probes in crevices, the warbler finches have the quick flitting movements of a true warbler and the ground finches hop about on the ground. Originally all like finches, the Galapagos finches have become like tits, woodpeckers or warblers, in accordance with their adopted lifestyles. The formation of a number of new species from a single parent species, as a result of adaptation to a variety of habitats, is called adaptive radiation.

1.2 Historical outline

Although the scientific study of animal behaviour has its origins in the work of eighteenth century naturalists such as Gilbert White (1720–1793) and Charles Leroy (1723–1789), it is Charles Darwin (1809–1882) who is

Box 1.1 *(cont)*

regarded as the father of the scientific study of animal behaviour. Darwin influenced the development of ethology in three main ways. First and foremost, his theory of natural selection set the stage for consideration of animal behaviour in evolutionary terms, a key aspect of modern ethology. Second, Darwin's views on instinct can be regarded as a direct forerunner of those of the founders of classical ethology. Third, Darwin's observations on behaviour were important, especially those that stemmed from his belief in the evolutionary continuity of man and other animals. In his *Descent of Man and Selection of Relation to Sex* (1871), for instance, Darwin writes:

'We have seen that the senses and intuitions, the various emotions and faculties such as love, memory, attention, curiosity, imitation, reason, etc., of which man boasts, may be found in an incipient or even sometimes a well developed condition in the lower animals.'

In his book *The Expression of the Emotions in Man and Animals* (1872), Darwin elaborates on this theme:

'With mankind some expressions, such as bristling of the hair under the influence of extreme terror, or the uncovering of the teeth under that of furious rage, can hardly be understood, except on the belief that man once existed in a much lower and animal-like condition.'

Darwin's friend and disciple George Romanes energetically continued Darwin's work on animal behaviour, and his *Animal Intelligence* (1882) was the first general treatise on comparative psychology. However, Romanes was not very critical in his evaluation of evidence. He credited animals with mental abilities, such as reasoning, and with feelings such as jealousy. This led to a revolt, led by Conway Lloyd Morgan, whose views on objectivity inspired the behaviourist school of psychology. In his *Introduction to Comparative Psychology* (1894) Morgan enunciated his famous canon:

'In no case may we interpret an action as the outcome of the exercise of a higher psychical faculty, if it can be interpreted as the outcome of one which stands lower in the psychological scale.'

This attitude resulted in a vast improvement in the control of experiments and the assessment of evidence.

This more sceptical approach to animal behaviour laid the foundations for the behaviourist school of psychology, initiated by John Watson (1913). The behaviourist view is that psychology is the study of behaviour itself, rather than mental events. In its strong form, behaviourism rejects all reference to inner processes in the explanation of behaviour. This view was very influential in the development of American psychology. It led to considerable improvements in experimental technique and the interpretation of behavioural evidence. However, the behaviourist approach was criticized, particularly by the European ethologists Konrad Lorenz and Niko Tinbergen, for being too sterile and divorced from reality. The ethologists thought that animal behaviour should be studied in the natural environment and not in the laboratory. Lorenz emphasized the importance of acute observation and Tinbergen showed that meaningful experiments could be carried out under natural conditions.

The modern approach to animal behaviour includes many features derived from both the behaviourist and the early ethological views. In addition, modern ethology draws upon the physiological tradition, with its emphasis on the explanation of behaviour in terms of the activity of the nervous system. In this book we will study all three approaches to animal behaviour.

1.3	## How this book is organized

This book is divided into three major parts. In the first part we look at the evolutionary approach to behaviour, in the second to causal mechanisms of behaviour, and in the third part we examine complex behaviour using both these approaches. The purpose of this structure is to give you a full understanding of one mode of explanation before moving on to another. Too often students remain confused between mechanistic and design based explanations, and are unable to follow discussion of topics in which the two are inevitably mixed. Accordingly, treatment of such topics is reserved for the third part of the book.

There are 28 chapters, organized into groups of three. Each group of chapters covers a major aspect of the study of animal behaviour, and is preceded by a profile of an eminent scientist who played an important role in the development of the subject. The aim is to give a feeling for the continuity of the science to which you, now reading this book, may contribute in the future.

The arrangement of chapters in groups makes it a simple matter to skip over a particular group if you have already been well grounded in a given field of study. Many topics are introduced at an elementary level early in the book and then taken up at a more advanced level later. This arrangement is to encourage flexibility in the use of the book by students coming to the study of animal behaviour from different backgrounds. In addition to this arrangement of chapters into groups concentrating on particular topics, the book is also designed to introduce you to the three main ways of studying behaviour.

Animal behaviour can be studied in the natural environment and in the laboratory. In the natural environment the animal is free to express its full range of behaviour, and this is discovered primarily through observation. Direct visual observation is the traditional approach, but in recent years this has been supplemented by indirect methods which rely on technological developments such as audio-recording, radio-telemetry, etc. The natural behaviour of animals can also be investigated by experiments involving selective interference in the environment. We see examples of this in Chapters 6 and 23.

In the laboratory the scientist has considerable control over the animal's environment and can thus design careful tests of particular hypotheses about behaviour. This approach is particularly important in investigating the sensory capabilities of animals, as we see in Chapters 12 and 13, and in the study of animal learning (Chapters 17 and 18). In addition to behavioural investigation in the laboratory, physiological experiments are making an increasing contribution to our understanding of behaviour. The physiological approach to behaviour has long been advocated but only in recent years has it made a substantial contribution to behaviour studies. We see examples of this approach in Chapters 11 and 13. Animal behaviour is very complex and a wide range of practical and theoretical approaches is necessary for a good understanding to be achieved. It is this multi-disciplinary approach that makes animal behaviour such an exciting subject to study.

The evolution of behaviour

In the first part of this book we look at animal behaviour from an evolutionary viewpoint. Since the time of Charles Darwin it has become increasingly possible to explain the role of behaviour in the survival and reproduction of the animal. Just as an animal's eyes, ears, legs or wings can be viewed as mechanisms designed to enable the animal to cope with its particular mode of life, so too can the mechanisms controlling behaviour. Thus we can ask why animals have particular behaviour patterns, and expect to find answers in terms of the evolutionary history of the species.

This evolutionary type of explanation requires some knowledge of the genetics and development of behaviour, of the theory of natural selection in relation to the ecology of the animal, and of evolutionary theory and its relevance to the social behaviour of animals. The first three sections of this book are devoted to these three subjects.

Genetics and behaviour

In this group of three chapters we enter the field of behaviour genetics. Chapter 2 deals with elementary Mendelian genetics and the cellular basis of heredity. Chapter 3 outlines some of the issues involved in the study of the development of behaviour, including the problem of innate behaviour. Chapter 4 is devoted to behaviour genetics and includes discussion of the effects of mutations, polygenic inheritance and the heritability of behaviour.

Gregor Mendel (1822–1884)

Born in peasant circumstances, Gregor Mendel entered the Church and was ordained at the age of 25. He trained to be a teacher but failed examinations to qualify as a high-school teacher of natural history. In 1856 he became a pupil in a school of lower rank, where he was much liked by his pupils. Mendel started his researches in 1858 in the garden of a monastery at Brunn, Moravia. His results and theory were presented in a paper read to the Brunn Society of Natural Science in 1865 and published in the proceedings of the Society in 1866. Mendel's work remained unknown to the scientific world at large until it was discovered in 1900 more or less simultaneously by three independent workers – Correns, von Tschermak, and de Vries. Its importance was quickly realized and Mendel's results were soon confirmed and extended. The main body of Mendel's research was carried out with ordinary garden peas. He was particularly interested in characteristics of the peas that could occur in either of two contrasting forms. Thus, the seeds could be red or white, the pods could be inflated or constricted, and green or yellow, the flowers could be axial or terminal, and the stems could be long or short. Mendel sowed every seed separately. He cross-fertilized the resulting plants by hand and kept different hybrids in different plots. Instead of trying to trace individual lineages, as was typical of the time, he had large numbers of plants and simply counted the various types of offspring. On the basis of these studies, Mendel formulated his revolutionary laws of inheritance. In 1866 Mendel wrote to the famous biologist Karl von Nägeli, who was engaged in breeding experiments with hawkweed. Mendel outlined his ideas, asked for help, and offered to work on hawkweed. However, von Nägeli was not impressed with Mendel's theories and suggested that Mendel grow more peas. Mendel had already recorded observations on some 13 000 specimens of pea plant and he tried instead to repeat von Nägeli's work on hawkweed. Hawkweed is small, difficult to fertilize by hand, and unsuitable for genetical work. Mendel persisted for a time, but made no progress. Von Nägeli offered to grow some of Mendel's pea seeds at the Botanic Gardens in Munich, and Mendel sent him 140 packets. Von Nägeli never planted the seeds and did not refer to Mendel in his major work of 1884. It is ironic that von Nägeli, who had been the first to observe and describe chromosomes under the microscope in 1842, never realized the importance of Mendel's work. The connection between chromosomes and Mendelian heridity was not established until 1914.

By 1875 the monastery at Brunn had become embroiled in a tax dispute, and Mendel, who was prelate, had no more time for research. He died in 1884.

Genes and chromosomes

2.1 **Natural selection and behaviour**

2.2 **Mendel's laws**

2.3 **The cellular basis of heredity**

2.4 **Genetic variation**

2.5 **The genetic material**

When Darwin's theory of natural selection was published, the question of how characters were inherited and how variation among the offspring was maintained became of acute scientific interest. Von Nägeli, famous for his work on the cellular structure of tissues, carried out breeding experiments on hawkweed. William Bateson made extensive studies of inheritance and variation, published in 1894. Dutch botanist Hugo de Vries discovered that variants of evening primrose plants arose from normal plants and seemed to breed true. He called these sudden jumps **mutations**. Darwin published *The Variation of Animals and Plants under Domestication* (1868), in which he put forward his theory of pangenesis – that is, each cell in the body throws off minute 'gemmules' containing information about itself. 'These multiply and aggregate themselves into buds and the sexual elements' (p. 481). Not until Mendel's work was discovered in 1900 could the science of genetics be reconciled with the theory of evolution by natural selection and with the growing science of cell physiology.

In this chapter we see how Mendel's work led to our modern understanding of the cellular basis of heredity, of genetic variation and of the nature of the gene. Genetics is of fundamental importance to the study of behaviour, because much animal behaviour is influenced by the animal's genetic make-up. By understanding genetics we can gain insight into the nature of this influence.

2.1 Natural selection and behaviour

Natural selection operates upon the physical characteristics, or **phenotype**, of the individual, including its behaviour. A mouse that did not attempt to escape from predators would be less likely to survive than one that did, and a mouse that did not take advantage of opportunities to obtain food would be similarly disadvantaged. Thus, we can attribute survival value to behaviour patterns, just as we can to the morphological properties of animals.

Moreover, the balance between different behaviour patterns may also be important. For example, the mouse that pays too much attention to food may fail to notice an approaching predator, while the mouse that is too fearful of predators may miss important feeding opportunities. Such considerations lead to two important questions about the relationship between natural selection and behaviour:

1. How can we determine or demonstrate the survival value of particular behaviour patterns?

2. Is there a best way for an animal to divide its time among the many activities possible in a given set of circumstances?

We address the first of these questions in Chapter 6. The second question is a little more complicated and is reserved for Chapter 9.

The effectiveness of natural selection in changing the nature of a population of animals depends upon the degree to which the phenotypic characteristics are inheritable. Although Darwin knew that there are usually variations among the offspring of particular parents and that such variations were essential for his theory, he could not really say how the variations occurred. In 1858 a Moravian monk, Gregor Mendel, started experiments on plant breeding. He described his results in 1865 but in a relatively obscure publication that was not brought to the attention of the scientific world until 1900, after Darwin's death. Mendel demonstrated that heredity is not blending but particulate. Offspring inherit discrete particles that we now call genes. After its rediscovery in 1900, Mendel's work led to the establishment of the genetical theory of natural selection by R.A. Fisher and others in the 1930s. This theory provides the mechanism for natural selection in that it accounts for the variation within a population of reproducing animals.

In the study of animal behaviour, it is important to know to what extent particular behavioural characteristics are inheritable. Not only will this knowledge enable us to estimate the extent to which behaviour traits are subject to selection, but also it highlights the distinction between innate and acquired behaviour, a currently controversial topic with far-reaching implications for human philosophy and politics.

Chapter 4 introduces the subject of behaviour genetics. For our present purposes it is sufficient to understand that the effectiveness of natural selection in influencing the evolution of behaviour depends upon the extent to which the behaviour is under genetic control. The evolution of behaviour is complicated by the fact that natural selection is not always the only important mechanism. In some animals, and especially in humans, behaviour characteristics can evolve by cultural means. Individuals can learn from each other in a variety of ways (see Chapter 27), and information can thus be passed from one generation to the next. The behaviour of birds and mammals is the result of a complex interaction of genetics and experience. To understand this interaction, ethologists must be familiar with the elements of genetics.

2.2 Mendel's laws

Mendel carried out most of his experiments with garden pea plants. In one typical experiment, he crossed plants with red flowers and plants with white flowers. All the offspring had red flowers. He then allowed these offspring to breed freely among themselves and obtained 705 plants with red and 224 plants with white flowers. Two aspects of these results are important. First, the first generation of offspring has no representatives of the white-flowered parent, but the second generation does. Second, the red and white flowers of the second generation appear in a 3:1 ratio. Darwin also experimented with peas and noticed the 3:1 ratio. Mendel, however, established this and other important ratios in numerous repeated experiments. Moreover, he realized the significance of the disappearance of one parental type (white) in the first generation of offspring. Mendel proposed that each pea plant possessed two hereditary factors for flower colour and for each of the other characters. One factor dominates the other so that the flowers have only one colour when two different factors are present. Thus, white-flowered plants contain two white factors, while red-flowered plants may contain two red factors or a red factor and a white factor. This type of reasoning enabled Mendel to account for all the observed ratios of offspring types and to formulate his revolutionary laws of inheritance.

On the basis of his breeding experiments with pea plants, Mendel came to a number of conclusions, noted in the following paragraphs.

1. Inheritance is particulate and the genetic contribution from each parent is equal. The genetic material from one parent cannot mix with, or contaminate, that from the other. Mendel's particulate factors are now called **genes**.

2. Each external character contributing to the structure and appearance of the individual is controlled by a pair of genes, one inherited from each parent. These external characters are called the **phenotype**.

3. Each gene may exist in two or more alternative forms, now called **alleles**. The total collection of genes within the individual is called the **genotype**. Thus, three kinds of combination are possible when there are two alleles. For a gene that takes two forms, A and a, the three possible genotypes that an individual might have are AA, aa and Aa. When an individual carries two copies of the same allele, as in the first two cases, the individual is said to be **homozygous** for the character in question. When an individual carries one copy of each allele, as in the third case, the organism is said to be **heterozygous**.

4. One allele may be dominant over another, in which case the non-dominant allele is called recessive. **Dominant genes** control the nature of the phenotype when they occur in homozygous or heterozygous combination. **Recessive genes** control the phenotype only when they occur in homozygous combination, as illustrated in Figure 2.1.

5. Of the genes making up a pair in a parent only one will be copied and inherited by an offspring. The offspring receives its other copy from its other parent. The result of this is that there may be one phenotype in the first filial or (F1) generation (the first generation following a specific cross) but more

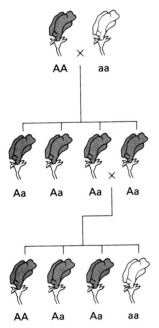

Fig. 2.1 Mendelian table showing crosses between red- and white-flowered peas. Red (A) is dominant over white (a).

than one in the second filial (F2) generation, as illustrated in Figure 2.1. **The law of segregation** (sometimes called **Mendel's first law**) results in typical Mendelian ratios of phenotypes, like the 3:1 ratio depicted in Figure 2.1.

6. When two or more pairs of genes segregate simultaneously, the distribution of any one is independent of the distribution of the others. This is called the **law of independent assortment** (or **Mendel's second law**). The result is that phenotypic characters that are paired in one generation need not be paired in subsequent generations, as shown in Figure 2.2.

Mendel's conclusions apply to the inheritance of characters in many species of plant and animal, but they have not been found to be universally applicable. One source of departure from the classical Mendelian picture is variability in the relationship between the hereditary constitution, or genotype, and the observed character, or phenotype. For most of the characters studied by Mendel, the dominance of gene over its allele was complete. However, we now know that there is a range of possible gene expression, as indicated in Figure 2.3, and that dominance is not always complete.

Fig. 2.2 Mating of two F1, individuals in one of Mendel's hybrid pea crosses. Round (R) is dominant over wrinkled (r), and yellow (Y) is dominant over green (y).

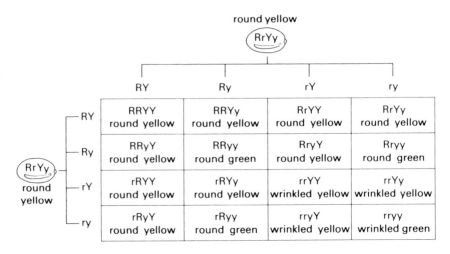

Fig. 2.3 Graphical representation of four different types of gene expression (After McClearn and DeFries, 1973).

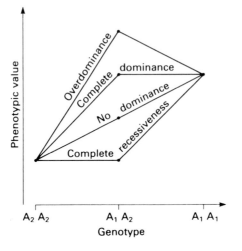

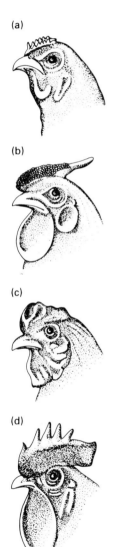

Another complication of the simple Mendelian picture is that a given gene may influence many different characters. This phenomenon is known as **pleiotropism**. The first demonstration of behavioural pleiotropism was by Sturtevant in 1915. Working with the fruit fly *Drosophila*, he found male flies appeared to mate preferentially with mutant females showing either white-eye or yellow-bodied phenotypes. He found conversely that the females of various types preferentially mated with non yellow-bodied and non white-eyed males. These results are due to a pleiotropic effect whereby the white-eye and yellow-body mutations both have the side-effects of lowering the mutant fly's overall activity level. Because the normal active females tend to flee the advances of males, the mutant females are more commonly inseminated. Conversely, because the males must approach females to initiate mating the inactive mutant males had lower success than their normal male counterparts. In other words the mutant genes were pleiotropic in that they influenced more than one character. In a situation where mating was at random, the mutant flies would be selected against on account of their lower sexual activity.

Gene interactions can also produce deviations from simple Mendelian principles. When two or more different genes interact with each other in determining a single character, the ratios obtained from matings may differ from the classical Mendelian ratios. For example, the form of the comb in domestic fowl is determined by the two genes R and P. The dominant gene R produces the rose comb (see Figure 2.4), and its recessive allele r produces the single comb, where homozygous. The dominant gene P produces the pea comb, while the recessive p produces the single comb. When R and P occur together, they interact to produce the walnut comb, which neither could produce on its own.

Another form of gene interaction is the case of **complementary genes**, which are mutually dependent: neither gene can produce its phenotypic effect without the other gene. Sometimes one gene masks the effect of another, a condition known as **epistasis**. Epistasis is similar to dominance, but whereas dominance involves different alleles, epistasis involves different genes.

A **modifier gene** may alter the expression of another gene. Thus, for instance, human eye colour is controlled largely by a single pair of genes. B is responsible for brown eyes and b for blue eyes. The genotypes BB and Bb both produce brown eyes, and bb gives rise to blue eyes. However, due to the action of modifier genes, many variations exist on these two basic eye colours. A modifier gene can switch on the phenotypic effects of other genes.

Some phenotypic characters, like height in humans, are influenced by a large number of genes. The result is that height varies continuously among different members of a population. If only a few genes were involved, we would expect to find that some particular heights are very much more prevalent than others.

Fig. 2.4 Comb types in chickens: (a) pea comb, (b) rose comb, (c) walnut comb, (d) single comb.

2.3   The cellular basis of heredity

The idea that plants and animals are composed of cells developed gradually and was first explicitly stated in 1839 by Schwann, who pointed out the correspondence in the cellular structure and growth of plants and animals. Albrecht von Koelliker was the first to recognize, in 1840, that sperm and ova were cells; it soon became established that all living tissue was made up of cells.

The first recorded observations that cell division included the nucleus and its chromosomes were made in 1842 by von Nägeli, but a detailed account of how the cell nucleus divides awaited improvements in the techniques of microscopy. The first correct account of the behaviour of chromosomes during cell division was made in 1882 by Walther Felmming, who observed the process in the cells of salamander larvae. By 1885 Eduard von Beneden had shown that the chromosomes remained unaltered from one cell division to the next. He discovered that the number of chromosomes is fixed for a given species. In 1856 Pringsheim was the first to see a sperm enter a female cell, but it was Oscar Hertwig in 1876 who realized that, where two nuclei were seen within a fertilized egg, one must have come from the sperm. Thus, he proposed that it is the chromosomes that carry the genetic material, which is provided by both parents at the time of fertilization. A clear account of cell division and fertilization was now possible, and the stage was set for a rapid assimilation of Mendel's work on heredity when it was unearthed in 1900.

The somatic cells of most animals are **diploid** – that is, each cell has two copies of each type of chromosome. Human somatic cells have 46 chromosomes or 23 different pairs, and the fruit fly *Drosophila* has eight chromosomes forming four different pairs. Normal cell division involves **mitosis** and produces diploid cells. Mitosis is characteristic of the formation of all forms of cells except germ cells, which give rise to the male and female gametes, the sperm and egg. Since sperm and egg combine to form a single cell with two chromosome sets (the diploid zygote) each gamete cell carries only one set of chromosomes, and is said to be **haploid**. The formation of the germ cells, therefore, involves a process by which the diploid number of chromosomes is reduced to the haploid number, by a process called **meiosis**.

Meiosis involves two separate cell divisions. The first division achieves a reduction in the number of chromosomes, while the second division is similar to normal mitosis, except that the daughter cells are haploid. Complete meiosis thus results in the division of a single diploid cell into four haploid cells. In male animals, all four haploid cells differentiate into spermatozoa, as illustrated in Figure 2.5. In females, the cells produced by the first cell division are usually of unequal size. The smaller cell is called the **first polar body**, and this may or may not undergo the second division. The larger cell divides into a **second polar body** and a large cell that differentiates into the **ovum** (Figure 2.5). The polar bodies do not survive. Thus, meiosis in males usually produces four spermatozoa of equal size, while in females it produces a single large ovum and two or three non-functional polar bodies.

Fig. 2.5 Schematic representation of spermatogenesis and oogenesis in an animal.

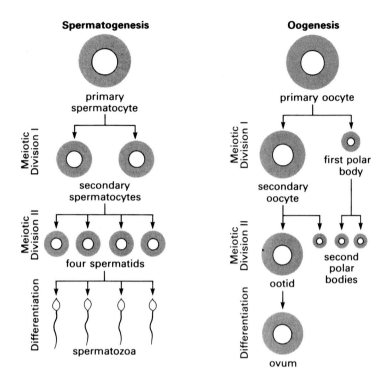

2.4   Genetic variation

The American biologist Thomas Hunt Morgan realized that the phenomenon of mutation, discovered by de Vries, provided a means of investigating genetics. After pilot studies on mice, pigeons and rats, he settled on the fruit fly *Drosophila* as a suitable experimental animal. The fruit fly reaches maturity in twelve days and breeds rapidly. Thirty generations can be bred in twelve months, providing an abundant source of experimental material. Morgan attempted to induce mutations by subjecting the flies to extremes of temperature, radioactivity, etc. At first he had no success, but in 1910 he discovered a male fly with white eyes, in contrast to the red eyes of normal flies. He bred this solitary specimen with normal females and obtained only red-eyed flies in the first generation. In the second generation, however, both red-eyed and white-eyed flies appeared.

Morgan soon identified many more mutants. He discovered that white eye occurred only with yellow wing, never with grey wing. Similarly, black body appeared only with vestigial wing and ebony body only with pink eye. This could be explained if the three pairs of characters each occurred on a different chromosome. Since *Drosophila* had three large chromosome pairs, this interpretation seemed reasonable. Furthermore, in 1914 bent wing appeared, followed by eyeless and shaven, all linked together, but unlinked to other mutations. Morgan postulated that these forms were due to mutations on the fourth, small chromosome of *Drosophila* (see Figure 2.6).

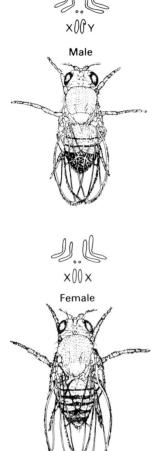

Male

Female

Fig. 2.6 *Drosophila melanogaster* and its chromosomes.

Some mutations like white eye appeared only in males. Morgan showed that one chromosome in the male is slightly larger than the corresponding chromosome in the female. The additional material carries the sex-determining factor plus some other factors that therefore are found only in the male. The chromosomal basis of heredity was now established.

One day yellow wing and white eye were discovered unlinked. Morgan postulated that a chromosome had broken during meiosis and that the pairs had rejoined in a different arrangement. He proved to be correct, and the phenomenon is now called **crossing over** (see Figure 2.7). Crossing over is an example of a chromosome mutation, and many other types are now known. The phenomenon of crossing over proved to be particularly useful in determining which parts of the chromosome normally carried particular genes. Thus, by studying numerous cases of crossing over, it is possible to calculate the relative positions of different factors and to establish the order of genes on the chromosome (see Figure 2.7).

Mutations can also occur within a single gene. The chemical structure of the gene is normally very stable and resistant to changes induced by normal environmental variables such as temperature extremes and toxic chemicals. However, Morgan's student H. Muller discovered that the rate of mutation could be enhanced greatly by strong radioactivity, which is not normally present in the natural environment.

Gene mutations occur at random, in the sense that there is usually no way of knowing which gene in which individual will mutate. Most mutations are deleterious and usually lethal in the homozygous form. For example, there is a mutant form of the domestic fowl, called **creeper**, that has very short crooked legs. When two creeper fowl are crossed, the offspring include normals and creepers in a 1:2 ratio. This is not a straightforward Mendelian ratio, but a quarter of the fertilized eggs fail to hatch and these are homozygous for the creeper gene, a lethal combination. Thus, the true ratio is 1:2:1 normal, creeper, lethal; a classical Mendelian ratio.

We now have three known sources for the genetic variation that Darwin recognized as so important for his theory of evolution by natural selection:

■ Mendelian variation, due to new combinations of characters as described by Mendel's laws

■ chromosome mutation

■ gene mutation.

2.5 The genetic material

Mendel did not use the term **gene**. He wrote about the observable (phenotypic) characters of living organisms and postulated the existence of heredity elements, or factors, that behaved in a particulate manner. When Mendel's work was brought to the attention of the scientific world in 1900, scientists already suspected that the hereditary material was carried by the chromosomes. Intensive research confirmed the generality of Mendel's law, although exceptions were recognized, as mentioned earlier. In 1905 William Bateson

Fig. 2.7 Schematic representation of crossing over (After Stern, 1973).

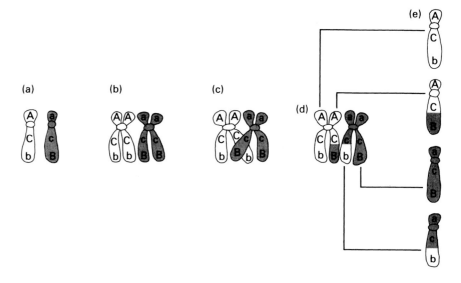

(a)   (b)   (c)   (d)   (e)

gave the name **genetics** to this field of research, and the term gene was applied to the Mendelian elements by Danish botanist Wilhelm Johannsen in 1909. Johannsen also made the important distinction between genotype and phenotype.

Morgan regarded the genes as the smallest units of recombination, arranged on the chromosome like beads on a necklace. Linked genes could recombine by crossing over if a break occurred in the chromosome between the two genes. The gene was also regarded as a functional unit of control over the phenotype and as the unit of mutation. Thus, a gene mutation was the smallest genetic change that could change the phenotype. For the first part of the century, the concept of the gene was a unified one, but this was to be undermined by advances in understanding the chemical nature of the gene. Through the work of Oswald Avery in 1944 and James Watson and Francis Crick in 1953 it was discovered that the genetic code is carried by molecules of DNA (deoxyribonucleic acid).

DNA molecules consist of two strands (Figure 2.8), each made up of a string of smaller molecules, called **nucleotides**. These are the elements of the genetic code. DNA does two things. It reproduces itself, providing the hereditary connection between generations, and it provides the information necessary to the production of thousands of different **proteins**. Just as the DNA is composed of a string of smaller nucleotides, so a protein is composed of a string of smaller **amino acids**. Each possible combination of three nucleotides from the four nucleotide types specifies one of the twenty amino acid types (there are also nucleotide triplets, or **codons**, indicating the end of the protein and other messages). The string of DNA itself is immobile within the nucleus so that it must first be copied into a matching **messenger RNA** (ribonucleic acid) which can move out of the nucelus into the cytoplasm where the amino acids are stored. In the cytoplasm a string of amino acids is matched to the string of nucleotide triplets in the RNA. When the amino acid chain or **polypeptide** is complete it separates from the RNA and takes up its

Fig. 2.8 Schematic representation of the DNA double helix. Each helical chain is made up of alternating deoxyribose sugar and phosphate. The two chains are joined by weak links between the nucleotide bases adenine (A), thymine (T), cytosine (C) and guanine (G).

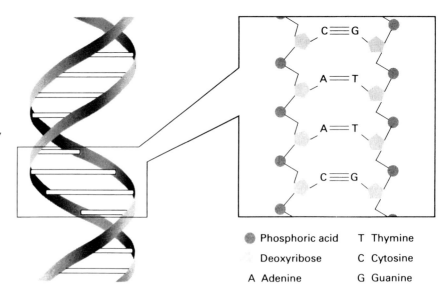

● Phosphoric acid   T Thymine

    Deoxyribose   C Cytosine

A Adenine   G Guanine

role as a functional protein. The gene, then, is simply that stretch of DNA which includes the entire nucleotide sequence for one protein (or one RNA active in gene regulation without translation into protein). In the chromosomes of any complex animal there are tens of thousands of genes as well as long pieces of DNA which do not code for any known product.

Gene mutations can induce various types of disruption of the DNA molecule. In particular, it has been shown that **deletions** or **substitutions** in particular nucleotides can occur. If such a change happens near the beginning of a gene, then the triplets are misread all along the line and a completely different, non-functional protein, results. If the addition or deletion occurs near the end of a gene, then the first part of the protein is coded correctly, the rest being altered in some way. If the alterations did not affect the active sites of the molecule, then some function might remain. Another type of mutation is base substitution – that is, the exchange of one nucleotide for another. This may code for a new triplet coding for a different amino acid, or it may produce a spacer triplet. Mutations may also occur in regulator genes that control the activity of other genes.

Chromosome mutations result in recombination of genes such that the genes appear in a different order on the chromosome, or appear on a different chromosome altogether. This type of mutation does not alter any single gene but may affect the way in which proteins are produced, since the order of genes on the chromosome is an important aspect of gene expression.

The rate at which any particular gene undergoes mutation is very low. However, there is a large number of different genes within an individual, and the number in the gene pool is vast. Thus mutations frequently occur within a population, pure chance determining which individual is affected. The vast majority of mutations are deleterious and cause death at an early stage of development. Only a few allow the animal to remain viable. These increase the genetic variability upon which natural selection depends, but they do not determine the direction of evolution.

Points to remember

- Natural selection depends upon genetic variation in the population. Those variants that are best suited to the environment tend to have more offspring. The variation comes about as a result of meiosis, chromosome mutation and gene mutation.

- During meiosis there is a reassortment of the genetic material in accordance with Mendel's laws. This can result in changes in the genetic combinations from one generation to the next.

- Chromosome mutations also occur during meiosis, but their occurrence is not systematic. They are errors in meiosis, which occur at random and result in novel genetic combinations.

- Gene mutations are changes in the chemical make-up of the genes. They occur at random and are generally deleterious. However, a few result in beneficial innovations which endow the individual with an evolutionary advantage.

Further reading

Ayala, F.J. and Kiger, J.A. (1984) *Modern Genetics*, 2nd edn. Benjamin/Cummings, Menlo Park, CA.

CHAPTER

Development of behaviour

Behaviour patterns result from the complex interactions of external stimuli and internal conditions. However, any behaviour pattern is constrained by the way in which information is processed by the animal. The internal information processing systems are established during the course of development from the fertilized egg, to the embryo, to the adult animal, a process called ontogeny. Through the study of ontogeny we can discover the ways in which genetic and environmental (or learned) information interact to give rise to the behaviour of the animal.

3.1 Ontogeny

Almost all animals are composed of the same basic materials; the differences between species result from differences in the way in which these basic materials are put together. This regulatory function is poorly understood but it is known that genes control development by producing proteins which regulate the complex organization of embryological processes.

The regulatory proteins are only effective insofar as the developing organism is sensitive to the messages they convey. A regulatory molecule does not act by constructing a system according to a map, rather it causes a specific reaction in cells which possess receptors for it on their surfaces. The response of a developing embryo to a regulatory gene product thus depends upon the embryo's internal organization at the time the regulator is released. In addition, the regulator is only functional under a certain range of environmental conditions. As an example we first look at some aspects of development in a species with a relatively simple behavioural repertoire.

Aplysia is a shell-less marine snail which can attain a body weight of up to 10 lb (4.5 kg). As is typical of snails, *Aplysia* is a true hermaphrodite. Fertilization is internal and mating chains commonly form in which up to a dozen individuals mate with one another, each animal inseminating one animal as it is being inseminated by another. Following fertilization, *Aplysia* lays over a million eggs which are connected in a long strip. The egg strand

Fig. 3.1 The egg-laying behaviour of *Aplysia*. The string of egg cases is expelled from the reproductive duct in the side of the body. The animal grasps the egg string in its mouth (left) and fixes it to the substrate (right) (Adapted from Scheller and Axel, 1984).

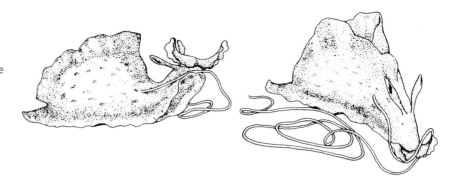

is expelled from the reproductive duct initially through contraction of duct muscles. When the strand is long enough, the snail picks up the end in its mouth and performs a series of head waving motions which help to pull the eggs free of the reproductive duct (Figure 3.1). A special mucus gland in the animal's mouth secretes adhesive onto the egg strand which is then wrapped over itself to make a more compact egg mass. Finally, the snail uses its head to press the egg mass firmly onto a solid substrate (Figure 3.1) where the eggs adhere until hatching. The entire series of egg processing actions is a tightly stereotyped pattern of behaviour. During the sequence the animal's heartbeat and respiratory rates rise and it stops all feeding and locomotion.

Richard Scheller and Richard Axel (1984) found that the behaviour sequence was the result of gene action at a number of levels, acting both during development and from minute to minute. The cells which give rise to *Aplysia*'s entire nervous system are descendants of a few cells in the embryo's body wall. Early in development a subset of these pre-nerve cells already produces a specific protein called ELH (**egg-laying hormone**). It is thought that these may all be daughter cells of a single cell in the very early embryo which has the ability to use the appropriate gene to make this hormone. All other cells in the embryo possess this gene as well, but it is never functional. The ELH-producing cells divide further and migrate to their final locations in the developing organism where they differentiate into adult nerve cells. In the adult snail, therefore, certain nerve cells have inherited the ability to produce ELH while others have not.

By the time of maturation of the snail, all of these systems are in place. At a specific moment another hormone is released which causes the ELH-producing cells to go into full production. ELH then causes neighbouring nerves to fire (that is it acts as a neurotransmitter) as well as circulating in the blood and causing specific muscle fibres to contract (acting as a hormone). This response results in the coordinated behaviour pattern observed as egg extraction and deposition. In this sequence we see the importance of embryogenesis, genes, neurotransmitters, hormones, nerves and muscles all coordinating to generate a behaviour pattern critical to the reproduction of the snail.

In *Aplysia*, the action of the genes is of direct importance in guiding the production of a stereotyped behaviour pattern. While genetic influences underlie all behaviour at some level (for instance the embryogenesis which lays down the body structure) there are also varying degrees of environmental influence which can modify the behavioural results of development. The same genes may have different phenotypic effects when the animal is subjected to different environmental influences during development. For example, the crustacean *Gammarus* normally has a red eye colour. E.B. Ford and Julian Huxley (1927) discovered a single mutation that affects the rate at which eye pigment is deposited during a certain stage of ontogeny. If the mutant is raised below a certain temperature, the eyes become red. At higher temperatures they become a chocolate brown, and at intermediate temperatures there will be intermediate colours. Given the vast complexity of the biochemical processes involved in gene expression and in the growth and differentiation of cells (see Ham and Veomett, 1980), it is not at all surprising that the medium in which these processes occur will affect the course of their development. To a certain extent, built-in stabilizing, or regulatory, mechanisms serve to correct deviations and to control the speed and direction of the developmental processes. It is even possible for a structure or behaviour pattern to develop by different routes, a phenomenon known as **equifinality**. Nevertheless, deviations from the general pattern of development are inevitable. While in some cases these may be lethal, in others they may give rise to differences among individuals that have the same genetic make-up.

Some early ethologists tended to think of behaviour as being directly determined by genes. Thus Konrad Lorenz (1965) proposes the metaphor of a blueprint analogous to that for the construction of a building. The genetic blueprint represents a plan for the construction of the adult animal. Recognizing that bricks, mortar and a work force are necessary for the construction of a building, Lorenz draws a sharp distinction between the conditions necessary for the translation of the blueprint into a building and the information contained in the blueprint upon which the characteristics of the finished building depend. Critics of this view (e.g. Lehrman, 1970) point out that, whereas an architectural blueprint is isomorphic with the structure it represents, this cannot be the case with a genetic blueprint. While there is a one-to-one correspondence between the measurements marked on an architectural blueprint and those that appear in the finished building, there is no such correspondence in biological ontogeny. Even if it is correct that the genes code for certain key enzymes, this is far removed from the idea of a blueprint. The idea that information provided by the genes can be separated from that provided by the environment during development was criticized by Hebb (1953), who argued that it is as meaningless to ask how much a given piece of behaviour depends upon genetic factors and how much upon environmental as it is to ask how much the area of a field depends upon its length and how much upon its width.

The major emphasis of research into the ontogeny of behaviour has been to demonstrate the variety of ways in which patterns of behaviour develop. There appears to be a spectrum of processes, ranging from those that appear to be relatively uninfluenced by environmental factors to those that are heavily dependent upon experience.

The genetic determinism of the early ethologists has given way to the recognition that genetic and environmental influence are inextricably bound up together in ontogeny – not, as Lorenz (1965) has suggested, like the relationship between the blueprint and the materials necessary for the construction of a building but more along the lines of a process, called epigenesis, by which each developmental event sets the stage for, but does not dictate, the next. As suggested by Brown (1975), epigenesis may be summarized as follows. Starting from the beginning of development with the fertilized egg or zygote, P_1, its phenotype at the next stage of development, P_2, will be determined jointly by the genes that are active in guiding its growth and differentiation during the intervening interval, G_1, and by the environment in which the development takes place, E_1: zygote + genes + environment phenotype at next stage of development, or:

$$P_1 + G_1 + E_1 \rightarrow P_2$$

The phenotype at the next stage of development, P_3, will be determined by the way in which P_2 has been changed by the genes, G_2, and environmental influences, E_2, affecting development between P_2 and P_3:

$$P_2 + G_2 + E_2 \rightarrow P_3$$

This notation points up the fact that all three elements, genes, environmental conditions and starting phenotype are aspects of normal development. Development can only proceed to the next stage if the phenotype of the developing animal is appropriate, if the correct gene products are available, and if environmental conditions fall within certain ranges.

During the early stages of development ($P_1 + G_1 + E_1 \rightarrow P_2$) the environmental component E_1 consists mainly of biochemical factors that surround the early embryo. At a later stage of development ($P_2 + G_2 + E_2 \rightarrow P_3$) the environmental component E_2 may consist of the post-hatching environment. After hatching or birth an animal's environment includes information taken in by its sense organs as well as biochemical conditions based upon feeding level, etc. In this situation learning can take place and can permanently influence the animal's behavioural phenotype, P_3.

It is interesting to explore the relationship between learned and innate, or non-learned, behaviour by comparing the baby bird of a **precocial** species (a species in which hatching occurs late in development) to the baby of an **altricial** species (a species in which hatching occurs early). If two such birds are compared at the same stage of physical development, such as the point at which the first feathers appear, we find that the precocial bird is still inside the egg, while the altricial bird has already hatched. By the time the precocial bird hatches, the altricial juvenile has had the opportunity to learn a great deal, while the precocial bird has not yet begun to learn. What we find, interestingly, is that the precocial hatchling may have innately developed capabilities which the altricial chick has had to learn. For example, the altricial white-crowned sparrow learns the song of its species while a nestling. By contrast, chickens are precocial, passing early infancy while still in the egg, but they have the innate ability to produce normal vocalizations, even though never previously exposed to them.

We may be inclined to think of what happens inside an egg as being purely maturational, in the sense that the relevant environmental factors are predetermined. We should not forget, however, that the normal environment of the white-crowned sparrow is also predetermined in the sense that the nestlings normally hear the song of their own species during the critical period of song learning.

Later still in development, $P_3 + G_3 + E_3 \rightarrow P_4$. By this stage, the effects of postnatal experience will be observable in both altricial and precocial species. Behaviour that clearly is affected by experience is not likely to be called innate. However, there is little difference in principle here from the situation at earlier stages of development. The extent to which the environmental factors are predetermined depends largely upon the variability of the environment. Inside the egg such variability is usually small, and we are therefore inclined to regard such behaviour as innate. Outside the egg, however, the variability may also be small, depending upon the ecological circumstances typical of the species. Thus, the postnatal behaviour of a precocial animal may be just as predetermined as the prenatal behaviour of an altricial animal. If it is evolutionarily important that the development should result in a particular end product, as might well be the case with the song of a passerine bird, then we might expect the environmental influences to be so contrived as to ensure that outcome.

3.2 Environmental influences on behaviour

Some forms of behaviour do not appear until a particular stage of development is reached. Some of these seem to develop without any obvious form of practice. For example, pigeons start to flap their wings and fly erratically at a particular age, and their flying ability appears to improve with practice. However, Grohmann (1939), in a classic experiment, reared a group of pigeons in tubes so that they could not move their wings. Another group of the same age was allowed to develop without restraint. When the unrestrained pigeons had reached the stage at which they could fly satisfactorily, those that had been restrained were freed. Grohmann discovered that they also were able to fly immediately upon being released. Chickens that are featherless as a result of mutation develop normal wing-flapping vestibular reflexes, even though these are completely ineffective (Provine, 1981).

At first sight it would appear that the behaviour is independent of environmental factors, but we must not assume that the ability to fly develops irrespective of any ontogenetic eventuality. Rather, it is contingent upon some essential conditions, although our knowledge of these is sketchy (Ham and Veomett, 1980). A similar example can be seen in the vocalizations of pigeons and domestic fowl. These vocalizations are highly stereotyped and appear at certain stages of development. They are not dependent upon auditory experience but upon certain hormonal conditions that normally occur as part of the overall process of maturation.

Many movement patterns appear to develop in the absence of practice or example (Hinde, 1979). However, study of the developmental history of some behaviour patterns has suggested that they involve fragmentary and incomplete movements that might influence the course of development (e.g. Kuo, 1932, 1967; Kruijt, 1964; Anthoney, 1968). As an example, we can consider Kuo's (1932) work on the development of behaviour in chick

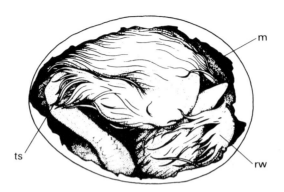

Fig. 3.2 The hatching position of the chick embryo approximately 1–2 days before emergence from the shell. m, membrane; ts, tarsal joint of leg; rw, right wing (From *The Oxford Companion to Animal Behaviour*, 1981).

embryos. Kuo placed windows in the eggshell and was able to observe the behaviour of the embryo directly (see Figure 3.2). He found that the embryo is constantly subject to stimulation both from its own activity and from outside the egg. Various movements can be observed, including initial passive movements of the head caused by the beating heart. Active head movements appear within a few days, and these may be accompanied by opening and closing of the beak. By the seventeenth day, movements similar to pecking are apparent. Some scientists believe that the chick learns to peck in this way (e.g. Lehrman, 1953; Schneirla, 1965). Others (e.g. Thorpe, 1963; Lorenz, 1965) ridicule this idea.

Whether or not true learning occurs, there is little doubt that the developing embryo responds to stimuli from both inside and outside the egg and that these stimuli may influence development. Thus, Margaret Vince (1969) showed that in some species, stimuli produced by the embryo can influence the development of other eggs in the clutch and can lead to synchronization of hatching. Gilbert Gottlieb (1971) shows that duck embryos can respond to maternal calls five days before hatching and that such experience may influence the ducklings' subsequent behaviour. However, other workers provide evidence that movement in the chick embryo develops in the absence of sensory stimulation and that the movement patterns that occur later in ontogeny are not necessarily influenced by earlier movements (for review, see Gottlieb, 1970; Oppenheim, 1974).

Readers should remember that the young of different species develop in very different ways. While ducklings and goat kids are mobile as soon as they are hatched or born, blackbird nestlings and kittens are helpless at this stage. We are not surprised to discover that the postnatal development of the latter is influenced by experience. Should we be surprised at prenatal influences in the case of animals which are still inside the egg or womb at the same overall stage of development? There seems little difference in principle between the protection the duckling receives while in its egg and that the young blackbird enjoys in the nest. Similarly, there would seem to be little difference in principle between the effects of experience in the two cases. For each species, certain types of experiences are essential for normal development. For example, adult cat vision is abnormal if the coordination of vision from the two eyes of a kitten is disrupted during the first few months of life, by inducing an artificial squint or by covering each eye on alternate days so that the two eyes never work together (Hubel and Wiesel, 1965). The exact nature of the experience that is necessary for normal development varies considerably from one species to another.

3.3 Sensitive phases of development

Some animals appear to be preprogrammed to learn about certain aspects of the environment during particular sensitive phases of their development. Language learning in humans is an example of this phenomenon. The learning is preprogrammed in the sense that it will occur in the absence of any obvious reward or punishment, in contrast to other types of learning. For example, the young of many precocial species (in which the young are born at an advanced stage of development) show a fairly indiscriminate attachment to moving objects. Thus, newly hatched mallard ducklings, separated from their mother, will follow a crude model duck, a person or even a simple box moved slowly away from them (see Figure 3.3). Some stimuli are more effective than others in eliciting the following response (see Chapter 20). In the natural environment the most effective stimuli are normally provided by the mother, and approaches to the mother are often rewarded by body contact and warmth or by food that the mother uncovers. In the laboratory the attachment to a model can be enhanced by food rewards.

The more an animal develops an attachment to one object, the less interested it is in others. This process of learning, through which attachment to the mother normally develops, is called **imprinting**. It occurs during a particular sensitive phase of development, which varies according to the species and the circumstances. Imprinting may have long-term effects, beyond the attachment to a parent or foster parent. In many mammals such

Fig. 3.3 A duckling following its 'mother' in an apparatus designed to test aspects of imprinting (After Gottlieb, 1971).

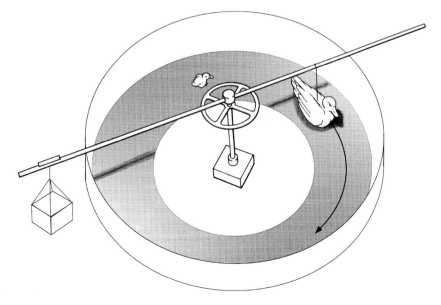

early experience affects subsequent social adjustment. In a number of bird species imprinting has been shown to affect subsequent sexual behaviour, as we see in Chapter 20. In short, imprinting is a process of learning that occurs at a particular stage of development and that affects subsequent behaviour toward parents, peers or sexual partners. If the object of attachment is a cardboard box, then the duckling will become attached to the box as to a parent. If male zebra finches (*Taeniopygia guttata*) are raised by Bengalese finches (*Lonchura striata*), then they court Bengalese finch females when adult (Immelmann, 1972, 1985).

Although this type of learning may be influenced by rewards, it is not dependent upon them or upon any particular consequences of the behaviour. The learning is preprogrammed to take place as part of the normal process of development and in whatever circumstances pertain at the time.

A similar arrangement applies to the song learning of some passerine birds. The first attempts at song, called subsong, usually occur in the young bird's first spring or first autumn, some months after hatching. The subsong resembles the adult song in length, pitch and tonal quality, but it is lacking elements and motifs typical of the adult song and is usually rather variable and imprecise (see Figure 3.4).

If white-crowned sparrows (*Zonotrichia leucophrys*) are reared in isolation, they develop the subsong but fail to develop the normal adult song. If exposed to the normal song of adult males when about 10 to 90 days old, male white-crowned sparrows subsequently (i.e. when about 8 months old) will develop the normal adult song. However, if exposed to a normal song before they are 8 days old, and not thereafter, they will not develop the adult pattern. Similarly, if exposed only after the age of about 100 days, male white-crowned sparrows will not develop normal song. Thus, there appears to be a sensitive period, between the ages of 10 and 90 days, when it is necessary for young male white-crowned sparrows to hear adult male song if they are eventually to learn to sing that song (Marler and Mundinger, 1971). Similar sensitive periods have been found for other species.

Fig. 3.4 Development of song in the canary (*Serinus canaria*). (a) Subsong: The sound spectrogram shows the ill-defined phrasing and lack of tonal purity. (b) Full song, by contrast, shows regular phrasing and notes which are relatively pure and free of harmonics (From *The Oxford Companion to Animal Behaviour*, 1981).

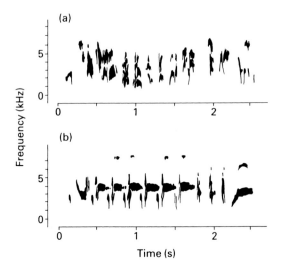

If white-crowned sparrows are surgically deafened before being exposed to normal song during the sensitive period, they do not develop normal song. If they are isolated after having heard normal song during the sensitive period, then they do produce normal song. Thus, the birds that hear the song during the sensitive phase remember it even though they do not sing until they are about 8 months old. However, if they are deafened after hearing normal song during the sensitive period but before they sing, then they do not produce normal adult song (Konishi, 1965). They may be able to hear themselves sing and to make use of their auditory memory in developing a song that matches the song they heard during the sensitive phase. If white-crowned sparrows are deafened after their song has been rehearsed and has crystallized, then their song is not affected. This means that during song rehearsal and perfection, the song is encoded in another kind of memory.

In nature, birds may be exposed to more than one song tutor, such as a parent and a neighbour. Experiments with Zebra finches (*Taeniopygia guttata*) indicate some selectivity in this respect. Young males were provided with the opportunity to listen to two song tutors, their father and another male. Sons developed a song pattern which was very similar to that of the father, although the song copies were not identical in all details (Bohner, 1983). These findings suggest a certain bias in the choice of song tutor, and a degree of infidelity in copying which leads to the development of individual differences among males. Zebra finches, and some other birds, will not learn from tape recordings, so there must be more to song learning in such species than simply hearing the appropriate sounds.

Numerous experiments show that social interaction with the tutor is an important aspect of song learning (Catchpole and Slater, 1995). In a number of species young birds have been found to possess more songs in early development than they subsequently use. This suggests another mechanism in addition to the auditory template (Box 3.1). Marler (1970) distinguishes between 'memory-based learning' and 'action-based learning', the latter being important at the stage of song production. Marler and Nelson (1992) go so far as to suggest that some species select from 'pre-encoded' songs. In the field sparrow, for example, two or more song types are sung initially, but as the young bird becomes territorial the number is reduced to one, which matches that of his most actively singing neighbour. Thus it appears that many birds may acquire a variety of songs, from which they select a few for use during interactions with neighbours.

3.4   Juvenile behaviour

The animal develops throughout the whole of its life, and must be well adapted to its environment throughout its life history. Thus behavioural development is not simply a matter of constructing adult behaviour patterns. The juvenile animal will often show behaviour that never occurs during its later life. This behaviour is usually tailored to suit the young animal's needs.

For example, the alarm behaviour of herring gull chicks is quite different from that of the adults. When alarmed, the chicks move a short distance from

Box 3.1 The auditory template model

The studies of Konishi (1965) and Marler (1970) led to the development of the auditory template model. According to this model, the young bird hatches with a crude 'template' which defines the approximate characteristics of the song of the species. During the sensitive phase of song learning, when the bird hears the songs of many species, only those that match the template are memorized. The crude template thus becomes an exact template: a representation of the song(s) that the bird will sing as an adult.

When the bird first starts to sing it hears its own song and compares this sound with the template. Usually, the bird initially emits a rather quiet and variable subsong which develops into a louder plastic song. Within some weeks of starting to sing, the plastic song is shaped into the full song characteristic of the species.

(After Catchpole and Slater, 1995.)

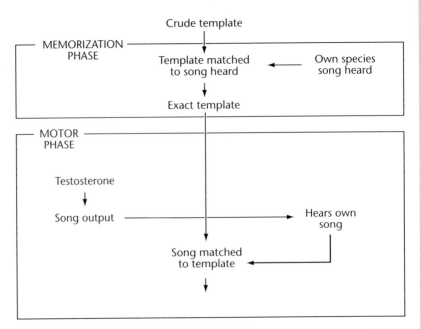

the nest and crouch motionless and silent among the vegetation. The adults, on the other hand, fly away from the nest uttering alarm calls. Chicks that run around during an alarm are very vulnerable to predation from crows and other gulls. By remaining motionless they increase their camouflage and reduce the risk of predation.

The extent to which the behaviour of the juvenile resembles that of the adult varies considerably from species to species. In some, such as the wildebeest (*Connochaetes*), where the young animal runs with the herd within a few minutes of being born, the juvenile lifestyle is very similar to that of its parents. In others, such as the larval forms of butterflies and frogs, the life of the juvenile is quite different from that of the adult.

3.5 Innate behaviour

The question of what it means to say that behaviour is innate has long been a subject of controversy. The term 'innate' has come to have different meanings which you must learn to distinguish. The early ethologists referred to innate behaviour as behaviour that is determined by heredity and that is part of the animal's original make-up and thus independent of the experience of the individual. Thus, Lorenz (1939) writes about characteristics of behaviour as being 'hereditary, individually fixed, and thus open to evolutionary analysis'. Tinbergen (1942) refers similarly to instinctive acts as being 'highly stereotyped coordinated movements, the neuromotor apparatus of which belongs, in its complete form, to the hereditary constitution of the animal'. These views were criticized soundly by Lehrman (1953) and others as being too rigid, as incorporating a naive idea genetic determinism and of being semantically confusing. A rigid distinction between innate and learned behaviour is unsatisfactory because many aspects of behaviour are influenced by both genetic factors and the experience of the individual.

The idea that genes determine behaviour is naive because the genes cannot possibly contain detailed instructions for particular aspects of behaviour. The genes may influence the processes of development in various ways, but these processes are also affected by environmental factors. As we have seen, the genes can never dictate the course of ontogeny without reference to the environmental medium in which the development occurs. For the student of behaviour it is convenient to use the term as a short-hand for 'behaviour that develops without obvious environmental influence'. This is the way it is used in this book.

Points to remember

■ The development of the individual from embryo to adult involves a continual interaction between the animal's genetic make-up and its environment. The interaction is one in which each phase of development sets the stage for the next, a process called epigenesis.

■ Environmental influences upon development are most important just after birth, or hatching, but may occur at any stage of the developmental process. The circumstances in which the parents raise their young are usually designed to protect the young from unfavourable environmental influences.

■ Many species have periods of development during which they are sensitive to particular kinds of environmental influence. What the animal learns during these sensitive periods usually affects it for the rest of its life.

■ Juvenile animals often have characteristic behaviour which enables them to respond in an appropriate way to environmental occurrences such as the appearance of a predator or the provision of food by a parent. This typical juvenile behaviour is lost in adulthood.

■ It is convenient to use the term innate for behaviour that occurs without obvious environmental influence, provided it is recognized that environmental factors influence the development of all behaviour to some extent.

Further reading

Bateson, P.P.G. and Klopfer, H. (eds) (1982) *Perspectives in Ethology*, Volume 5, *Ontogeny*. Plenum, New York.

Behaviour genetics

Although Charles Darwin was interested in hereditary factors in relation to behaviour, the scientific pioneer in this field was another grandson of Erasmus Darwin, Francis Galton. *On the Origin of Species* inspired Galton to devote the rest of his life to the study of the inheritance of mental characteristics. In 1869, Galton published *Hereditary Genius: An Inquiry into its Laws and Consquences*. He argued that people of outstanding mental ability are to be found more often among the relatives of similar people than among the population at large. Not having any satisfactory way of measuring mental ability, Galton relied upon an index of reputation, 'the reputation of a leader of opinion, of an originator, of a man to whom the world deliberately acknowledges itself largely indebted' (1869). He examined the pedigrees of some 300 families containing eminent judges, statesmen, military commanders, literary men, scientists, poets, musicians, painters, etc. His results showed that eminent social status was most likely to appear in close relatives, and as the degree of relationship became more remote, so the likelihood of eminence decreased.

Galton was aware of the fact that eminent people would share social, educational and financial advantages. He argued that the reputation of eminent people is an indication of their natural ability and not due to environmental factors. To support his argument, Galton pointed out that many eminent men had humble family backgrounds. He also judged that the adopted kinsmen of Roman Catholic popes, who enjoyed great social advantages, were a less distinguished group than the sons of eminent men.

Galton devoted considerable effort to improving the means of assessing mental characteristics. He developed procedures for measuring smell, touch and visual acuity; judgements of length, weight and of the vertical and reaction time and memory span. In statistics, he pioneered the concepts of correlation, median and percentile. In 1883, he introduced the twin-study method of distinguishing between the effects of nature and of nurture. He studied 35 pairs of twins who were very similar at birth and who were raised in similar conditions. He noted that their behavioural similarities persisted even after they had developed separate adult lives. He also studied

20 pairs of twins who were dissimilar at birth and raised in similar environments:

'There is no escape for the conclusion that nature prevails enormously over nurture when the differences of nurture do not exceed what is commonly to be found among persons of the same rank of society and in the same country. My fear is that my evidence may seem to prove too much, and be discredited on that account, as it appears contrary to all experience that nurture should go for so little.' (1883).

The study of the genetic inheritance of behavioural traits developed rapidly following Galton's lead, but it has remained controversial to this day. In the early part of this century, the main challenge came from the rise of behaviourism. Watson (1930), in particular, persuaded many psychologists that genetics was irrelevant to behaviour development, and the strong environmentalist view adopted by the behaviourists remained influential until the 1960s. The field of behaviour genetics may be said to have become firmly established by 1960 when Fuller and Thompson published their *Behaviour Genetics*. This book relates the history of psychological studies of human behaviour and intelligence from the beginning of the century and reviews the evidence for genetic influences upon behaviour. Despite the overwhelming weight of evidence, many sociologists and psychologists remained opposed to the idea of genetic influences in behaviour and the debate continued (see Hirsch, 1963, 1967). Even today some theories that are based upon genetic arguments remain controversial. In this chapter we review the evidence concerning the influence of genetic factors upon behaviour.

4.1   Single genes and behaviour

Behavioural traits under the control of single genes can provide a clear demonstration of genetic analysis of behaviour. A classic example is Bastock's (1956) study of mating success in the fruit fly *Drosophila melanogaster*. She crossed a sex-linked yellow mutant with wild stock for seven generations. This ensured that the wild stock was genetically similar to the yellow stock, except in the region of the yellow gene. Bastock found that males with the yellow mutant were less successful in mating with wild-type females than were wild-type males. The yellow males had an altered courtship pattern that reduced their mating success. They were less stimulating to the females because their courtship contained a smaller proportion of wing vibration.

Rothenbuhler (1964) carried out an elegant genetic analysis of the nest-cleaning behaviour of honey-bees (*Apis mellifera*). The larvae are sometimes killed by a disease called American foulbrood. To maintain a hygienic environment within the hive, the worker bees normally uncap, or open, the comb cells that contain diseased larvae and remove them. Some strains of bees, called 'unhygienic', do not follow this procedure. When these are crossed with normal hygienic bees, all the offspring are unhygienic, indicating that this is a dominant character. When these hybrids were back crossed with the parental hygienic strain, Rothenbuhler obtained the following results. Out of a total of 29 back-crossed colonies he found that nine uncapped the infected cells but did not remove the diseased larvae, six did not uncap the cells but would remove the larvae if the cells were uncapped by the experimenter and eight would not uncap cells and remove larvae.

These results show that different genes must control the uncapping and the removal behaviours. The results can be explained on the basis of two pairs of alleles, of which the unhygienic alleles are dominant. Thus, worker bees with Uu or UU will not uncap the infected cells, and those with Rr or RR will not remove larvae.

The male bees, called drones, are haploid, having only one set of chromosomes. Among his 29 colonies, Rothenbuhler must have had four types of drones (UR, Ur, uR, ur) that he crossed with hygienic queens (uurr), as shown in Figure 4.1. According to this simple Mendelian scheme, the back cross should produce four genotypes in equal proportions, which does not differ significantly from the results Rothenbuhler obtained. No physical or physiological differences have been discovered among totally hygienic, partially hygienic or unhygienic workers, although there is some evidence that unhygienic workers do perform hygienic activities at a very low frequency and require a stimulus that is more powerful than normal. This suggests that the alleles U and u act as switches that release the uncapping behaviour, provided there is a certain threshold of stimulation.

Genes that act as switches, activating a group of other genes, are known in a variety of circumstances. For example, some butterflies have elaborate wing patterning that mimics that of other species, which are distasteful to predators. The development of the mimic patterns is under the control of

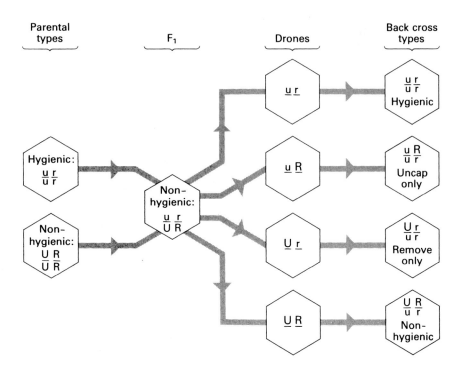

Fig. 4.1 Mendelian scheme proposed to account for resistance to American foulbrood in honey-bees (After Rothenbuhler, 1964).

many genes, but it appears that a single switch gene determines whether or not the pattern appears (Sheppard, 1961). Single-gene effects are also known in humans. An example is the lactase deficiency found in certain racial groups and discussed in Chapter 5. Lactase production seems to be controlled by a gene with three alleles L, l_1 and l_2. Both l_1 and l_2 are recessive to L and l_2 is recessive to l_1. Individuals with LL, Ll_1 or Ll_2 genotypes produce lactase both as adults and as children. Those with l_1l_1 or l_1l_2 do not produce lactase as adults, and those with l_2l_2 cannot produce lactase even in infancy. Adults with l_1l_1 or l_1l_2 genotypes can digest milk that has been soured or turned into yogurt or cheese.

4.2 Chromosome mutations

The arrangement and number of chromosomes can often be observed directly under the microscope. There are various known types of chromosome mutations, and some of these have identifiable effects upon the phenotype. The study of behavioural correlates of chromosome structure thus provides a useful way of investigating genetic influences upon behaviour.

A favourite animal for this type of study has been the fruit fly *Drosophila*. The *Drosophila* larvae have giant chromosomes in the salivary glands that can be prepared and observed relatively easily. The application of chromosomal analysis in *Drosophila* to the study of behaviour was pioneered by Jerry Hirsch and his co-workers. They investigated the tendency of *Drosophila melanogaster* to move toward or away from the direction of gravity (positive and negative geotaxis). The behavioural response is tested in a vertical plastic maze (Figure 4.2), through which the flies are attracted by the odour of food. Backward movement is discouraged by cone-shaped funnels at the junctions in the maze. The flies are introduced, in large numbers, in a vial on the left-hand side of the maze and collected from a series of vials on the right. In this way thousands of flies can be tested without being handled by humans. The flies recovered in the various right-hand vials are segregated into those that are strongly positively geotaxic, those that are strongly negatively geotaxic and those in between. By breeding from those flies that are collected from the extreme upper and lower vials, it is possible to produce strains selected for positive or negative geotaxis, as shown in Figure 4.3.

In one experiment, three populations of *Drosophila* were compared (Hirsch and Erlenmeyer-Kimling, 1962). One was selectively bred for positive geotaxis, one for negative geotaxis and the third an unselected control population. These were crossed with a special stock that carried various chromosomal inversions and marker genes. *Drosophila melanogaster* have four pairs of chromosomes, three large and one small. The marker genes were used to identify the three large chromosomes. They were dominant genes controlling phenotypic features by which their presence in the genotype is made visible. By means of a special mating design, females were produced that were either homozygous or heterozygous for the chromosomes to be investigated.

Fig. 4.2 A geotactic maze for *Drosophila*. The flies are introduced in the vial on the left and collected from the array of vials on the right (*Photograph: Jerry Hirsch*).

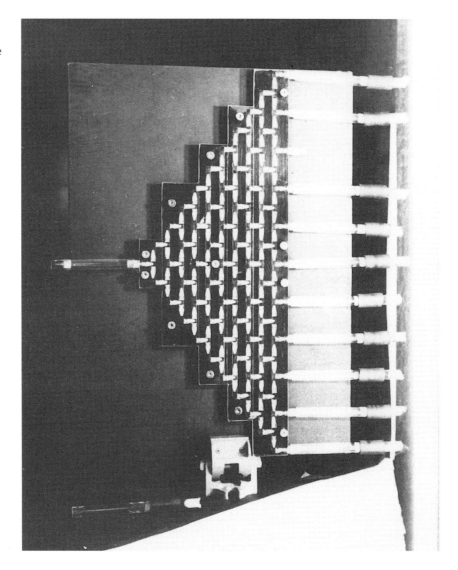

The three chromosomes were identified by marker genes as follows: chromosome X by bar eyes (B), chromosome II by curly wings (Cy) and chromosome III by stubble bristles (Sb). Tester females carrying these markers were mated with males from one of the stocks to be investigated. Of the progeny, only those carrying all three marker genes were used in the subsequent part of the experiment. These were back crossed with the original male population, producing eight possible genotypes. Each of the three major chromosomes is therefore heterozygous or homozygous for an S chromosome obtained from the sample line (s) to which the father belonged. From the eight classes of genotype, the individual effects of the chromosomes and their interactions can be studied.

Jerry Hirsch and Linda Erlenmeyer-Kimling (1962) found that, in an unselected population, chromosomes X and II contributed to positive geotaxis

Fig. 4.3 Geotactic scores obtained for three strains of *Drosophila* in the maze shown in Figure 4.2 (After Hirsch, 1963).

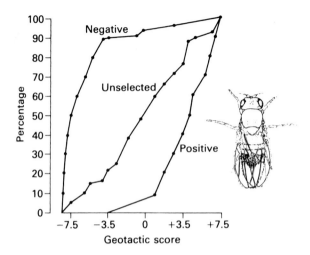

and chromosome III contributed to negative geotaxis. In a strain selected for positive geotaxis, there was little change on chromosomes X and II, but chromosome III now contributed to positive geotaxis. In a strain selected for negative geotaxis, the negative contribution of chromosome III was increased, while the positive effects of chromosomes X and II were reduced. When the effects on the three chromosomes are taken together, the magnitude of the effect is greater for negative geotaxis. This is not surprising because the total response to selection (see Figure 4.3) is greater for negative geotaxis. These results demonstrate that geotaxis behaviour is controlled by a number of genes, which are distributed over all three of the major chromosomes.

Chromosome analysis has been applied extensively to various types of *Drosophila* behaviour including mating speed and other aspects of courtship (Ehrman and Parsons, 1976). Chromosome inversions are common, and it has been found that in *Drosophila pseudo-obscura* that inversion heterozygotes have greater fitness than inversion homozygotes, due to the effects upon courtship behaviour. Chromosome anomalies in humans have been the subject of considerable study and are thought to be involved in a number of genetically induced behavioural disorders, including epilepsy, manic depression, mental retardation, and schizophrenia (McClearn and DeFries, 1973; Ehrman and Parsons, 1976).

4.3 Polygenic inheritance of behaviour

Many behavioural traits are influenced by a large number of genes as well as environmental factors. A variety of methods can be used in the genetic analysis of such complex situations. Their effectiveness depends largely upon the degree of heritability of the behavioural trait in question. However, some aspects of behaviour have high heritability, and with these we begin our discussion of polygenic inheritance.

Male crickets attract females over long distances by a calling song. The sound is produced by rhythmic opening and closing of specialized forewings that carry friction mechanisms. Each closing stroke of the wings produces a sound pulse, while the opening stroke is silent. The songs are remarkably stereotyped among members of a local population, but they differ considerably from one species to another. The differences occur primarily in the temporal patterning of the pulses, as illustrated in Figure 4.4. Although hybrids are not common in nature (Hill *et al.*, 1972), they can be produced in a laboratory. For example, Leroy (1964) obtained hybrids between the Australian field crickets *Teleogryllus commodus* and *Teleogryllus oceanicus*. Bentley and Hoy (1972) found that the songs of the F hybrids are distinctly different from either parental song. In particular, the intrachirp and intratrill intervals of the hybrids are intermediate between those of the parents. Figure 4.5 shows how the parental songs compare with the hybrids. Bentley and Hoy found that reciprocal hybrids differed from each other. The hybrid from female *T. oceanicus* by male *T. commodus* was similar to that of *T. oceanicus* in having a well-defined intertrill interval. The hybrid from female *T. commodus* by male

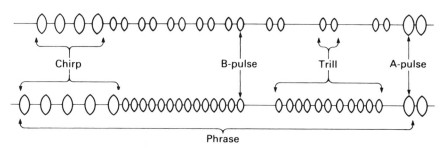

Fig. 4.4 Phrase structure of the calling song of the cricket *Teleogryllus*. Each phrase is composed of two types of pulse: A-pulses contained in the chirp portion of the phrase, and B-pulses contained in the trill. The song of *T. oceanicus* is shown above and that of *T. commodus* below (After Bentley and Hoy, 1972).

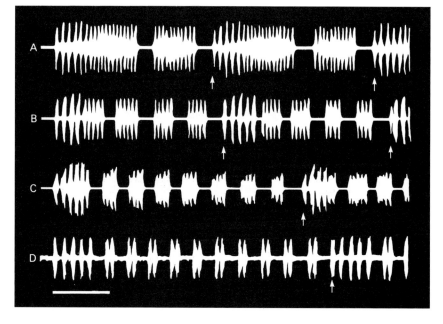

Fig. 4.5 Oscilllograms of calling songs of *T. oceanicus* (A), *T. commodus* (D), and their hybrids *T. oceanicus* female × *T. commodus* male (B) and *T. commodus* female × *T. oceanicus* male (C) (arrows mark the beginning of the phrases) (After Bentley and Hoy, 1972).

T. oceanicus lacks a well-defined intertrill interval, and its song is similar to that of the *T. commodus* parent. These differences are best seen in the inter-pulse–interval frequency histograms in Figure 4.6.

These results suggest that the inheritance of the song pattern is polygenic. There is no evidence of any simple dominance of any particular song feature, and the hybrid songs are intermediate between those of the parents. The hypothesis is supported by the fact that back-crossed hybrids also show inter-mediate inheritance (Bentley, 1971). There is also some indication that sex-linked factors are involved. The songs of the reciprocal hybrids differ in a way that suggests that the intertrill interval characteristic of *T. oceanicus* is present in one hybrid but not the other. Sex determination in crickets is XO, mean-ing that there is no Y chromosome. The male receives the X chromosome from his mother but nothing to match it from his father. Thus, it appears that the intertrill interval characteristic of the female *T. oceanicus* by male *T. com-modus* hybrid is determined by the X chromosome of the *T. oceanicus* parent. This trait is present neither in the female *T. commodus* by male *T. oceanicus* hybrids nor in their mothers.

Fig. 4.6 Interpulse interval frequency histograms of cricket calling songs. Each histogram represents the analysis of one individual song. Intratrill, intrachirp and intertrill intervals can be distinguished from these histograms (After Bentley and Hoy, 1972).

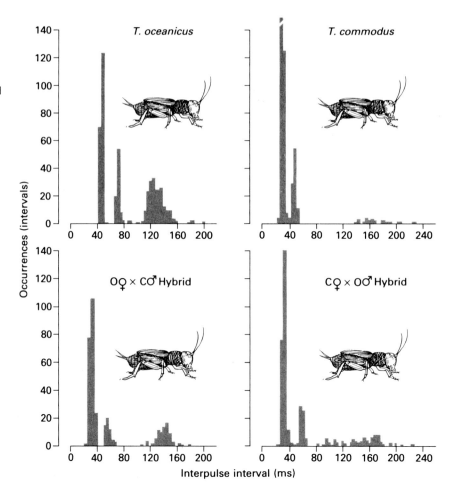

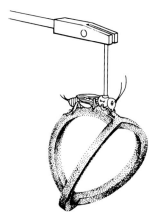

Fig. 4.7 Female field cricket on a Y maze. Her movements on the maze indicate her preferences among male calling songs (After Hoy, 1974).

The male songs characteristic of different species presumably are responded to by the corresponding female. It is, therefore, of interest to discover how hybrid females respond to the songs of their parents and siblings. Female crickets are mute and respond to the male song by walking toward the source of the sound. Hill *et al.* (1972) showed that free-walking *T. oceanicus* and *T. commodus* females can discriminate between the song of their own species and that of another species. To obtain a more quantitative measure of female preference, Hoy and Paul (1973) employed the Y maze illustrated in Figure 4.7. They presented recorded songs from loudspeakers placed on the right and left of the maze. In this way they were able to measure the relative attractiveness of any test song. Hoy and Paul found that *T. oceanicus* and *T. commodus* females make species-specific discriminations, just as Hill *et al.* had found in free-walking females. They also found that females that were female *T. oceanicus* by male *T. commodus* hybrid preferred the song of sibling hybrid males to that typical of either parent species. As can be seen from Figure 4.6, the song of the male *T. oceanicus* by female *T. commodus* hybrid is easily distinguishable from that of either parent (note that this is not so clear cut in the case of the alternative hybrid). The fact that the hybrid song is preferred by the female implies that the production of the song in the male and its reception and recognition by the female are somehow coupled genetically.

It is well known that different breeds of dog have different behavioural characteristics. In 1965, Scott and Fuller published some of the results of their extensive research into the genetic basis of such differences. In one study they compared the behaviour of the cocker spaniel, the basenji and their hybrids. Spaniels have long been bred as sporting dogs, obedient and devoted to their master (see Figure 4.8). Spaniels originally came from Spain, where they were used in hawking and in hunting birds with nets. The dog was trained to crouch close to the ground when it detected partridge or other birds. A net was then thrown across both the dog and the birds. When shotguns were invented, nets were no longer used and land spaniels were developed to stop and point when they located birds. Springer spaniels were developed to spring and flush birds. The original tendency to crouch remains only in the cocker spaniel.

Basenjis used to be widespread in Africa and were used for hunting by Pygmies and several other African tribes (see Figure 4.9). They are a general purpose hunting dog with a somewhat wary temperament. Basenjis bark little, if at all, although they do sometimes indulge in bouts of prolonged howling.

Scott and Fuller (1965) carried out extensive breeding programmes with these two breeds, including reciprocal hybridization and various types of back cross. In Box 4.1 the characteristics of the two breeds are summarized, together with the most likely modes of inheritance as judged on the basis of the results of the experiments. They came to the conclusion that the behavioural traits they investigated were controlled by one or two genes. The situation appears to be intermediate between simple Mendelian inheritance and polygenic inheritance.

It may seem surprising that such complex behavioural traits are controlled

Fig. 4.8 Use of spaniels in hawking. The dogs crouch in response to a handsignal given by their master (From Blome's *Gentleman's Recreation*, 1966).

Fig. 4.9 Pygmies returning from a hunting trip with a basenji. The nets are used to trap game which the dogs flush out of the bush (*Courtesy of J.P. Scott*).

by so few genes. However, the two breeds of dog have been isolated for a very long period of time and have been subjected to intense artificial selection. This seems to have led to genotypes that are homozygous for particular traits, so that there is little segregation within breeds. The genetics of dog behaviour is particularly interesting because the ability to learn particular types of behaviour is often the inherited trait.

4.4 Heritability of behaviour

'The question whether the nature or the nurture, the genotype or the environment, is more important in shaping man's physique and his personality is simply fallacious and misleading. The genotype and the environment are equally important, because both are indispensable . . . The question about the roles of the genotype and the environment in human development must be posed thus: To what extent are the differences observed among people conditioned by the differences of their genotypes and by the differences between the environments in which people were born, grew and were brought up?' (Dobzhansky, 1964).

Much of the work of Francis Galton was based on the idea that phenotypic similarities among relatives are due partly to similarities of genotype. However, Galton was not aware of the work of Mendel, and he assumed that inheritance was blending. He thought the hereditary material was continuous so that offspring were genetically intermediate between their parents. On the basis of such assumptions, Galton pioneered the technique of using

Box 4.1 Characteristics of basenjis and cocker spaniels and their hybrids

Basenjis (left) and parental cocker spaniels (right) used in the genetic studies of Scott and Fuller, 1965 (*Photograph: J.P. Scott*).

Basenji puppies tend to be wild in contrast to the tameness and friendliness of spaniel puppies. They avoid being handled and struggle against restraint. Handling tests showed that the F behaviour of basenjis is similar to parental basenji behaviour, suggesting that dominant genes are involved. The results of back cross experiments pointed to a single dominant gene controlling wildness, normally present in basenjis. The tameness of spaniels is controlled by a single recessive gene. To take another example, basenji females come into oestrus once per year, at the time of the autumnal equinox. Spaniels, like most European breeds, may show heat once every six months, at any time of year. The basenji type of oestrus cycle appears to be controlled by a single recessive gene.

Table 4.1 Characteristics of basenjis and cocker spaniels

Characteristic	Basenji	Cocker spaniel	Most likely mode of inheritance
Wildness and tameness			
Avoidance and vocalization in reaction to handling	High	Low	One dominant gene for wildness
Struggle against restraint	High	Low	One gene with no dominance
Playful aggressiveness at 13 to 15 weeks of age	High	Low	Two genes with no dominance
Barking at 11 weeks			
Threshold of stimulation	High	Low	Two dominant genes for low threshold
Tendency to bark a small number of times	High	Low	One gene with no dominance
Sexual behaviour (time of oestrus)	Annual	Semi-annual	Basenji type as a recessive gene
Tendency to be quiet while weighed	Low	High	Two recessive genes for high tendency

After Scott and Fuller, 1965.

correlations among relatives as a means of estimating heritability. In 1918, Ronald Fisher published a paper that showed that it was possible to predict the correlations to be expected among relatives, on the basis of Mendelian theory. Although Mendel's work had been brought to public attention in 1900, it had been applied only to discontinuous traits. Fisher showed how Mendelian theory could cope with continuous traits such as height and weight. It is interesting that his conclusions were similar to those of Galton, with many predictions of the two theories being almost identical.

The concept of **heritability** was first defined as the fraction of the observed variance which was caused by differences in heredity (Lush, 1940). In other words, when we examine a particular trait, such as bodyweight, in a population of animals, we are looking at a phenotypic value which is made up of both genotypic and environmental components. By looking at related populations in different circumstances, it is sometimes possible to estimate how much of the variability between individuals is due to environmental factors, and how much to genetic factors.

There are various statistical methods for estimating the heritability of behaviour. Some are based upon examining phenotypic variation in genetically identical individuals. The total phenotypic variability among genetically identical individuals is compared with the total phenotypic variability in a natural, genetically variable population. The comparison generates a ratio which contrasts the genetic versus environmental components of phenotypic variability. This technique has certain technical statistical problems, but it also makes clear the fallacy of calling one behaviour genetic and another environmental. If a new gene variant appears, or if the population is exposed to new types of environmental variation, the ratio may well be altered. A behaviour pattern which is 'genetically determined' in one environment (identical twins reared apart are found to have identical phenotypes) may be shown to have a strong environmental component in a different climate (identical twins reared apart are found to have divergent phenotypes). Conversely, a behaviour pattern which is 'environmentally determined' (genetically variable individuals all have identical phenotypes when reared in the same environment) can come to have an apparently strong genetic component in an environment in which the genetic variability is expressed. Because of these interactions, the determination of behavioural heritability, just as with the distinction between innate and learned behaviour, is always determinable only within very strict limits. We can say that a given behaviour is 80 per cent heritable across the range of environments and genotypes in a specific population, but we cannot generalize to other environments or populations.

Estimates of heritability may be subject to error from various sources, the most important of which are genotype–environment interaction and genotype–environment correlation. Interaction between genotype and environment introduces a variability that is not taken account of in normal heritability calculations. For example, Tryon (1942) was able to breed rats selectively for maze dullness and maze brightness. These rats were bred and tested in ordinary laboratory conditions. Cooper and Zubek (1958), however, raised some rats in ordinary conditions and others in impoverished conditions with bare surroundings or in enriched conditions with complex

Fig. 4.10 Mean number of errors made in a maze by 'bright' and 'dull' rats reared in enriched, normal and impoverished environments (After Cooper and Zubek, 1958).

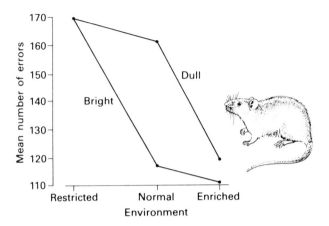

maze-like surroundings. When tested in a standard maze, the maze-bright and maze-dull strains of rats that were bred in ordinary conditions showed large differences in performance, as other workers had found previously. However, both strains of rats raised in a restricted environment performed equally badly in the test maze, while both strains raised in the enriched environment performed equally well, as illustrated in Figure 4.10. Although a definite genotypic difference exists between the two strains, this is only evident phenotypically in rats raised in particular environmental conditions.

Correlation between genotype and environment, another potential source of error in estimating heritability, can arise when individuals select particular environments or develop particular habits to compensate for genetic defects. Despite these difficulties, however, heritability measures have proved to be particularly useful in separating the effects of nature and nurture.

Human heritability studies

Estimates of heritability based on similarity among relatives are much used in human behaviour genetics. There are, however, a number of complicating factors. The first problem is that genetic dominance lowers the correlation among relatives. Under genetic dominance the parental contributions to a trait are not equal, and an extra source of variation is introduced. This is called the dominance variance. However, dominance variance has differential effects on parent–offspring correlations and sibling–sibling correlations. In fact, the difference between these two correlation coefficients should be exactly one-quarter of the proportion of the total variance due to dominance (Bodmer and Cavalli-Sforza, 1976).

The second problem is that, for certain traits, mating is likely to be non-random in humans. While we can assume that random mating occurs with respect to unobservable traits like enzyme levels, this is not the case with traits such as height and intelligence. Correlations between husband and wife are usually positive for such traits, indicating non-random, or assortative, mating such that people tend to take partners of similar height and I.Q. The husband–wife correlation for height is about +0.3, while for I.Q. it is about +0.4. These correlations may be influenced by social effects because positive correlations also exist between these variables and socioeconomic status. The tendency to choose a mate from the same social group, therefore, would lead to positive correlations between husband and wife. Assortative mating has the effect of increasing the frequency of homozygotes.

The third problem lies in estimating environmental variance in human studies. Environmental conditions cannot be controlled as they can in some animal studies. An approach adopted by Cavalli-Sforza and Bodmer (1971) is to divide the environmental variance into separate components. Initially, we have the variance among individuals within the family. This may result from age differences, birth order, sex differences, etc. Families of different sizes will usually experience different environmental effects such as nutritional level and degree of crowding.

Next we have variance among families within socioeconomic groups. Correlations between adopted children and their foster parents can give some estimate of the importance of such effects, but the results may be biased by the selection of parents by adoption agencies.

We then have the variance among socioeconomic groups. Cultural inheritance may give rise to correlations among relatives that are difficult to distinguish from those due to genetic factors. Cultural transmission from parent to child is confounded with biological inheritance to a considerable extent (Cavalli-Sforza and Feldman, 1974). This remains one of the outstanding problems of biology.

We may also have to consider the variance in environmental conditions accompanying racial differences. For example, the environmental conditions typical of black and white Americans show considerable variation. In part, these may be due to cultural differences and in part to socioeconomic factors.

Finally, we have the variance due to interaction between the genotype and the environment. In humans, such interaction may occur, for example, between age of maturity and the type of schooling available to different age groups.

Overall, we can specify the variance due to environmental factors by adding the component variances outlined above.

Cavalli-Sforza and Bodmer (1971) concluded that estimates of heritability are not adequate yardsticks of the relative importance of nature and nurture, except in particular circumstances. Changes in environmental conditions can invalidate even straightforward measures of heritability. For example, although the heritability of human stature is high, there have been large changes in the average stature of some human populations. These are due to environmental factors like improved nutrition. While the heritability measures provide an indication of the ratio of prevailing genetic differences to the ratio of prevailing environmental differences, they cannot be extrapolated to other populations, environments, or time periods.

Points to remember

- In this chapter we review some methods of investigating genetic influences upon animal behaviour. We see that in a few cases it is possible to demonstrate the effect of single genes; usually however, many genes are involved in even simple aspects of behaviour.

- Polygenic inheritance can be investigated by breeding experiments designed to study the effects of chromosome mutations upon behaviour or to study quantitative differences in the behaviour of different genetic strains.

- The heritability of behaviour is measured in terms of the variance in phenotypic traits within a particular population. The heritability of a trait is the fraction of the observed variance which is due to differences in heredity. It can be studied by looking at differences in closely related animals raised in different environments, and at differences among animals raised in the same environment.

Further reading

Bodmer, W.F. and Cavalli-Sforza, L.L. (1976) *Genetics, Evolution, and Man*. Freeman, New York.

Ehrman, L. and Parsons, P.A. (1976) *The Genetics of Behavior*. Sinauer Associates, Sunderland, MA.

Natural selection

The hawkmoth caterpillar (*Leucorampha*) mimicking the snake *Bothrops schlagelli* (*Photograph: Nicholas Smythe*).

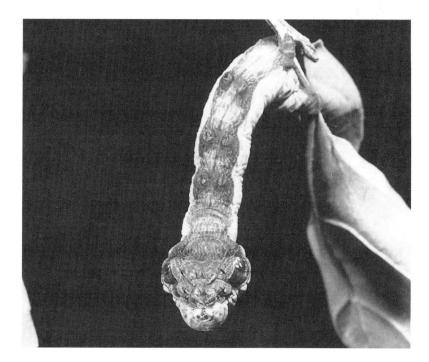

In this section we discuss the elements of evolutionary theory as they affect behaviour. The aim is to provide a basic understanding of some key concepts and to show how these apply to the study of behaviour. Chapter 5 introduces the theory of natural selection and some assessment of the evidence. It also outlines the physical and biotic pressures of natural selection that shape the evolution of animal behaviour. Chapter 6 gives an account of some of the classical experiments demonstrating the survival value of behaviour. It also provides an introduction to the various concepts of fitness that are so important to evolutionary biology. Chapter 7 introduces the concept of evolutionary strategy, including evolutionarily stable strategies (ESS).

Charles Darwin (1809–1882)

Charles Robert Darwin was born at Shrewsbury, England, in 1809. His father, Robert Darwin, was a physician, and his grandfather, Erasmus Darwin, was a noted biologist. Although interested in natural history, Darwin showed no early academic promise. He studied medicine briefly at the University of Edinburgh, and then transferred to the University of Cambridge to study for the ministry. In 1831 he was offered a place as a naturalist on HMS *Beagle*. The *Beagle* carried Darwin to Brazil, Argentina, the Falkland Islands, Tierra del Fuego, Chile, the Galapagos Islands, Tahiti, New Zealand, Tasmania, and Australia. He discovered both the riches of the tropical forest and the fossilized bones of long-extinct reptiles. He experienced earthquake and insurrection. Darwin kept extensive notes on animals, plants, fossils, geological formations, and coral reefs. He thought a great deal about the adaptiveness, geographical variation, and competitiveness that characterizes living things. Upon returning to England, Darwin initially lived in London but then settled in Kent, where he lived for the rest of his life. He died there in 1882.

Darwin's thinking was greatly influenced by Lyell's *The Principles of Geology*, which he took with him on the *Beagle*. This book presented the view that the history of the world is the result of natural laws and forces that are active and observable in the present-day world. It argued against the prevailing view that the world was shaped by a series of catastrophes, either natural or divinely inspired.

Darwin's voyage on the *Beagle* opened his eyes to the variability of species and to their remarkable adaptations to the environment. In 1838 Darwin read Malthus's *Essay on Population*, first published in 1798. In this essay Malthus pointed out that populations tend to increase geometrically, and maintained that food supplies could increase only in arithmetic ratio. He argued that population increase will inevitably be checked by starvation or some other form of 'vice and misery.' Darwin realized that competition for scarce resources would inevitably lead to the survival of those individuals with attributes that gave them an advantage in the struggle for life. He knew that domestic animals and plants could be changed over generations by artificial selection, and he began to form the idea that the characteristics of wild species were established through natural selection.

Darwin wrote a first draft of his theory in 1842 and another in 1844. He intended to marshal the evidence necessary for a much larger work, but in 1858 he received a letter from Alfred Russel Wallace outlining a theory of evolution by natural selection, which Wallace had conceived independently. Darwin and Wallace jointly published a preliminary sketch of their theory in 1858 and in 1859 Darwin published his book entitled *On the Origin of Species by Means of Natural Selection, or the Preservation of Favoured Races in the Struggle for Life*.

This book created a revolution in scientific thought and provided the foundation of modern biology. Darwin wrote a number of other books that have implications for the study of animal behaviour and that greatly influenced the development of ethology, the most important of which were *The Descent of Man and Selection in Relation to Sex* (1871) and *The Expression of the Emotions in Man and Animals* (1872).

5.1 | The evidence for evolution by natural selection

Although Darwin could not point to natural selection in action, he noted that artificial selection was practised by man in the domestication of animals. The rapid changes that occur in animals under domestication show that evolution can result from selection, in this case from selective breeding. He also put forward many other types of evidence to support his theory.

In Darwin's day most people were resistant to the idea of evolutionary change, preferring to imagine that the world has always remained much the same. However, during the late eighteenth and early nineteenth century geologists discovered numerous fossils and they realized that many of them were the remains of species that were no longer living. Initially it was thought that climatic catastrophes could account for the extinction of species and separate creation for the appearance of others. As more and more fossils were discovered, however, it became possible to trace changes in some species which had occurred over many thousands of years. Much of the fossil evidence was known to Darwin and was used by him as evidence that some kind of evolution had occurred (Romer, 1958).

Darwin also regarded the resemblances between living species as important evidence for evolution. For example, most vertebrates have the same basic bone structure in their forelimbs. This would not be expected if species were independently created, but it would be expected among species that had evolved from a common ancestor. Darwin pointed out that parts that were functional in some species were vestigial in others. Thus pigs walk on two toes and have two other vestigial toes that protrude from the leg well above the ground. In his *The Descent of Man* (1871) Darwin wrote:

'It is notorious that man is constructed on the same general type or model with the other mammals. All the bones in his skeleton can be compared with the corresponding bones in a monkey, bat or seal. So it is with his muscles, nerves, blood vessels and internal viscera. The brain, the most important of all the organs, follows the same law.'

Darwin realized that similarities between species were sometimes (though not always) due to common ancestry, while differences between closely related species were due to evolutionary adaptation to different environments. By comparing species one can often make important deductions about their biology, and this comparative approach is an important part of ethology.

During his voyage on the *Beagle* Darwin discovered that the geographical distribution of animals was hard to account for in terms of separate creation. The presence of subspecies or variants of a species on neighbouring islands, for example, is easy to account for in terms of evolution, but it is hard to see why a creator should choose to make such minor variations. During his voyage on the *Beagle*, Darwin visited the Galapagos archipelago situated some 600 miles off the west coast of South America. In the *Voyage of the Beagle* (1841) Darwin wrote:

'I never dreamed that islands about fifty or sixty miles apart, and most of them in sight of each other, formed of precisely the same rocks, placed under a quite similar climate, rising to nearly equal height, would have been differently tenanted ... It is the circumstance, that several of the islands possess their own species of the tortoise, mocking thrush, finches and numerous plants, these species having the same general habits, occupying analogous situations, and obviously filling the same place in the natural economy of the archipelago, that strikes me with wonder.'

In *On the Origin of Species* (1859) he wrote:

'How has it happened in the several (Galapagos) islands ... that many of the immigrants should have been differently modified, though only in a small degree. This long appeared to me a great difficulty: but it arises in chief part from the deeply-seated error of considering the physical conditions of a country as the most important for its inhabitants: whereas it cannot be disputed that the nature of the other inhabitants with which each has to compete, is at least as important, and generally a far more important element of success ... When in former times an immigrant settled on any one or more of the islands, or when it subsequently spread from one island to another, it would undoubtedly be exposed to different conditions of life in the different islands, for it would have to compete with different sets of organisms ... If then it varies, natural selection would probably favour different varieties in the different islands.'

The tendency for closely related species to diverge in characteristics that reduce competition between them is nowadays called **character displacement**. Darwin's finches, the Geospizini which inhabit the Galapagos Islands, are a well-studied example of this phenomenon (see Box 1.1).

5.2 Frequency distribution of phenotypes

Within a given population, each characteristic of the species varies between individuals, and one can plot the frequency distribution of the various values. For example, in a population of mice, the size of the litter might range from two to 16. If we were to plot the percentage of litters of each possible size on a graph like that illustrated in Figure 5.2, then we would have established the frequency distribution. If there were no variation among offspring and if individuals of all phenotypes had identical numbers of surviving offspring, then the form of the frequency distribution would remain unchanged from one generation to the next. Where a change in the frequency distribution persists for more than one generation, evolution may be said to have occurred. Changes can occur if some phenotypes leave fewer surviving offspring than others. Mice with a small litter size may or may not have the same number of surviving offspring as mice with a large litter size. The former may have more litters than the latter, or the survival rate may be lower in large litters than in small ones. However, a change in circumstances may favour litters of a particular size so that more mice from such litters survive and reproduce. Evolution of litter size will then begin.

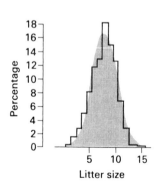

Fig. 5.2 Frequency distribution of litter size in mice (After Falconer, 1960).

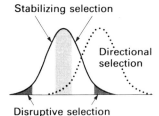

Fig. 5.3 Section of a normal population distribution favoured under disruptive, directional and stabilizing selection.

Natural selection can have a variety of effects upon the frequency distribution of phenotypes within a population. As a result of mutation, recombination and genetic drift, the variance of a character tends to increase from one generation to the next. This tendency is usually opposed by **stabilizing selection**, which acts relatively more severely at the extremes of a frequency distribution, as illustrated in Figure 5.3. When there is a differential selective pressure along a phenotypic gradient, **directional selection** occurs. This may result in a shift in the mean of the frequency distribution or in a skewed distribution without alteration of the mean, as in Figure 5.3. Sometimes there may be strong selective pressures against characters that are typical of the population, so that selection is stronger against animals near the mean than against extreme individuals. Such **disruptive selection** is rare, and it tends to result in a bimodal frequency distribution. In some cases it may lead to division of a species into two separate species.

The relative abundance of different genes can change at random from generation to generation by accident without the action of natural selection. This effect, known as **random genetic drift** (Wright, 1921) is greatest in small populations. Figure 5.4 illustrates how a particular genotype can be eliminated from a small population purely by chance. Variation due to chance fluctuation sometimes gives rise to unusual proportions of genotypes, especially in small, isolated populations (see Figure 5.5). For instance, in some isolated alpine villages the frequency of albinism is ten times higher than normal. Other isolated human communities are known to have abnormal incidences of colour blindness, deaf-mutes and certain types of mental deficiencies.

The rate of change of gene frequency induced by natural selection depends upon the **relative fitness** of the various genotypes. The relative fitness of a genotype reflects the difference in fitness between that genotype and the

Fig. 5.4 Schematic representation of genetic drift. In each generation each individual of the grey and white genotypes produces two identical offspring. Half the juveniles die and the population remains constant in size. However, which juveniles die is entirely random and not dependent on genotypes. The frequency of each fluctuates at random until by chance grey becomes extinct and white is fixed in the population (After Futuyma, 1986).

Fig. 5.5 Genetic variation as a function of human population density. The graph summarizes the result of early study in which the Italian province of Parma was divided into areas and the genetic variation between villages measured in each area (Adapted from Cavalli-Sforza, 1969).

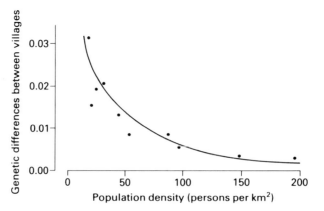

others represented in the same population. If the fitness of the genotype with the highest rate of increase is designated 1, the relative fitness of another genotype with, say, 85 per cent as great a rate of increase would then be 0.85. The difference, 0.15 is called the **coefficient of selection** against the inferior genotype.

Selection coefficients can be used to calculate the rate of change of phenotype in a population. An interesting example arises from the domestication of animals some 10 000 years ago. Before the domestication of animals, the only source of milk for human babies was the human mother's milk given during the first years of life. The child's stomach produces the enzyme lactase, which is necessary for the digestion of the lactose sugar present in milk. In later life the enzyme disappears, and among many present-day human populations, it is completely absent in adults.

People who lack lactase are intolerant to lactose and experience nausea, vomiting and abdominal pains if they drink milk. The retention of lactase in later life is an inherited trait, which is thought to be the result of a mutation that became advantageous in populations for which milk became a regular part of the diet. Presumably, the mutant was rare before the domestication of animals, when humans lived entirely by hunting animals and gathering plants. Thus, in some populations the lactose-intolerant genotype must have increased from an initially low frequency to the present high frequency among northern Europeans and other milk drinkers.

Let us assume that the human generation time is 30 years on average. On this basis, 300 generations have passed since the beginning of domestication. During this period, the proportion of adults that can tolerate lactose has risen from almost zero to 75 per cent among northern Europeans. If we assume a single gene is responsible for the trait, the gene frequency must have increased from about 0.001 per cent to the 50 per cent required to account for the present genotype frequency of 75 per cent. It is possible to calculate that a selection coefficient of 0.04 would be necessary to achieve this change. If the initial frequency of the lactose-tolerant gene was 1.0 per cent then a selection coefficient of 0.015 would be sufficient (see Figure 5.6). This means that a relatively rapid change, in evolutionary terms, can take place even if the difference in fitness between two alleles is less than 5 per cent.

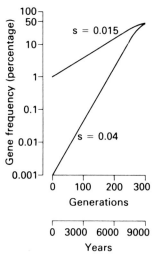

Fig. 5.6 Selection for lactose tolerance of adults. Assuming 300 generations have elapsed since cattle were first domesticated and it became advantageous to be able to digest milk, then the present gene frequency of 50 per cent can be explained in terms of a particular coefficient of selection(s) acting upon the assumed initial frequency of the gene (After Bodmer and Cavalli-Sforza, 1976).

It is possible to calculate (e.g. Futuyma, 1979), that an advantageous allele increases in frequency at a rate that is proportional to the selection coefficient and to the frequencies of both the advantageous allele and the disadvantageous allele. The rate of genetic change therefore is only high when both genes are common in the population. New mutations will initially increase in frequency very slowly. Moreover, complete replacement of one allele by another will take a very long time because the process is slow initially when the favourable allele is rare and, terminally, when the disadvantageous allele becomes rare. Deleterious recessive alleles are seldom completely eliminated by natural selection because they show themselves phenotypically only when homozygous. Natural populations, therefore, have many deleterious recessive genes that occur at low frequency and that create a kind of genetic reservoir.

The rate of evolution of a gene depends upon the phenotypic expression of the trait as well as the gene frequencies and the coefficient of selection. For example, a rare advantageous dominant gene increases in frequency much more rapidly than a rare advantageous recessive gene, as illustrated in Figure 5.7. However, a dominant gene will achieve complete fixation (replacement of the competing allele) more slowly than the recessive because the last few recessive genes occur in heterozygous form and are thus shielded from natural selection.

No animal species exists in the absence of other life, and the processes of adaptation of one species may change the conditions under which another species lives. This may induce compensatory adaptation in the affected species. Thus if a bird species adapts to the scarcity of its principal prey species by broadening its diet to include new insects, then a new selective agent is introduced into the ecology of the newly preyed-upon insect population. Because such interconnections exist in every ecosystem, we must always imagine the behaviour of an animal as embedded in a complex of ecological relationships.

5.3 Ecology and behaviour

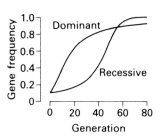

Fig. 5.7 A rare advantageous dominant gene increases more rapidly than a rare advantageous recessive, but it achieves fixation more slowly because the last few recessive genes are protected in the heterozygous form.

In considering the behaviour of animals in the natural environment, it is important to have some understanding of the consequences of the behaviour in terms of the survival of the animal. The consequences of a particular activity will depend largely upon the animal's immediate environment. If the animal is in a situation to which it is well adapted, the consequences of the activity may be beneficial. The same activity performed in a different environment may be harmful. In order to appreciate how the behaviour of animals has been shaped during evolution, we need to understand the adaptive relationship between the animal and its environment.

Ecology, or scientific natural history, is the study of the relationships between animals and plants in the natural environment. It is concerned with all aspects of these relationships, including the flow of energy in the environment, the physiology of the animals and plants, animal populations, animal behaviour, etc. In addition to specific knowledge of particular animals, the ecologist seeks to understand the general principles of ecological organization, and we are concerned with some of these here.

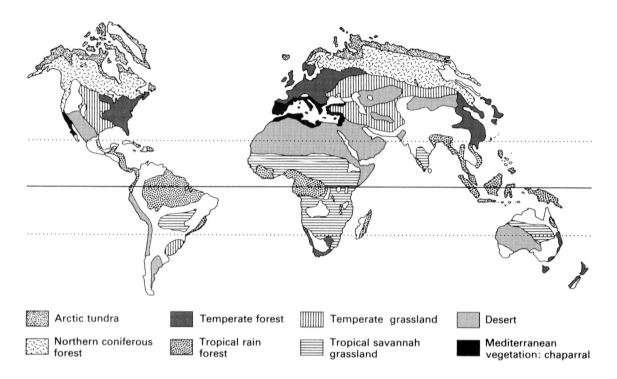

Legend			
Arctic tundra	Temperate forest	Temperate grassland	Desert
Northern coniferous forest	Tropical rain forest	Tropical savannah grassland	Mediterranean vegetation: chaparral

Fig. 5.8 The distribution of the main terrestrial biomes of the world.

During the course of evolution, animals become adapted to a particular environment, or habitat. Habitats are usually described in terms of the physical and chemical features of the environment. Plant communities depend upon the physical features of the environment, such as the type of soil and the climate. The plant communities provide a variety of possible habitats that animals are able to exploit. The association of plants and animals, together with the physical features of the habitat, constitute an ecosystem. There are about ten general types of ecosystem found in the world, and these are known as **biomes**.

Figure 5.8 shows the distribution of the major terrestrial biomes of the world. There are also marine and freshwater biomes. To take one example, the savannah biome occupies large areas of Africa, South Amercia and Australia. It consists of a combination of grassland and scattered trees and is found in tropical or subtropical parts of the world. Savannah is typified by seasonal rainfall. At the higher end of the rainfall range, the savannah grades into tropical forest, and at the lower end it falls into desert. The dominant trees of the African savannah are the acacias; in South America they are the palms and in Australia, the eucalyptus. The characteristic feature of the African savannah is the large variety of grazing ungulate species, which supports a variety of carnivores. In South America and Australia the same niches, or roles, are occupied by other species.

The association of animals and plants living in a particular habitat is called a **community**. The species that make up a community may be classified into producers, consumers and decomposers. The **producers** are the green plants that trap solar energy and convert it into chemical energy. The **consumers** are the animals that eat the plants or that eat herbivorous animals and thus depend indirectly upon plants for energy. The **decomposers** are usually fungi and bacteria that break down animal and plant material into a form that can be reused by plants.

The **niche** is the role the animal plays in the community in terms of its relationship both to other organisms and to the physical environment. Thus, a herbivore eats plant material and usually is preyed upon by carnivores. The species occupying a given niche varies from one part of the world to another. For example, a small herbivore niche is occupied by rabbits and hares in northern temperate regions, by the agouti and viscacha in South America, by the hyrax and mouse deer in Africa and by wallabies in Australia.

In 1917 the animal ecologist Joseph Grinell pioneered the study of niches in his research on the California thrasher (*Toxostoma redivivum*). This bird nests in dense masses of foliage one or two metres above the ground. This nesting habit is an aspect of the animal's niche. In mountain areas, the necessary vegetation is available only in that ecological community called the **chaparral**. The habitat of the thrasher, described in terms of the physical characteristics of the environment, is determined partly by the response of the thrasher population to the niche situation. This, in turn, depends upon the status of the thrasher in the ecological community. Thus, if height of the nest above ground were a critical factor in the avoidance of predators, there would be competition within the population for nest sites at the optimal height. If this factor were less critical, more individuals could nest in sub-optimal sites. The niche can also be affected by competition from other species, for nesting sites, food, etc. The habitat of the California thrasher is determined partly by the niche situation, by the distribution of the various shrub species that characterize the chaparral and by the population density of the thrasher. Clearly, if the population were small, only the best nest sites would be occupied, and this would affect the habitat of the species. Thus, the thrasher's total relationship to its environment, for which the term **ecotope** is sometimes used, results from a complex interaction of niche, habitat and population factors.

When animals of different species use the same resources or have certain preference or tolerance ranges in common, **niche overlap** occurs (see Figure 5.9). Niche overlap leads to competition, especially when resources are in short supply. The **competitive exclusion principle** states that two species with identical niches cannot live together in the same place at the same time, when resources are limited. The corollary is that, if two species coexist, there must be ecological differences between them.

As an example, let us consider the niche relationships of a group of bird species that comprises a 'foliage-gleaning guild', feeding on oak trees on coastal mountains in central California (Root, 1967). A **guild** is a group of species that exploits the same environmental resources in a similar manner. Such species overlap significantly in their niche requirements, and thus, they compete with each other. One advantage of the guild concept is that it focuses attention on all competing species in a given geographic area, without regard to their taxonomic relationship. To be considered as a member of the foliage-gleaning guild in the oak woodland, the major portion of the

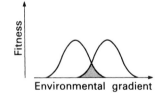

Fig. 5.9 Niche overlap. Along an environmental gradient, such as temperature, the fitness of an animal can often be represented as a bell-shaped curve. Niche overlap (shaded area) occurs along that part of the gradient shared by representatives of different species.

animal's diet must consist of arthropods obtained from the foliage. This is an arbitrary classification since a species may be a member of more than one guild. For example, the plain titmouse (*Parus inornatus*) belongs to the foliage-gleaning guild on account of its foraging habits, and it is also a member of the hole-nesting guild by virtue of its nesting requirements.

Although there are five species feeding on insects in the same area, each tends to eat insects from characteristic size distribution and taxonomic categories that were different from those of the other species. An overlap exists between the taxonomic categories of insects eaten by the five bird species but that each species tends to specialize on certain taxa. Similarly, Figure 5.10 shows complete overlap in size, but different mean values and variances at least in some cases. Root also found that the bird species fell into three general categories with respect to foraging technique:

- gleaning consists of taking prey from leaf surfaces while the bird walks on a solid substrate
- hovering involves taking prey from leaf surfaces while in flight
- hawking consists of taking insects on the wing

The proportions of time that the five bird species devote to these methods of foraging are illustrated in Figure 5.11. In this example we see the operation of ecological specialization in behaviour. The behaviour of each species influences that of the others such that the species of the guild are spread out over the available foraging sites and prey types.

Competition often results in dominance of one species over another, in the sense that the dominant species has priority in use of resources such as food, space and shelter (Miller, 1967; Morse, 1971). On theoretical grounds, we would expect that a species that becomes subordinate to another species should shift its use of resources in a way that decreases overlap with the dominant species. This typically involves the subordinate's reducing its use of certain resources, thus reducing its niche breadth. In some circumstances it may be possible for the subordinate species to expand its niche to include hitherto unutilized resources. This would imply either that the subordinate species is able to dominate another species in an adjoining niche or that it is able to make fuller use of its own fundamental niche.

If a subordinate species is to survive in competition with a dominant species, it must have a fundamental niche that is broader than that of the dominant species. Such cases have been documented for bees and new-world blackbirds (Orians and Wilson, 1964). As dominant species have priority in use of resources, subordinate species may be excluded from areas of the niche space when resources are limited in availability, are unpredictable or require considerable effort to find. This may mean that the fitness of subordinate species is reduced considerably in regions of overlap. In such cases we would expect subordinate species to be under considerable selective pressure to modify their fundamental niche by becoming specialists or by evolving a tolerance for a wider range of physical environmental conditions.

Fig. 5.10 Size distribution of intact prey in the stomachs of foliage-gleaning birds (After Root, 1967).

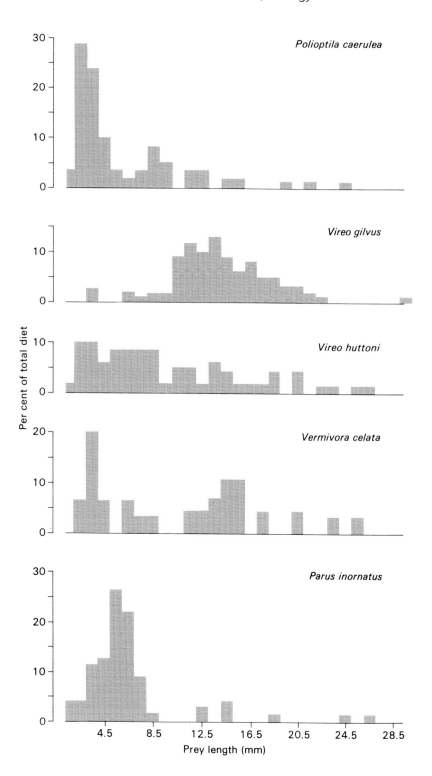

Fig. 5.11 Three types of foraging tactics by members of a foliage-gleaning guild are represented as the three sides of a triangle. The length of line perpendicular to a side of the triangle is proportional to the percentage of time spent in that foraging tactic. The sum of all three lines for each species equals 100 per cent (After Root, 1967).

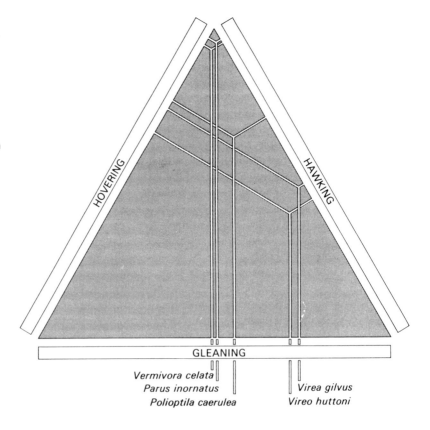

HOVERING

HAWKING

GLEANING

Vermivora celata
Parus inornatus
Polioptila caerulea
Virea gilvus
Vireo huttoni

5.4 The adaptedness of behaviour

Naturalists and ethologists have discovered numerous examples of the marvellous ways in which animals appear to be perfectly adapted to their environment. The problem with such accounts of animal behaviour is that they seem convincing precisely because the various details and observations fit together so well; that is, a good story may seem convincing simply because it is a good story. This is not to say that there is anything wrong with a good story. Indeed, in any correct account of behavioural adaptation, the variety of details and observations should fit together to make a coherent whole. The problem is that biologists, as scientists, should evaluate the evidence, and a good story is not necessarily good evidence. As in a court of law, the evidence must be more than circumstantial and must have some element of independent corroboration.

One way to obtain evidence about the adaptedness of behaviour is to compare related species that occupy different habitats. A classic example of this approach is Ester Cullen's (1957) comparison of the nesting habits of cliff-nesting kittiwakes (*Rissa tridactyla*) with those of ground-nesting gulls such as the black-headed gull (*Larus ridibundus*) and the herring gull (*Larus argentatus*). The kittiwake nests on cliff ledges that are inaccessible to predators and is thought to have evolved from ground-nesting gulls as a result of predation pressure. Kittiwakes retain some features of ground-nesting gulls, like the partial cryptic coloration of their eggs. The eggs of ground-nesting birds are usually well camouflaged as a protection from predators, but this cannot be the function of egg coloration in kittiwakes because each nest is marked conspicuously by white droppings. The adults and young of ground-nesting gulls are careful not to defecate near the nest and thus advertize its whereabouts. Thus, it seems most likely that the camouflage of kittiwake eggs is a vestigial indication of their ground-nesting ancestry.

Cullen (1957) studied a breeding colony of kittiwakes in the Farne Islands, off the east coast of Britain, where they nest on very narrow cliff ledges. She observed that the eggs were free from ground predators like rats and from aerial predators like herring gulls, which frequently prey on the eggs of ground-nesting birds. Kittiwakes feed mainly on fish and plankton and do not cannibalize the eggs and chicks of their neighbours as do many ground-nesting gulls. The kittiwakes seem to have lost many of the anti-predator adaptations of other gulls. For example, as well as failing to camouflage the nest, they rarely give alarm calls and they do not mob predators.

Kittiwakes have many special adaptations to cliff nesting. They are lightly built and have strong toes and claws that enable them to cling to ledges too small for other gulls to manage. Compared with ground-nesting gulls, kittiwake adults appear to have a number of behavioural adaptations to their cliff habitat. Their fighting behaviour is limited and stereotyped compared with that of ground-nesting relatives (Figure 5.12). They build fairly elaborate cup-shaped nests using both sticks and mud, whereas ground-nesting gulls build rudimentary nests of grass or seaweed, without using mud as cement. Young kittiwakes are different in many ways from the young of ground-nesting gulls. For example, they stay in the nest for a longer period and face the cliff wall much of the time. They take regurgitated food directly from the

Fig. 5.12 Red-legged kittiwakes (*Rissa brevirostris*) nesting on the cliff ledges off the Pribilof Islands, Bering Sea. Note the restricted form of fighting (*Photograph: Ronald Squibb*).

throat of the parent, whereas most gulls pick it up from the ground where it has been deposited by the adult. The young of ground-nesting gulls run and hide when alarmed, but kittiwake young remain in the nest. The young of ground-nesting gulls are cryptic in both appearance and behaviour: kittiwake young are not.

Comparison of species can provide evidence about the function of behaviour patterns in the following ways. When a behaviour pattern is observed in one species but not in the other, there may be correlated differences in the way that natural selection has exerted its effects upon the two species. For example, herring gulls remove broken eggshells from the region of the nest, and a possible function of this behaviour is that it helps to maintain the camouflage of the nest because the white interior of the eggshell is conspicuous. Evidence in support of this hypothesis comes from the observation that kittiwakes do not remove broken eggshells. As we have seen, kittiwakes are not subject to much nest predation, and their nests and chicks are not camouflaged. If eggshell removal functions primarily to maintain nest camouflage, then we would not expect to find it among kittiwakes. If, however, it serves some other function like prevention of disease, then we might expect to observe it in kittiwakes. Kittiwakes usually maintain very clean nests and throw away any foreign objects that appear there. Herring gulls do not normally bother to do this.

The type of evidence discussed here is strengthened if we can show that other related species, faced with similar selective pressures, show similar adaptations. An example is provided by Jack Hailman's (1965) study of the cliff-nesting swallow-tail gull (*Larus furcatus*) of the Galapagos. Hailman investigated various aspects of behaviour that appeared to be relevant to the danger of falling off the cliff. Swallow-tail gulls nest neither on such steep cliffs as kittiwakes nor so high above the ground. Thus, we might expect the relevant adaptations of the swallow-tail gull to be intermediate between those of the kittiwakes and typical ground-nesting gulls. Swallow-tail gulls are subject to greater predation than kittiwakes, and Hailman observed several aspects of behaviour that appeared to be correlated with this difference. For example, as mentioned above, kittiwake chicks defecate on the nesting ledge, thus rendering it conspicuous. The chicks of swallow-tail gulls, however, defecate over the edge of the ledge. In a number of characteristics, likely to be related to the amount of predation, swallow-tail gulls are therefore intermediate between kittiwakes and other gulls. Hailman similarly evaluated those behavioural characteristics of swallow-tail gulls that would appear to be adaptations to the amount of available room for nesting and the availability of nesting sites and nesting materials. He then set out to evaluate the evidence relevant to Cullen's (1957) hypothesis that the peculiarities shown by kittiwakes are the result of selective pressures accompanying cliff-nesting habits. He took 30 characteristics of swallow-tail gulls and divided them into three groups according to the degree of similarity to kittiwake behaviour. Taken as a whole, this comparison supports Cullen's hypothesis that the special features of the kittiwake are the result of selective pressures that accompany cliff nesting.

The comparative approach has proved to be a fruitful method of studying the relationship between behaviour and ecology. There have been several studies of birds (Crook, 1964; Lack, 1968) and also extensive studies of ungulates (Jarman, 1974) and primates (Crook and Gartlan, 1966; Clutton-Brock and Harvey, 1977). Though not without its critics (e.g. Clutton-Brock and Harvey, 1977; Krebs and Davies, 1981), the comparative approach can provide satisfactory evidence concerning evolutionary aspects of behaviour, provided that sufficient care is taken to avoid confounding variables and circular arguments. Hailman (1965) regards the comparative method as satisfactory only in cases where comparison of two populations of animals leads to predictions about a third population that has not been studied already at the time the predictions are formulated. In this way a hypothesis resulting from a comparative study can be tested independently of the data used in the study. Simply to discover that correlated differences exist in behaviour and ecology between two populations is not sufficient evidence to conclude that the characters reflect selective pressures arising from environmental differences between the populations. Differences due to confounding variables or to comparison at inappropriate taxonomic levels may be eliminated by careful statistical analysis (Clutton-Brock and Harvey, 1979; Harvey and Pagel, 1991).

Points to remember

■ Evidence for evolution by natural selection can be found by comparing living species with each other and with their fossil ancestors, and by studying the effects of geographical isolation. There are also some present-day examples of natural selection in action.

■ The frequency distribution of phenotypes may change as a result of selective pressure, and the rate of change depends partly upon the genetic make-up of the population and partly upon the phenotypic expression of the trait.

■ The main types of interaction among animals of different species are predation and competition. In obtaining food, each species occupies a particular niche. Species with similar niches in the same habitat compete for food and may also compete for other resources, such as nest sites. No two species can occupy the same niche in the same habitat.

■ Comparison of closely related species living in different habitats can often reveal those aspects of behaviour which are particularly important in adapting the animal to its environment.

Further reading

Dawkins, R. (1986) *The Blind Watchmaker*. Longman, Harlow.

Dennet, D.C. (1995) *Darwin's Dangerous Idea*. Simon & Schuster, New York.

Futuyma, D.J. (1979) *Evolutionary Biology*. Sinauer Associates, Sunderland, MA.

Harvey, P.H. and Pagel, M.D. (1991) *The Comparative Method in Evolutionary Biology*. Oxford University Press, Oxford.

CHAPTER

6 Survival value and fitness

In Chapter 5 we saw that the survival value of a hereditary trait within a population depends upon the extent to which the trait contributes to reproductive success, which depends partly upon the selective pressures inherent in the environment. A number of features of the environment could jeopardize reproductive success by leading to the death of the parent by starvation, predation, etc.; failure to breed as a result of competition for mates or nesting sites; or failure of the young to thrive due to lack of parental care, food or protection from predators.

In studying the way in which structural or behavioural traits contribute to survival and reproductive success, we can think of the animal as designed by natural selection to fulfil certain functions. In everyday language, the word function refers to the job that something is designed to do. Thus, we might say that the function of a bicycle is to transport a person from place to place. The biologist uses the word function in a slightly different way. Strictly speaking, as we shall see, the function of a trait is to increase, via natural selection, the genetic contribution to future generations.

However, biologists also use the word function in a less rigorous sense to indicate the role of the trait in the survival and reproductive success of the individual animal. For example, a biologist might say that the function of incubation behaviour is to keep the eggs warm up to the point of hatching. Implicit in this statement would be the understanding that birds that do not incubate adequately have low reproductive success. That is, that incubation behaviour has evolved by natural selection, to ensure that the eggs hatch. We have the same type of logic in ordinary language when we say that the function of a bicycle is to transport people, or that a bicycle is designed to transport people.

The idea that the function of incubation behaviour is to keep the eggs warm is only one among a number of possibilities. Incubation behaviour, in some cases, may serve to shield the eggs from the hot sun or to protect the eggs from predation (Drent, 1970). Moreover, we can ask about the costs and benefits of incubation behaviour. Thus, while the benefits may be counted in terms of maintaining the eggs at the correct temperature and protecting them from predators, the incubating bird may suffer costs in terms of its own vulnerability to predators and its lost opportunities for feeding. In addition, there will be other consequences of incubation that may or may not affect future survival and reproductive success: the eggs are kept dry and the owner's presence on the territory is maintained. For the time being, however, we confine ourselves to the more straightforward ways of investigating the survival value of behaviour.

6.1 Experimental studies of survival value

We have seen that natural selection acts upon the phenotype of an individual and that its effectiveness in changing the nature of a population depends upon the degree to which the phenotypic characteristics are controlled genetically. Thus, the effectiveness of natural selection depends upon the genetic influence that an individual can exert upon the population as a whole. Obviously, an individual that has no offspring during its lifetime will exert no genetic influence upon the population, however great its ability to survive in the natural environment. Evolutionary biologists distinguish between two aspects of survival value. First, the survival value of the traits of a particular individual may be estimated, as we see in Chapter 10. Thus individuals with traits of high survival value are said to be well adapted to the environment in that they efficiently obtain food, avoid predators, etc. Second, the survival value of a trait within a population depends upon the extent to which the trait contributes to reproductive success.

The term **survival value** is akin to the concept of **fitness**. Fitness is a measure of the ability of genetic material to perpetuate itself in the course of evolution. It depends not only upon the individual's ability to survive but also upon its rate of reproduction and the viability of its offspring. In this book we use the term survival value to designate factors relevant to the survival of the individual and the term fitness in reference to long-term reproductive success.

Estimating the survival value of behaviour is partly a matter of conjecture, since we have no direct way of measuring the selective pressures of the past. As we have seen, one of the problems in interpreting the results of a comparative study is that differences among species are the result of events that occurred long ago. However, we can ask questions about present-day selective pressures, and one way of doing this is to conduct experiments.

Eggshell removal in the black-headed gull

A classic example of the experimental approach is the study by Niko Tinbergen and his co-workers (1962) on eggshell removal in the black-headed gull (*Larus ridibundus*). Many birds dispose of the empty eggshell after the chick or nestling has been hatched. This may be done in a variety of ways, but usually the eggshell is trampled into the nest, eaten by the parent or picked up and carried away (Nethersole Thompson, 1942). The black-

Fig. 6.1 Black-headed gull (*Larus ridibundus*) removing an eggshell from its nest (*Photograph: Niko Tinbergen*).

headed gull removes the eggshell by picking it up in its bill (Figure 6.1) and flying some distance away before dropping it. This is done invariably within a few hours of hatching. Although eggshell removal takes only a few minutes, by leaving the nest unguarded the parent is undoubtedly exposing its chicks to predation. It would seem, therefore, that there must be considerable survival value attached to eggshell removal for the parent to take such a risk.

Tinbergen and his co-workers considered various possible ways in which disposal of the eggshells might benefit the gulls: for example, the sharp edges of the shell might injure the chicks; the shell might slip over an unhatched egg, thus trapping the chick in a double shell; the shells might interfere with brooding; the inside of the shell might become a breeding ground for infectious organisms; or the white inside of the shell might attract the attention of predators.

The results of their previous work gave the researchers some clues. First, black-headed gulls remove not only eggshells but also many other objects of equivalent size, even those that do not resemble eggshells in any obvious respect – almost any foreign object. Second, this tendency occurs not only at the time of hatching but also for a number of weeks before and after hatching. Thus, it seems unlikely that the response is connected with injury or disease in newly hatched chicks. Third, as we have seen, the kittiwake never disposes of its broken eggshells. These observations combine to suggest that the prime function of eggshell removal in black-headed gulls is protection from predators by maintaining the camouflage of the nest. The kittiwake is not subject to nest predation, but presumably it would dispose of eggshells if they were injurious to the chicks. The black-headed gull maintains the response long before and after hatching, which makes sense for anti-predator behaviour but not for behaviour directly related to the health of the chicks.

In order to test the hypothesis that eggshell removal serves to maintain the camouflage of the nest, Tinbergen and his team conducted a series of experiments. They laid out eggs in a widely scattered pattern (about 20 metres apart) in an area of land outside the gull colony. This site was chosen to resemble the gullery, without provoking interference from the gulls' defensive behaviour. In some experiments hen eggs were laid out, some of which were painted to look like gull eggs. In others, gull eggs were laid out, camouflaged with bits of vegetation, some with a broken eggshell nearby. After the eggs were laid out, a watch was kept from a hide, or blind, and a count was kept of the number of eggs that were discovered and taken by typical nest predators such as carrion crows (*Corvus corone*) and herring gulls (*Larus argentatus*). The results showed that white hen eggs were taken much more quickly than artifically camouflaged hen eggs. Similarly, black-headed gull eggs painted white were more vulnerable than unpainted black-headed gull eggs. Eggs with a broken eggshell placed nearby were also much more vulnerable to predation than those not so marked.

Tinbergen and his co-workers also discovered that carrion crows soon learn to associate broken eggshells with the presence of eggs. First, they showed that the crows ignore empty eggshells put out alone. Then they paired broken eggshells with real eggs, some of which were taken by crows. When empty eggshells subsequently were put out alone, the crows paid particular attention to them and searched in their vicinity. Overall, the results of the experiment strongly suggest that the presence of a broken eggshell near the nest of a black-headed gull makes the nest more vulnerable to predators because it makes the nest easy to spot from the air. Moreover, if nests are distinguishable by the presence of broken eggshells, then crows and other predators will soon learn to use them as a cue to the presence of eggs or chicks. It is interesting, however, that the parent black-headed gull does not remove the eggshell immediately but waits about an hour before doing so. Tinbergen suggested that this delay allows time for the chicks to dry. They are then more robust, better camouflaged and less easily swallowed by a predator.

Spine function in sticklebacks

The experimental approach to the study of survival value may be combined with the **comparative approach**, for example, the study by Hoogland and co-workers (1957) of the function of the spines of sticklebacks. Two species of stickleback occur in European fresh water. They are the three-spined stickleback (*Gasterosteus aculeatus*) and the ten-spined stickleback (*Pygosteus pungitius*). They are preyed upon by perch (*Perca fluviatilis*) and pike (*Esox lucius*). The spines of sticklebacks are thought to give some protection against predators, and to test this possibility Tinbergen and his co-workers carried out experiments in which they compared sticklebacks with fish of similar size, but no spines, such as minnow (*Phoxinus phoxinus*), roach (*Rutilus*

Three-spined stickleback

Ten-spined stickleback

Minnow

Roach

Rudd

Crucian carp

Fig. 6.2 Fish used in experiments on prey capture by pike (After Hoogland *et al.*, 1957).

rutilus), rudd (*Scardinius erythrophthalmus*) and crucian carp (*Carassius carassius*), as illustrated in Figure 6.2.

The experiments were conducted in large aquariums and involved observations of the predatory behaviour of perch and pike and the reactions of the prey fish. The researchers observed that pike and perch usually attempt to take their prey head first but that they are not always successful. They can swallow minnows tail first without difficulty, but they cannot cope with a stickleback in this way and usually spit it out and try to catch it again immediately. When attacked, sticklebacks raise their spines and attempt to keep their head away from the predator. Once caught, sticklebacks sometimes would manage to escape, especially if the predator attempted to manoeuvre its prey in order to swallow it head first (see Figure 6.3).

When presented with mixed schools of minnows and sticklebacks, both pike and perch ate the minnows more quickly than the sticklebacks (see Figure 6.4), partly the result of the predators' preferring to attack prey without spines and partly a result of sticklebacks being caught and then rejected or being caught and then escaping. Three-spined sticklebacks survived longer than ten-spined sticklebacks (Figure 6.5), presumably because their spines are much larger.

In some tests, sticklebacks were offered to perch and pike after their spines had been cut off. Such fish survived much less well than intact sticklebacks. This supports the main conclusion gained from these experiments, which is that the spines of sticklebacks do indeed provide some protection against predators. The differences between the three-spined and ten-spined stickleback are large, and the tests showed that the three-spined sticklebacks survived longer than the ten-spined sticklebacks. How, then, are the ten-spined sticklebacks, or any fish without protective spines, able to survive in nature? Comparison of the habitats and behaviour of three-spined and ten-spined sticklebacks gives some clues. The male three-spined stickleback builds its nest on a substratum relatively free of weeds, and it has bright red nuptial coloration. The male ten-spined stickleback builds its nest among thick weeds and has black coloration during the breeding season. Thus, the ten-spined stickleback apparently compensates for its relative lack of protection by its cryptic habitat, coloration and behaviour.

6.2 Assessment of mortality

The comparative and experimental studies of survival value are indirect ways of gaining evidence about the action of natural selection. A more direct approach is to measure the mortality that results from a particular selective pressure. For example, Hans Kruuk (1964) studied the predators and anti-predator behaviour of the black-headed gull (*Larus ridibundus*).

As discussed earlier these gulls nest on the ground in large colonies, where they are vulnerable to attack from a variety of predators. Some predators such as other gulls, crows and hedgehogs prey only on eggs and chicks.

Fig. 6.3 Drawing made from film showing the variety of positions of a stickleback in the mouth (After Hoogland *et al.*, 1957).

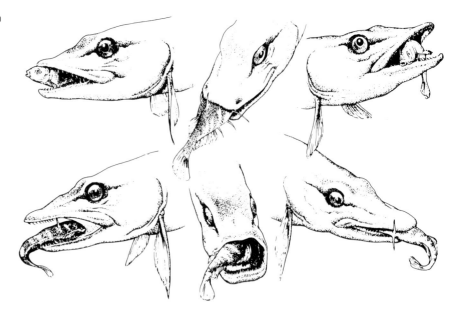

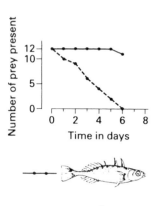

Fig. 6.4 The effect of predation by pike on a mixed shoal of twenty three-spined sticklebacks (solid line) and twenty minnows (dashed line) (After Hoogland *et al.*, 1957).

Others including hawks, stoats, foxes and humans are a danger to both the adults and their brood. The peregrine falcon attacks only the adult birds. Kruuk (1964) found that the reactions of the gulls depend upon the type of predator. They would flee from the peregrine falcon, but in reaction to humans, foxes, stoats and hawks their behaviour was ambivalent. The gulls flee the nest but launch aerial attacks at the predators. They do not flee from other gulls, hedgehogs or crows but attack them if they come near the nest.

These differences in reactions to predators are highly adaptive. If the gull leaves the nest, it is leaving its eggs or chicks protected only by their camouflage. This does not matter when the predator is a peregrine falcon that preys only on the adult birds. The gulls can remain at the nest and protect their brood from hedgehogs and ferrets, which are a threat only to the eggs and chicks. By vigorously attacking these predators, the gulls can deter them not only on a particular occasion but also possibly from visiting the colony in the future. Kruuk (1964) found that crows tend to avoid the colony, though hedgehogs remain persistent, especially at night. In reacting to foxes and humans, the gulls are faced with a problem. To flee the nest endangers the brood, but not to flee endangers the adult. A fine balance of costs and benefits exists in this type of situation, and the gulls seem to make some good compromises.

In addition to direct responses to predators, the gulls can protect themselves in various indirect ways. Ian Patterson (1965) showed that choice of nest site and of egg-laying date were important factors affecting reproductive success. Patterson noted that the majority of black-headed gulls in his study area nested in the densely packed colony where the nests are about one metre apart. Some, however, nested more than a hundred metres from the colony. He showed that there was a marked difference in breeding success between birds that nested within the colony and those that nested outside it. Obviously, there must be some advantages in nesting within the colony. To

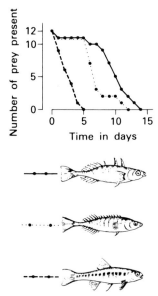

Fig. 6.5 The effect of predation by pike on a mixed shoal of twenty three-spined sticklebacks (solid line), twenty ten-spined sticklebacks (dotted line) and twenty minnows (dashed line) (After Hoogland *et al.*, 1957).

investigate these, Patterson followed the case histories of some 800 marked nests. He found that nests in the centre of the colony had a higher success rate than those near the edge in terms of the number of young fledged. Patterson thus used naturally occurring variation to see whether differences in behaviour were correlated with differences in breeding success.

Black-headed gulls tend to synchronize their egg laying, but again there is natural variation. Some birds lay much later than the majority, as illustrated in Figure 6.6. Patterson found that breeding success was correlated closely with laying date. The most successful birds were those that laid in synchrony with the majority, while those that laid later had much greater mortality among their eggs and chicks. Predation was by far the most common cause of chick and egg losses (Patterson, 1965). The greater success of birds that synchronize their egg laying is probably due to the fact that this leads to an overabundance of food for the predators (Kruuk, 1964). It is difficult for predators to exploit a sudden and short-lived increase in potential food when the predator population is adjusted to a low level of food availability. Thus, gulls that lay in synchrony have a lower probability of being attacked by a predator than those that are providing a larger proportion of the available food by laying at a different date.

Patterson (1965) found that breeding success was not correlated with nest density. This means that the greater success of birds nesting within the colony was due not to closer packing but to the total number of neighbours that would be alerted by a predator. This conclusion is supported by Kruuk's (1964) observation that the gulls often joined together in combined attacks on predators so they often were deterred from entering the highly populated parts of the colony.

This work shows that the black-headed gulls that synchronize their egg laying and that can establish themselves in the centre of the colony benefit in terms of reproductive success. Therefore, they have more offspring than those birds that nest on the edge of, or away from, the colony or that lay their eggs later than the majority. Since these offspring presumably inherit the traits of colonial behaviour, these traits are maintained in the population through the action of natural selection. To confirm this hypothesis it would

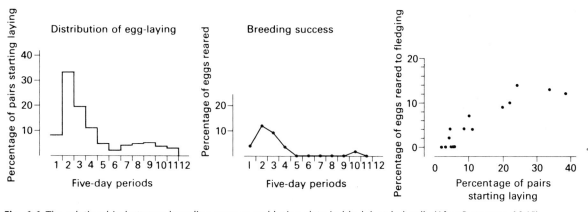

Fig. 6.6 The relationship between breeding success and laying date in black-headed gulls (After Patterson, 1965).

be necessary, ideally, to show that the offspring of gulls that show the typical colonial behaviour have a greater probability of surviving to maturity. This would be difficult to do, however, because black-headed gulls do not reach sexual maturity until they are three to four years old.

Sometimes it is possible to measure the number of offspring surviving to maturity. The females of the side-blotched lizard (*Uta stansburiana*) hold territories, and in one study (Tinkle, 1969) it was found that those that held larger territories had a greater number of mature progeny than those with small territories. The territories were similar in quality, so the main reason for the correlation was probably the amount of food in each territory.

6.3 Darwinian fitness

We have seen that the extent to which a genetic trait is passed from one generation to the next, in a wild population, is determined by the breeding success of individuals of the parent generation and the value of the trait in enabling the animals to overcome natural hazards such as food shortage, predators and sexual rivals. Such environmental pressures can be looked upon as selecting those traits that fit the animal to the environment.

The concept of fitness has proved to be something of a problem for the reason elucidated by the historian William Dampier (1929):

'Herbert Spencer's phrase for natural selection, the survival of the fittest standing alone begs the question What is the fittest? The answer is: The fittest is that which best fits the existing environment. That which is fit survives, and that which survives is fit.'

Darwin discussed the survival of the fittest without defining fitness rigorously. 'Darwinian fitness', however, is now widely recognized by biologists as a measure of capacity to produce offspring. The apparent circular reasoning in the phrase can be broken if we recognize that the fittest is not 'that which best fits the existing environment' but that 'the fit are those who fit their existing environments and whose descendants will fit future environments' (Thoday, 1953). Thus, in defining fitness, we are looking for a quantity that will reflect the probability that, after a given lapse of time, the animal will have left descendants.

In any population of animals there is variation among individuals, and consequently, some have greater reproductive success than others. Individuals that produce a higher number of viable offspring are said to have greater Darwinian fitness. An individual's fitness depends upon its ability to survive to reproductive age, its success in mating, the fecundity of the mated pair and the probability of survival to reproductive age of the resulting offspring.

The fitness of a genotype in a Darwinian sense can be measured by means of numbers of its progeny, different generations being counted at the same stage of the life cycle. The fitness of an individual must take its age into account because the potential for reproduction changes with age. In this respect, Ronald Fisher (1930) was the first to draw attention to the importance of the concept of **age-specific reproductive value**. This is an index of the extent to which the members of a given age group contribute to the next

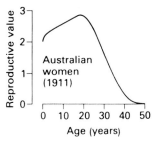

Fig. 6.7 Reproductive value plotted against age for Australian women about 1911 (After Fisher, 1930).

generation between now and when they die. In the example of Australian women, illustrated in Figure 6.7, we can see that reproductive value increases up to the age of twenty and then declines. The initial increase is probably due to childhood mortality. This means that the average ten-year-old Australian girl is less likely to have children than the average twenty-year-old woman (because some ten year olds will die before child-bearing age). Conversely, the average twenty-year-old woman has more child-bearing potential left than the average woman of thirty.

The age at which an animal should ideally become sexually mature and capable of reproduction is a matter of evolutionary life-history strategy. In unpredictable environments natural selection usually favours early maturity and large numbers of offspring which are left to fend for themselves. In more stable environments it is a better strategy to mature late and have few offspring which are well cared for.

Reproduction may expose an animal to risks, thus reducing the chances of subsequent survival and reproduction. In Figure 6.8 the data do not provide conclusive evidence of the risks involved in reproduction because they compare differences among rather than within species. However, Loschiavo's (1968) data on the length of life and number of eggs laid by the beetle *Trogoderma parabile* (Figure 6.9) do provide more direct evidence of risks to

Fig. 6.8 Total fecundity per reproductive season plotted against probability of surviving to a subsequent reproductive year for fourteen lizard populations (After Tinkle, 1969).

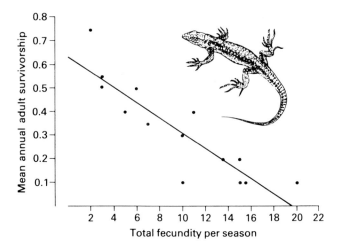

Fig. 6.9 The effect of reproductive effort on the subsequent life span of the beetle *Trogoderma parabile* (Data from Loschiavo, 1968).

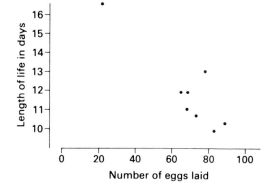

the individual. We must distinguish here between the evolutionary strategy represented by a particular species and the factors affecting individuals within a species. On the one hand, high reproductive output may be such a drain on the energy resources of the individual that its subsequent survival is endangered – for example, many birds lose a considerable amount of weight during incubation. On the other hand, species with short life expectancy, perhaps due to predation, can be expected to have high fecundity to compensate, while longer-lived species may find that relatively low fecundity increases overall fitness, especially if the resources available for the offspring are limited.

In general, the more time and energy a parent expends upon a particular offspring, the fitter that offspring will be. There is often an inverse relationship between the total number of offspring produced and their average fitness. As an example, let us consider briefly the work of David Lack and his co-workers (1954, 1966) on the great tit (*Parus major*) in Marley Wood near Oxford. These studies show that annual fluctuations in the breeding population are due primarily to variations in the juvenile mortality before the winter. Survival of the young was greater for broods reared early rather than later in the season, largely due to a decline in food availability toward the end of the season. Chris Perrins (1965) found that the average weight of the nestlings in the great tit broods declined with increasing brood size (Figure 6.10). This effect was more marked in seasons when food for the female at egg-production time was in short supply. Perrins also found that the larger nestlings had a greater chance of survival. To be able to rear a brood early in the season, the parents must be successful in establishing a territory during the spring. Behavioural factors like aggressiveness are of importance here. In general, it appears that parental effectiveness, including the ability to establish a territory, the tendency to produce a clutch of the optimal size and experience in rearing the nestlings, is an important factor determining fitness in this species. To maximize fitness the best reproductive strategy must be a

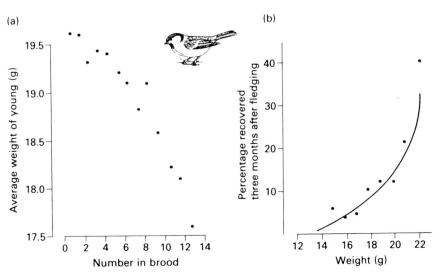

Fig. 6.10 Average weight of young great tits (*Parus major*) as a function of the number in the brood (a), and subsequent survival in relation to weight at fledging (b) (After Perrins, 1965).

compromise between having a large number of progeny and attaining a high individual fitness for each offspring.

The studies of Lack and his co-workers show that an optimal clutch size exists from which the greatest number of offspring survive to breed in the next generation. The smaller clutches produce fewer offspring, while young birds from the larger clutches leave the nest at a lighter weight and have a smaller chance of survival. Here we have an example of stabilizing selection in which the intermediates in a population leave more descendants than the extremes. In a stable environment, genetic recombination increases the variability among the individuals of a population during each generation, but stabilizing selection reduces the variability back to the level of the previous generation. In a changing environment, on the other hand, the average individuals may not be the fittest members of the population. Under such conditions directional selection can occur, such that the population average shifts toward a new phenotype better adapted to the altered environment.

The components of fitness such as viability, fecundity and fertility can be described in various ways. For example, Prout (1971) distinguishes between larval and adult components of fitness. The larval component consists of the probability of survival to adulthood, while the adult component is divided into a male and a female sub-component. The female sub-component includes viability and fecundity, and the male sub-component includes viability and virility. Mating ability is the most important aspect of virility, and this aspect of fitness is, therefore, connected with sexual selection (see Chapter 8).

6.4 Inclusive fitness

We have seen that an animal's individual fitness is a measure of its ability to leave viable offspring. The process of natural selection determines which characteristics of the animal confer greater fitness. However, the effectiveness of natural selection depends upon the mixture of genotypes in the population. Thus, the relative fitness of a genotype depends upon the other genotypes present in the population, as well as upon other environmental conditions.

The concept of fitness can be applied to individual genes by considering the survival of particular genes in the gene pool from one generation to another. A gene that can enhance the reproductive success of the animal carrying it will thereby increase its representation in the gene pool. It could do this by influencing the animal's morphology or physiology, making it more likely to survive climatic and other hazards, or by influencing its behaviour, making the animal more successful in courtship or raising young. A gene that influences parental behaviour will probably be represented in the offspring so that by facilitating parental care, the gene itself is likely to appear in other individuals. Indeed, a situation could arise in which the gene could have a deleterious effect upon the animal carrying it but increase its probability of survival in the offspring. An obvious example is a gene that leads the parent to endanger its own life in attempts to preserve the lives of its progeny. As we shall see, this is a form of **altruism**.

Both Fisher (1930) and Haldane (1955) realized that the fitness of an individual gene could be increased as a result of altruistic behaviour on the part of animals carrying the gene. However, Hamilton (1964) first enunciated the general principle that natural selection tends to maximize not individual fitness but **inclusive fitness**; that is, an animal's fitness depends upon not only its own reproductive success but also that of its kin. The inclusive fitness of an individual depends upon the survival of its descendants and of its collateral relatives. Thus even if an animal has no offspring its inclusive fitness may not be zero, because its genes will be passed on by nieces, nephews and cousins.

In normal diploid animals, each parent contributes a copy of one of its two sets of genes to each of its offspring (see Chapter 2). The chance that any given gene in a parent will appear in one of the progeny is therefore one half. By the same logic the chance that a single gene in a parent will be inherited in common by two of the offspring is a half. The chance that one gene in a grandparent will be transmitted to a given grandchild is a quarter, and the chance that two first cousins both carry copies of a gene present in a grandparent is one-eighth.

We can now define the **coefficient of genetic relatedness** (**r**) as a measure of the probability that a gene in one individual will be identical by descent to a gene in a particular relative. An equivalent (Dawkins, 1979) and alternative measure is the proportion of an individual's genome that is identical by descent with the relative's genome. Note that the genes must be identical by descent and not simply genes shared by the population as a whole. This means that **r** is the probability that a gene common to two individuals is descended from the same ancestral gene in a recent common relative.

Inclusive fitness is sometimes equated with a simple weighted sum based on the animal's various coefficients of relationship. Thus, it is sometimes seen as the sum of the animal's individual fitness and that of the relatives discounted in proportion to the coefficient of relationship. This measure counts all the animal's offspring, and although this may be a useful measure for practical purposes, it does not accord with Hamilton's (1964) original definition. Thus inclusive fitness has often been misdefined (Grafen, 1982). It should exclude those of the animal's own offspring that exist because of help received from others, and include those offspring of relatives whose existence is the result of the animal's help being offered to the relative (Grafen, 1984).

6.5 Fitness in the natural environment

Implicit in our discussion of inclusive fitness are two alternative ways of describing natural selection. In population genetics, the unit of selection is the gene, and replication of the gene is the quantity maximized by natural selection. While this approach has an appealing logic that can sometimes be used to illuminate aspects of animal behaviour (e.g. R. Dawkins, 1976), it is not really convenient for those interested in the behaviour of individual animals. An equivalent approach is to regard the individual animal as the unit of selection and inclusive fitness as the quantity maximized by natural selection (Dawkins, 1978). For practical purposes, the reproductive success of an individual is a good guide to its inclusive fitness (Grafen, 1982).

The fitness derived from behaviour in the natural environment depends upon the selective pressures that are operating. In theory, every aspect of an animal's behaviour will make a difference to its fitness, either because of the direct consequences of the behaviour or simply because one type of behaviour precludes another that might be more, or less, beneficial. In practice, the natural environment is so complex that it is extremely difficult to measure the changes in fitness that derive from behaviour. There are two main approaches to this problem. One approach is to attempt to specify in detail the various costs and benefits that are associated with behaviour in the natural environment. This approach is discussed in Chapter 24. The other approach is to measure directly the changes in reproductive success that result from particular aspects of behaviour.

In Section 6.1 we saw that the survival value of particular aspects of behaviour like eggshell removal by the black-headed gull could be estimated from experiments carried out in the field. While such experiments are valuable in providing an understanding of the function of particular features of morphology and behaviour, they do not bear directly upon the question of fitness. The concept of survival value is usually applied to relatively short-term questions like the survival of the young in a particular breeding season. Ideally, estimates of fitness should relate to the long-term survival of particular genetic traits or the long-term consequences, in terms of reproductive success, of particular behaviour patterns.

Sometimes differences in reproductive success can be attributed to fairly simple aspects of behaviour. For example, Nick Davies and Tim Halliday (1977) found that the reproductive success of a pair of toads (*Bufo bufo*) was influenced by both the size of the female and the degree to which the mating pair (Figure 6.11) were matched for size. Thus, a larger female produces more eggs, and more eggs are fertilized when the mating pair are well matched for size. A male that mates with a female just slightly larger than himself should have the greatest reproductive success, while a female should prefer larger males that can ward off rivals. Thus, both the male and female should each prefer a mate larger than itself. Observed pairings lie between these two theoretical extremes, with the females usually being larger than the males (see Figure 6.12).

The lifetime mating success of a natural population of the damsel fly *Enallagma hageni* has been studied in northern Michigan (Fincke, 1982). These flies breed in a shallow pond separated from a lake by a sand beach 10 to 20 metres wide. Feeding occurred in a marshy open field adjacent to the breeding site. On sunny days male flies move from the feeding area to the pond at about 9.30 a.m. Up to 1.00 p.m. most of the males present at the pond are unmated. The number of mated pairs (clasped together in a tandem formation) increases until 3.00 p.m. and remains high until about 5.30 p.m. On a particular day, nearly all females are found in tandem pairs, but usually more than half the males fail to mate. Since there are more males than

Fig. 6.11 Pair of toads
mating. The male is repelling
a rival (Drawing from Davies
and Halliday, 1977.
Photograph: Tim Halliday).

females in the breeding area, competition among males is intense. The males
intercept females on their way to the pond, and nearly all successful males
are paired before arriving at the water. Competition among males takes the
form of interference with tandem pairs. Such harassment sometimes results
in the uncoupling of a tandem pair and displacement of one male by
another.

Surprisingly, Fincke found no correlation between the body size of males
and their mating success. Apparently, a small male is as capable as a large
male at flying in tandem with a female. Although large males may be more
successful at takeovers than small males, this advantage is probably offset by
the fact that harassment of tandem pairs exposes single males to predation
near the water. Fincke found that the larger males tended to spend a greater

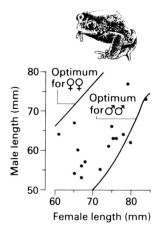

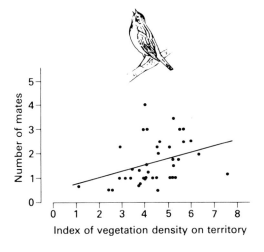

Fig. 6.12 Optimal and observed pairings of toads (*Bufo bufo*) of various sizes. Optima are calculated on the assumption that the number of fertilized eggs produced in a season is maximized. The female optimum curve represents a minimum, because no experimental pairings with males larger than females could be obtained. The observed pairings are represented by the solid dots (After Davies and Halliday, 1977).

amount of time near the water, suggesting that interference with mated pairs was their preferred strategy. However, the alternative strategy of searching for unmated females is probably more successful overall.

Mating success among males varies considerably from day to day, 39 per cent of males mating on average. Lifetime mating success is also very variable. Only 59 per cent of males in the population mate at least once, and this measure is highly correlated with longevity. The major cause of death at the breeding site is predation from spiders, dragonflies and frogs. Thus, it seems that the most important component of lifetime breeding success in this species is to avoid predators and live to mate another day.

Many animals defend territories during the breeding season. In some cases, this is a form of spacing out that has to do with the acquisition of mates (Davies, 1978). Territories sometimes differ in quality, and males that defend the better territories can be expected to have greater reproductive success. In the three-spined stickleback, for example, males with large territories are more successful in attracting females (van den Assem, 1967). Females that deposit their eggs in large territories have greater reproductive success because fewer eggs are lost to predation by other males. Rival males from neighbouring territories try to interfere with the nest, but this is harder to do when the territory is large (Black, 1971; Wooton, 1976). Territory quality may depend not only upon size but also upon food or nest-site availability as shown in Figure 6.13. In cases where animals compete for mates, territory or habitat (Partridge, 1978), the successful may be fitter in various respects. They may have greater reproductive success not only because they acquire better mates or territories but also because they are fitter in other respects. This is a perennial problem with the assessment of fitness. In the natural environment there are so many interacting factors that it is virtually impossible to account for all of them.

Fig. 6.13 Number of mates attracted by male dickcussel (*Spiza americana*) as a function of the vegetation density on the territory (After Zimmerman, 1971).

Points to remember

- The survival value of particular aspects of behaviour can be investigated by experiments performed under natural conditions.

- Investigations of the causes of mortality within a population can often provide useful evidence about the selective pressures.

- Fitness in the Darwinian sense is measured by the number of progeny, counted at a particular stage of the life cycle.

- The inclusive fitness of an individual depends upon the survival of its descendants and of its collateral relatives. Even if an animal has no offspring, its inclusive fitness may not be zero because its genes will be passed on by nieces, nephews and cousins.

- In studying fitness in the natural environment, it is usually necessary to use an index of mating success or reproductive success. Such indices only approximate fitness because they take no account of the viability of the offspring.

Further reading

Dawkins, R. (1982) *The Extended Phenotype*. Freeman, New York.

Hurst, L.D., Atlan, A. and Bengtsson, B.O.C. (1996) Genetic conflicts. *Quart. Rev. Biol.* **71**, 317–364.

Krebs, J.R. and Davies, N.B. (1981) *An Introduction to Behavioural Ecology*. Blackwell Scientific Publications, Oxford.

CHAPTER

The evolution of adaptive strategies

7.1 Evolutionary strategies

7.2 Evolutionarily stable strategies

7.3 Digger-wasp strategies

Evolutionary biologists are interested in explaining how a state of affairs observed today is likely to have come about as a result of evolution by natural selection. To account for the establishment of a particular genetic trait, they imagine a time before the trait existed. Then they postulate that a rare gene arises in an individual, or arrives with an immigrant, and that individuals carrying the gene exhibit the trait. They then ask what circumstances will favour the spread of the gene through the population. If the gene is favoured by natural selection, then the individuals with genotypes incorporating the gene will have increased fitness. The gene may be said to have invaded the population. This chapter considers the implications of this kind of argument for studies of animal behaviour.

7.1 Evolutionary strategies

To become established, a gene not only must compete with the existing members of the gene pool but must also resist invasion from other mutant genes. Indeed, evolutionary biologists, when speculating about a particular evolutionary situation, often impose upon themselves the test of postulating an invasion by hypothetical mutants in order to see whether a particular theory will stand up to the competition. Before going into the detail of such procedures, let us look at a few examples.

A high degree of similarity between an animal and its visual background may result from selection for features that enable animals to avoid predators or predators to lurk undetected and ambush a suitable prey. This type of strategy sometimes takes the form of an astonishing resemblance between an animal and the plant with which it normally associates. For example, the leaf-like grasshopper (*Arantia rectifolia*) shown in Figure 7.1 can rest with its wings held close together so it resembles a complete leaf, or it can rest on a leaf and hold its wings flat so it appears to be part of the leaf (Edmunds, 1974). It takes up these camouflaged postures as a protection against predation. Conversely, the mantis (*Phyllocrania paradoxa*) may be disguised similarly (Figure 7.2), but its camouflage is partly defensive and partly aggressive, since it sits motionless waiting to seize its prey.

Fig. 7.1 The leaf-like grasshopper *Arantia rectifolia* (*Photograph: Malcolm Edmunds*).

Fig. 7.2 The praying mantis (*Phyllocrania paradoxa*) mimicking a dead leaf (*Photograph: Malcolm Edmunds*).

In addition to resembling the background or vegetation, animals may also resemble parts of other animals. This is often a feature of the deimatic (Edmunds, 1974) displays by which animals attempt to scare off their enemies, for example, the caterpillar of the hawkmoth (*Leucorampha*). This animal normally rests upside down beneath a branch or leaf. When disturbed, it raises and inflates its head, the ventral surface of which has conspicuous eye-like marks and whose general patterning resembles the head of a snake, as illustrated on p. 52. Many moths and butterflies have eye spots on their wings that they reveal suddenly when disturbed, with the possible effect of frightening the predator.

True mimicry is the resemblance of one animal, called the **mimic**, to another animal, called the **model**, so that the two are confused by a third animal, usually a predator. Various types of mimicry exist, of which the most exemplary is **Batesian mimicry**. In this type, the mimic resembles a model that is noxious or distasteful to predators. There is no advantage to the model, but experiments have shown that mimics definitely can benefit as a result of predators becoming confused between model and mimic. For example, birds soon learn to avoid the salamander *Notophthalmus viridescens*, which is unpalatable. They readily eat the similarly coloured salamander *Pseudotriton ruber*, provided that they have not previously experienced the unpalatable *Notophthalmus*. Thus *Pseudotriton* gains some protection from its resemblance to *Notophthalmus*, provided that the birds are already familiar with *Notophthalmus* and that they are not so experienced as to be able to discriminate between the two salamanders (Howard and Brodie, 1971). Some other examples of Batesian mimicry are illustrated in Figure 7.3.

Sometimes a number of noxious species share the same warning coloration. This is called **Mullerian mimicry** and is of advantage to all participant species because once a predator has learned to avoid one species, it will avoid all the mimics. Figure 7.3 illustrates some examples of Mullerian mimicry. A remarkable case of aggressive mimicry is illustrated in Box 7.1

In the case of Batesian mimicry the anti-predator markings on one species, the model, are imitated by another species, the mimic. This will be of advan-

---

Box 7.1 Aggressive mimicry by the sabre-toothed blenny

The black-and-white-striped cleaner wrasse (*Labroides dimidiatus*) lives among the Pacific coral reefs. Other fish recognize it because of its conspicuous coloration and its advertising dance or display. Large fish permit the cleaner wrasse to approach and remove parasites from the body surface and the inside of the mouth. The relationship is symbiotic because the wrasse benefits by obtaining food and the host fish benefits by having its parasites removed. However, the cleaner wrasse has a mimic that resembles it closely, the sabre-toothed blenny (*Aspidontus taeniatus*). Other fish often mistake it for a cleaner wrasse and allow it to approach them. The sabre-toothed blenny then quickly bites a piece from the fin of the host fish and escapes.

(a) The cleaner wrasse (*Labroides dimidiatus*) with its host fish a coral cod (*Cephalopholis miniatus*) and (b) its mimic the sabre-toothed blenny (*Aspidontus taeniatus*) (From *The Oxford Companion to Animal Behaviour*, 1981).

(b)

(a)

Fig. 7.3 Some examples of Mullerian and Batesian mimicry. (a) Mullerian mimicry among West African Heteroptera. The reduviid *Phonoctonus fasciatus* normally lives in colonies of one or more of the pyrrhocorids *Dysdercus fasciatus*, *D. superstitiosus* and *D. voelkeri*. *Phonoctonus lutescens* normally lives in colonies of *Odontopus*, sometimes with *D. voelkeri* as well. All of these bugs are black, red (shown stippled) and greyish orange (shown white). (b) Batesian mimicry. The wasp *Vespula vulgaris* and two mimics, the beetle *Clytus arietis* and the hoverfly *Helophilus hybridus*. In all three the body is dark brown or black with yellow bands. (c) The distasteful white ermine moth (*Spilosoma lubricipeda*) and its Batesian mimic the buff ermine (*S. lutea*). The abdomen of both is yellowish with darker markings, while the wings are white or cream with black spots (After Edmunds, 1974).

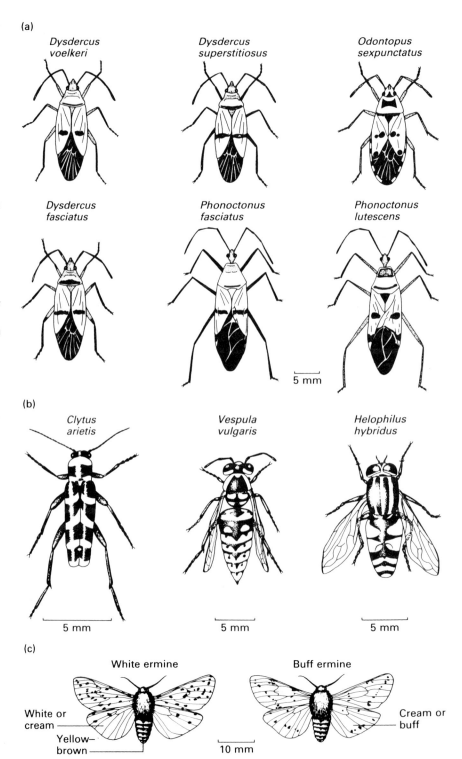

Fig. 7.4 Blue jay eating a toxic monarch butterfly and then vomiting (*Photograph: Lincoln Brower*).

Danaus plexippus

Limenitis archippus

|⎯⎯⎯⎯⎯⎯|
20 mm

Fig. 7.5 The monarch butterfly *Danaus plexippus* and its mimic the viceroy *Limenitis archippus*. The stippled areas are orange (After Edmunds, 1974).

tage to the mimic, provided that predators are unable to discriminate the mimic from the model and provided that the mimics do not become too common in relation to the models. If the model is distasteful to predators, the mimic will benefit, because predators are likely to avoid both model and mimic, while the mimic does not have to bear the cost of bearing the deterrent chemical. In some situations, a model may be mimicked by a number of species, and there is a danger predators will take more models than they might otherwise. For example, the large and conspicuous monarch butterfly (*Danaus plexippus*) of the eastern United States is distasteful to birds, which learn to avoid it after one or two experiences (Brower, 1958). As a caterpillar, the monarch feeds on species of milkweed, which are poisonous to many vertebrates. The caterpillars accumulate toxins, which are retained in the pupae and which are thus present in the butterfly. Predators that eat the butterflies usually vomit (Figure 7.4). The emetic monarch butterfly is mimicked by the viceroy (*Limentis archippus*), which resembles it closely (Figure 7.5). The viceroy is edible, and, as Brower (1958) has shown, jays will eat them readily if they have no experience of the monarch butterfly. Once having eaten a monarch, however, the jays avoid both monarchs and viceroys. The emetic monarch also has other mimics, so there is a possibility that, in a particular area, the number of mimics may be large in relation to the number of models. In such a situation, predators will not so quickly learn to avoid the monarch and its mimics, since their first experiences may be with edible butterflies, and thus, the level of predation on both model and mimic is likely to be higher.

In attempting to trace the evolution of such a situation, the biologist must try to calculate the consequences of the possible combinations of frequencies of different genotypes in the populations of models and mimics. For example, what is likely to happen if some monarchs evolve the habit of eating non-toxic plants and are no longer distasteful to predators? On the one hand, we might expect such individuals to gain an advantage by being perfect mimics of the unpalatable members of their species, without having to pay the price of being restricted to eating toxic milkweed, on which basis the genotype should increase in frequency. On the other hand, if the edible monarchs become too common, predators presumably would start to eat them. It is possible to calculate, on the basis of certain assumptions, the chances of survival of the edible monarchs as a function of the proportions of emetic and edible types in the population. The assumptions concern the degree of predation and the number of trials it takes a predator to avoid a monarch after having eaten an emetic one. In fact, it has been discovered that there is considerable variation in the toxicity of milkweeds and in the palatability of the monarchs raised on them (Brower, 1969). Thus, the natural populations do contain some monarchs that eat particular species of milkweed and that are palatable to avian predators, a form of mimicry called **automimicry** (Brower, 1969). In some localities the automimics outnumber the models 3:1, and yet predators are deterred from eating them.

If the change to a non-toxic diet is genetically based and not merely fortuitous, then such a change can be regarded as an **evolutionary strategy**. Thus evolutionary biologists sometimes find it useful to think of genes as employing a strategy to increase their numbers at the expense of other genes, even though, in reality, the whole exercise is conducted passively by natural selection.

In considering evolutionary strategies that influence behaviour, we have to visualize a situation in which changes in genotype lead to changes in behaviour. It is not necessary to suppose that the behaviour is genetically determined in a direct manner, since there are many routes by which genetic changes can influence behaviour. It is convenient, however, to refer to genes for behavioural traits, provided we realize this is merely shorthand. Thus, by 'gene for sibling care', we mean that genetic differences exist in the population such that some individuals aid their siblings, while others do not. Similarly, by 'dove strategy', we mean that animals exist in the population that do not engage in fights and that this trait is passed from one generation to the next.

At first sight it might seem that the most successful evolutionary strategy will spread through the population and eventually supplant all others. While this may sometimes be the case, it is not always so. In many situations it may not be possible to say what is the best strategy, because the effectiveness of a strategy may depend upon the behaviour of other animals; that is, competing strategies may be interdependent in that the success of one depends upon the existence of the other and the frequency with which the other is represented in the population. For example, the strategy of mimicry has no value if the warning strategy of the model is not effective.

Although it has long been realized that alternative strategies could exist, it is only recently that it has been possible to give a satisfactory explanation in terms of evolutionary theory. The key of this type of problem is the concept of 'the evolutionarily stable strategy', developed by John Maynard Smith, Geoff Parker, and others (see Maynard Smith, 1982).

7.2 Evolutionarily stable strategies

An **evolutionarily stable strategy** (ESS) is a strategy that cannot be bettered by any feasible alternative strategy, provided sufficient members of a population adopt it. This is another way of saying that the best strategy for an individual depends upon the strategies adopted by other members of the population. Since the same applies to all individuals in the population, a true ESS cannot be invaded successfully by a mutant gene. As an example, let us consider the question of fighting and assessment of rivals. Animals that employ strategies to avoid unnecessary fighting will usually be at an evolutionary advantage. This idea can be formalized in the following way.

Suppose we imagine that two types of strategy are represented in a population: the **hawk strategy** – which is to fight to kill or injure an opponent, even though there is a risk of injury to oneself; and the **dove strategy** – which is to threaten and display but to avoid serious fighting. These two strategies are extreme examples of those that might occur in real life. Suppose we now assign fitness-increment payoffs to the consequences of an encounter between the two animals. Let the winner of the contest score +50 and the loser zero. Let the cost of wasting time in a display be −10 and the cost of injury be −100. We have four possible types of encounter in a population containing both hawks and doves, the average payoffs from which are illustrated in Table 7.1. We can see that in an encounter between hawk and hawk, each stands to lose because while there is a 50 per cent chance of winning a given encounter (a gain of 50), there is also a 50 per cent chance of losing (a loss of 100 through injury). On average, each hawk will gain a payoff of −25 as a result of fighting another hawk. When a hawk meets a dove, the hawk always wins 50 while the dove gains zero, because by avoiding a fight, it avoids injury. When a dove meets a dove, each threatens the other and wins half the contests without fighting. The average payoff is therefore +15.

It is easy to see that a hawk could invade a population made up entirely of doves. While the doves would average +15 in contests against each other, an invading hawk would get +50 in every contest with a dove and, thus, would

Table 7.1 The game between hawk and dove (after Maynard Smith, 1976). The payoffs are as follows: winner +50, injury −100, loser 0, display −10. The matrix of average payoffs to the attacker appears below

Attacker	Opponent	
	Hawk	Dove
Hawk	$\frac{1}{2}(50) + \frac{1}{2}(-100)$ $= -25$	+50
Dove	0	$\frac{1}{2}(50-10) + \frac{1}{2}(-10)$ $= +15$

Note that when a hawk meets a hawk we assume that on half of the occasions it wins and on half of the occasions it suffers injury. Hawks always beat doves. Doves always immediately retreat against hawks. When a dove meets a dove we assume that there is always a display and it wins on half of the occasions.

be at an advantage. In a population with only hawks, the average payoff is −25, so that a dove could easily invade, since a dove gains zero in a contest with a hawk, a better score than −25. Thus, we see that neither all-dove nor all-hawk populations are proof against invasion; neither one is an ESS. However, it is possible that a mixture of hawks and doves could provide a stable situation when their numbers reach a certain proportion of the total population. Let the proportion of hawks be h and the proportion of doves be $(1 − h)$. The average payoff for a dove (D) can be calculated from the probability of meeting a hawk or another dove and the payoff from each type of encounter. Thus:

$$D = 0h + 15(1 − h)$$

Similarly, the average payoff for a hawk (H) is:

$$H = −25h + 50(1 − h)$$

When D is equal to H there will be a stable equilibrium. If $D = H$, then $h = 7/12$ and $(1 − h) = 5/12$. Therefore, an ESS will exist in which 7/12 of the population are hawks and 5/12 are doves. An alternative possibility is that individuals behave as hawks for 7/12 of their encounters and as doves for 5/12, the strategy being chosen at random on each occasion. This is called a **mixed strategy**. In either case, equilibrium is achieved only when the dove strategy is deployed on 5/12 of all contests and the hawk strategy on 7/12. Each strategy is at an advantage when it is relatively rare. Its representation in the population then increases until the equilibrium point is reached again. No invasion of hawks or doves can upset the ESS.

Another strategy is conceivable in this case, a so-called **bourgeois strategy**, where the individual behaves like a hawk when it is the owner of a territory and like a dove when it is an intruder into the territory of another. If we assume that a bourgeois is an owner for half its encounters and an intruder for the other half, then the payoffs will be as illustrated in Table 7.2. If the population were made up entirely of bourgeois, then the average payoff per contest would be +25, more than could be gained by invading doves or hawks, who would get +7.5 and +12.5 respectively, in contests against bourgeois. Thus the bourgeois strategy is an ESS. It seems that the

Table 7.2 The hawk, dove, bourgeois game (after Maynard Smith, 1976). The payoffs, as in Table 7.1, are: winner +50, injury −100, loser 0, display −10. The matrix of average payoffs to the attacker appears below.

Attacker	Opponent		
	Hawk	Dove	Bourgeois
Hawk	−25	+50	+12.5
Dove	0	+15	+7.5
Bourgeois	−12.5	+32.5	+25

Note that when bourgeois meets either hawk or dove we assumed it is winner half the time and therefore plays hawk, and intruder half the time and therefore plays dove. Its payoffs are therefore the average of the two cells above it in the matrix. When bourgeois meets bourgeois on half the occasions it is winner and wins while on half the occasions it is intruder and retreats. There is never any cost of display or injury.

stable strategy is the conditional strategy – fight hard if the territory owner, but be prepared to retreat if an intruder.

These examples, first formulated by Maynard Smith (1976), are gross simplifications of real-life situations. Nevertheless, they represent a considerable advance in evolutionary theory because they show that this coherent approach to a complex situation can in principle provide answers to long-standing problems. The basic principles of ESS may have to be elaborated to account for particular natural situations, but this is true of any scientific law that has reasonable generality. Bourgeois strategies can be observed in nature, as illustrated in Box 7.2.

The dove strategy is often shown by animals that carry potentially dangerous weapons. For example, the oryx (*Oryx gazella*) (Figure 7.6) has sharp pointed horns that could inflict mortal wounds. These may be used in defence against predators, but in contests among oryx, the horns are used in a purely ritualized manner. It is against the rules to stab a rival in the side. Similarly, rattlesnakes settle their contests with ritualized trials of strength in which one attempts to pin the other to the ground (Figure 7.7). They do not use their poisonous bite against rivals. The hawk strategy is not observed commonly among animals, but it may occur in contests over a valuable resource like the opportunity to mate. In species in which access to females

Box 7.2 The bourgeois strategy of the speckled wood butterfly

Male speckled wood butterflies (*Pararge aegeria*) compete for mating territories, which are patches of sunlight. Contests consist of brief upward flight during which the contestants spiral around each other. The loser is the first to quit. The owner always wins (Davies, 1978). This is an example in which the bourgeois strategy is followed by all members of the population.

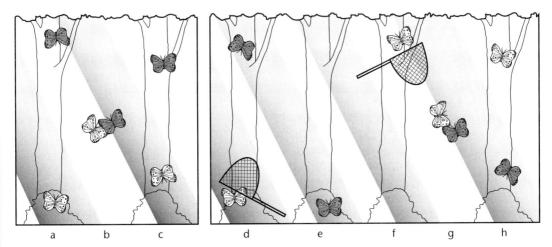

(a) A speckled wood butterfly basking in a patch of sunlight is challenged by an intruder, and a contest results (b), which the resident always wins (c). If the resident is removed (d), another butterfly becomes resident (e). If the original resident is then released (f), a contest results (g), but the new resident wins (h) (After Davies, 1978).

Fig. 7.6 Ritualized technique of fighting in the oryx (*Oryx gazella*) (From *The Oxford Companion to Animal Behaviour*, 1981).

Fig. 7.7 Technique of fighting in the rattlesnake (*Crotalus*). The snakes do not bite each other, but each attempts to pin the other to the ground (From *The Oxford Companion to Animal Behaviour*, 1981).

is difficult to attain, or is short-lived, we might expect to see hawk-like fighting because the payoff may be the one chance of a lifetime to contribute genetically to future generations. For example, male fig wasps (*Idarnes*) engage in lethal combat for the opportunity to mate with females inside the figure. The males have large mandibles and can bite another wasp in half. Bill Hamilton (1979) found one fig that contained 15 females, 42 males that were dead or dying from injury, and 12 uninjured males. In the musk ox (*Ovibos*) up to ten per cent of adult males may die per year as a result of fights over females (Wilkinson and Shank, 1977). Serious injury may occur among red deer, though only after fairly prolonged assessment routines (see Chapter 8).

7.3 Digger-wasp strategies

Female great golden digger wasps (*Sphex ichneumoneus*) (Figure 7.8) lay their eggs in underground burrows that they have provisioned with katydids (long-horned grasshoppers) as food for the larvae. Jane Brockmann studied the female wasp's behaviour in detail. She maintained almost continuous records of the nest-related activities of 68 individually colour-marked females at three different field sites over a total of six breeding seasons (Brockmann and Dawkins, 1979). Brockmann discovered that the females obtain a burrow either by digging one for themselves or by entering an already dug burrow. It takes a female an average of 100 minutes to dig a burrow. She then provisions it with stung and paralysed katydids (a process that may take a few days), lays a single egg, and seals up the burrow prior to starting the cycle again. There is a five to fifteen per cent chance that her burrow will be entered by another female wasp, who also provisions the burrow. Both wasps will be fully engaged in provisioning the same burrow and will not have another burrow at the same time. Because both wasps spend most of their time hunting, it may be some time before they meet. When they do meet they fight, and one wasp is usually driven away. Only one wasp eventually lays an egg in the brood cell.

Fig. 7.8 A female digger wasp (*Sphex ichneumoneus*) at the entrance to its burrow (*Photograph: Jane Brockmann*).

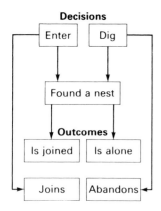

Decisions

Enter | Dig

Found a nest

Outcomes

Is joined | Is alone

Joins | Abandons

Fig. 7.9 The mixed strategy of the female digger wasp. The consequences of the decisions to dig a new burrow or enter an existing one are shown in this figure. Digging may result in abandoning the burrow, or in founding a nest where the wasp may remain alone or be joined by another. Entering an existing burrow may result in obtaining a nest, in which she may or may not be joined by another wasp, or in joining an already occupied nest (After Brockmann *et al.*, 1979).

A female wasp has two strategies open to her. She can undertake to dig her own burrow and run a small risk of being invaded by another wasp, or she can enter an already dug burrow, saving herself the work of digging but running the risk that the burrow is being used by the owner. The best strategy will depend upon that adopted by the majority of the other female wasps in the vicinity. If nearly all other females dig their own nests, then it is better to adopt the entering strategy, since there would be plenty of empty burrows and few other wasps exploiting them. On the other hand, if the majority of other wasps are employing the entering strategy, competition will be fierce and it will be better to dig one's own burrow. Thus it seems that the ESS lies between the two extremes.

It is possible to measure the success of the two strategies in terms of the number of eggs laid in a given period of time (Brockmann *et al.*, 1979). Digger wasps employ a mixed strategy, so instead of comparing the success of individuals, the results of the decisions to enter or dig have to be compared. On the basis of the hypothesis that the alternative decisions constitute a mixed ESS, we would expect the number of eggs laid per unit time to be the same whether the wasp decides to enter an existing burrow or to dig her own. On the basis of Brockmann's observations, Brockmann *et al.* (1979) calculated the success attributable to all possible outcomes of the two decisions, as illustrated in Figure 7.9. They discovered no significant difference between the two strategies on the basis of the number of eggs laid. This conclusion supports the hypothesis that entering and digging are components of a mixed ESS.

As we have seen, when two females share a burrow, they usually end up fighting. They rear up, lunge with open mandibles, and wrestle with each other. The duration of the fights varies between two and sixteen minutes. A fight ends when one wasp, the loser, leaves the area. Out of 23 fights observed, in eighteen cases the loser never returned to the nest, while in the remaining five cases it returned many hours later (Dawkins and Brockmann, 1980). The winner gains the use of the burrow, but the value of the prize depends upon the number of katydids it contains. A burrow containing four katydids is ready for egg laying, a prize worth fighting for. A burrow containing no katydids would still be worth some effort because the winner is saved the trouble of having to dig a new burrow. On this basis we would expect each wasp to fight to an extent that is related to the payoffs. It would not be worth fighting very hard for an empty burrow because the effort and risk involved could amount to more than that required to dig a new burrow.

The problem is that the situation is the same for both wasps. If both wasps know how many katydids the burrow contains, then we can expect them to fight equally hard. Would such a situation be evolutionarily stable? Dawkins and Brockmann (1980) argue that it would not. If the burrow is valuable, both wasps would fight for a long time but would surrender at about the same time, the winner being determined by chance. A less valuable burrow would result in a less prolonged fight, but both participants would still bear a substantial cost. Such a situation could be invaded by a gambler strategy: 'on a random half of your encounters give up immediately without a fight; on the remaining half of your encounters, persist indefinitely until your rival surrenders' (Dawkins and Brockmann, 1980). In a population dominated by

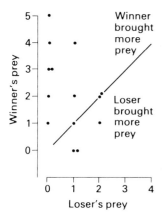

Fig. 7.10 Number of katydids brought by the winner of each fight between female digger wasps, plotted against the number brought by the loser (After Dawkins and Brockmann, 1980).

rational wasps that fight according to the total value of the burrow, the gambler would be at an advantage because she would win half her fights on average, but would pay nothing for the fights lost. However, the gambler strategy alone probably would not be an ESS.

Dawkins and Brockmann analysed the circumstances surrounding observed fights between two female wasps. For each of 23 fights they ascertained: the duration, which wasp dug the burrow, how long each wasp had been associated with the nest, which wasp was larger, the number of katydids each wasp had placed in the nest, and which wasp won. They found that the winner was not more likely to be the larger wasp or the one that dug the burrow or the one that most recently visited the burrow. Eleven wasps fought more than once, but there was no significant individual effect on the distribution of wins and losses. However, Dawkins and Brockmann did discover that the wasp that had placed the most katydids in the nest was more likely to be the winner (Figure 7.10). This is a surprising result because it appears to be an example of the so-called 'Concorde effect' (see Dawkins and Carlisle, 1976); that is, future behaviour is decided on the basis of past investment instead of on future prospects. Decisions about the Concorde airliner were reputedly based upon such considerations, but we can hardly expect natural selection to act in such a way (R. Dawkins, 1976).

The duration of the fight was highly correlated with the number of katydids the loser had supplied. A fight ends when the loser breaks off, so the loser effectively determines the duration of the fight. Dawkins and Brockmann show convincingly that the loser breaks off the fight at a time determined by her investment in the nest. This also is an example of the Concorde effect. The best strategy is always to base decisions upon the expected future payoff, but the female digger wasps do not seem to do this.

It appears that the wasps are not able to evaluate the contents of the burrow, but presumably the ability to do so could evolve. Apparently, they can tell how many katydids they have caught, though it is not known on what basis. As we have seen, the best policy for the individual is to fight in proportion to the total number of katydids in the nest, although this is probably not an ESS. However, it is not readily apparent that the Concorde strategy is necessarily the one most likely to be an ESS. Dawkins and Brockmann merely suggest that it might be. However, they do point to an important principle in evolutionary theory: the strategy that is best for the individual may not necessarily be the one that evolves, since it may not be an ESS.

Points to remember

- Evolutionary strategies are the passive result of natural selection which give the appearance of a ploy employed by genes to increase their numbers at the expense of other genes.

- An evolutionarily stable strategy (ESS) is an evolutionary strategy that cannot be bettered by any feasible alternative strategy, provided sufficient members of the population adopt it. Where the best strategy for an individual depends upon the strategies adopted by other members of the population, the resulting ESS may be a mixture of strategies.

Further reading

Hurst, L.D., Atlan, A. and Bendtsson, B.O. (1996) Genetic conflicts. *Quart. Rev. Biol.* **71**, 317–364.

Maynard Smith, J. (1982) *Evolution and the Theory of Games.* Cambridge University Press, Cambridge.

Parker, G.A. (1984) Evolutionarily stable strategies. In Krebs, J.R. and Davies, N.B. (eds) *Behavioural Ecology*, 2nd edn. Blackwell Scientific Publications, Oxford.

Evolution and social behaviour

Profile of Sir Ronald Fisher

Bushmen going hunting (*Photograph: Jen and Des Bartlett: courtesy of Bruce Coleman Ltd*).

In this section we are concerned with the ways in which evolutionary theory can illuminate the social behaviour of animals. Chapter 8 offers an introduction to the theory of sexual selection as conceived by Darwin and provides some necessary background for understanding sexual strategy and its effects on social behaviour, with explanations of altruism in terms of kin, selection, and reciprocal altruism and cooperation among animals. Chapter 9 is concerned with behavioural ecology. Chapter 10 is devoted to evolutionary aspects of primate social behaviour. The aim is to provide a good grounding in the modern applications of evolutionary theory and a prelude to further reading in sociobiology.

Sir Ronald Fisher (1890–1962)

Ronald Aylmer Fisher was born in London to a prosperous business family. As a child he was academically precocious, especially in mathematics. In 1909, after attending Harrow School, Fisher entered Cambridge University, where he studied mathematics. He was, however, deeply interested in biology, particularly genetics. He was a prominent member of Cambridge University Eugenics Society, and retained that interest throughout his life. After graduating, Fisher took a job as a statistician in London. When war broke out in 1914 he volunteered for active service, but was rejected because of his poor eyesight. He became a schoolmaster, teaching physics and mathematics in various schools. It was during this period that he began to make important contributions to statistics and to evolutionary theory. For example, he tackled the problem of reconciling the particulate nature of Mendelian genetics with the continuously varying characteristics (such as height in humans) measured by biometricians. The prevailing view was that such characters were the result of blending, and not Mendelian, inheritance. In a paper published in 1918, Fisher showed that the biometrical results follow logically from Mendelian principal cases where a number of gene pairs contribute to the character in question.

After the war, Fisher took up a post as statistician at Rothamstead Experimental Station, where he remained for fourteen years and where he made many very important contributions to statistics, biometry and experimental design. In 1929 Fisher was elected a Fellow of the Royal Society in recognition of his work in statistics.

While at Rothamstead, Fisher worked at home on various projects, including genetics. He made a number of important contributions to genetics, eugenics and evolutionary theory, culminating in his major work *The Genetical Theory of Natural Selection* published in 1930. In this work Fisher brought together the subjects of genetics and evolution by natural selection, and developed his fundamental theorem: The rate of increase in fitness of any organism at any time is equal to its genetic variance in fitness at that time. Fisher argued that since variation in populations is maintained by mutations, the rate of mutation occurrence determines the speed of evolution, while natural selection determines its direction. Many of Fisher's formulations were incompletely explained and some were not well understood by his fellow geneticists. Some were independently discovered by subsequent workers. It was said of Fisher that he was a geneticist of such prescience that the genius of his conclusions is still unfolding today. As we shall see, Fisher anticipated some of the fundamental ideas of modern sociobiology.

In 1933 Fisher was elected to the Chair of Eugenics at University College London. In 1949 he became Balfour Professor of Genetics at Cambridge. He was knighted in 1952. Fisher retired in 1957 and died in Adelaide, Australia, in 1962.

Sexual and social behaviour

In 1871 Darwin published *The Descent of Man*, in which he considered the subject of sexual selection (to which he had referred in his *On the Origin of Species* (1859)) as a form of natural selection. According to Darwin (1871), sexual selection depends on the advantage which certain individuals have over others of the same sex and species solely in respect of reproduction. Darwin reasoned that females make a definite choice of sexual partner and that males have acquired particular adornments and courtship behaviour 'not from being better fitted to survive in the struggle for existence but from having gained an advantage over other males, and from having transmitted this advantage to their male offspring alone'. Darwin realized that there are two ways in which a male can gain advantage over other males. First, they can compete directly with one another by fighting or by some form of ritualized combat, sometimes called **intrasexual selection** (selection within a sex). Second, males can compete indirectly in attracting females by special displays and adornments, sometimes called **intersexual selection** (selection between sexes).

8.1   Sexual selection

Male rivalry is exemplified by the red deer *Cervus elaphus*, a species native to Europe and common in Scotland. The stags grow antlers each year, and in the autumn rutting season, they directly challenge each other for ownership of females, which have no antlers and which are herded into harems by the successful males. The females have little or no choice of sexual partner because the males defend their harems against possible rivals. Tim Clutton-Brock and his co-workers have made an intensive study of this species on the Isle of Rhum. Males challenge each other initially by roaring (Clutton-Brock and Albon, 1979), starting slowly and then speeding up. The challenger usually retreats if the harem owner can roar faster. These roaring matches are thought to help the stags assess each other, since a stag has to be in good condition to be able to roar well. If the challenger is able to match the roaring of the harem owner, then the two approach and walk parallel to each other, which enables the rivals to assess each other more closely, particularly with respect to body size. Many contests end at this stage, but some escalate into fighting proper. The stags interlock antlers and push against each other. The fighting is quite dangerous, and during its life there is about a 25 per cent chance that a stag will be injured permanently from fighting. Usually the larger stag wins, although one that is handicapped by a previous injury or exhausted by the strenuous rutting season may be supplanted by a younger stag.

The older, larger stags normally have larger antlers, and these confer an advantage in fighting. Apart from this advantage, however, the antlers contribute little to survival. Since stags do not possess antlers outside the breeding season, and since females do not grow them at all, it is unlikely that antlers are an important protection against predators. Moreover, for their growth, each year the antlers require large dietary quantities of materials such as phosphorus and calcium salts, so growing them must constitute a burden on the stag's metabolism. It seems, therefore, that with respect to antlers, at least, the advantages of intrasexual selection outweigh the pressures of natural selection.

Competition between two stags involves both **fighting** and **assessment**. Fighting is risky because it can lead to injury and may distract the animal from other important aspects of behaviour. For example, while a harem-owning stag is engaged in a fight, other males may attempt to steal some of his unguarded females. Although winning a fight can bring considerable benefits, these may be offset by the risk of loss of assets or of future fighting potential. It is not surprising, therefore, that natural selection has led to the evolution of modes of assessment of the fighting potential of rivals. There is not much point in engaging in a fight that is certain to be lost, and animals that avoid unnecessary fighting will be at an evolutionary advantage. The question of assessment is directly relevant to sexual selection and merits some discussion here.

One way to assess the likelihood that a rival will win a fight is to look for direct indications of fighting potential, such as body size and weapons. This may seem simple enough. However, natural selection may favour cheaters. If an animal exaggerates its body size, for instance, it may deter rivals from fighting. Figure 8.1 shows some examples of this. In a similar manner, natural selection may favour weapons exaggerated beyond the point of usefulness. For example, fossils of the extinct giant deer (*Megaloceros giganteus* which lived in Europe during the Ice Age) have been found with antlers measuring more than 3 metres in span (Figure 8.2). Some scientists (e.g. Gould, 1978) think these animals became extinct because the intrasexual advantage of their large antlers was offset by their disadvantage in terms of natural selection when the climate became colder, since the antlers probably had to be regrown every year, a huge cost when food resources became scarce.

One way for a male to circumvent possible cheating is for him to base his assessment of his competitor on some factor well correlated with real body

Fig. 8.1 Body size is exaggerated in these courtship displays of the Siamese fighting fish (*Betta splendens*) and frigate bird (*Fregata magnificans*) (From *The Oxford Companion to Animal Behaviour*, 1981).

Fig. 8.2 The extinct giant deer *Megaloceros giganteus*.

size and physical fitness, like the roaring of red deer mentioned earlier. Similarly, two males may take part in ritualized trials of strength that do not involve real fighting and injury. Such contests are found in many species. Thus some beetles engage in pushing matches, which the larger individual usually wins. Buffalos (Sinclair, 1977) and bighorn sheep (Geist, 1971) charge each other and clash head on (Figure 8.3). Darwin thought that the evolution of weapons in species where they appear in the males but not in the females was due to their usefulness in fighting sexual rivals. However, some biologists take the view that the effectiveness of weapons is primarily psychological, serving to threaten and intimidate the rival. The way to assess the likelihood that a rival will win a fight is to gauge his aggressive

Fig. 8.3 Two male bighorn sheep in a contest of rivalry over females (*Photograph: L. Lee Rue*).

Fig. 8.4 Two male elephant seals fighting (*Photograph: Burney Le Boeuf*).

motivation or tendency to attack. For example, many animals undergo changes in posture that reflect their motivational state (see Chapter 21).

Male rivalry may be influenced by the behaviour of females. Among elephant seals (*Mirounga angustirostris*), for example, male rivalry is intense and often involves fighting (Figure 8.4). When a male attempts to copulate with a female, she protests loudly, thus attracting the attentions of other nearby males, who attempt to interfere (Cox and Le Boeuf, 1977). A male is likely to be successful in copulating only if he is dominant and can ward off his rivals. The female intensifies the competition among males by her protests and ensures herself a dominant male. Thus, while the female does not directly exercise a choice, her behaviour indirectly has that effect. If the female does not protest, the copulation is less likely to be interrupted and a low-ranking male will have a greater chance of success.

Female choice

Female choice is the aspect of sexual selection that most fascinated Darwin. He maintained that the advantage of behaviour evolved through sexual selection lies primarily in the satisfaction of female choice. He did not say, however, why such female preference might arise or be maintained within a population. Many examples exist of male adornment and courtship behaviour that has evolved as a result of sexual selection (see Figure 8.5). Males that succeed in attracting females by virtue of their special features are likely to father more offspring than less attractive males and thus pass on their features to the next generation. The problem is that, although it is clear how males benefit by acquiring attractive features, it is not so clear why females

Fig. 8.5 Male and female African Paradise birds (*Vidus paradisea*). Outside the breeding season the male looks similar to the female (After Halliday, 1980).

should benefit by choosing males with features that may be irrelevant to survival and that may even be a disadvantage in the face of natural selection. For example, the enormous tail of the peacock (*Pavo cristatus*) is costly to produce, in terms of food and metabolic load. It is unwieldy and likely to hinder escape from predators – it is a handicap. Why, then, do pea hens not prefer to mate with males that carry less of a burden on their chances of survival?

Amos Zahavi (1975) suggested that females prefer males precisely because they carry a **handicap** and therefore must be robust individuals. The handicap is an advertisement for male quality. If a peacock can survive despite the encumbrance of his large tail, he must be a worthy male, since his good qualities will be passed on to the next generation. This suggestion has been criticized (e.g. Halliday, 1978; Maynard Smith, 1976) in that even a modest handicap, if inherited, places a burden on the next generation that outweighs any possible correlated advantages. A female might do well to choose a male that had survived despite an injury or other non-inheritable handicap, although this could not explain the evolution of male adornments. The handicap principle may operate in cases where the male's attractive features are not fixed genetically but are a direct indication of male quality (Zahavi, 1975; Halliday, 1978; Hoelzer, 1989). It is also relevant in cases where apparent handicaps reveal desirable traits (Pomiankowski, 1987), or where sex ornaments are condition-dependent (Andersson, 1986).

Ronald Fisher (1930) provided the most popular explanation of intersexual selection. He pointed out that females that mate with attractive males will tend to have attractive sons, provided the attractive characteristics are inherited. These sons, in turn, will be successful in attracting females and in reproducing themselves. Therefore, a female that chooses to mate with a male on the basis of his sexual attractiveness is likely to have more grandchildren than a female that mates with a less attractive male. It does not really matter what male feature is attractive to females, provided that it is not too strongly opposed by the forces of natural selection. Fisher suggested that initially, females are attracted to male characteristics that have survival value and that these characteristics became exaggerated during evolution through the action of intersexual selection. For example, female birds might show a preference for males with well-maintained plumage, since this would indicate not only a direct survival value in flying efficiently but also that the male could afford the time and trouble to maintain his plumage in good order. Once such a female preference was established in a population, almost any arbitrary feature of the plumage could become exaggerated and evolve into a superplumage irrelevant to the survival of the individual male bird. Fisher recognized that the interaction of male attractiveness and female preference should lead to an escalating evolution of a particular fashion that eventually would be checked by natural selection.

Darwin assumed that the female selects her sexual partner while the males compete for her attentions. Though this is typically true throughout the animal kingdom, there are exceptions. Usually the female pays the greater cost in the process of reproduction. She provides the developing embryos with nutrition and may devote much of her time to their care. Before making such a large investment, the female can be expected to exercise a certain amount of caution. In particular, it is important that she should mate with a male of the correct species that is fully capable of carrying out his functions and likely to endow her offspring with high survival value.

8.2 Sexual strategy

Darwin (1871) was convinced that many of the differences between the sexes (sexual dimorphism) were the result of sexual selection. It is obvious that we should expect the sexes to differ in species where there is intense male rivalry. Features such as large body size, weapons, and aggressiveness will assist males in the competition for females, but there is no advantage in females possessing such features. Therefore, the females of such species will tend to lack weapons and to be smaller and less aggressive than the males.

In species where female choice is important, Darwin's argument is that males evolve particular adornments and behaviour that attract females and that are therefore maintained through sexual selection. Often, however, the differences between the sexes will be due to natural selection rather than sexual selection. In many species, for instance, the females possess special features adapted for care of the young.

Recently there has been a change of emphasis. Rather than attempting to account for differences between the sexes in terms of sexual or natural selection, we can assume that the sexes are fundamentally asymmetrical, and then ask what evolutionary consequences follow from this basic fact.

The fundamental feature of sexual reproduction is that it involves fusion of two gametes to form a zygote that develops into a new individual. The gametes come from the male and female parents and usually take the form of a small sperm and a large egg, respectively. Originally, we may imagine gametes were of the same size, as with some protozoa, such as *Paramecium*. Early in evolution there was probably some genetic variation in gamete size, and then the larger gametes produced larger zygotes, which had a better chance of survival because of their greater food reserves. Therefore, we might imagine the larger gametes to have been favoured by natural selection. If this were the case, then immediately there would have been selection for small gametes that could find a large partner to fuse with them, effectively parasitizing on its food reserves. Thus small gametes could survive only if they became highly mobile and able to discriminate between large and small mating partners. This theory of the evolutionary origin of eggs and sperm, first suggested by Geoff Parker *et al.* (1972), is widely accepted as an explanation of the fact that virtually all present-day sexually reproducing multicelled animals produce eggs and sperm (Krebs and Davies, 1981).

Males produce small sperm in large numbers and females produce large eggs in relatively small numbers. Although these may cost roughly the same amount of energy to produce, the disparity in numbers of gametes means that sperm must compete for the chance to fertilize eggs. With relatively little cost per female, a male can increase his reproductive success by fertilizing many females. The reproductive success of a female, however, depends upon the number and viability of the eggs she produces. A female can follow the strategy of producing many eggs, and releasing them in conditions where at least some have a chance of fertilization and ultimate survival. Alternatively, she may produce fewer eggs and devote more resources to each one, thereby increasing each one's chance of survival, in which case we would expect the female to be highly discriminating in her choice of mate, since she has much to lose from an unsatisfactory partnership. Examples of both extreme types of strategy can be found in the animal kingdom. Most species follow some kind of intermediate position, and the best strategy for a species is determined largely by its ecological circumstances.

A.J. Bateman (1948), working on the fruit fly *Drosophila melanogaster* discovered that the reproductive success of males was closely related to the number of matings achieved, while that of females was not so related. He explained his results in terms of the difference in cost between sperm cells and egg cells. Given that the energy resources available to male and female are roughly equivalent, males can produce many small sperm, while females are limited to fewer eggs of a viable size. The females have more resources invested in each egg, and can be expected to be selective in their choice of mate, while the males collectively have more sperm available than there are eggs to fertilize, and so there is likely to be competition among males for access to females.

The main evolutionary problem for males is to attract females in the face of competition from other males. The main evolutionary problem for females is to select, as sexual partners, males that will endow the offspring with the greatest chance of survival and reproduction. Initially it would seem, the female has the advantage because she can refuse to cooperate. As mentioned above, the female usually exercises the choice in a sexual encounter. A male can do little if the female is unreceptive. It is possible that the male could enforce copulation, but this is usually very difficult because in most species the female must take up a specific posture before copulation can take place. Forced mating, however, has been observed in some animals including mallard ducks and scorpion flies (*Panorpa*). Usually, the male scorpion fly presents the female with a nuptial gift during courtship. This gift is often a dead insect obtained from a spider's web and copulation occurs while the female is eating the gift. Sometimes the male forces mating without presenting the female with a gift (Thornhill, 1980). The male benefits from a successful forced mating because he does not have to run the risk of finding a nuptial gift. Apparently 65 per cent of adult male scorpion flies die as a result of being caught in the spider's web. The female loses because she does not receive a gift, which she normally uses to provide energy for her eggs. However, the success rate from forced mating appears to be small, so it may be a strategy of last resort.

Males usually benefit from a given sexual encounter if they mate, whereas females do not always benefit and may even suffer a reduction in fitness (Parker, 1979). Therefore, it is a good strategy for females to be coy. George Williams (1966) has described courtship as a contest between male salesmanship and female sales resistance. A female would not need to be coy if she could directly gauge the male's fitness and likely future behaviour. All the female has to go on, however, is the male's appearance and present behaviour. The female has to induce the male to reveal his true nature, or she has to arrange matters so that his interests coincide with hers. In both cases, the female's best policy is to be cautious and to prolong the courtship. The first task for the female is to ensure that her potential mate is of the correct sex and species. This is also important for the male, but as we have seen, the female has much more to lose by making a mistake. Natural selection favours individuals that have distinctive markings and behaviour by which they can be identified as members of a particular species. These features play a prominent role in courtship, and they are called **isolating mechanisms** because they promote reproductive isolation among species by reducing the possibility of hybridization.

A female must try to ensure that she mates with a male that is the same species as herself, and is fully mature and sexually competent. By prolonging the courtship she is more likely to mate with males that are fully motivated and sexually vigorous. In the European smooth newt (*Triturus vulgaris*) (Figure 8.6), the intensity of the male's display is correlated with the number of spermatophores (sperm capsules) he produces in the sexual encounter (Halliday, 1976). During courtship, the male deposits the spermatophore on the substratum of the pond, and the probability that the female will pick it up successfully increases during the course of the encounter (Halliday, 1974). To be able to deposit three spermatophores, the male must remain under water without breathing for much longer than he normally would (Halliday, 1977a). Thus, it is possible that in prolonging courtship, the female is providing a test of male fitness.

The assessment of the male by the female is facilitated if the male can give an indication that he is a good choice. Some researchers have suggested that **courtship feeding** may be important in this respect. Courtship feeding, in

Fig. 8.6 A male European smooth newt (*Triturus vulgaris*) (*Photograph: Tim Halliday*).

which the male presents food or food-like objects to the female, is observed in many species. In the herring gull (*Larus argentatus*) the female begs for food as a juvenile would, and the male regurgitates food that the female then eats. In most species the quantity of food transferred from male to female is small in relation to the female's normal daily intake, but there is some evidence that it does contribute to egg production in gulls (Salzer and Larkin, 1990). In the pied flycatcher (*Ficedula hypoleuca*), however, it is about half what a nestling of equivalent weight would receive. In some species the courtship feeding is highly ritualized, and there is no exchange of real food. In the common tern (*Sterna hirundo*) courtship feeding influences clutch weight and it may be a predictor of the male's future performance in providing food for the young (Nisbet, 1973, 1977). Virginia Niebuhr (1981) found that courtship feeding was a reliable predictor of chick feeding by the male herring gull. It also provides an indication of the male's tendency to incubate and to protect the chicks after hatching. For these reasons courtship feeding would appear to provide a good indication of a male's parental abilities.

As we have seen, males usually compete for opportunities to mate with females. Such competition can take many forms, the most obvious being **aggressive competition**. If a male can prevent other males from gaining access to females, then he should have little difficulty in persuading the females to mate with him since they have little alternative. In some species there is competition not for access to females but for fertilization. This is sometimes known as **sperm competition**. For example, male dungflies (*Scatophaga stercoraria*) compete for females and may manage to displace one another during copulation. Geoff Parker (1978) showed that when two males mate with the same female, the sperm of the second male fertilize most of the eggs. Parker irradiated males with cobalt, which prevents eggs fertilized by these sperm from developing. If an irradiated male mates with a particular female after a normal one has, only 20 per cent of the eggs hatch, whereas 80 per cent hatch when a normal male mates after an irradiated male. Somehow, the sperm of the second male displace most of those of the first.

In mammals, when two males copulate with the same female, the male with the larger number of ejaculates usually ensures paternity of a larger proportion of the female's litter (Dewsbury, 1984). Among primates, sperm competition is likely to be important in species where females are routinely mated by more than one male. Such species have larger testes for their body size than do harem-holding or monogamous primates (Harvey and Harcourt, 1984).

In species where sperm competition is possible, it is not surprising to find males taking precautions against it. Many mammals copulate in private, away from possible interference. The male dungfly sits on top of the female after copulation and guards her until the eggs are laid. Some insects cement up the genital opening of the female after they have copulated. Such mating plugs are thought to prevent subsequent matings in some water beetles, butterflies and moths (Wilson, 1975). Effective mating plugs can be achieved by prolonged copulation. Male houseflies remain in the copulatory position for about an hour, although most of the sperm are transferred during the first fifteen minutes. Some moths copulate for a full day. Following copulation in the ceratopogonid fly, *Johannseneilla nitida*, the female eats the male with the exception of his genitalia, which remain in place and serve as a mating plug (Wilson, 1975).

If a male is unable to gain access to females in competition with other males, he may deploy an **alternative strategy**, stealing copulations from a dominant male. For example, male bullfrogs (*Rana catesbeiana*) compete for territories in ponds where the females come to lay their eggs. The females prefer some parts of ponds to others, and these are usually those parts where the water is warm but where the vegetation is not too dense. In such conditions the eggs develop quickly in a tight ball and are relatively safe from attack by leeches (Howard, 1978). The males compete vocally and physically for the best territories. The older, larger males usually win, and small, young males may end up with no territory. The younger males adopt the alternative strategy of sitting silently near a calling male and attempt to intercept females that he attracts. Young elephant seals may attempt to join the harem of a dominant male by behaving like a female. They then sneak copulation when the large bull is preoccupied with a rival (Le Boeuf, 1974). Young red deer stags may attempt to steal copulations from the harem of a dominant stag.

Many species contain males that are never likely to become dominant or territorial. They may adopt the alternative strategy permanently. For example, male ruffs mating at a lek adopt alternative strategies (Figure 8.7). The light-coloured satellite males are tolerated by the territorial males, probably because they attract females to the lek. Each satellite male lurks near a particular territory and steals copulations while the owner is otherwise engaged. The two types of male are genetically distinct, and males do not change from one strategy to the other during their lifetime.

There are some species in which the male invests more in the reproductive process than the female. In such species, according to theory, we would expect to find greater variability in reproductive success among females than among males, and we would expect to find the male being more careful in choice of mate than the female. A number of examples are reviewed by Trivers (1985). In the three-spined stickleback (*Gasterosteus aculeatus*) the male establishes a territory that he defends against other males. He builds a nest and courts any ripe females that enter his territory. Once the female has laid her eggs in the nest, the male fertilizes them and then vigorously drives

Fig. 8.7 Ruffs displaying at their lek. This is the species with the most marked individual variations in male plumage. Males with dark ruffs and tuffs are territorial, while the white one is a satellite male. The bird without a ruff is a female (From *The Oxford Companion to Animal Behaviour*, 1981).

the female out of his territory. Care of the eggs and young is entirely the responsibility of the male. Although there is competition among males in establishing territories and attracting females, the males are discriminating as to who is allowed into the territory. Only females evidently ready to spawn are permitted to approach the nest. Thus, choice of mate is partly the prerogative of the male. It is most important for the male to guard against raids by other males seeking to steal eggs and destroy the nest (Wooton, 1976). Thus, any mistakes made in identifying intruders into his territory could have serious consequences. The female stickleback is also able to exercise choice, and some researchers have shown that females prefer to spawn with males whose nests already contain eggs (Ridley and Rechten, 1981).

8.3 Altruism

Altruistic behaviour benefits other animals at some cost to the donor. In evolutionary biology, altruism is defined by reference to its effects on survival prospects without reference to any motivation or intention that may be involved. The possibility that animals may have altruistic or selfish intentions is, of course, of interest (see Chapter 26), but it is not relevant to consideration of altruism from an evolutionary point of view. This distinction is sometimes forgotten.

In the strict sense, altruism is an evolutionary possibility only if it is defined in terms of individual fitness. An altruistic act increases the individual fitness of the recipient while it decreases the individual fitness of the donor. In certain circumstances, as we see later, natural selection may favour such behaviour. In terms of inclusive fitness, however, this is not the case. Thus, natural selection would not favour behaviour that benefitted another animal at the expense of the donor's inclusive fitness because inclusive fitness is maximized by natural selection (see Chapter 6).

Natural selection will favour altruistic behaviour under two main circumstances:

1. If the benefit in fitness to the recipient of an altruistic act exceeds the cost (decrement in fitness) to the donor by more than their coefficient of relationship.

2. If one individual benefits another at little cost and this situation is later reciprocated.

This second aspect of natural selection was proposed first by Trivers (1971) as a form of altruism that could occur between unrelated individuals.

There have been various other explanations of apparently altruistic behaviour. The most notorious of these, called **group selection**, claims that natural selection will favour behaviour that reduces the fitness of the donor if it benefits the group or species as a whole. However, evolutionary biologists have been unable to demonstrate how such a situation could arise during evolution, and many consider it to be impossible (e.g. R. Dawkins, 1976; Maynard Smith, 1964).

The main problem is to prevent cheating. For example, let us imagine a population of rabbits in which the members do not warn each other of approaching danger. Suppose that a gene is introduced that promotes thumping on the ground when a rabbit senses danger. The thumping serves to alert other rabbits, but it also attracts the attention of the predator. Thus, it would seem that neighbouring rabbits benefit from the warning while the thumper endangers itself. If we assume that the thumper manages to pass on this thumping gene before he is eaten by a predator, then we have a sub-population of thumping rabbits within a population of non-thumpers. Because thumping is of benefit to the group, the group selection argument maintains that the thumping group will evade predation better than the non-thumping group. Although the thumpers endanger themselves by attracting the attention of the predator, the nearby rabbits escape. Thus, the predator gets only one rabbit, whereas a stealthy predator in a non-thumping population might be able to pick off prey one at a time and end up with more than one rabbit per visit.

However, this situation is not evolutionarily stable because a rabbit in the thumping group that did not possess the thumping gene would not endanger itself when it detected a predator, yet it would benefit from the warnings given by other members of the group. This rabbit would be a cheat and would have an advantage over other members of the group. The greater reproductive success of the cheating rabbits would mean that the thumping gene would gradually be eliminated from the thumping population. There have been various attempts to circumvent this type of theoretical argument, but none has won much support among evolutionary biologists (see Grafen, 1984).

Natural selection favours genes that promote altruistic behaviour toward individuals that are genetically related to the altruist. John Maynard Smith (1964) coined the term **kin selection** to distinguish this type of selection from group selection. Some authors (e.g. Wilson, 1975) erroneously define kin selection as a special case of group selection. Kin selection is a special consequence of gene selection (R. Dawkins, 1976).

The degree to which altruistic behaviour should be extended toward other individuals will depend upon the probability of the gene being represented in those other individuals (e.g. upon the **coefficient of relationship**, r, between the two animals involved). This does not mean that the individual altruist has to calculate its relatedness to each possible recipient. J.B.S. Haldane (1955) joked that 'on the two occasions when I have pulled possibly drowning people out of the water (at an infinitesimal risk to myself) I had no time to make such calculations'. He realized, of course, that the animal behaves as if it had made the calculations. Nevertheless, the animal somehow has to direct its altruistic behaviour toward its kin rather than toward other animals. There are two main ways by which this can be achieved. The first is by kin recognition, which is discussed below. The second is simply a result of living near one's relatives. For example, in the case of rabbits discussed previously, the thumping gene may spread if the thumping rabbit tends to be surrounded by kin; that is if the gene for thumping primarily benefits rabbits who also carry some of the thumper's genes, then the inclusive fitness of the thumper will be increased by thumping behaviour. The cheat who does not carry the thumping gene may benefit initially, but once surrounded by its own non-thumping relatives, it will be at a disadvantage. Thus, it will be difficult for a population of thumping rabbits to be invaded by cheats, provided that related individuals tend to stay close to each other.

The alternative strategy is for altruists to recognize their kin and to confine their altruistic behaviour to them. Such **kin recognition** is known to occur in some species and is often due to early experience. Beldings ground squirrels are born underground, but mother and offspring do not learn to recognize each other until the young emerge above ground at three and a half weeks old. Only then will there normally be any chance of confusion (Trivers, 1985). Ground squirrels that are cross-fostered in nature (raised by by a foster family following the death of their mother) treat their foster family as if it were their own. This occurs only if the cross-fostering occurs before the squirrels are weaned and emerge above ground (Sherman, 1981). Underlying this effect of early association there is also a genetic effect. If squirrel pups are reared apart and then tested together, siblings are less aggressive than unrelated squirrels.

Most ground squirrel litters are fathered by more than one male, and littermates may be full siblings or half siblings. Observations in nature show that female squirrels act more altruistically toward full siblings than half siblings. The ability of the female to make this discrimination must be due to some kind of **phenotypic matching**, an ability to compare one's own phenotype with that of others. Other examples are discussed by Trivers (1985).

Altruism towards kin can be regarded as selfishness on the part of the genes responsible, because copies of these genes are likely to be present in relatives. Altruism could also be regarded as a form of gene selfishness if by being altruistic an individual could ensure that it was a recipient of altruism at a later date. The problem with the evolution of this kind of altruism is that individuals that cheated, by receiving but never giving, would be at an advantage.

It is possible that cheating could be countered if individuals were altruistic only toward other individuals that were likely to reciprocate. For example, Craig Packer (1977) observed that when a female olive baboon (*Papio anubis*) comes into oestrus, a male forms a consort relationship with her. He follows her around, waiting for an opportunity to mate, and guards the female from the attentions of other males. However, a rival male may sometimes solicit the help of a third male in an attempt to gain access to the female. While the solicited male challenges the consort male to a fight, the rival male gains access to the female. Packer showed that the altruism shown by the solicited male is often reciprocated. Those males that most often gave aid were those that most frequently received aid.

This type of **reciprocal altruism** obviously provides scope for cheating. An individual that receives aid may refuse to reciprocate at a later date. However, if opportunities for reciprocal altruism arise sufficiently often, and if the individuals involved are known to each other, then a non-cooperative individual can be identified easily and discriminated against. Thus, for natural selection to favour reciprocal altruism, the individuals must have sufficient opportunities for reciprocation, they must be able to recognize each other individually and remember their obligations and they must be motivated to reciprocate. These conditions are found in primitive human societies, and reciprocal altruism has played an important role in human evolution (Trivers, 1971)(see Box 8.1).

8.4 Parental care

Parental care is a form of altruism. In spending time and energy in aiding its offspring, the parent is increasing their fitness to the detriment of its individual fitness in that it is favouring current offspring at the expense of possible future offspring. The degree of parental care varies considerably from species to species. It depends upon the number of offspring produced, the type of mating system involved and the aid given to the offspring by animals other than the parents.

The female usually spends more time and energy in caring for the young than does the male. The unequal contribution of the parents in caring for the young can be seen as an evolutionary contest in which each sex, while having an interest in the offspring in common with the other, also has an interest in minimizing the cost to itself of the necessary parental care. Thus, if a gene in a male can so influence the behaviour of the animal that the burden of parental care is shifted toward the female, then the gene is likely to become more frequent in the population. Conversely, if a gene in a female has influenced her behaviour so that the male is required to raise his contribution, without any extra risk to the offspring, then that gene is likely to spread.

Robert Trivers (1972) used the term **parental investment** to indicate the effort, in terms of time and resources, put into rearing an individual offspring. Trivers argued that 'where one sex invests considerably more than the other, members of the latter will compete among themselves to mate with members of the former.' The degree of competition depends upon a number of factors, however. Obviously, much depends upon the relative numbers of males and available females, which in turn depend upon the ratio of males to females in the population (the sex ratio) and the type of mating system typical of the species.

Parental investment can be defined as any investment by the parent in an individual offspring that increases the offspring's chance of surviving (and hence its reproductive success) at the cost of the parent's ability to invest in other offspring. Although undoubtedly useful, this definition suffers from the implication that past investment can influence future behaviour. For example, suppose a mother has a choice of saving the life (by investment) of one of two offspring of different ages. The one she neglects is bound to die, but if she can save only one, which should it be? In her lifetime, the mother has limited resources to invest. She stands to lose a higher proportion of her life's investment if she neglects the older offspring; thus, it might seem that she should save the older and let the younger die (R. Dawkins, 1976).

Box 8.1   Reciprocal altruism among the !Kung bushmen

The !Kung bushmen of the Kalahari desert provide a good example of the importance of reciprocal altruism in hunter–gatherer societies. The women bring in about two-thirds of the protein and carbohydrate by gathering vegetables and fruit. The men spend many hours hunting for game which provides essential amino acids, and minerals (Lee, 1972). The food supply is very variable and times of plenty may be followed by periods of hardship. The !Kung work to obtain calories and nutrients which are shared among all members of the group, plus any visitors. The consumption of each person, therefore, can be assumed to be the same, in relation to their requirements.

In San society, the successful provide food for the unsuccessful on the understanding that the situation may be reversed. The successful never boast about their achievement, but are always very modest and self-effacing. This is typical of hunter–gatherers and contrasts with the boastful behaviour of people who practise redistributive exchange. Marvin Harris (1985) maintains that the typical reciprocal exchange occurs in environments where hunting follows the law of diminishing returns; over-exploitation of the game is not to be encouraged, so the different hunting groups do not compete, and do not advertise their prowess. The San carry out their duties without coercion, as a result of social obligation. Shirkers are maligned, but suffer no material loss. Whether foraging or doing camp chores, the individual is working for payment that is delayed. The payment comes in the form of shared-out food. If the payment had come in the form of money, it would have to be spent on that same shared-out food, because there is nothing else to buy. In the animal kingdom, the normal consequences of foraging are food preparation and eating. By inserting an additional stage in this sequence, the San have taken the first step towards a monetary economy (see Chapter 24). It is worth noting that something very similar occurs among those animals that obtain food and then transport it home to feed to their young.

When !Kung women go gathering fruit and vegetables, they have to carry their children with them (*Photograph: Gerald Cubitt: courtesy of Bruce Coleman Ltd*).

However, we would expect natural selection to design animals to behave so as to maximize their future reproductive success irrespective of past invest-ment. Dawkins and Carlisle (1976) pointed out this fallacy in some applica-tions of the notion of parental investment.)

The coefficient of relationship between parent and offspring is often only one-half. Consequently there is a conflict of genetic interest between parent and offspring. The selective pressure on a parent is to avoid too much invest-ment in a particular offspring, because this would decrease the total number of offspring surviving. The selective pressure on the offspring is to stop asking for investment by the parent when the cost of the investment is more than twice the benefit that it receives. This is due to the offspring being identically related to itself, but only half related to its full siblings. This situation usually results in the offspring demanding investment for longer than the parent is inclined to give it, as shown in Box 8.2.

In summary, parental care is a form of altruism. Parents invest in their off-spring at the expense of their own survival and chance of future reproduc-tion. The role of the sexes in parental care varies from species to species in accordance with ecological factors which are only partly understood.

8.5 Cooperation

Cooperation among animals usually involves some form of altruism. In co-operation among members of different species, called **symbiosis**, the relationship is reciprocal. For example, many aphids gain protection by asso-ciating with ants, while the ants benefit by obtaining food from the aphids. Thus, when a garden ant (*Lasius niger*) encounters a bean aphid (*Aphis fabae*), it strokes the aphid with its antennae. This induces the aphid to exude honeydew, a sugary liquid that is a byproduct of digestion, that the ant then consumes.

Anemone fish, *Amphiprion*, are able to gain protection from predators by swimming among the tentacles of sea anemones without being harmed. Each fish has to develop an immunity to the anemone sting, and it is highly prob-able that the anemones could evolve an effective deterrent to anemone fish. However, the fish do not harm the anemones, and some fish defend their anemone against predators like the butterfly fish (*Chaetodon*), which bite the ends of anemone tentacles. The anemones also benefit from the anemone fish's habit of bringing its food into the anemone, some of which is eaten by the anemone. Thus, the relationship between the two animals is reciprocal. The anemone benefits by obtaining food and by gaining some protection from predators. The anemone fish gains protection for itself and for its eggs and fry, which are able to develop among the anemone tentacles undisturbed (see Figure 8.8).

Cooperation among members of the same species often involves some form of altruism. The cooperative hunting of wild dogs, lions and hyenas usually occurs among relatives (see Figure 8.9). Wild dogs (*Lycaon pictus*) prey upon animals much larger than themselves, such as zebra and wildebeest. They select a single quarry and chase it over a long distance. The hunt

Box 8.2 Trivers' (1974) model of parent–offspring conflict

Conflict during the period of investment over the amount of investment. The benefit (*B*) or the cost (*C*) of investment at a given moment is plotted as a function of the amount of parental investment (such as amount of milk). The parent is selected to maximize the difference between the two functions (*B* and *C*); that is, to invest *p*. The offspring maximizes (*B* − *rC*) (here, *r* = ½); that is, it is selected to receive investment *o*. (RS = reproductive success; IF = inclusive fitness.) (From Trivers, 1974).

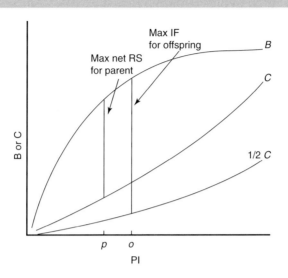

Submissive posture of herring gull chicks near their parent. Although fully grown, these chicks of two months are still dependent on their parents for food; their parents are now ambivalent about feeding them and tolerating them nearby. As a result of this conflict, the chicks have adopted a very submissive posture (the very opposite of attack): head withdrawn into the body, wings relaxed, body small. Similar postures at this stage are found in other gull species.

involves cooperation in the choice of prey and in bringing the animal down. During the chase, there may be exchange of leaders, thus sharing the burden over long distances. A dog in the rear will sometimes cut corners in an attempt to head off prey. The food is shared among the members of the pack, and upon arrival home, adults often regurgitate food for the young. Some adults may stay behind to guard the juveniles, and these animals are also fed by the returning pack. Thus, individual dogs do not act purely in their own interest but are altruistic toward other members of the pack.

The altruistic behaviour involved in cooperative hunting is fairly straight-forward and not difficult to account for in terms of kin selection. More

Fig. 8.8 Anemone fish (*Amphiprion*) are not stung by anemones to which they have become acclimatized. They can then gain protection from predators by hiding among the tentacles (*Photograph: Tim Halliday*).

Fig. 8.9 Wild dogs pulling down a wildebeest after running it down (*Photograph: Hans Kruuk*).

difficult problems for the evolutionary theorist arise from **cooperative breeding**. For example, the Mexican jay is a communally breeding species that lives in flocks of four to fifteen individuals. It is a non-migratory species, and each flock communally defends a territory in oak or pine forest. A mated pair builds a nest, and the female lays her eggs only in her own nest. Each flock may have between one and four nests. When the eggs hatch, the nestlings are fed not only by their parents but also by other members of the flock. About 50 per cent of the food an infant receives is provided by birds other than its parents. These helpers at the nest are behaving in an appar-

Box 8.3   Cooperative breeding in acorn woodpeckers

Acorn woodpecker
outside its nest
(*Photograph by Ian Tait*).

Acorn woodpeckers (*Melanerpes formicivorus*) frequently breed in groups containing mate-sharing males and joint-nesting females (called cobreeders). Cobreeders are usually closely related and thus gain direct fitness benefits by raising their own young, and indirect fitness benefits by raising the young of relatives.

Walter Koenig of the University of California (Koenig, 1990) has performed experiments to determine whether indirect fitness benefits are sufficient to induce normal care of the young among male cobreeders whose opportunity of parentage had been reduced or eliminated by temporarily removing them during egg laying. He found that dominant males always destroyed the eggs in the nest that they had been removed from, while subordinate males did not do this when they were returned to the nest after a period of imprisonment.

Specifically, when renesting is a likely possibility, dominant cobreeding acorn woodpeckers, whose opportunity of parentage is reduced or eliminated, destroy the ongoing nest and force a renest, while subordinates do not. When renesting is unlikely to be successful late in the season, then neither dominants or subordinates destroy the ongoing nest, but help to raise the offspring already in the nest, the offspring of another male.

It is possible that the direct fitness benefits accruing to dominants by forcing a renest outweigh the indirect fitness benefits of feeding non-descendant offspring, while the reverse is true for subordinates. An alternative hypothesis is that destroying the nest is in the best interest of all detained birds, but that subordinates fail to do so because they are denied access to the nest by dominants. Koenig's observations support the former hypothesis.

ently altruistic manner, because at considerable cost to themselves, they are helping the offspring of another individual (see Box 8.3).

It is necessary to explain why a mutation that led to failure to aid another individual's offspring should not spread through the population and eventually destroy the basis for the cooperative behaviour. In other words, the non-breeding birds must benefit in some way. There are various possibilities:

■ they may gain experience in caring for young

■ they may benefit from belonging to a group

■ they may increase their inclusive fitness by aiding relatives

■ they may inherit part of the parental territory

The main disadvantage is that the helper may lose or postpone the opportunity to breed itself. However, the chances of successfully breeding independently are offset by the difficulty of finding a mate, establishing an independent territory and breeding successfully as a novice.

Many of these features are illustrated by Ulrich Reyer's study of the pied kingfisher, a common fish-eating bird that occurs along rivers and freshwater lakes in Africa. He studied two marked populations in Kenya. The pied kingfisher has two types of helper at the nest, primary ones that help their own parents and secondary helpers that help birds other than their parents (Reyer, 1980). These helpers are always males, probably because there are many more males than females in populations of pied kingfishers. The primary helpers are the one- or two-year-old offspring of the parents. They remain with their parents throughout the year, forming a permanent co-operative group. Primary helpers usually become breeders or secondary helpers. These are two to three years old, and thereafter become breeders. They join a breeding group, to which they are unrelated, a few days after the young are hatched.

The primary helpers feed the young and protect them against predators. In addition, they chase away competitors for nest sites, and regularly feed the male and female breeders prior to egg laying. They do not take part in nest building, incubating or brooding. They do not copulate and do not appear to be physiologically capable of reproduction (Reyer *et al.*, 1986). The secondary helpers are more capable of fertilizing eggs. They are potential rivals to the parental male and are often attacked by him, and by the primary helpers (Reyer, 1986). By offering fish the secondary helpers reduce the probability that they will be attacked. The fish appears to be payment for being accepted as a group member. After the young have hatched the primary helpers carry more food to the nestlings, whereas the secondary helpers carry more to the parental female. The primary helpers increase their inclusive fitness by helping their siblings, whereas the secondary helpers benefit mainly by finding a mate and reproducing themselves.

It is easy to see why the primary helpers are tolerated by the parents. They are closely related, they are not a threat to the parents, and they assist in feeding the young. It is less easy to see why the secondary helpers are tolerated. They are unrelated, they are potential rivals to the male parent, and they provide much less help than do the primary helpers. Reyer (1980) found that more helpers are recruited in unfavourable conditions than in favourable conditions. There were rarely more helpers per pair than was necessary for maximal breeding success. Secondary helpers, in particular, were not tolerated when breeding conditions were favourable. The accessibility and quality of food, the distance between the breeding colony and the fishing grounds, and the frequency of disturbance by people, were all factors that contributed to breeding success and to the number of helpers recruited by breeding pairs.

Reyer (1984) attempted to calculate the inclusive fitness of birds following different strategies over a two-year period. He concluded that breeding is superior to helping, and helping is superior to doing nothing. The primary helper strategy is more costly, but the inclusive fitness of these helpers is increased as a result of the enhanced survivorship of their parents' new brood. The secondary helper strategy is less costly and these helpers may increase their individual fitness by increasing their chances of mating with the resident female. They also become familiar with the locality and come to be tolerated by neighbours, both of which probably increase their eventual breeding efficiency.

In general, altruistic behaviour can be distinguished from other types of social interaction on theoretical grounds in terms of the apportionment of benefit between the two participants. Bill Hamilton (1964) suggested the terminology shown in Figure 8.10 to describe the four main types of interaction. This classification can be usefully applied to the problem of helpers at the nest. If the relationship between a breeding bird and its helper is truly cooperative, then we would expect both to benefit in terms of individual fitness. This could happen in situations where the consequences of living in a group are of benefit to both the breeding pair and the helpers. Thus, a group might be able to hold a better territory than a pair. There might also be benefits in combating predation. The help given by non-breeding birds could be a payment for joining a group (Gaston, 1976), or it could be a form of apprenticeship (Emlen, 1978), the helper eventually taking over the role of the female, or mating with the female (Reyer, 1984).

In cases in which the helper is altruistic, the breeding birds should benefit while the helper loses individual fitness. The inclusive fitness of the helper is increased in cases in which it is aiding its relatives. This strategy could become established by kin selection. The alternative strategy, for the juvenile bird to mate and set up independently, might not be viable at the outset. Thus, it might pay the helper to wait for a vacant territory.

The selfish helper is one that gains from associating with a breeding pair but that may decrease the pair's reproductive success. Among some woodpeckers (Skutch, 1969) and the white-winged chough (Rowley, 1965), novice helpers tend to be inefficient. The extra birds on the territory could deplete the food resources, and the extra activity around the nest may attract predators. Why should the parents tolerate the presence of helpers if they do not benefit from it? A possible reason might be that they have a genetic interest in the welfare of their kin and that their tolerance of helpers is an extended form of parental care.

If neither the breeders nor the helpers benefit, then the relationship is spiteful. The helpers could be tolerated by the parents, even though their presence was detrimental, for the reason discussed in the last paragraph. The helper may gain nothing in the short term, but it may be able to sabotage the efforts of the breeding pair and eventually take over the territory (Zahavi, 1974, 1976).

Fig. 8.10 Types of behavioural interaction (After Hamilton, 1964).

		Fitness change to recipient	
		Gain	Loss
Fitness change to donor	Gain	Cooperative	Selfish
	Loss	Altruistic	Spiteful

Points to remember

■ Males which have a (genetically based) mating advantage over other males have more offspring to which the advantage is transmitted. This is called sexual selection.

■ Males may have an advantage over other males as a result of male rivalry or of female choice. Male rivalry involves ritualized combat, bluff, assessment of the rival's fighting potential and, sometimes, actual fighting.

■ Females that choose to mate with males as a result of their sexual attractiveness will tend to have attractive sons, provided the attractive features are inherited. Therefore, a female that chooses a mate on the basis of sexual attractiveness is likely to have more grand-children than a female that mates with a less attractive male.

■ Fundamental differences between the sexes result from the fact that males produce many small sperm while females produce few large eggs. This means that the female invests more in the outcome of each fertilization. This difference in initial parental investment means that male and female are likely to pursue different evolutionary strategies.

■ The main evolutionary problem for males is to attract females in the face of competition from other males. The main evolutionary problem for females is to attract sexual partners that will endow their offspring with the greatest chance of survival and reproduction.

■ Courtship involves a contest between male salesmanship and female sales resistance. Coyness helps the female to assess the suitability of the male as a potential mate and may also encourage the male to invest in the future offspring.

■ Sexual strategies adopted by males may include aggressive competition, sperm competition, and sneaked copulations. Different strategies may be employed at different stages of the life cycle.

■ Natural selection will favour altruistic behaviour if the benefit in fitness to the recipient of an altruistic act exceeds the cost (decrement in fitness) to the donor by more than their coefficient of relationship. It may also favour altruism in which one individual benefits another at little cost and this situation is later reciprocated.

■ Natural selection will favour genes that promote altruistic behaviour toward individuals that are genetically related to the altruist. This form of selection is known as kin selection.

■ Parental care is a form of altruism because the parent diminishes its own fitness by investing time and energy in the care of its current offspring. This increases their fitness at the expense of possible future offspring because the parent has a limit to the rescues that it can invest during its reproductive life.

■ Cooperation among animals usually involves some form of altruism. The (symbiotic) cooperative relationships between members of different species are usually reciprocal. Cooperation among members of the same species may also be based upon kin selection.

Further reading

Gould, J.L. and Gould, C.G. (1989) *Sexual Selection*. Scientific American Libary, W.H. Freeman, New York.

Grafen, A. (1984) Natural selection, kin selection and group selection. In Krebs, J.R. and Davies, N.B. (eds) *Behavioural Ecology*, 2nd edn. Blackwell Scientific Publications, Oxford.

Ryan, M. (1977) Sexual selection and mate choice. In Krebs, J.R. and Davies, N.B. (eds) *Behavioural Ecology*, 4th edn. Blackwell, Oxford.

Trivers, R. (1985) *Social Evolution*. Benjamin/Cummings, Menlo Park, CA.

Behavioural ecology

Behavioural ecology, so named because 'the way in which behaviour contributes to survival and reproduction depends on ecology' (Krebs and Davies, 1993) is a branch of evolutionary biology. As we saw in Chapter 7, evolutionary biologists are primarily interested in explaining how a state of affairs observed today is likely to have come about as a result of evolution by natural selection. The evolutionary persistence of a trait, such as a particular aspect of behaviour, depends upon its contribution to the survival and reproduction of the individual carrying the trait. Therefore, when we see an animal behaving in a particular way, we can ask how the behaviour contributes to survival and reproduction under the ecological circumstances. Notice that this type of question is not aimed at an explanation of behaviour in terms of mechanisms possessed by the animal. In terms of Tinbergen's four types of answer to 'why' questions (see Chapter 1), we are dealing here with type 4, 'Why do animals of a particular species characteristically behave in particular ways in particular situations?'

An individual's success at surviving and reproducing depends critically upon its behaviour, and natural selection will tend to design animals as efficient foragers, efficient predator avoiders, efficient copulators, efficient parents, etc. (Krebs and Davies, 1993). The best thing to do, at any particular time, depends upon the behavioural alternatives available, and the consequences of each alternative.

In this chapter we show how these can be analysed, concentrating upon the main concepts employed in behavioural ecology.

9.1   Energy and foraging

All behaviour uses energy which the animal must replace through its own behaviour. Food is the prime source of energy in animals. Food may be gained in various ways, all of which may be listed under the heading of foraging, as shown in Figure 9.1. Natural selection favours efficient foragers, and most animals are extremely adept at searching for, and harvesting, food.

Different species employ different foraging methods, some searching for food, some lying in wait for prey, some grazing, etc. Thus some species have

Fig. 9.1 Modes of food gain in animals.

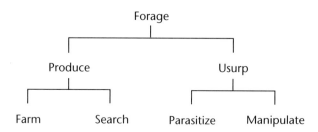

a high rate of energy expenditure while foraging, but spend relatively little time foraging, while other species have a low rate of energy expenditure but spend a lot of time foraging. The rate at which energy is obtained depends upon the availability and accessibility of the food. These determine the rate of return on foraging.

Foraging efficiency is usually a matter of **trade-off** between competing priorities. These may include energy gained versus energy spent (see Box 9.1), energy gained versus risk of predation, energy gained versus losses to rivals (see Box 9.2) etc. Moreover, such trade-offs apply not only to energy produced by foraging, but also to that usurped. Thus the food-finding abilities of one species can become part of the foraging strategy of another species. In other words, the foraging investment of one species (the producer) is parasitized by another species (the scrounger). For example, lapwings foraging on earthworms are pirated by black-headed gulls. Often the gulls are involved in expensive aerial chases. They can avoid these if they attack the lapwings just as they obtain their prey. When taking large worms, the lapwings adopt a tell-tale crouching posture, which the gulls can use as a cue as to the best time to attack. In one study it was observed that attacks against crouching lapwings were four times as successful as attacks against those that did not crouch (Barnard and Stevens, 1981). Lapwings can reduce their vulnerability to attack by concentrating on the smaller worms nearer the surface, and thus obviating the crouching required to obtain the larger, deeper, worms. Lapwings may even discard the large worms (which are the gulls' principal target) after having extracted them at great expense from the ground (Thompson and Barnard, 1984). The kleptoparasitic strategies and counter-strategies will often give rise to an evolutionary arms race (Dawkins and Krebs, 1978; Dawkins, 1982; Barnard, 1984), which may or may not reach stable equilibrium.

Many animals are able to select among various kinds of food according to their physiological requirements (see Chapter 15). In order to obtain food animals have to expend energy. They may also have to spend valuable time, spend physiological commodities such as heat and water, and risk exposing themselves to predation. Given freedom of choice, the animal should trade off among the alternative possible behaviours and perform the one that is most beneficial in terms of fitness. However, the animal does not always have freedom of choice because of **constraints** operating on, and within, the system.

Animals must expend energy in order to forage, and there may be circumstances in which the energy available to spend is limited. Similarly, the

Box 9.1 An example of a foraging trade-off

Reto Zach (1979) discovered a simple trade-off in the foraging behaviour of crows that feed on shell-fish on the west coast of Canada. The crows hunt for whelks at low tide, usually selecting the largest ones. When they find one they hover over a rock and drop the whelk so that it breaks open to expose the edible inside. By dropping whelks of various sizes from different heights, Zach discovered that the number of times a whelk has to be dropped in order to break is related to the height from which it is dropped. The crows have to expend energy in flying up to drop a whelk, so

Zach calculated the total amount of upward flying that a crow would have to do to break a whelk from a given height. He showed that the lowest flying cost is incurred if the whelks are dropped from about 5 metres. Thus, there is a trade-off between the cost in terms of the number of drops required to break the whelk and the height of drop. Calculations based upon this trade-off reveal that the optimal dropping height is indeed about 5 metres. Thus, it appears that the crows somehow have been programmed to exploit this particular feature of the foraging situation.

Number of times a whelk shell needs to be dropped to break it, when dropped from different heights (After Zach, 1979).

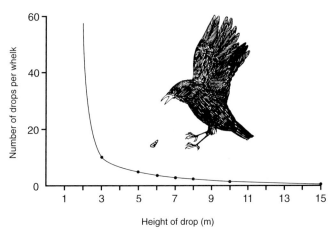

Total amount of upward flight (number of drops × height of each drop) is minimal at the height (arrow) usually used by crows (After Zach, 1979).

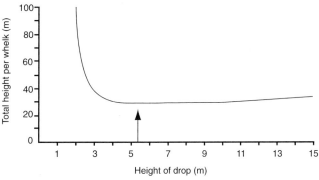

expenditure of a shopper in a supermarket may be limited by the amount of money available. Such limitations on expenditure constrain the behaviour of the forager, or shopper, in ways that are discussed further in Chapter 24.

A shopper with a limited amount of time available is constrained by the rate at which items can be selected and paid for. Similarly, foraging animals

Box 9.2   Singing versus foraging in great tits

In the great tit (*Parus major*) song is an important aspect of territorial defence and advertisement (Krebs, 1971). The major daily episode of singing during early spring occurs at dawn, starting when the sun is just below the horizon. This is typical of many birds; hence, the dawn chorus. It is something of a puzzle because most small birds lose up to ten per cent of their body weight overnight in winter and therefore ought to be very hungry when they wake up. Why do they not forage rather than sing first thing in the morning? There seem to be two main reasons. First, the advantage of territorial defence just after dawn is particularly high because at this time birds without territories probe the defences to exploit any vacancies that have occurred since the previous day (Kacelnik and Krebs, 1983). Second, the benefits of foraging are low because foraging efficiency is limited by the light intensity. Alex Kacelnik (1979) determined by laboratory experiments that great tits cannot forage efficiently below a certain level of illumination and showed that this level normally occurred between 40 and 80 minutes after dawn, depending on the weather. Thus, the birds probably gain more benefits from singing first and foraging second than they would by foraging first and singing later.

Luminance of great tit feeding sites in relation to time of day, for sunny and overcast weather. P and S are the luminance values at which measures of profitability and searching efficiency, as determined in the laboratory, respectively reach 95 per cent of their maximum value (After Kacelnik, 1979).

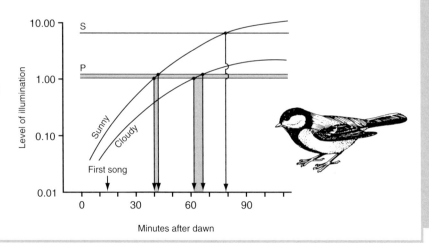

are often constrained by the **handling time** required to recognize, capture and process each food item. An animal searching for prey will encounter prey items at a rate that depends upon the **availability**, or density of prey in the environment. Upon encountering an item, the animal will take some time to recognize it as prey. It will then either attack the prey, taking some time to pursue and capture it, or it will resume searching. An attack may, or may not, be successful, but success involves a further period of time during which the prey is consumed. As shown in Figure 9.2, the handling time is made up of the recognition time + the pursuit and killing time + the consumption time. Thus, in addition to the availability of the resource, the returns that an animal obtains per unit time spent foraging partly depend upon the **accessibility** of the resource. The accessibility includes both the handling time and the energy that the animal must spend in obtaining the prey.

When a predator encounters a prey item, it has to pay a cost in terms of the time taken to catch and eat the prey. This cost is offset by the net energy

Fig. 9.2 Time taken to obtain prey. The forager must first take time to recognize an item as prey, then it must decide whether to attack. Pursuing and killing the prey takes further time, and if the attack is successful, time is taken to consume the prey. (From Stephens and Krebs, 1986).

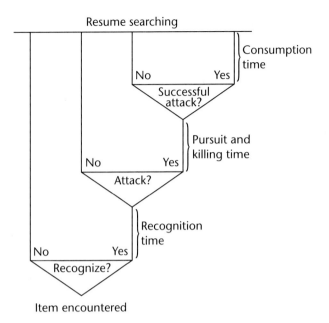

value of the prey item, which is the gross value minus the energy expended in handling and digesting the food. The **profitability** of the prey is the net energy value divided by the handling time. When an animal is able to choose among prey items, we might expect it to choose the most profitable prey. For example, in one study, bluegill sunfish (*Lepomis macrochirus*) were allowed to hunt for water fleas (*Daphnia*) in a larger aquarium (Werner and Hall, 1974). The investigators found that the fish showed no preferences among small, medium and large *Daphnia*, when these were available at low densities. When the *Daphnia* were abundant, however, the fish chose the largest and most profitable *Daphnia* and ignored the small ones. This is the result that would be expected if the fish were maximizing profitability, but we still do not know how the animal makes the right choice.

One possibility is that the fish take whichever *Daphnia* appears largest at each choice (O'Brien *et al.*, 1976). A nearby small *Daphnia* may appear to be larger than a larger one that is further away. As the density of prey increases, the probability that there will be a large *Daphnia* nearby also increases. By selecting whichever prey item appears larger, the fish could behave in the manner predicted by the profitability model. Thus, while the profitability model specifies what ought to happen in a given situation, the take-the-largest model proposes a rule that the fish might actually use.

Animals usually consume their food on the spot, but if an animal has a nest, or comes from a colony, it may take food home. Such animals usually make an outward journey, spend some time searching, and then make a return journey. This type of foraging is called **central place foraging**. The distance between the central place and the foraging area is usually called the **travel distance**. Orians and Pearson (1979) made a number of predictions about the foraging behaviour relative to the travel distance:

1. Provided food availability remains constant, the optimal load carried by the animal should increase with travel distance.

2. Time spent at the foraging site should increase with travel distance.

3. The return travel should be shorter than the outward travel, because the animal weighs more on its return trip and should minimize the time spent carrying the load. An example can be seen in desert ants that have meandering outward paths and straight return paths (see Figure 9.3).

The first and second hypotheses have been tested by Hegner *et al.* (1982), studying breeding colonies of the white-fronted bee-eater (*Merops bullockoides*) in Kenya. When travel distance increased from 25 to 575 metres, there was an increase in search bout duration and in the load carried back to the nest.

Central place foragers may incur extra predation risks by carrying food home. Lima *et al.* (1985) studied grey squirrels (*Sciurus carolinensis*). These animals maximize energy profitability if they consume food immediately upon finding it, but they risk predation during the handling time. This risk can be reduced if the squirrel carries the food to the safety of a tree, but this incurs extra energy expenditure. It was found that the squirrel's tendency to carry a food item to a safe place decreases with distance to cover (travel distance) and increases with the size of the food item (handling time). Thus there is a trade-off between energy profitability and predation risk.

The central place for a group of foraging animals can act as an **information centre**. Foraging trail-laying ants are an obvious example, as are the honey-bees discussed in Chapter 23, and various species of colonially breeding birds (Brown, 1988; see Bell, 1991, for a review).

Fig. 9.3 Outward and homeward (bold) paths of an individually foraging desert ant, *Cataglyphis fortis* (inset) (From Wehner, 1997).

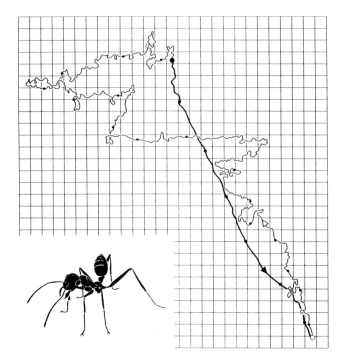

9.2 | **Territorality**

Many animals defend patches of ground against other individuals, usually of the same species. This often takes the form of aggression against intruders, as in many species of birds. For instance, pairs of tawny owls (*Strix aluco*) inhabit a fixed, exclusive area of woodland for the whole of their adult lives, and they defend this against all other tawny owls by loud calling and chases. However, in many instances animals maintain exclusive areas by less overt behaviour. Even though they may seldom meet, many carnivorous mammals avoid each other by smelling the scent deposits of other individuals. Therefore, the best indicator that an area is divided into territories is the finding that the occupying individuals or groups of animals are spaced out more than would be expected from a random occupation of suitable habitats.

Various ecological factors may favour territorality. Three important factors are resource quality and distribution in space, resource distribution in time, and competition for resources (Davies and Houston, 1984). An animal faced with resources, such as food or mates, that are spread over a large area, cannot economically defend a territory that includes sufficient resources. On the other hand, for an animal faced with clumped resources territorial defence may well be worthwhile. Zahavi (1971) showed that when high quality food was presented (by the experimenter) in small clumps, individual pied wagtails (*Motacilla alba*) defended them, but when the same food was sparsely distributed, the birds did not attempt to defend any territory.

Some species are typically faced with food that is distributed in time. Rufous hummingbirds (*Selasphorus rufus*) migrate south along the western mountains of the USA. They fly between alpine meadows, where they may defend feeding territories for a few days or weeks. A number of workers have found a close relationship between flower density and territory size in these birds (see Box 9.3).

Gill and Wolf (1975, 1977) studied golden-winged sunbirds (*Nectarinia reichenowi*) in Kenya. These birds defend winter territories consisting of patches of *Leonotis nepetifolia* flowers. Like the hummingbirds, territory size varied greatly, but the number of flowers defended varied less. Gill and Wolf found that the nectar levels in the flowers were higher in defended patches than in undefended patches. By excluding competitors the birds benefitted, because nectar levels could increase without being depleted by other birds. If the rate of nectar renewal was low, the birds abandoned their territories, because they could not satisfy their daily energy requirements. As the rate of renewal increased it became more and more worthwhile for the birds to expend some energy in defending the territory, but if the rate of renewal became very high, the birds ceased to defend territories, and tolerated intruders, because any depletion caused by them was quickly made up. It was not worthwhile to spend energy on defending such a rapidly renewed resource.

Competition for resources occurs in two main ways, by **exploitation** and by **exclusion**. Exploitation of the resources in a habitat depletes the resources, and although there will be some rate of renewal of the resource, over-exploitation will lead to diminishing returns. For example, a single

Box 9.3   Territory size in hummingbirds

Kodric-Brown and Brown (1978) studied hummingbirds in Arizona. They found a relationship between territory size and flower density (graph (a)), which suggests that the birds were changing their territory size to always defend about the same number of flowers. If flowers were removed experimentally, individual birds expanded their territories accordingly.

The study by Gass *et al.* (1976) of hummingbirds in north-west California, found that the birds varied their territory size depending upon flower density and flower species. Two common plants were the Indian paintbrush (*Castilleja miniata*) and the red columbine (*Aquilegia formosa*), the latter producing four times as much nectar as the former. The study showed (graph (b)) that territory sizes were adjusted to provide about the same amount of energy regardless of species composition and flower density.

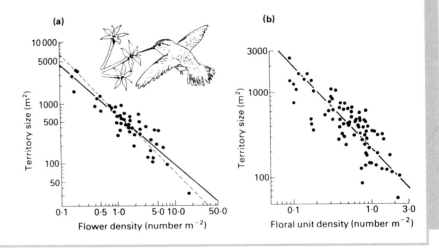

animal will home in on that area of the habitat where food availability is greatest. If it is joined by a competitor, the food availability will be lessened. As more and more competitors arrive, even though there is no territoriality or fighting, there will come a point where a new arrival will find this part of the habitat less profitable than another which was initially less rich. Thus competitors will adjust their distribution in relation to density of available resources, so that each individual gains the same rate of return on time spent foraging (or whatever activity is at issue). This theoretical pattern of distribution of competitors among resources is called the **ideal free distribution** (Fretwell, 1972) to reflect the idea that animals are ideal in having complete information about the availability of resources, and are free to go where they do best.

Milinski (1979) tested the ideal free model experimentally, using sticklebacks and water fleas. Water fleas (*Daphnia*) were dropped into a tank from a pipette at each end. At one end they were dropped at twice the rate of the other end. There were six sticklebacks in the tank and they were free to go to either end, there being no resource defence. Milinski found that four fish went to the richer end and two to the poorer end as predicted by the model.

When the food sources were reversed, the fish redistributed themselves accordingly. Similarly, Powers (1984) found that armoured catfish (*Ancistrus spinosus*) distributed themselves in a stream in Panama in an ideal free manner. In sunlit pools the algae upon which the fish fed grew six times as fast as in shady pools. The catfish were six times as numerous in sunny pools, as predicted by the model.

Competition by exclusion occurs whenever an animal successfully defends a resource. Territory owners may use a variety of means to defend their territories. For example, in song-birds there appears to be a three-tier system of defence. Song acts as a long-range signal that deters potential trespassers; visual displays are used at intermediate range to deter actual trespassers; and finally, if the intruder persists, it is attacked and chased.

Keep out signals have been studied experimentally in a number of species (Davies, 1978). Krebs (1971) showed that when he removed male great tits (*Parus major*) they were replaced by other males within a few hours. These new birds came from poor quality territories outside the wood, or were non-territorial floaters. They must have been regularly monitoring the wood to detect empty spaces so quickly. When males were removed from their territories and replaced by loud speakers that broadcast either great tit song or a control sound (a note on a tin whistle) invading birds settled quickly in the areas with the control sound, but took much longer to occupy the empty territories where great tit song was played (Krebs, 1977). In mammals scent acts as a keep out signal, and many species mark their territories with urine, faeces, or with scent produced by specialized glands. European badgers (*Meles meles*) live in groups and defend a communal territory, marking the boundaries with droppings deposited in latrines. Wolves (*Canis lupus*) mark the edge of their pack territories with urine, and experiments with red fox (*Vulpes vulpes*) show that males can be inhibited from trespassing onto a territory by the smell of the owner's urine scent alone.

The circumstances under which animals defend territory may be finely tuned economically (see Box 9.4), and as we have seen, territorial defence is sometimes worthwhile and sometimes not. There is a considerable literature on optimal territory size and the economics of defence (e.g. Davies and Houston, 1984; Lendrem, 1986; Krebs and Davies, 1993), which should be familiar to students of behavioural ecology.

9.3 Group living

In most species, some communication and cooperation is essential for reproduction. Courtship, mating and caring for the young all involve some expenditure of time and energy for the benefit of other individuals. Sexual and parental behaviour can therefore be regarded as rudimentary forms of social behaviour.

Some species, such as polar bears, lead a relatively solitary existence, while others congregate in vast herds. The size of a group, however, is not a good indication of the degree of sociality. Aggregations may occur as a result of individuals being independently attracted to a particular environmental

Box 9.4 Territorial management in pied wagtails

Pied wagtails (*Motacilla alba*) defend winter feeding territories along river banks, and in order to collect sufficient food during the short daylight hours at this time of the year they have to spend over 90 per cent of their time searching for food. They feed by walking along the river edge and picking up insect food which is washed up onto the bank. As they walk along they deplete the food supply temporarily, so that if they visit the same stretch again soon afterwards they experience a low feeding rate. The longer they leave the stretch the more time there is for more food to wash up onto the bank, and thus the greater the wagtail's feeding rate.

In a classic study, Davies and Houston (1981) found that each wagtail defends a territory of about the same length of river, and exploits the food systematically by walking up one bank of the territory boundary, crossing the river, and then walking back down the other side until it completes the circuit. The length of the territory defended is just sufficient to enable the food supply to be renewed to a profitable level during the time it takes the bird to make one circuit.

Pied wagtails do not defend their territories when the renewal rate is very high or very low. When high the territory owner's feeding rate is not affected by the presence of intruders and it is not worthwhile expending time and energy on territorial defence. When it is very low the owners abandon the territory and go off to feed elsewhere. At intermediate levels of food abundance the owner may defend the territory alone, or may allow another bird (usually a juvenile) to share it. This 'lodger' competes with the owner for food, but also helps the owner to defend the territory. Davies and Houston showed that the owner takes in a 'lodger' when the increased feeding rate obtained by spending less time in territorial defence (because the other bird is helping out) outweighs the losses that result from sharing the food supply with another bird. The 'lodger' is evicted if the food supply decreases, and it is more profitable for the owner to defend the territory alone.

feature, such as a food source, or simply being attracted to each other as a form of defence. Butterflies of the genus *Heliconius* have conspicuous warning coloration and are distasteful to birds. They aggregate in large numbers in response to a chemical attractant (pheromone). This is a form of defensive behaviour, because a predator that samples one quickly learns to avoid others of a similar colour.

The main environmental influences upon group size are food and predators (Krebs and Davies, 1993). Comparison of closely related species (see Chapter 5), attempting to find out how differences in behaviour affect differences in ecology, can be a powerful method of analysis. John Crook was one of the first to apply this approach to social behaviour.

Crook (1964) worked on some 90 species of weaver bird (Ploceinae). These small birds live throughout Asia and Africa. Although similar in appearance, the different weaver bird species vary markedly in their social organization. Some defend large territories in which they build camouflaged nests, while others build their nests in conspicuous colonies. Crook found that the species that live in forests tend to be solitary and insectivorous and to have camouflaged nests in large defended territories. They are monogamous and have little sexual dimorphism. Species living in the savannah are usually seed eaters that live in flocks and nest colonially. They are polygamous, the males being brightly coloured and the females dull.

Crook argued that food is difficult to find in the forest and that it is necessary for both parents to feed the young. Therefore, the parents have to stay together as a pair throughout the breeding season. The insects which the forest birds feed on are widely dispersed and a pair of birds must defend a large territory to ensure an adequate supply of food. The nests are camouflaged and the adult birds are dull coloured so that their visits to the nest do not so easily alert predators as to its whereabouts.

In the savannah, seeds may be abundant in some places and absent in others. This is usually called a patchy food distribution. Foraging in such conditions is more efficient if the birds form flocks that then search over a wide area. Nesting sites that offer some protection from predators are scarce in the savannah, and many birds nest in a single tree. The nests are bulky to insulate against the heat of the sun, and the colonies tend to be conspicuous. To gain some protection from predators, the nests are usually built high up in a spiny acacia or similar tree (see Figure 9.4). The female can feed the young by herself because food is relatively abundant. The male invests little in the young and is free to court other females. Males compete for nest sites within the colony, and the successful males may each attract several females while other males fail to breed.

In the colonial village weaver (*Textor cucullatus*), for example, the males steal nest material from each other. They therefore have to remain near to the nest to protect it. To attract females the male performs an elaborate display while hanging from the nest. If the male is successful in his courting, then the female enters the nest. Such nest displays are typical of colonial weaver birds and contrast with the courtship of the forest-living species. In these, the male chases the female, courts her away from the nest and then leads her to it. In the colonial polygamous situation, sexual selection is strong

Fig. 9.4 A colony of the village weaver (*Ploceus cucullatus*). Note the large number of nests positioned relatively free from predation (*Photograph: Nicholas Collias*).

and it is not surprising that there is considerable sexual dimorphism. Some weaver birds live in grassland where nests are vulnerable to predators. These species tend to space out their nests in scattered colonies. Like the savannah species, they feed in flocks on patchily distributed food, but the males are not usually so brightly coloured.

A specific advantage of living, or feeding, in a group is increased **vigilance** against predators. Most predators rely on an element of surprise to catch their prey, so constant vigilance is a good insurance against predation. However, the solitary animal cannot afford to spend all its time watching out for predators, because some time must be spent feeding, etc. In a group of animals, individuals can to some extent rely upon the vigilance of other members of the group. For example, Bertram (1980), observing ostriches (*Struthio camelus*), found that the feeding bird raises its head to scan for predators at random time intervals. This makes it impossible for a stalking lion to determine how much time it has to creep forward undetected, while the bird is feeding with its head down. Individual ostriches spend more time scanning for predators when alone than when in a group. The overall vigilance of the group (proportion of time that at least one bird has its head up) increases slightly as group size increases. This is what would be predicted if each individual looks up independently of the others.

The cost of vigilance is partly a matter of the time it takes, and this is offset by the benefit gained in detecting predators. The efficiency of predator detection depends partly upon the effort put in by the individual animal (see Figure 9.5), partly upon technique, such as scanning at random intervals, and partly upon the efficacy of the warning signals given to other members of the group. As we have seen (Chapter 8) alarm signals may themselves incur costs in attracting the attention of a predator to the individual giving the alarm. Thus the group situation is one which is open to cheating. It might

Fig. 9.5 Grazing and vigilant postures of the pink-footed goose: (a) grazing posture, (b) head-up posture, (c) extreme head-up posture.

pay an individual to save time and reduce risks by reducing the time spent vigilant when in a group situation. It appears that some species have mechanisms that reduce this possibility. For example, pigeons feeding in a flock give a special 'intention' signal when they are about to depart (see Chapter 21). If they depart without giving this signal, then all the other birds fly off in alarm (Davis, 1975). Thus the vigilant pigeon, by flying away without giving a 'goodbye' signal, is warning other members of the flock, but it gains an advantage by flying away first.

Another advantage of group living is the **dilution effect**. By joining a group, an individual dilutes the impact of a successful attack by a predator, because there is a chance (depending on group size) that another animal may be the victim (see Box 9.5). For example, Duncan and Vigne (1979) studied semi-wild horses in the Camargue, a marshy delta in the south of France. During the summer the horses are attacked by tabanid flies, and they cluster together in large groups. Measurements showed that horses in large groups are less likely to be attacked than horses in small groups. Experiments in

Box 9.5 Dilution effect in water skaters

Water skaters (*Halobates robustus*) are insects that move about on the water surface. They are preyed upon by small fish such as *Sardinops sagax*. The fish snap at the insects from below, so it is unlikely that the vigilance of the insects is affected by their group size. Foster and Treherne (1981) found that the attack rate by fish did not vary with group size. Therefore, attacks per individual insect decline with group size. In the graph, the predicted line is what would be expected if the decline were entirely due to the dilution effect. This line is very close to the observed line (from Krebs and Davies, 1993.)

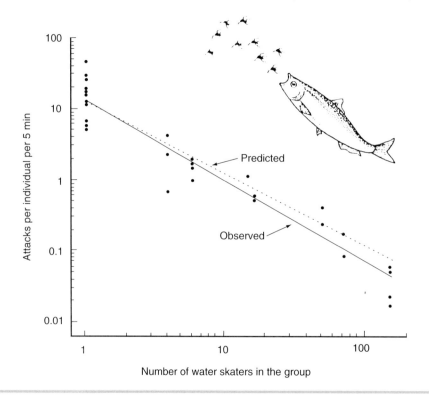

which horses were transferred from one group to another confirmed that living in a large group conferred an advantage because of the dilution effect.

Other possible advantages of group living include **group defence** against predators (Kruuk, 1964; see also Chapter 6); **group hunting**, which may enable predators to catch difficult prey (Major, 1978; Pusey and Packer, 1997); and enhanced offspring viability through **cooperative breeding** (see Emlen, 1997, for a review).

Disadvantages of group living arise partly from competition for food or mates. Goss-Custard (1976) studied foraging in the redshank (*Tringa totanus*) which feeds on coastal mudflats (see also Chapter 24). At night these birds feed in tight flocks, whereas during the day they feed in more scattered groups. At night the main prey are snails (*Hydrobia*) which the birds detect by touch. During the day they feed on small shrimps (*Corophium*) which they detect visually. Undisturbed shrimps live with their tails sticking out of the mud surface, but when disturbed vanish into the mud. Experiments with captive birds showed that redshank footsteps trigger this retreat behaviour. Therefore, redshank feeding close together are likely to disturb each other's shrimp prey, to their mutual disadvantage. At night this **disturbance effect** does not occur, because the snail prey are not detected visually and do not react quickly to disturbance. Comparison of shorebird species reveals a similar story. Some, such as the knot (*Calidris canutus*) live in dense flocks and feed by touch, whereas others like the ringed plover (*Charadrius hiaticula*) feed by sight and forage as solitary individuals, or in loose flocks (Goss-Custard, 1970).

Competition for mates may take the form of direct rivalry (see Chapter 10) or cuckoldry by neighbours. As an example of the latter, Bray *et al.* (1975) found that among colony-nesting red-winged blackbirds (*Agelaius phoeniceus*) the mates of vasectomized males laid fertile eggs, presumably fathered by other males. Other hazards of group nesting are cannibalism and parasitism. The carrion crow is known to cannibalize the eggs and chicks of neighbours, not for the nutrition they might provide, but in order to prevent a pair from breeding. Herring gulls (*Larus argentatus*) are opportunistic feeders on neighbours' eggs and chicks, indeed for some it develops into a feeding specialization.

Parasitism can be a major disadvantage of group living. Brown and Brown (1986) studied colonially nesting cliff swallows (*Hirundo pyrrhonota*). They found that the nestlings are often attacked by a blood-sucking ectoparasite, the swallow bug *Oeciacus vicarius*. Larger colonies of birds have more bugs per nest, and the growth rate of nestlings in parasitized nests is severely retarded. By fumigating some nests with insecticide, it was found that fumigated nestlings were about fifteen per cent heavier than the non-fumigated controls, at ten days old.

The advantages and disadvantages of group living may seem to cancel each other out. For example, Brown (1988) compared nestling grown in cliff swallow colonies of different sizes (without fumigation), and found little difference. The disadvantage of increased parasitism seems to be cancelled out by the role of the colony as a foraging information centre. However, we might well ask whether or not there is an optimal colony size.

The notion of an optimal group size has been discussed by some workers (e.g. Pulliam and Caraco, 1984; Krebs and Davies, 1993), but there are two reasons why it may not be a viable notion (Krebs and Davies, 1993):

1. The individuals within a group may benefit from group living to different extents. For example, in flocks of starlings, the birds in the middle spend less time scanning for predators than those at the edge (Jennings and Evans, 1980). It may be that dominant individuals acquire the better situations (such as the middle of the flock), leaving subordinates to take up the less favoured positions (such as the edge). It would pay a subordinate to tolerate such a situation so long as it could not do better elsewhere (Vehrencamp, 1983). Thus observed groups may be a compromise between the optima characteristic of different types of individual.

2. Optimal group sizes may not be stable. If there were a group of the optimal size, then it would pay an outside individual to join the group, pushing the group above the optimal size (Sibly, 1983). As shown in

Fig. 9.6 Sibly's model of optimal and stable group size. Although seven is the optimal group size, a solitary individual will gain fitness by joining any group smaller than fourteen. Thus the optimal group size is not stable (After Sibly, 1983).

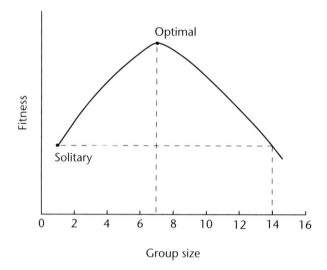

Figure 9.6, the fitness payoff for foraging alone will be less than that for joining a group that is much larger than the optimum. Therefore an outsider should join the group rather than remain a solitary forager. In nature, one should expect to find stable groups rather than optimal groups, and observed groups will often be larger than the optimum.

Overall, one can point to various advantages and disadvantages of group living, but it is difficult to combine these into a coherent model, because the advantages and disadvantages affect different individuals to differing extents. Moreover, individuals are free to join and leave groups, and they will be designed to make such decisions on the basis of factors that reflect the overall payoff in terms of fitness.

Points to remember

- Behavioural ecology is a branch of evolutionary biology, and is concerned with explaining how a state of affairs observed today is likely to have come about as a result of evolution by natural selection.

- Foraging efficiency is usually a matter of **trade-off** between competing priorities. These may include energy gained versus energy spent, energy gained versus risk of predation, and energy gained versus losses to rivals.

- The most important ecological factors favouring territoriality are resource quality and distribution in space, resource distribution in time, and competition for resources.

- The main environmental influences upon group size are food and predators.

Further reading

Krebs, J.R. and Davies, N.B. (1993) *An introduction to Behavioural Ecology*, 3rd edn. Blackwell, Oxford.

CHAPTER

Primate social behaviour

In preceding chapters we have seen how many aspects of behaviour can be viewed in terms of evolutionary strategy. These include foraging, mating, and parental behaviour. In Chapter 9 we looked at the advantages and disadvantages of group living. In this chapter we move to the most complex aspect of animal behaviour – primate social behaviour – and ask if the principles that apply to other species also apply to primates. In Chapter 22 we extend this question to humans.

10.1   The primates

The first primates are thought to have to appeared some 70 million years ago (Figure 10.1), and evolved into diverse groups. Present day primates fall into four main groups: the relatively primitive primates (prosimians), the New World monkeys (ceboids), the Old World monkeys (cercopithecoids), and apes and man (hominoids). The exact classification of modern primates is subject to changes in academic opinion. That given by Passingham (1982) is shown in Table 10.1.

From the fossil record, it appears that the Old and New World primates diverged some 40 million years ago, and the apes split off from the Old World monkeys about 20 million years ago (Figure 10.1). The prosimians are the relatively little-changed descendants of the earliest primates. The early prosimians were arboreal fruit eaters, with binocular vision and relatively large brains compared with those of other contemporary mammals. Modern prosimians are now confined to isolated areas (e.g. Madagascar), or to nocturnal niches, and the number of species is much less than it was in the early period of primate evolution. It is thought that, as higher primates evolved, they outcompeted the prosimians. The prosimian brain, in both its fossil and modern forms, is well developed in those areas involved in visual and auditory perception, visual and auditory memory, and the integration of these senses with each other. Thus it is probable that the early prosimians lived in

Table 10.1 Classification of living primates (after Passingham, 1982)

Order	Suborder	Infraorder	Superfamily	Family	Subfamily	Genus	Common name
Primates	Prosimii	Lemuriformes	Lemuroidea	Lemuridae	Lemurinae	*Lemur*	Common lemur
						Hapalemur	Gentle lemur
						Lepilemur	Sportive lemur
					Cheirogaleinae	*Cheirogaleus*	Dwaf lemur
						Microcebus	Mouse lemur
				Indridae		*Indri*	Indris
						Avahi	Avahi
						Propithecus	Sifaka
				Daubentoniidae		*Daubentonia*	Aye-aye
		Lorisiformes	Lorisoidea	Lorisidae		*Loris*	Slender loris
						Nycticebus	Slow loris
						Arctocebus	Angwantibo
						Perodicticus	Potto
				Galagidae		*Galago*	Bushbaby
		Tarsiiformes	Tarsioidea	Tarsiidae		*Tarsius*	Tarsier
	Platyrrhini		Ceboidea	Callithricidae	Callithricinae	*Callithrix*	Marmosets
						Leontideus	Tamarin
					Callimiconinac	*Callimico*	Goeldi's marmoset
					Aotinae	*Aotus*	Night monkey
						Callicebus	Titi
					Pithecinae	*Pithecia*	Saki
						Chiropotes	Saki
						Carcajao	Uakari

Suborder	Infraorder	Superfamily	Family	Subfamily	Genus	Common name
Anthropoidea			Cebidae	Alouattinae	*Alouatta*	Howler
				Cebinae	*Cebus*	Capuchin
					Saimiri	Squirrel monkey
				Atelinae	*Ateles*	Spider monkey
					Brachyteles	Woolly spider monkey
					Lagothrix	Woolly monkey
	Catarrhini	Cercopithecoidea	Cercopithecidae	Cercopithecinae	*Macaca*	Macaque
					Cynopithecus	Black ape
					Cercocebus	Mangabey
					Papio	Baboon
					Theropithecus	Gelada
					Cercopithecus	Guenon
					Erythrocebus	Patas monkey
				Colobinae	*Presbytis*	Common langur
					Pygathrix	Douc langur
					Rhinopithecus	Snub-nosed langur
					Simias	Pagai Island langur
					Nasalis	Proboscis monkey
					Colobus	Gueraza
		Hominoidea	Hylobatidae		*Hylobates*	Gibbon
					Symphalangus	Siamang
			Pongidae		*Pongo*	Orang-utan
					Pan	Chimpanzee
					Gorilla	Gorilla
			Hominidae		*Homo*	Man

Fig. 10.1 Primate radiation. The geological timescale given on the left is based on rough estimates (After Passingham, 1982).

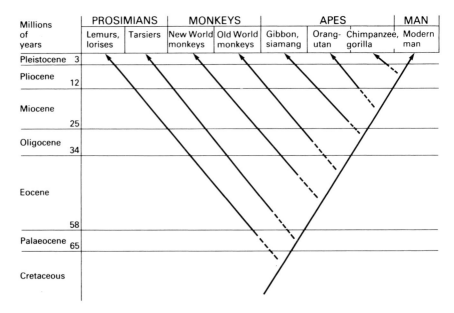

groups and had highly developed social communication. As we see below, sophisticated social behaviour is characteristic of the primates.

Of the higher primates, one group, the New World monkeys, evolved in South America. Many species, such as the spider monkey, are highly specialized for arboreal locomotion, swinging by the arms and tail through the high branches of trees. The Old World primates evolved in Africa, and comprise two groups, the Cercopithecoidea (Old World monkeys) and the Hominoidea (apes and man). From the fossil record, it seems that these two groups split about 35 million years ago. The Old World monkeys include both tree-living species, such as the colobines, and ground-living species, such as baboons. The hominoids are divided into three families, the lesser apes (gibbons and siamangs), the great apes (gorilla, chimpanzee and orangutan), and the hominids, of which the only living species is *Homo sapiens*.

The anatomical and biochemical evidence suggests that man is more closely related to the chimpanzee and gorilla than to other living hominoids. The social behaviour of man's immediate ancestors was probably not unlike that of chimpanzees. The prehominids, like chimpanzees, were forest animals capable of life at the forest fringe and open woodland (Pilbeam, 1972). As indicated in Figure 10.2, the transition from forest to savannah habitat probably occurred more than once in hominoid evolution. The emergence of the hominids is characterized by changes in dentition indicating a shift from fruit eating towards ground feeding on tough vegetable items.

10.2 Primate life histories

'In the game of life an animal stakes its offspring against a more or less capricious environment. The game is won if its offspring live to play another round. What is an appropriate tactical strategy for winning this game? How

Fig. 10.2 Ecological shifts within various higher primate lineages (After Pilbeam, 1972).

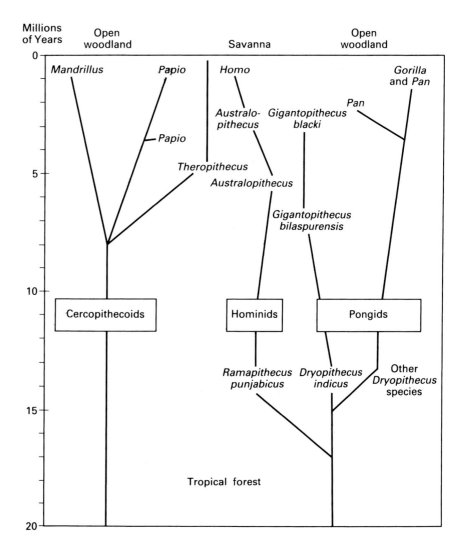

many offspring are needed? At what age should they be born? Should they be born in one large batch or spread over a long lifespan? Should the off-spring of a particular batch be few and tough or many and flimsy? Should parents lavish care on their offspring? Should parents lavish care on them-selves to survive and breed again? Should the young grow up as a family, or should they be broadcast over the landscape at an early age to seek their for-tunes independently?' (Horn, 1978, p. 411).

Some of the answers to these questions depend upon the type of environ-ment inhabited by the species we are considering. In highly variable and unpredictable environments the causes of mortality are often unrelated to the genetic differences among individuals, and competition among individuals is low. In a competitive vacuum, the best reproductive strategy is to produce as many progeny as possible. These can often thrive even if they are small and inexpensive to produce. In stable and predictable environments those individuals able to cope with high population density and strong competition will be at a selective advantage. For the young to survive they have to be relatively strong and capable, and are usually more expensive to produce.

Those species that are subject to environmental depletion and are generally at low population density are said to be subject to **r selection**. Those that live in high density populations are subject to **K selection.** These terms were introduced by MacArthur and Wilson (1967), and are derived from the equations of population growth (Futuyma, 1979). r selection is typified by unpredictable environments, density-independent mortality, and varying population size well below the carrying capacity of the environment. K selection is more typical of predictable environments, density-dependent mortality and fairly constant population size near the carrying capacity of the environment.

Given that an animal has limited reproductive resources, there are two basic types of life-history strategy:

■ the r strategy which is to produce many, relatively unprepared offspring, each with a low chance of individual survival

■ the K strategy, which is to produce few well-prepared offspring which have a high chance of individual survival

Thus r selection favours rapid development, small body size, reproductive productivity and short life. K selection favours slow development, large body size, reproductive efficiency and long life. The primates are primarily K selected, although there is some variation among species in these respects. Thus all primates, being mammals, suckle their young, so that their investment in individual offspring is considerable. Harvey *et al.* (1987) have highlighted the extreme life history patterns of primates. They compare the mouse lemur (*Microcebus murinus*) with the gorilla (*Gorilla gorilla*).

The mouse lemur, the smallest living primate, probably produces three–six young during the year following her own birth, and continues to reproduce at this rate throughout her fifteen year lifespan. The female gorilla, on the other hand, usually produces her firstborn after ten years, and thereafter gives birth once every four or five years throughout her 40 year lifespan. Such differences between species have evolved as adaptations for exploiting different ecological niches. Each niche is associated with a particular optimum body size, determined in part by the available food supplies. The larger primate species tend to be diurnal, terrestrial vegetarians, while the small species are nocturnal, arboreal and omnivorous. Paul Harvey and co-workers (1987) looked systematically at primate life history variables, such as body size, gestation period, lifespan, etc. and carried out a large quantitative comparative study.

They discovered that brain size is more closely correlated with several life-history variables than is body size. They suggest that the brain, being slow to grow, may be the pacemaker of development. The brain is a relatively expensive organ, and it has been suggested (Martin, 1981) that the size of the neonatal brain that a mother can nourish sets a constraint on adult brain size. High female adult body size, and a long gestation period, is necessary for the production of large-brained offspring (see Chapter 11). Primates that produce relatively large-brained neonates have relatively little brain growth after birth. Humans are the single exception (Harvey *et al.*, 1987). If there is to be considerable postnatal development of the brain, then a prolonged period of parental care will be required. This has implications for the mating system of the species concerned.

10.3 Primate mating systems

A successful sexual strategy (see Chapter 8) results in a particular kind of mating system, which has a profound effect upon the social organization of the species. Thus, a male may maximize his fitness by mating with many females and fathering a large number of offspring. Such **polygyny** is a good strategy provided the offspring survive to reproductive age. Their survival depends upon the ecological circumstances and upon the degree of parental investment by both male and female. A successful polygynous male has little time to care for his offspring. Either they must be self-sufficient or the female must be able to care for them unaided. Polygyny will usually result in a few successful males and a large number of unsuccessful males. The harem may then become the basic feature of the social structure, as in many antelopes and deer. Males compete for possession of females and young males have to wait for a chance to steal copulations or to challenge a dominant male.

If we compare birds and mammals, we find that polygyny is much more common among mammals. Female mammals are specialized for parenthood, providing milk and elaborate maternal care. Male mammals are less well equipped, though they can help by providing food and guarding against predators. Male birds, in contrast, are just as well equipped to look after their progeny as female birds. **Monogamy**, in which each adult mates with only one member of the opposite sex, is often a necessary strategy when both parents are required to raise the young. More than 90 per cent of the bird species of the world appear to be monogamous. In some the monogamy is perennial in that a mated pair remains together for life, either on a seasonal or a continuous basis. Many migrant birds are monogamous during the breeding season but live separately for the rest of the year. In swans and geese, a close association between partners may be maintained even outside the breeding season.

In general, the formation of pair bonds is associated with parental care. In **promiscuous** species, in which both males and females mate with more than one member of the opposite sex, there are no pair bonds and parental care is minimal. In other species, the mating system is related partly to the ecological circumstances and partly to the needs of the young. This is well illustrated by comparing species of primate.

Although monogamy is not common among mammals, it does occur in fourteen or more primate species (Clutton-Brock and Harvey, 1977). In most primates the males show little parental care, though they defend the family group in time of danger. In the monogamous species a much greater degree of male parental care is evident (Passingham, 1982). For example, marmosets (*Callithrix*) and tamarins (*Leontideus*) usually produce twins. The male carries these about (Figure 10.3), and returns them to the mother only for nursing. Similarly, male titi monkeys (Figure 10.4) spend more time with the infants

Fig. 10.3 Marmoset father carrying twins (*Photograph by Ron Garrison © Zoological Society of San Diego*).

Fig. 10.4 Pair of titi monkeys (*Photograph by Ron Garrison © Zoological Society of San Diego*).

than does the mother. In humans, the helpless baby is a burden on the mother because it cannot cling to her of its own accord. A toddler cannot keep up an adult walking pace and has to be transported by the mother. In primitive human societies, this degree of dependence can restrict the frequency with which a woman can have children because of the work required to provision the family.

Ecological factors, particularly food availability, may have a profound influence upon mating systems. Many prosimian species, including lemurs, loris and tarsiers, have a widely dispersed food supply. They are usually insectivorous, arboreal and nocturnal. They have a solitary way of life and have social contact only during courtship and mating. Monogamous species such as the white-handed gibbon (*Hylobates lan*) defend territories, like many monogamous birds. In this way they protect their food supply. The young receive considerable care and remain with the parents until they become sexually mature.

The social organization of gelada baboons (*Theropithacus gelada*) is very similar to that of certain antelopes and appears to be based upon existence in a habitat that is subject to marked seasonal variation in food availability. Among impala (*Aepycerus melampus*), for example, there are bachelor herds made up of single males and harem herds made up of females under the control of a single male. Gelada baboons inhabit the mountain grassland of Ethiopa. During the dry season there are distinct types of herds, the harem herds, the all-male groups, and groups of juveniles. The all-male groups are the most widely dispersed, and this has the effect of reducing competition for food between the groups. During the rainy season, when food is more plentiful, the groups aggregate in large herds, within which the harems remain intact.

Primates with a more stable food supply tend to form multi-male groups. For example, the mountain gorilla (*Gorilla beringei*) has a small home range and feeds on the leaves and stems of forest plants. There animals live in small groups, each led by a silver-backed male, but often including other males. They do not defend territories and may overlap with other groups.

10.4 Evolution and primate social behaviour

In accounting for primate social behaviour in evolutionary terms, we are seeking an explanation in terms of the genetic fitness of the participants. As we have seen in previous chapters, much of what animals do can be seen as the result of individual selection. Thus improvements in foraging efficiency, predator avoidance, etc. are likely to lead to greater longevity and increased reproductive success. This is sometimes also true of social interactions. Thus aggressive behaviour in competition for a limited resource may be enhanced by individual selection. Difficulties arise in accounting for sexual behaviour and apparently altruistic behaviour (see Chapter 8). Explanations of these aspects of behaviour may have to be given, not only in terms of individual selection, but also in terms of inclusive fitness and sexual selection.

There are many aspects of social behaviour that remain difficult to account for in evolutionary terms. This is partly because social behaviour can be very complex, especially among the primates, and partly because many of the issues are difficult to resolve empirically, especially among long-lived species. Nevertheless, there are certain general questions that we can ask:

■ Why do primates live in groups?

■ Why do individuals decide to team up and form groups with some particular individuals and not others?

There are four main factors that might explain why primates live in groups. These are:

■ protection against predators

■ defence of resources

■ foraging efficiency

■ improved care-giving opportunities (Dunbar, 1988)

Protection against predators is enhanced by group living as a result of both improved detection and increased deterrence. Predators searching at random are less likely to find prey if they are clumped into groups, and the vigilance of the prey is increased by group living (e.g. van Shaik *et al.*, 1983). Vigilance is less time-consuming for the individual if it is a member of a large group. Capuchin monkeys living in small groups spend significantly more time in visual scanning of the environment than do those living in a large group (De Ruiter, 1986). This is a claim that has also been made for humans (Wirtz and Wawra (1986).

That predators can be deterred by group action by primates has often been observed. There have been a number of reported incidents in which dogs, leopards, cheetah and even lions have been driven off by the large males belonging to a group of baboons or chimpanzees (Dunbar, 1988). There is also evidence that primates respond to high predation risk by forming larger groups. For example, van Schaik and van Noordwijk (1985) found that macaques living on a predator-free island formed significantly smaller groups than those living on the mainland where predators were common.

Defence of resources, especially of food sources, is an important aspect of group living in many animals. Most of the evidence concerning primates comes from the observation that, in some species, the groups do defend territories (see Dunbar, 1988). In addition, among non-territorial species, group size is an important factor in a group's ability to displace other groups from key resources. In some species population size is known to be limited by food availability, and competition for food is an important aspect of reproductive success (Dittus, 1977).

Improved foraging efficiency and care-giving appear to be less important effects of group living in primates compared with some other species (Dunbar, 1988). In some large carnivores foraging efficiency improves significantly with increased group size (Schaller, 1972; Kruuk, 1972; Macdonald, 1983), but there is little evidence that this is the case among primates, apart from hominids. Help with rearing has been an important factor promoting sociality in birds, but it is rare in mammals. In primate groups juveniles may be protected from harassment by adults that are allies of their parents, and adoption of infants has been reported in some species. These benefits could, however, be as effectively provided by a group of two or three adults as by a large group. It seems unlikely, therefore, that care of the young has played an important part in the evolution of group-living in primates.

Set against the benefits of group living are certain disadvantages. Larger groups require more coordination and cooperative behaviour, at the expense of extra time and energy. For example, increase in group size in long-tailed macaques is directly reflected in the length of their daily journeys, thus incurring extra costs in energy expenditure and risk of predation (van Schaik *et al.*, 1983). Probably the most important disadvantage of large groups is the increased competition amongst individuals. This may range from harassment of subordinates to outright aggression. Such interactions are well documented among primates (see Dunbar, 1988), and they have important evolutionary implications.

Social dominance

In many species of primate individuals actively compete over a variety of resources, including food, water, grooming partners and resting places. Joan Silk (1987) has reviewed a number of lines of evidence that indirectly suggest that competitive success is related to reproductive success. The fecundity of females is greatly influenced by habitat quality, especially food supply. This is primarily because nutritional level affects fertility. If food is limited, high nutritional status may result from success in competitive encounters over food. A contributory factor here is social dominance. In one group of wild vervet monkeys, for example, high-ranking females weighed more than lower-ranking females of similar body size (Whitten, 1983).

Females of several species of Old World monkey (e.g. macaques, baboons, and vervet monkeys) form **dominance hierarchies** in which individuals related through the female line tend to occupy adjacent ranks. Thus offspring acquire dominance ranks just below those of their mothers and female siblings rank in reverse order of their ages. Maintenance of dominance rank is primarily dependent upon support from female relatives (Gouzoules and Gouzoules, 1987).

If high social dominance among females is evolutionarily advantageous, then we might expect such females to have greater reproductive success than lower-ranking females. Silk (1987) is unable to find clear-cut evidence of this, partly because there have so far been insufficient long-term studies. It appears that dominance rank is more clearly associated with reproductive success in some populations than in others. In some cases the benefits of high rank are offset by disadvantages. For example, Gouzoules *et al.* (1982) found that high-ranking Japanese macaque females produced their firstborn at a younger age and produced a larger number of infants than did lower-ranking females. On the other hand the high-ranking females suffered higher infant mortality. On balance, the high-ranking females did enjoy greater reproductive success than lower-ranking females in this population.

Competition among primate males for limited resources such as food is likely to be influenced by social dominance rank, as in females. Nutritional state will usually have a lesser effect on reproductive success in males than in females, because the parental investment of males is much lower than females (see Chapter 9). More important for males, at least in multi-male groups, is competition for access to receptive females. We might, therefore, expect dominance rank to be positively correlated with access to receptive females and related to male reproductive success.

Male dominance rank is correlated with reproductive success across a wide range of mammal species (Dewsbury, 1982), but the evidence for primates is equivocal (Silk, 1987). This is partly because there is little genetic paternity data, and few lifetime studies of reproductive success in wild populations. Dominance relationships seem to play an important part in the life of some primate species but not others. The reasons for this are not fully understood, but it may well be that altruism towards kin and the formation of temporary coalitions are more important than social dominance in some primate species.

Altruism

As we saw in Chapter 8, for altruistic traits to evolve, individuals with genes promoting altruistic behaviour must interact with other individuals that carry the same genes. For this to occur, they must recognize their kin, or live among them. Among primates both these situations are common. Kin-related groups may persist as social units over many generations (Silk, 1987), and kin recognition through the maternal line is well documented (Gouzoules and Gouzoules, 1987).

A common form of altruism among cercopithecine females is intervention in aggressive encounters. The individual who intervenes on behalf of another expends energy and risks injury. The evidence (e.g. Kaplan, 1978) shows that such interventions occur more commonly in aid of maternal kin, and that greater risks are run for close kin. The effect of this type of altruism, in populations containing a number of related females, is that **coalitions**, or mutual-aid pacts, are formed.

Other forms of altruism are mutual grooming and food sharing. When one monkey grooms another it expends a certain amount of time and energy, while the recipient benefits by the removal of dirt and parasites. Patterns of grooming have been studied in many types of primate group, and it has been found to occur mainly among maternal kin (Gouzoules and Gouzoules, 1987). This strongly suggests that kin selection has shaped the evolution of grooming among primates. Grooming interactions among non-kin are also observed, and these may have political implications, as we see below.

Food sharing is the direct transfer of food items from one individual to another. It may occur as a result of stealing, especially from a subordinate by a dominant animal; or as a result of donation, usually from mother to off-spring. In an analysis of the distribution of provisioned bananas by chim-panzees at Gombe, Bill McGrew (1975) found that 86 per cent of all transfers were among kin, virtually all from mother to offspring. The remainder were mainly from adult males to adult females. These may indicate some reciprocal altruism, since the males sometimes buy sexual favours in this way.

Reciprocal altruism, as we saw in Chapter 8, can evolve as a form of cooperation among non-kin. Some primate studies suggest that reciprocal altruism is an important feature of social interaction, providing an explanation that is complementary rather than competing with that offered by kin selection theory (Seyfarth, 1983). There are indications that individuals are more likely to receive aid from others if they have previously groomed them (Seyfarth and Cheney, 1984), and that grooming another individual may reduce the risk of being harassed by higher-ranking group members (Silk, 1982). Seyfarth and Cheney (1984) carried out a series of field experiments to test the hypothesis that grooming non-kin increases the probability of future support in aggressive encounters. In each experiment, two individuals, A and B, were selected for a pair of matched trials. A tape recorded vocalization used by animal A to solicit support from others in a coalition was played in the vicinity of animal B, either after A had groomed B, or following a period during which no grooming had occurred between A and B. The duration of B's visual response to the vocalization was compared under the two conditions. The results showed that 'grooming between unrelated individuals increases the probability that they will subsequently attend to each others' solicitations for aid' (Seyfarth and Cheney, 1984).

Evolutionary theory, relating to individual selection, kin selection and reciprocal altruism, goes some way to account for primate social behaviour, such as grooming, competition over resources, sharing food and giving aid. However, our understanding of primate social behaviour remains very primitive and incomplete. As we see in Chapter 26, studies of primate communication indicate a vast potential for sophisticated social and political behaviour, which we have only recently begun to explore.

10.5 Social structure

Social structure results from the behaviour of individuals. Each individual is unique, having a different genetic make-up, and different history, from every other individual. In a large group of animals there will be individuals of similar age, with similar genomes and similar past experiences. Natural selection will act on similar individuals in a similar manner, but at the same time each individual will react to the forces of nature in its own way, developing its own tactics and forming its own relationships with other individuals. Robert Hinde (1976) suggested that social structure emerges from the pattern and quality of relationships among individuals. It is the scientific observers, not the animals, who see structure in the network of relationships that characterizes a social group. In this respect non-human primates may differ from humans, because people can recognize that they belong to a particular social structure, or institution.

Dominance relationships among individuals have long been regarded as the basis of social structure, but family relationships are also important. Kummer (1968), studying the social organization of hamadryas baboons,

showed that a baboon group could be seen as a nesting hierarchy of interacting units, the smallest being a family unit made up of an adult male, one
or two adult females and their young offspring. The head of each family unit
will have some dominance relationship with the heads of other family units.
Each family unit changes with time, sometimes being formed by young males
and females and maturing with age. Kummer (1978) emphasized that
baboons do not merely react to each other's signals, or latest move, but that
their social behaviour is purposive, undertaken with long-term motives in
view. Each baboon attempts to monitor and manipulate the behaviour of
others to further its own career.

We can imagine a population consisting of a large number of individuals
who have relationships of varying intensity with each other. Some individuals will interact frequently, others rarely. The relationships will vary in quality, some being exploitative or antagonistic, others mutualistic. Where a set
of relationships, among a set of individuals, is reciprocal and stable, a social
group can be recognized. Such groupings can be created at many different
levels within the same population at the same time. Figure 10.5 illustrates
this for the gelada baboon.

It is thought that the ancestors of modern baboons were forest-living and
fruit-eating. They would probably have lived in small single-male groups
with an inherent tendency to form multi-male groups under the appropriate
demographic conditions (Dunbar, 1988). When they began to occupy more
open terrain, they would become more suceptible to ground predators, and

Fig. 10.5 Association
patterns among individual
animals in a population of
gelada baboons (After
Dunbar, 1988).

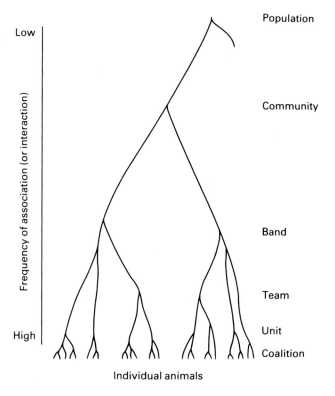

there was an increase in group size to counteract this threat (see above). While the family-group structure would be retained within the larger band, the increased group size would inevitably lead to increased social tension, and increased daily travelling and foraging time. The social tensions would probably have led to the development of kin-based coalitions, making it more difficult for females to migrate between groups.

Modern savannah-living baboons, such as the olive baboon (*Papio anubis*), usually live in troops made up of a number of males and females. The dominant males have privileged access to females, food and water. They cooperate in maintaining their social position against rivals, and in protecting mothers and infants from interference by other members of the troop. The dominant males also cooperate in defence against predators. Their habitat tends to attract large predators and male–male cooperation is a considerable advantage in such circumstances.

Comparative study (e.g. Crook, 1966; Crook and Gartlan, 1966; Kummer, 1968; Rowell, 1966; Wrangham, 1987) suggests that in harsh environments baboon troops are dominated by a single male, whereas in rich environments the troops have a number of cooperating dominant males. The hamadryas baboon (*Papio hamadryas*) of Ethiopia is typical of the former, while the yellow baboon (*Papio cynocephalus*) represents the opposite extreme. The anubis (*Papio anubis*) and chackma (*Papio ursinus*) baboons are intermediate in these respects (see Figure 10.6). Male rivalry is not very intense among the yellow baboons and there is little sexual dimorphism. The hamadryas baboon is subject to considerable sexual selection and the dominant males are distinguished by their large cape. Their society is based upon two types of bond, that between the male and his females and that between males of different one-male groups. Breeding males may steal females, and the males within a group tend to form defensive alliances within which there is a strong inhibition against mutual aggression. In contrast, the females of gelada baboons do not transfer between groups, and the breeding males are not a threat to each other. They form alliances only against bachelor males.

In contrast to the baboons, the forest-living apes live in small groups. A summary of some of their socio-ecological characteristics is given in Table 10.2. Two distinct grouping patterns are evident: the relatively stable groups of the monogamous gibbons and the polygamous gorillas, and the less stable groups of the orangutans and chimpanzees. Some of the pressures leading to monogamy have been discussed above. In many respects, these are similar to those affecting monogamous birds. Gorillas live in permanent stable groups. Each group is dominated by a silver-backed male, and male rivalry is often intense. Both males and females emigrate from the group. The females first breed at about ten years of age, males at about fifteen years. Each group thus contains many immature individuals. The harem-like pattern of gorilla society is facilitated by the fact that the members of the group do not have to disperse widely to obtain food, since they can exploit the abundant herb layer within the tropical forest.

Orangutans are arboreal and live on fruit. They tend to forage separately, as this is the best way to exploit the dispersed food supply. The males are larger than the females; some become so large that they can no longer con-

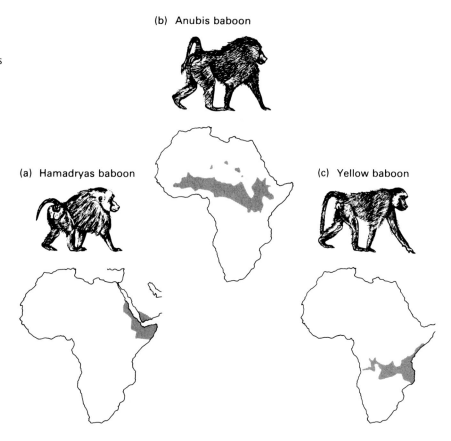

Fig. 10.6 Distribution of baboons in Africa. The male hamadryas baboon (a) has a large cape and many females per dominant male. The anubis baboon (b) is intermediate, and the male yellow baboon (c) has a small cape and few females per male (After Dorst and Dandelot, 1970).

tinue their arboreal existence. There is intense male rivalry, but the large males cannot guard their females. Consequently the orang utans have area-based polygamy, in which the males defend territories, and females have relatively free choice among males.

Chimpanzees live in groups of up to 30 adults. Each has its own ranging area, which varies in size according to the habitat quality. The male members of the group defend a large common range, while the females have their own individual ranges within the common range. Chimpanzees spend their time in small parties whose composition is constantly changing. The size of the party often depends upon food availability, and up to 30 individuals may congregate at a particularly good food source.

The size of a female range means that a male would not be able to defend a territory that included more than one female. This may be the reason for the formation of male coalitions that can defend the ranging areas of a number of females. In this way they can maximize priority of access to receptive females.

Levels of complexity

Studies of non-human primates have led to a recognition of successive levels of complexity in social structure, as illustrated in Figure 10.7. At the simplest

Table 10.2 Behavioural ecology of the apes (after Dunbar, 1988)

Species	Body weight (kg)[a]		Diet[b]	Group size			Range size (mean, km²)	Day journey (km)	Habitus	Mating system
	Male	Female		Mean	Females					
Whitehanded gibbon	6.1	5.5	F	3	1		0.50	1.5	Arboreal	Monogamous
Agile gibbon	5.9	5.9	F	4	1		0.22	–	Arboreal	Monogamous
Kloss's gibbon	5.8	5.8	F	3	1		0.07	–	Arboreal	Monogamous
Siamang	10.7	10.7	L	3	1		0.36	0.7	Arboreal	Monogamous
Chimpanzee	43.0	33.2	F	2–12[c]	1		5–13	3.9	Terrestrial	Multi-male
Bonobo	45.0	33.2	F	8–17[c]	3–6		22	2.4	Terrestrial	Multi-male
Orangutan	57.0	37.0	F	1	1		0.7–5.2	0.5	Arboreal	Area polygamy
Gorilla	170.0	80.0	H	10	2.5		4–12	0.4	Terrestrial	One-male

[a] Weight data for Asian apes from Clutton-Brock and Harvey (1977).
[b] F, frugivore; L, folivore; H, herbivore.
[c] Mean number of individuals travelling together.

Fig. 10.7 Successive levels of complexity in social phenomena (After Hinde, 1987).

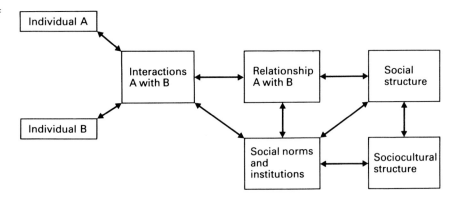

level we have the behaviour of individuals including interactions between individuals. At the next level we have relationships involving a series of interactions between individuals known to each other. Such relationships contribute, to some extent, to the emergence of social norms. For example, two individuals that have a long-standing relationship (such as a monogamous pair) normally behave in the same way when in the same situation. Such norms contribute to social stability. At another level we have relationships between sub-groups (such as family groups), which contribute to the social structure of the group as a whole. In the human case this includes sociocultural structure, involving beliefs about the society and the formation of institutions (Hinde, 1979).

The emergence of social institutions can be illustrated by food sharing. Most animals consume their food as soon as it is obtained, but they may also hoard it, or donate it to others. Food sharing occurs, in the form of courtship-feeding and feeding the young, in many species. Food sharing among unrelated individuals has been reported to occur in vampire bats (Wilkinson, 1984) and in some primates (Silk, 1987, and see above). Humans may sometimes eat food as soon as it is obtained, but they do not usually live such a hand-to-mouth existence. Exchange, the practice of giving and receiving valuable objects and services, is characteristic of humans, especially when it comes to food.

The products of human foraging and agriculture are distributed by means of various types of exchange. Of these the most primitive is reciprocal exchange. This is characterized as giving of labour products and services that is not contingent on any definite receipt of goods and services in return. Rather, the giving is based on a general understanding that there will be some eventual reciprocity. This form of exchange is characteristic of some hunter–gatherer societies, but some form of reciprocal exchange occurs in all human societies, especially among relatives and close friends.

Reciprocal exchange is open to abuse by cheats, or freeloaders, who receive favours from others, but do not reciprocate. In some cases, such as the giving of food, clothing, etc. to children, no reciprocation is expected, but a roughly symmetrical reciprocity is expected in exchanges among adults. In societies practising reciprocal exchange, freeloaders are subject to subtle forms of disapproval. It is usually part of the ethos of reciprocal exchange to deny that any balance is being calculated. We are all familiar with the ritualized exchange of presents that takes place at certain times of year. Such occasions have become institutions that are characteristic of our society.

A summary of the different types of exchange, as classified by Harris (1985), is given in Figure 10.8. In societies more complex than family, or hunter–gather, groups some form of coercive exchange is the norm. Redistributive exchange has much in common with reciprocal exchange, but also involves a small amount of coercion. The produce of several individuals is brought to a central place, sorted and counted, and shared among producers and non-producers alike. On a small scale, this is little different from reciprocal exchange, but some extra organization and coordination is required if large quantities of material are to be collected in one place, and distributed at the same time. The organizational role is sometimes taken by a prominent provider who has an interest in increasing production, and in receiving recognition for his efforts. This is called egalitarian redistribution, and it differs from reciprocal exchange primarily in encouraging productivity.

In other cases the redistribution is done by an individual who takes no part in the production, and usually retains the largest share of the produce. To fulfil such a role the redistributor must be able to exercise some form of coercion, or political power, implying some kind of stratified social structure. A certain amount of control over productivity is characteristic of stratified redistribution.

Exchange among unrelated (non-kin) groups usually takes place at some kind of market place. In the absence of money, goods and services are bartered. There is bargaining between the opposing parties, each trying to maximize their profitability. If the barter exchanges are not direct, then some sort of accounts must be kept. For this reason, and because of the danger of

Fig. 10.8 Summary of modes of exchange in humans (From McFarland, 1989b).

Means of exchange	Definition	Control	Social	Ecology
Reciprocal exchange	Whoever produces shares	Cheats disliked	No thanks given or expected	Production subject to diminishing returns
Egalitarian redistribution	Whoever produces gives	Accounts kept	Producers boast and receive gratitude	Productivity encouraged
Stratified redistribution	Whoever produces must contribute	Policed	Political control of producers	Productivity controlled
Barter market	Producers bargain	Accounts kept	Some trust required	Few commodities
Price market	Producers sell	All-purpose money	Trade with strangers	Many commodities in mixed economy

possible use of force by one, or both, parties, there must either be a degree of mutual trust, or some form of policing. The barter market is an institution which can work satisfactorily if relatively few commodities are involved in the exchange.

A society which has some form of all-purpose money can operate a price market. The producers sell their goods in exchange for money, and use money to purchase other goods or services. The main advantage of the price market is that it can handle many commodities in a mixed economy. Another advantage is that it facilitates trade with strangers, with whom no elaborate trust arrangements exist. Traditionally, money has been regarded as a portable, recognizable material. It has a certain legality which makes it divisible and convertible, and confers wide generality of use. Money is an institution that is characteristic of modern human society.

Points to remember

■ The primates are primarily *K* selected, although there is some variation among species in these respects.

■ A prolonged period of parental care is necessary for prolonged postnatal development of the brain. This has implications for the mating system of the species concerned.

■ In the primates, the formation of pair bonds is assoicated with parental care. Ecological factors, particularly food availability, may also have a profound influence upon mating systems.

■ There are a number of lines of evidence which suggest that competitive social success is related to reproductive success.

■ Evolutionary theory, relating to individual selection, kin selection and reciprocal altruism, goes some way to account for primate social behaviour, such as grooming, competition over resources, sharing food and giving aid.

■ Social structure emerges from the pattern and quality of relationships among individuals. It is the scientific observers, not the animals, who see structure in the network of relationships that characterizes a social group. In this respect non-human primates may differ from humans, because people can recognize that they belong to a particular social structure, or institution.

Further reading

Dunbar, R.I.M. (1988) *Primate Social Systems*. Croom Helm, London.

Smuts, B., Cheney, D.L., Seyfarth, R.M., Struhsaker, T.T. and Wrangham, R.W. (eds) (1987) *Primate Societies*, University of Chicago Press, Chicago, IL.

PART

Mechanisms of behaviour

In the second part of this book we look at animals as machines. We focus our attention on the individual animal and ask how it adjusts to changes in the environment. The type of explanation we are seeking is a mechanistic one. How are we to explain, in terms of immediate causes, changes in the animal's behaviour?

Ideally, we might wish to describe behavioural mechanisms in terms of the physiological hardware responsible – the events in the brain and other parts of the nervous system. However, our knowledge of physiology is not yet sufficiently developed to enable us to do this. Even if it were, physiological detail may not be the most appropriate vehicle for explaining complex behaviours. Accordingly, we look at alternative ways of describing behavioural mechanisms: the physiological approach, the more abstract control-systems approach and the psychological approach. All three have their advocates, but they should not be regarded as rivals, because they complement each other, being suited to different aspects of behaviour.

Animal perception

A bat catching a moth (Photograph: Frederick Webster).

In this group of three chapters we deal with the nervous and sensory systems of animals. Chapter 11 deals with the basic elements of the nervous, sensory and muscular systems. The nervous systems of invertebrates and vertebrates are then reviewed. The aim is to give a general picture of the physiological apparatus that is characteristic of species throughout the animal kingdom.

Chapter 12 discusses the major sensory systems, including chemo-reception, hearing and vision. Three different ways of looking at perception are also discussed. Chapter 13 is concerned with ecological aspects of sensory systems, particularly the special sensory adaptations possessed by animals that operate in unfavourable environments.

Johannes Müller (1801–1858)

Johannes Peter Müller, the son of a poor shoemaker, introduced experimental physiology into Germany and is largely responsible for our picture of the body as a machine. Müller had many famous pupils, including du Bois Reymond, Helmholtz, Henle, Koelliker, Remak, Reichert and Virchow. He was involved in many pioneering aspects of biology, especially sensory physiology and marine zoology. Müller published the monumental *Handbuch der Physiologie des Menschen für Vorlesungen*, which was translated into English by W. Baly and published in 1827 as *Elements of Physiology*. Müller's most important contribution to the study of behaviour is his doctrine of 'specific nerve energies'. Previously it was believed that each different environmental event or stimulus affected sensory nerves in a manner specific to that stimulus. Müller discovered that a given nerve always produces the same type of sensation regardless of its mode of stimulation. Thus light falling upon the eye produces visual sensation, but so does mechanical stimulation such as a blow to the eye, and so does electrical stimulation of the optic nerve. If our ear could be attached to the optic nerve, auditory stimulation then would give rise to visual sensation. Thus it is not the sense organ that is important, but the nerves that relay sensory messages to the parts of the brain that receive visual, auditory, tactile and olfactory information and translate this into the relevant sensations. The sense organs are responsible for transforming the various forms of environmental stimulation, such as light, heat and mechanical energy, into an electrical potential that can be registered by the receptor cells, that is, the nerve cells that are part of the sense organ and that connect to other cells of the nervous system. The doctrine of specific nerve energies provides a key organizing principle for sensory physiology. It banishes all previous speculations as to the role of sense organs in perception, making it clear that these are primarily transducers of one form of energy to another. Although a certain amount of stimulus filtering may take place at the periphery, it is the brain that sorts and categorizes incoming information, purely on the basis of the intensity of stimulation of the peripheral sensory nerves. As it is now known, partly through the work of Müller's pupils, du Bois Reymond and Helmholtz, the nerve impulses that travel along a particular nerve axon have a measurable size and speed of propagation that are characteristic of that nerve cell, which means that the nerve cell can transmit messages about the intensity of stimulation only. Other information about the nature of the stimulus can be gained by the brain only by integrating the information from many nerve cells.

Neural control of behaviour

In this chapter we discuss the neural control of behaviour, starting with an outline of general principles. We then review the types of nervous system that occur throughout the animal kingdom, and see how the gross organization of the nervous system is related to behaviour.

11.1   Nerve cells

Nervous systems are made up of nerve cells, called **neurons**, which are specialized for transmitting information to one another. Each cell has a body, containing the nucleus, and a number of branching protrusions as shown in Figure 11.1. Usually, there are many short branches, called **dendrites**, and a single long protrusion, called the **axon**. The dendrites connect with other

Fig. 11.1 One nerve cell forming a synaptic connection with another.

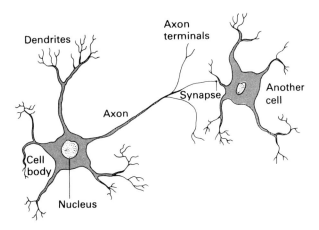

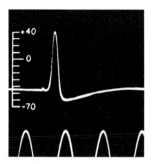

Fig. 11.2 Action potential spike recorded from the giant axon of a squid (From Hodgkin and Huxley, 1945).

nearby neurons, while the axons convey messages over relatively long distances.

The membrane of a neuron is usually polarized. That is, there is an electrical potential across it – called the **resting potential** in the inactive neuron – which provides a stable state of readiness, similar to that of an electrical battery which stores energy releasable upon demand. The resting potential results from unequal concentrations of K^+ ions on the inside and outside of the cell. When the cell is at rest the interior is negative relative to the outside. When the cell is depolarized its membrane potential is reduced towards zero, as shown in Figure 11.2. When the membrane potential is made more negative, the cell is said to be hyper-polarized.

If the resting potential is depolarized beyond a certain threshold then an **action potential** is propagated along the membrane. The action potential is a transient potential caused by systematic changes in the relative proportions of Na^+ and K^+ ions on each side of the membrane (see Figure 11.3). The action potential passes down the axon as an electrical wave. It is always of the same amplitude (height) and this is usually determined by the diameter of the axon. Larger axons propagate larger action potentials than small axons, and at a higher speed.

After each action potential has passed there is a refractory period during which the membrane recovers its normal ionic equilibrium and its normal resting potential. Since no other action potential can be generated during the refractory period, the refractory properties of the axon determine the upper maximum at which action potentials can be generated.

When as action potential occurs the neuron is said to 'fire'. The action potential often appears as a spike on an oscilloscope rigged up to measure the membrane potentials by means of electrodes inserted into nervous tissue. The neuron fires in an all-or-none manner (there is either a full spike or there is none) at a maximum frequency determined by its refractory properties. The frequency of firing depends upon the intensity of stimulation of the neuron. Thus the neuronal message is coded in terms of frequency (see Figure 11.4).

The membranes of axons and dendrites do not make physical connections with other neurons, but come very close at junctions called **synapses**. Usually, very small quantities of chemical **neurotransmitters** are released at

Fig. 11.3 Propagation of the action potential as a result of passage of K^+ and Na^+ ions through the axon membrane.

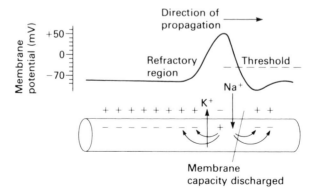

Fig. 11.4 Extracellular recording of nerve impulses from a neuron in the visual system of a cat. Note the increase in frequency when the stimulus is turned on (From Guthrie, 1980).

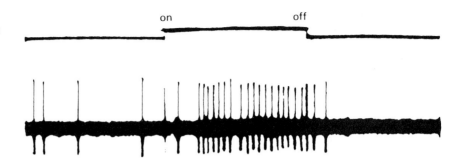

the synapse, and influence the resting potential of the recipient membrane, and hence the readiness of the receiving neuron to generate action potentials.

Neurons can be stimulated by other neurons, by injury, or by sensory receptors. The principles are the same in each case. The stimulation causes a change in the membrane potential and when this reaches a threshold an action potential is triggered. We now look at how this process is brought about by sensory receptors.

11.2   Sensory receptors

Sensory receptors are specialized nerve cells responsible for the transduction and transmission of information. Like ordinary nerve cells, they have a cell body, dendrites and one or more axons. Receptors are specialized according to the type of environmental energy to which they react. For example, **photoreceptors** contain pigments that are chemically modified by light and that give rise to an electrical potential when this occurs. **Mechanoreceptors** undergo electrochemical changes as a consequence of deformation of the cell membrane. This transduction of energy usually takes place in the cell body, and a characteristic of all receptor cells is that the environmental energy is converted into a graded electrical potential, called the **generator potential**. This potential is usually proportional to the intensity of stimulation of the receptor. When the generator potential reaches a certain threshold level, it triggers an action potential that travels along the axon of the receptor cell. This is the transmission part of the sensory process, and the information is usually coded such that the more intense the stimulus, the higher the frequency of action potentials.

In the absence of stimulation, the generator potential falls gradually to its resting level. When it falls below the threshold level, no further action potentials are generated. When stimulation is resumed, there may be a short delay (the refractory period) while the generator potential rises from its resting level to the threshold level. When stimulation is intermittent, the generator potential will rise and fall rhythmically, giving rise to bursts of action potentials. If, however, the intermittent stimulation is of high frequency, the generator potential may not have time to fall during the gaps in stimulation, and the production of action potentials will be continuous. This explains why we are not able to distinguished between a constant stimulus and an intermittent stimulus of very high frequency. This flicker fusion phenomenon holds with all senses, but it is most obvious in vision. The fact that a rapidly flickering light produces the same visual sensation at a constant light makes television and movie pictures possible.

The action potentials conveying sensory information are no different from any other nerve impulses. Their magnitude is determined by the size of the neural axon and their frequency by the intensity of stimulation. Each type of receptor sends impulses, either directly or indirectly, to a particular part of the brain. The sensations experienced depend not upon the type of receptor or the messages they send but upon the part of the brain that receives the message. The brain is also responsible for the localization of the sensation. In the case of pain, for example, the nerve fibres from the hand go to one part of the brain, those from the arm go to another and so on. The pain experienced by the brain is referred to that part of the body from which the message came. An illustration of this phenomenon comes from reports of people who have had limbs amputated and who complain of pain sensations that seem to come from the missing or phantom limb. Irritation of the cut nerve endings sends impulses to those parts of the brain that were concerned with the amputated limb. The messages are interpreted by the brain as coming from the lost limb, and the sensations experienced depend upon which nerve is irritated. Sensations of heat, cold or touch also may appear to come from the phantom limb.

11.3   Muscles and glands

The nervous system is responsible for controlling behaviour and, to a certain extent, for controlling the animal's internal environment (see Chapter 15). This control is exercised by commands to muscles and glands.

Muscles are made up of complex protein molecules that are capable of contraction and relaxation. Nerve endings connect with muscles by means of synapses similar to those by which neurones connect with each other. The nerve impulses arriving at the neuromuscular junction set up electrical potentials which cause the muscle to contract. Muscular relaxation results from lack of stimulation. When the muscle contracts it becomes shorter, provided it is not prevented from doing so by being fixed at each end. A muscle can lengthen when it is relaxed, but only if it is stretched by the action of other muscles or by some extraneous force. Thus muscles are usually arranged in opposing groups which act against each other. In some invertebrates, such as annelid worms, muscular contraction may be resisted by the hydrostatic pressure caused when the muscles squeeze part of the body cavity. This pressure causes the muscles to lengthen when they relax. In other invertebrates, such as arthropods, the muscles are housed inside a hard exoskeleton which provides the necessary leverage for opposing sets of muscles (see Figure 11.5). In vertebrates the internal skeleton provides the leverage and the muscles are arranged so that they pull against each other (see Figure 11.6). One set of muscles relaxes when the other contracts.

Fig. 11.5 Mechanical arrangement of muscle and skeleton in an insect's leg. The muscles are housed inside the skeleton, and in (a) muscle a in the flexor (bending the limb) and muscle b in the extensor (straightening the limb). In (b), where the muscles span the joint, the arrangement is the opposite. Muscle a is the extensor, and muscle b is the flexor.

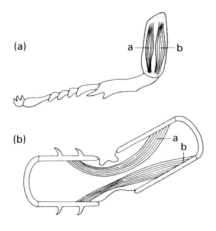

Fig. 11.6 Mechanical arrangement of muscle and skeleton in a human arm. The muscles are arranged outside the skeleton. Muscle a is the flexor and muscle b the extensor.

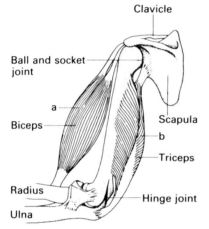

Some glandular secretions are under nervous control. In vertebrates these include the salivary glands, which produce saliva, the adrenal medulla, which produces adrenalin, and the posterior pituitary gland, which produces a number of important hormones. The secretions of these glands can influence behaviour indirectly through their influence upon the animal's internal state, as we see at the end of this chapter.

11.4 The somesthetic system

It is important for an animal's brain to receive information about the state of the body. The positions of limbs, pressure on internal organs, the temperature of various parts of the body and many other factors are monitored by the **central nervous system** (CNS) via interoreceptors located at strategic points. This system, responsible for bodily sensations, is called the **somesthetic system**.

Numerous types of sensory receptors exist in the skin, skeletal muscles and viscera of vertebrates. Some examples are shown in Figure 11.7. Invertebrates also have a wide range of receptors. Humans have five types of

Nerve plexus around hair
(touch)

End bulb of Krause
(cold)

Ruffini ending
(warmth)

Pacinian corpuscle
(deep pressure)

Meissner's corpuscle
(touch)

Free nerve ending
(pain)

Fig. 11.7 Some sensory receptors found in the skin and the senses with which they are associated (After Keeton, 1972).

skin receptor, giving rise to sensations of touch, pressure, cold, heat and pain. The pain receptors are numerous, being 27 times as common as cold receptors and 270 times as common as heat receptors. Some skin receptors show rapid sensory adaptation. In response to a step change in stimulation, the frequency of nerve impulses rises rapidly and then declines to its resting level. This means that the receptor is a good indicator of changes in intensity of stimulation but a poor indicator of the absolute level of intensity. This is an advantage in cases where the skin receptors give early warning of environmental changes that are likely to affect the body, like changes in temperature.

Receptors deep inside the body serve a wide variety of functions including detection of changes in blood pressure, the tension of muscles, the amount of salt in the blood, etc. We are not directly aware of the information produced by the majority of interoceptors. They do not give rise to sensations. Sometimes their effects combine to produce sensations of hunger, thirst, or nausea, but these are due to complex processes in the brain, which do not always refer the sensation to particular parts of the body. This is presumably because the action that has to be taken in response to hunger and thirst is much more indirect than action that is taken in response to peripheral touch or temperature change.

The orientation of animals in relation to gravity and external stimuli like light depends partly upon information about the spatial relationship of the various parts of the body. In mammals this information comes from the vestibular system and from receptors in the joints, muscles and tendons. The joint receptors provide information about the angular position of each joint (Howard and Templeton, 1966). In the tendons of mammals are the Golgi tendon organ receptors, which are sensitive to tension. They send messages to the spinal cord and are involved in a simple reflex that acts to oppose increases in muscle tension.

Within the muscles are the muscle spindles that are sensitive to changes in muscle length (Figure 11.8). The muscle spindles consists of modified muscle fibres that have a spiral nerve ending, called the **primary** (or **annulospiral**) **ending**, wrapped around its middle. When the muscle increases in length, the muscle spindle is stretched and the primary endings send fast messages to the spinal cord. There may also be **secondary flower-spray endings** that send slower messages. Many mammalian spindles have both primary and secondary endings, while others have only primary endings (Prosser, 1973). These spindles form part of a simple reflex that acts to oppose increases in muscle length.

The muscle spindles are contained within a **fusiform connective tissue**, and their muscle fibres are called **intrafusal fibres**. Ordinary muscle fibres are called **extrafusal fibres**. These are innervated by **alpha motor neurons**, which have their cell bodies in the spinal cord. In mammals the intrafusal fibres are innervated by smaller **gamma motor neurons**, which keep the spindle in a tonic state of activity so that less muscle stretch is required to activate the spindle. Since muscle spindles are arranged in parallel with the extrafusal muscle fibres, they tend to become slack when the muscle contracts. The gamma motor neurons can command the intrafusal fibres to take up the slack so that the spindle remains in a state of readiness (Figure 11.9).

Fig. 11.8 Mammalian muscle spindle.

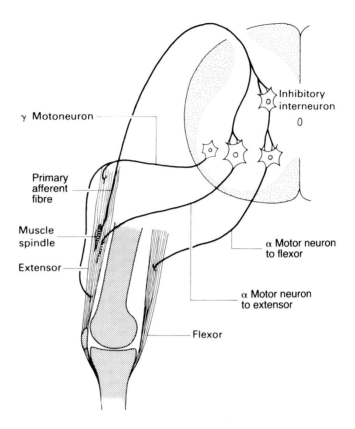

Fig. 11.9 The innervation of the muscle spindle.

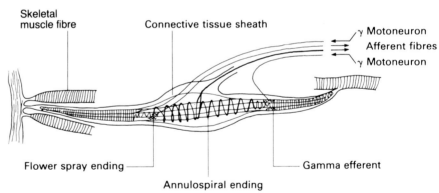

The muscle spindles of birds resemble those of mammals, with the intrafusal and extrafusal fibres in parallel. In the lizard *Tiliqua*, however, the muscle spindles appear to be in series with the extrafusal fibres (Prosser, 1973). In reptiles and amphibians there is no gamma motor neuron system, and both intrafusal and extrafusal fibres are supplied by alpha motor neurons. Fish have no muscle spindles, but they do have receptors in their fibre that are sensitive to their angular velocity.

Arthropods have numerous kinds of stretch receptors, of which there are two general types:

- those that lie between exoskeletal elements and that respond to vibrations in the cuticle

- those that are attached to tendons and that can signal stretch and pressure changes (Prosser, 1973)

Thus, crabs have receptors that signal both position and movement at a joint, and blowflies have stretch receptors in the gut that inhibit feeding when the gut is distended.

Most biologists would agree that one of the main evolutionary trends within the animal kingdom has been the increasing elaboration of the nervous system. In tracing the evolution of sensory processes, therefore, it seems sensible to use the complexity of the nervous system as a guide. However, we have very little direct evidence about the nervous system of animals in the past because soft nervous tissue is seldom preserved in fossilized form. Indirect evidence sometimes can be gained from skeletal remains, especially where well-preserved vertebrate skulls are available. Most of our conclusions about the evolution of the nervous system derive from present-day representatives of animal types that are known to have changed little over millions of years.

11.5   Nervous systems of invertebrate animals

A nervous system may be defined as an organized constellation of nerve cells and associated non-nervous cells (Bullock, 1977). This definition includes receptors but not effector organs such as gland and muscle. Throughout the animal kingdom, nerve cells have common attributes that readily distinguish them from other cells. These attributes include the graded electrical potentials that occur in receptor cells and at synapses and the impulse-like action potentials that carry information along the axons. Although not universal, these are the typical properties of neurons that are described above. Important exceptions to the general pattern occur in some primitive animals and in the protozoa.

Protozoa, being single-cell organisms, cannot possess a true nervous system. Although they have undoubted sensory capabilities and exhibit rudimentary behavior, they do not appear to have specialized intracellular organelles that conduct excitation (Bullock, 1977). Instead, they seem to be organized along principles similar to those governing the physiology of neurons. Thus, the protozoan is like a receptor cell equipped with effector organnelles.

The coelenterates contain the simplest animals that have a true nervous system. Nerve cells communicate with each other via recognizable nerve impulses and synapses. Coelenerates have no CNS, although nerve cells may sometimes be organized into simple ganglia. The nerve cells are often organized into **nerve nets** that permit diffuse conduction throughout the body, as in the case of *Hydra* (Figure 11.10).

Hydra has receptor cells responsive to touch and to chemical stimuli. Photoreceptors also occur in some coelenterates. The receptors pass infor-

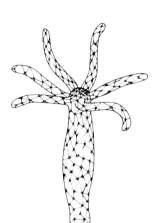

Fig. 11.10 The nerve net of *Hydra* (From *The Oxford Companion to Animal Behaviour*, 1981).

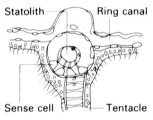

Fig. 11.11 A statocyst (gravity sensor) from a medusa (jellyfish). These are complexes of tissue and hence true sense organs.

mation across synapses to other nerve cells, but the nerve impulses are very slow compared to those of vertebrates. The nerve net coordinates the movements of the animal in a way that is not understood fully. The movements are slow and the behavioural repertoire is very limited. Learning is not known to occur, except in the form of simple habituation. If the mouth of *Hydra* or of a sea anemone is touched repeatedly, it closes in a reflex manner at first, but the response gradually disappears with repeated stimulation. This habituation is known to be due to sensory adaptation and is not normally regarded as true learning.

Among the coelenterates, the medusae (jellyfish) show two important features of more advanced nervous systems. These are representatives of what were probably the first ganglia and the first sense organs. The ganglia are found in the marginal bodies, which innervate photoreceptors and statocysts. The marginal bodies contain four or more different types of neuron that interconnect with each other (Bullock, 1977). The **statocysts** are devices for detecting the direction of gravity. They contain a small pebble-like object, the **statolith**, whose position in a cavity can be detected by mechanoreceptors in the wall of the cavity, as illustrated in Figure 11.11. The receptors provide information about gravity because of the arrangement of the associated non-nervous tissue. This makes the statocyst a true sense organ. It is interesting that this fairly sophisticated sensory device is found in action with a relatively complex feature of the nervous system, the ganglia found within the marginal bodies. An association between specialized features of the nervous system and particular sense organs is a common phenomenon in the evolution of sensory processes.

A further advance in nervous system organization is seen in the flatworms (Platyhelminthes). Unlike the coelenterates, but like most other invertebrates, the flatworms show bilateral symmetry and possess a head and tail. Sensory receptors tend to be concentrated in the head region instead of scattered over the body. The nervous system also shows concentration at the head in the form of an anterior ganglion that constitutes a simple brain. From the anterior ganglion, two nerve cords run down the body, and these are linked by nerves in the form of a ladder, as illustrated in Figure 11.12. Nerve fibres pass from the cords to all parts of the body in a net-like pattern. The anterior ganglion and nerve cords together form the CNS, while the network of nerve fibres constitutes the **peripheral nervous system**. This distinction is common to most invertebrates and all vertebrates and is seen in flatworms in its most primitive form. As a rule, the CNS contains most of the effector nerve cell bodies, while the peripheral nervous system contains the sensory receptors.

In flatworms such as *Planaria*, sensory cells on the head respond to touch, temperature and the chemical composition of the water. They also have two eyes consisting of photoreceptors grouped together. The messages from the sensory cells are transmitted to the anterior ganglion. Conduction of nerve impulses along the nerve cords is faster than in nerve nets, and the behaviour of flatworms is correspondingly more varied and definite than that of coelenterates. They are quick to detect the presence of food and to approach it. They avoid strong light and noxious chemicals and seem to be capable of

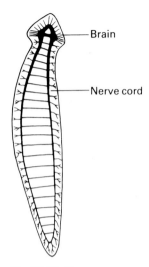

Fig. 11.12 The nervous system of a flatworm (From *The Oxford Companion to Animal Behaviour*, 1981).

rudimentary learning. If placed in a simple T maze, planarians can learn to turn to one side rather than the other to avoid being tapped with a rod. Learned responses to light stimuli have also been claimed, but these have proved to be controversial.

The more advanced invertebrates follow a general pattern based on the annelids, or segmented worms. These have a fairly complex nervous system, which is typified by that of the earthworm *Lumbricus*, illustrated in Figure 11.13. The head region is characterized by a pair of supra-oesophageal ganglia, above the pharynx, connected to another pair of ganglia below. These suboesophageal ganglia are the first of a chain of ganglia along a central nerve cord, one ganglion for each segment of the worm. Many annelids have a system of giant fibres containing neuronal axons of large diameter and high conduction velocity. These are involved in rapid escape or withdrawal reactions to danger. Most annelid receptors are single elements that include chemoreceptors, mechanoreceptors and photoreceptors. Proper sense organs appear in some species, examples being the taste buds in the skin of *Lumbricus* and the eyes of the free-swimming polychaete *Alciope*. Annelids have a fairly restricted repertoire of behaviour patterns and are capable of rudimentary learning.

The gross plan of arthropods is similar to that of annelids, with a central chord and a pair of ganglia in each segment, plus cross connections called **commissures**. There is a dorsal anterior brain with circumoesophageal connections to the central cord. In primitive arthropods, the segmental pattern can be seen clearly, but in more advanced forms there is considerable fusion of ganglia, as illustrated in Figure 11.14. This fusion of ganglia is characteristic of the evolution of the nervous systems of invertebrates and is associated with increasing complexity of the sensory systems and of behaviour. Giant fibre systems occur in many arthropods including shrimps, lobsters, scorpions and some insects. They normally mediate a rapid tail click or jumping movement, which forms part of the animal's escape behaviour.

Arthropods have evolved a greater variety of types of receptor than any other group, including vertebrates (Bullock, 1977). Table 11.1 gives a

Fig. 11.13 The nervous system of an earthworm (From *The Oxford Companion to Animal Behaviour*, 1981).

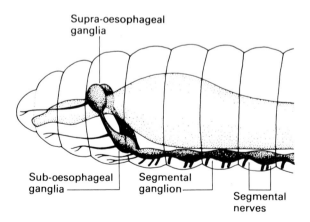

Table 11.1 Some types of arthropod skin receptors

Structure	Morphological character	Function (known or presumed)
Sensilla trichoidea	Sensory hairs and setae	Mechanoreceptors, proprioceptors, sound, contact chemoreception, humidity, olfactory in different places
Sensilla chaetica	Sensory spines and bristles	Mechanoreceptors, proprioceptors
Sensilla squamiformia	Sensory scales	Mechanoreceptors
Sensilla basiconica	Short, thick hairs; few to many neurons	Mechanoreceptors, contact chemoreceptors, olfactory, humidity, osmotic, temperature receptors
Sensilla coeloconica	Sunken cuticular cones	Olfactory or humidity receptors
Sensilla ampullacea	Sensory tubes	Olfactory receptors
Sensilla campaniforma	Cuticular domes	Directional strain gauges
Sensilla placodea	Cuticular plates and pore plates	Unknown; abundant on bee antennae

Fig. 11.14 Cephalization in Crustacea. Above: the brain and ventral cord of a fairy shrimp. Below: the brain and cord of a crab.

summary of the main types. The sensory neurons of these receptors have their cell bodies close to the sensory surface and not grouped into sensory ganglia. Some have few sensory neurons, and others have many, as illustrated in Figure 11.15

Many types of mechanoreceptor, including statocysts, are common in crustaceans. **Chordotonal organs** are a distinct kind of internal mechanoreceptor that usually serve as proprioceptors responding to mechanical

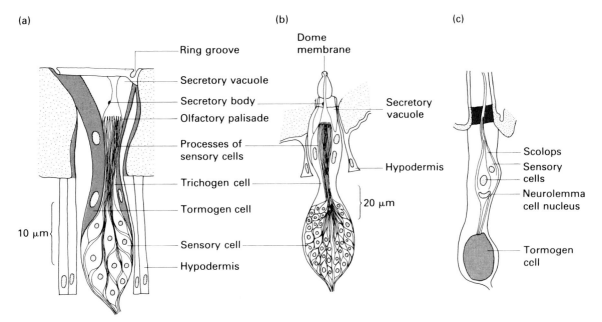

Fig. 11.15 Examples of sense organs in anthropods: (a) olfactory plate of a hornet (Sensilla placode), (b) olfactory cone of a hornet (Sensilla coeloconica), (c) antenna hair of a moth (Sensilla trichoidea) (After Bullock, 1977).

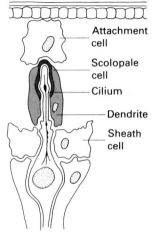

Fig. 11.16 Chordotonal mechanoreceptor of an insect (After Gray, 1950).

displacement (Figure 11.16). They also may be involved in detection of wind direction, velocity of water flow, flight speed, direction of gravity and various types of vibrations. Mechanoreceptors are also found in ligaments and muscles, where they provide information about limb position and movement.

The photoreceptors of arthropods are of particular interest, and some of these are illustrated in Figure 11.17. In general the arthropods display a wide range of nervous complexity and demonstrate many evolutionary trends, including a progression from the segmented annelid pattern to the large brains that result from many fused ganglia. Arthropods have a remarkably wide repertoire of behaviour patterns. Some species, such as the honey-bee which we discuss in detail in Chapter 23, have a complexity of behaviour that rivals that of the vertebrates. Many arthropods are capable of simple learning, though this seems to be tried to specific situations.

These evolutionary trends are even more apparent among the molluscs, which have achieved the highest degree of sophistication of the CNS found in invertebrates. Among the gastropods, for example, there is an enormous range of complexity in nervous system organization. Thus, *Haliotis* (the abalones) have an annelid-like nervous system, while the land snail *Helix* has considerable fusion of ganglia, as illustrated in Figure 11.18. The most highly developed nervous systems are found in the cephalopods. These are the most active molluscs. They hunt mainly by vision and are capable of complex behaviour, including recognition of complex objects, and rapid learning.

The most studied representative of the cephalopods is probably the octopus. The octopus brain contains about 170 million nerve cells, compared with about 100 000 for the larger crustaceans. It is made up of about 30 different lobes, many of which have distinct functions (see Figure 11.18). The optic lobes make up more than half the nervous tissue of the brain.

Fig. 11.17 The compound eye of a fly. Each eye is made up of numerous ommatidia. Each ommatidium is connected to a nerve cell which signals the intensity of light from the direction in which the ommatidium points. A single ommatidium does not produce an image, but as adjacent ommatidia point in slightly different directions, the CNS receives information about the distribution of light intensity over the visual field.

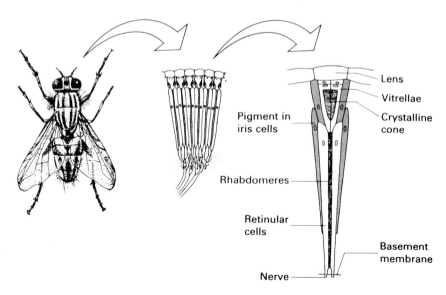

| *Haliotis* | *Helix* | Octopus |

Fig. 11.18 Examples of molluscan nervous system. The abalone (*Haliotis*) has a ladder-like system. The land snail (*Helix*) shows considerable cephalization, and the octopus has the most highly developed brain found among the invertebrates.

They are connected to a pair of large eyes that are the most advanced among the invertebrates and that rival those of vertebrates (see Chapter 12).

The molluscs vary considerably in the complexity of their behaviour, but the cephalopods are the most advanced. This group has received considerable study from behavioural scientists on account of their remarkable learning abilities.

11.6 The vertebrate nervous system

The organization of the vertebrate nervous system is distinct from that of invertebrates, though it is not always more complex. The vertebrate CNS is developed from a dorsal neural tube to form a brain and single dorsal nerve cord, rather than the brain and dual ventral nerve cords characteristic of invertebrates. The evolution of the vertebrate brain is reflected in the embryonic development of the CNS in the individual, as illustrated in Figure 11.19.

The pattern of embryological development of the brain is remarkably constant throughout the vertebrates (Laming, 1981). As illustrated in Figure 11.19, the neural tube develops into three distinguishable parts:

■ the prosencephalon (forebrain)

■ the mesencephalon (midbrain)

■ the rhombencephalon (hindbrain)

These three hollow vesicles are traditionally associated with the olfactory, visual and auditory senses respectively. Each major division develops a secondary outgrowth: the telencephalon (cerebrum), optic tectum and cerebellum respectively, as illustrated in Figure 11.20. The CNS comprises the brain and spinal cord, which are enclosed within the bones of the skull and vertebral column. The peripheral nervous system consists primarily of the

Fig. 11.19 Embryonic development (from left to right) of the human brain showing the major divisions of the brain.

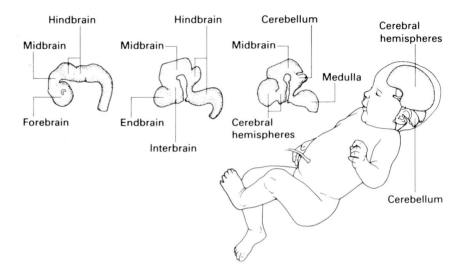

Fig. 11.20 One hemisphere of the rat brain (from the inside after section), showing the subdivisions of the forebrain, midbrain and hindbrain.

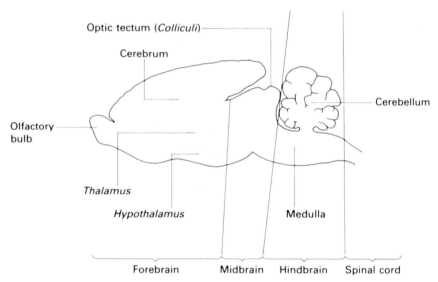

inflow (afferent nerves) and outflow (efferent nerves) from the CNS. The **somatic nervous system** carries sensory information to the CNS and commands from the CNS to the skeletal muscles responsible for bodily movement. The **autonomic nervous system** supplies the internal organs with two types of innervation that have antagonistic effects (see Box 11.1). The functioning of the autonomic nervous system is described in further detail in Chapters 15 and 28.

It is fair to say that most of the interesting evolutionary trends within the vertebrates have to do with the CNS. The changes in the peripheral nervous system are more a reflection of the animal's particular anatomy and its adaptations to its particular niche, than to any evolutionary 'progress'. The more ecological aspects of the senses are discussed in Chapter 13. Two important evolutionary trends that are worth discussing here are those of increasing

Box 11.1 Plan of the autonomic nervous system

The sympathetic system is protrayed on the left, and the parasympathetic on the right (After Russell and Woodburne, 1965).

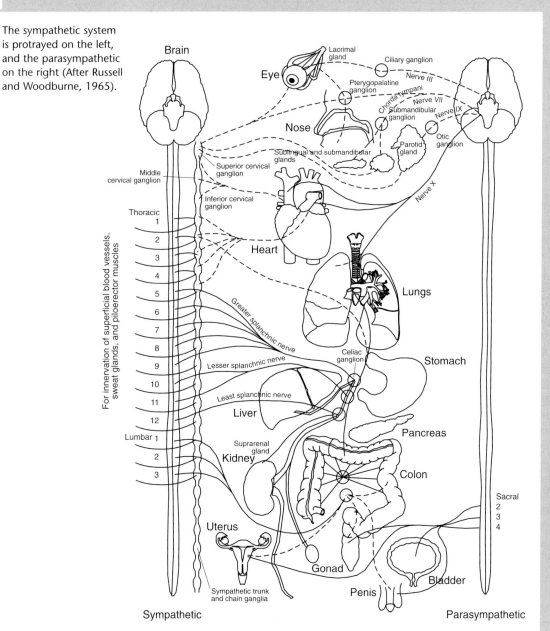

The **sympathetic nerve pathways** have an emergency function and become active under conditions of exertion or emotion. They have effects such as accelerating heart rate, dilating air passages to the lungs, reducing intestinal activity and increasing the blood supply to the muscles and the brain. The **parasympathetic pathways** serve a recuperative function, restoring the blood supply to normal and counteracting the effects of sympathetic activity.

brain size, and increasing tendency for behaviour to be controlled by the higher centres of the brain.

Vertebrate brains vary considerably in size. However, direct comparison of brain size does not enable us to draw conclusions about intelligence or about the role of the CNS in controlling behaviour. Larger animals tend to have larger brains because the larger the body, the greater the number of sensory fibres entering the brain and the more fibres leaving the brain to control the muscles. Therefore it is the size of the brain in relation to the body that we must use as a measure. When we take account of differences in body size, we find that the brains of fish, amphibians and reptiles are of similar sizes.

The brains of birds and mammals are normally much larger. Among the mammals, the brains of rodents and insectivores are relatively small, while those of ungulates and carnivores are much larger. The biggest brains are found among the primates and the sea mammals. The brains of the lower primates (prosimians) differ little from those of other mammals, but monkeys and apes have larger brains than any other land mammals. Even bigger brains, in relation to body size, are found among the dolphins and whales, but it is difficult to know what conclusions to draw in comparing animals specialized for life on land and at sea. The human brain is three times as large as would be expected for a non-human primate of the same body size. Among humans, however, considerable variation exists. The brain of a normal person can be as small as 1000 cubic centimetres or nearly as large as 2000 cubic centimetres (Coon, 1962). The norm for modern *Homo sapiens* is about 1450 cubic centimetres. It is interesting that the fossil skulls of archaic man (*Homo neanderthalensis*), who lived some 45 000 to 75 000 years ago, indicate a slightly larger cranial capacity than that of modern man.

The size of the brain gives a rough indication of the number of nerve cells in the brain. Larger brains have more nerve cells, and these cells tend to be larger and more loosely packed than those of smaller brains. In a large brain, each neuron tends to have a more complex set of dendrites and is open to influence from a greater number of other neurons. When we compare the brains of different vertebrates, after taking body size into account, can we see a definite evolutionary trend toward larger brains, which implies greater and more sophisticated control of behaviour?

The brains of animals have evolved as adaptations to their ecological niche, just as have other organ systems (Jerison, 1973). Animals that have specialized sensory systems and forms of behaviour must have correspondingly specialized brain mechanisms. Contrary to popular belief, there is no progressive increase in brain size when animals in the sequence fish, reptile, bird, mammal are compared with each other (Jerison, 1973). Some fish brains are larger than those of reptiles of equivalent weight, and some bird brains are relatively larger than those of some mammals.

In charting the evolution of the brain and intelligence (see Chapter 27), we see no smooth progression from primitive to advanced animals (Hodos and Campbell, 1969; Hodos, 1982). Jerison (1973) calculated an encephalization quotient by relating the brain size of each species to the size expected for an average mammal of the same body weight (see Figure 11.21). This measure shows some marked differences among different groups, but some anomalies

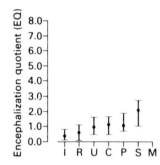

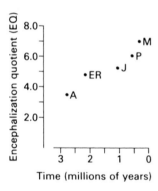

Time (millions of years)

Fig. 11.21 Above: emcephalization quotient for a variety of mammals. I, insectivores; R, rodents; U, ungulates; C, carnivores; P, prosimians; S, simians; M, man. Below: EQ in hominids plotted against time. A, *Australopithecus africanus*; ER, *Homo* skull from Kenya; J, *Homo erectus* from Java; P, *Homo erectus* from Peking; M, modern man (After Passingham, 1982).

occur. Thus, among the primates, some small monkeys are placed well above other primates. When brain size is related not to body size but to the size of the medulla (a more direct measure of the inputs and outputs of the brain), then the great apes appear to have the larger brains (Passingham, 1975). Encephalization quotients for fossil hominids (Figure 11.21) show that brain size has increased over the past three million years (Passingham, 1982).

11.7 | Vertebrate hormones

The nervous system has a parallel system of feedback and control of behaviour in vertebrates. This is the **endocrine** (hormonal) **system**. Hormones are chemicals (usually peptides or steroids) that are produced by endocrine glands and released into the blood. Each hormone has specific functions and often affects specific target organs. The endocrine glands turn their secretions on and off at appropriate times. The control of hormone secretion is achieved either by direct nervous action or by the action of other hormones. There is usually a series of events that provide negative feedback that regulates the level of the hormone in the blood. Hormones circulate in very small amounts but most hormone molecules have a life of less than one hour in the blood. They have to be secreted continuously if their presence is to be effective.

The most important hormones that influence behaviour are indicated in Table 11.2. Some of these, such as follicle stimulating hormone (FSH) and luteinizing hormone (LH) have indirect effects, priming other endocrine glands. Others have a more direct effect. There are three main ways in which hormones can affect behaviour:

1. They can influence effectors, such as special structures involved in behaviour.

Table 11.2 Important hormones in mammalian behaviour

Source	Hormones	Principal effect
Kidney	Angiotensin	Stimulates vasoconstriction and causes a rise in blood pressure. Stimulates thirst
Testes	Testosterone	Stimulates development and maintenance of male secondary sex characteristics and behaviour
Ovaries	Oestrogen	Stimulates development and maintenance of female secondary sex characteristics and behaviour
	Progesterone	Stimulates female secondary sex characteristics and behaviour and maintains pregnancy
Adrenal medulla	Adrenaline	Stimulates 'fight or flight' reactions
Anterior pituitary	Follicle stimulating hormone	Stimulates growth of ovarian follicles and seminiferous tubules of the testes
	Luteinizing hormone	Stimulates secretion of sex hormone by ovaries and testes
	Prolactin	Stimulates milk secretion by mammary glands
Posterior pituitary	Oxytocin	Stimulates release of milk by mammary glands and stimulates contraction of uterine muscles

2. They can influence peripheral sensory receptors and so modify the input to the brain.

3. They can affect the brain directly.

An example of type 1 can be seen in the mating behaviour of the African clawed toad (*Xenopus laevis*). During the breeding season the male develops special nuptial pads on his forelimbs. These consist of a thick mass of small spines set into the skin, which enable the male to grip the smooth and slippery pelvic area of the female during mating. The appearance of the pads depends upon the hormone testosterone which is released from the testes at the appropriate time. At the end of the breeding season testosterone production ceases and the pads disappear.

An example of the influence of hormones upon peripheral sensory receptors can be seen in the parental behaviour of pigeons. Pigeons and doves feed their young on 'pigeon's milk', which is a proliferation of the lining of the crop. This substance is regurgitated in response to the begging behaviour of the young. Its formation is under the control of the hormone prolactin. Daniel Lehrman (1955) showed that sensory stimulation of the enlarged crop is essential for inducing parental feeding. If the crop sensitivity is reduced by means of a local anaesthetic, then the parental feeding is reduced. Prolactin is thus instrumental in inducing parental feeding behaviour by its effect on the crop.

The direct action of hormones on the brain is the most important mode of hormonal influence upon behaviour. It was first demonstrated (Harris *et al.*, 1958) by implantation of minute quantities of artificial oestrogen into certain regions of the brain of female cats. This could induce sexual behaviour, even though the level of oestrogen in the blood was well below the level normally required for such behaviour.

Hormonal influences upon behaviour may be slow and prolonged, or quick-acting and short-lived. The role of hormones in reproductive behaviour, and their interactions with day length and seasonal effects, is discussed in Chapter 16. As an example of the prolonged action of hormones, we can point to the influence of hormones on the development of behaviour. In the male zebra finch, for example, there is growing evidence that the development of many morphological and behavioural features occurs under hormonal control (Prove, 1983). Thus behaviour patterns which are known to be dependent upon testosterone in adult males (Prove and Immelmann, 1982), such as song, courtship and copulation, first occur at times of high testosterone production. The onset of the sensitive phase for song learning is accompanied by increasing oestrogen production and the areas of the brain that are responsible for the control of song are differentiated under the control of oestrogens (Gurney and Konishi, 1980).

As an example of the rapid effects of hormones, let us consider the suckling of goats (Figure 11.22). The luteinizing hormone (LH) stimulates the ovary to produce oestrogens and progesterone, which are responsible for the development of the mammary glands at puberty, and for maintaining them in a state of readiness for milk secretion. During pregnancy there is a rise in the level of prolactin, and this stimulates milk secretion. When the kid

Fig. 11.22 Diagram of hormones involved in suckling.

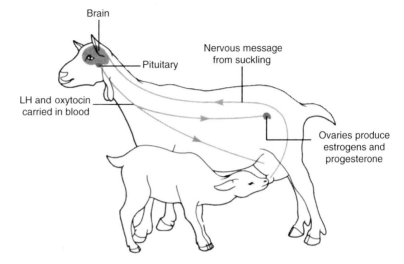

Brain

Pituitary

Nervous message from suckling

LH and oxytocin carried in blood

Ovaries produce estrogens and progesterone

suckles the mechanical stimulation induces the hypothalamus of the brain to release oxytocin from the posterior pituitary gland. The rise in blood oxytocin triggers milk letdown. The oxytocin response is very rapid and the kid will normally receive milk within the first 40 seconds of suckling.

Points to remember

■ Animal behaviour is controlled by the nervous system, which is made up of specialized cells called neurons. These operate on the same principles throughout the animal kingdom.

■ The nervous system receives information from the environment through sensory receptors. These are nerve cells specialized as energy transducers. They may provide information about the animal's external environment and about its internal state.

■ Behaviour is effected through the action of muscles and glands. The muscles are responsible for animal movement, and the glands produce substances which may play a variety of behavioural roles, such as communication with other animals, protection against predators and adherence to the substrate.

■ Muscular activity is controlled and coordinated by means of the somesthetic part of the nervous system. This provides feedback to the central nervous system about the position, tension and length of joints, tendons and muscles. The nervous commands to the muscles are modulated by this feedback of information.

■ The nervous systems of invertebrates vary in plan from one phylum to another. Within a phylum there is usually a trend, from simple primitive forms to more complex advanced forms, which involves increasing cephalization and concentration of neurons and sensory receptors in the anterior region of the animal. In some invertebrates this has resulted in a complex brain rivalling that of some vertebrates.

■ The vertebrate nervous system has a basic plan common to all. It has evolved primarily by accretion of new mechanisms rather than modification of old ones. This means that it is possible to trace the evolution of the vertebrate nervous system in the developing embryo. The most distinctive feature of the vertebrate nervous system is the highly developed brain and associated sense organs.

■ The vertebrate endocrine system produces hormones, some of which have important behavioural roles. Hormones may influence behaviour by acting directly on the brain, by affecting peripheral organs or by influencing the production of other hormones.

Further reading

Bullock, T.H. (1977) *Introduction to Nervous Systems*. Freeman, New York.

Young, D. (1989) *Nerve Cells and Animal Behaviour*. Cambridge University Press, Cambridge.

CHAPTER

Sensory processes and perception

The ability of animals to respond to changes in the external environment depends upon sensory processes designed to detect such changes. These range from simple discrimination of differences in temperature, or light intensity, to the recognition of complex patterns. In this chapter we look at the major sensory modalities of animals, and explore the nature of perception.

12.1   Chemoreception and thermoreception

Chemoreception is the capability of identifying chemical substances and detecting their concentration. It exists even among very primitive forms of life. In a technical sense, virtually every nerve cell is a chemoreceptor in that it reacts specifically to substances released by other nerve cells. The mechanisms of chemoreception involve the recognition of specific molecules by receptor sites on cell membranes. Whether this recognition occurs on a basis of chemical action, molecule shape or both is not fully understood. Thus, we do not know what sugar and saccharine have in common that makes them both taste sweet to blowflies, rats, monkeys and humans.

Chemoreceptors may be **exteroceptors** or **interoceptors**. The former detect the presence of chemicals in the external environment, while the latter detect substances circulating in the body fluids, such as carbon dioxide, nutrients and hormones. Both taste and smell depend upon chemoreceptors. In the traditional sense, smell is concerned with the detection of low concentrations of airborne substances, while taste results from direct contact with relatively high concentrations of chemical substances. In both cases, however, the chemicals are presented to the receptor in solution, and the distinction is difficult to justify in some animals, like those that live in water. Nevertheless, in many animals there is a neurological distinction in that some nerves are concerned with olfaction, or the detection of low concentrations, while others convey **gustatory** messages from different receptors specialized for detecting high concentrations of chemicals. In the blowfly, for example, chemoreceptors on the antennae detect small quantities of airborne substances, and chemoreceptors on the tarsi (feet) are capable of detecting salt, sugar and pure water. In vertebrates, the sense of taste is relayed via the facial (VII) and glossopharyngeal (IX) cranial nerves, while the sense of smell is transmitted by the olfactory nerve (I).

One of the most intensively investigated invertebrate olfactory systems is the perception of courtship pheromone by the silkworm moth (*Bombyx mori*). A **pheromone** is a chemical, or mixture of chemicals, that is released into the environment by an organism and that causes a specific behavioural or physiological reaction in a receiving organism of the same species. Pheromones are thus chemical messengers, and they are involved in what is probably the most primitive form of communication. The first chemically defined sex pheromone was that of the silkworm moth, and this chemical (a polyalcohol) has become known as bombykol. It is secreted by an abdominal gland of the female, and by this means females can attract males over a distance of several kilometres. The chemical structure of bombykol is known, and it has been synthesized and presented to male moths or used to stimulate isolated antennae whose receptors were monitored with electrodes (Schneider, 1969; Payne, 1974). A full physiological response is shown only to bombykol, while a lesser response occurs to certain very close chemical relatives of bombykol. This demonstrates very great specificity. Even more remarkable is the finding that a single molecule of bombykol is sufficient to trigger a full response (Kaissling and Priesner, 1970). The behavioural response of the male silkworm moth is to fly upwind upon detecting bombykol molecules, and when he finds the female he copulates with her. This type of orientation to chemical stimulation is common in insects.

Insects have various types of olfactory sensilla, as illustrated in Figure 12.1. These usually have numerous minute pores in the surface, which terminate in fluid-filled tubules. Dendrites of the receptor cells extend into the sensilla, and the receptor axons travel directly to the brain. Airborne pheromone molecules enter the pores of the sensillum and pass through the pores into the fluid-filled interior, where they come into contact with the receptor membrane.

In vertebrates the olfactory receptors are primary sensory neurons with dendrites that extend as cilia into a mucous layer, as illustrated in Figure 12.2. The axons of these neurons go to the olfactory bulb where they synapse with secondary neurons whose axons form the olfactory tract that enters the forebrain. The taste receptors consist of sense cells that are usually arranged in clusters called papillae. The sense cells are in close contact with sensory nerve fibres that remain fixed, while the taste cells are replaced every few days.

The basic tastes in mammals are acid, bitter, salt and sweet. In humans, different parts of the tongue are sensitive to different tastes. Acid or sour tastes affect the sides of the tongue towards the back, and bitter tastes affect the extreme back of the tongue. Salty tastes affect the sides towards the front, and sweet tastes affect the tip of the tongue (see Figure 12.3). The flavour of

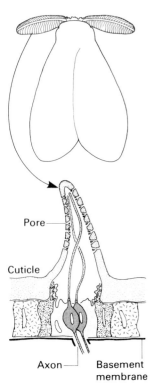

Pore

Cuticle

Axon   Basement membrane

Fig. 12.1 Schematic diagram of the olfactory sensillum of a moth.

Fig. 12.2 Schematic representation of the olfactory mucosa of a rabbit. Three receptor cells are surrounded by supporting cells.

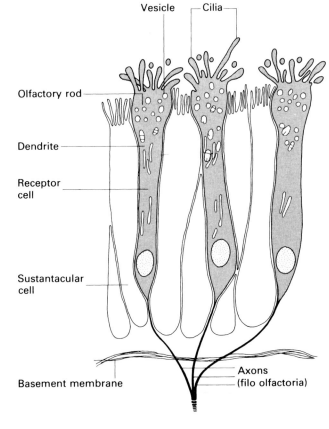

Vesicle — Cilia —
Olfactory rod
Dendrite
Receptor cell
Sustantacular cell
Basement membrane
Axons (filo olfactoria)

Sweet   Sour

Salty   Bitter

Surface of tongue   Sensory hair

Supporting cell   Sensory neurons

Receptor cell

Fig. 12.3 Distribution and structure of human taste buds.

food depends upon both taste and smell. This can be demonstrated easily by asking a person to close his or her eyes and tell the difference between small pieces of apple and onion placed on the tongue. Most people find the discrimination easy if they are allowed to breathe through the nose but impossible if they are asked to hold the nose while taking the test.

There have been many attempts to identify basic odours equivalent to the four basic tastes. The most widely accepted classification is that of John Amoore (1964), illustrated in Table 12.1. While an undergraduate at Oxford University, Amoore noted great dissimilarities in the chemical structure of substances that smelled alike. He examined over 600 organic compounds that

Table 12.1 Primary odours with chemical and more familiar examples

Primary odour	Chemical example	Familiar substance
Camphoraceous	Camphor	Moth repellant
Musky	Pentadecanolactone	Angelica root oil
Floral	Phenylethylmethyl ethyl carbinol	Roses
Pepperminty	Menthone	Mint candy
Ethereal	Ethylene dichloride	Dry-cleaning fluid
Pungent	Formic acid	Vinegar
Putrid	Butyl mercaptan	Bad egg

had well-described smells, and in 1952 he published his stereochemical theory of olfaction, in which all odours are described in terms of the seven different primary odours given in Table 12.1. Amoore's theory classifies chemicals in terms of their molecular shape and size. He postulates seven basic types of receptor, each with characteristic receptor sites into which certain shaped molecules can fit. Thus, camphoraceous-smelling molecules are roughly spherical, while musky-smelling molecules are disc-shaped. Amoore's theory has received some support as a result of subsequent research (Amoore, 1964), but it remains controversial.

Somewhat akin to chemoreception is **thermoreception**. This probably occurs in most animals, but it has been studied in relatively few. Nerve endings sensitive to temperature are known to occur in a variety of insects. For example, the cockroach *Periplaneta* has thermoreceptors on the legs that perceive ground temperature and some on the antennae that perceive air temperature. Fish have thermoreceptors in the skin, lateral line and brain and are very sensitive to temperature changes. Catfish have been shown to respond to changes in temperature of less than 0.1 °C. Many reptiles have a well-developed temperature sense, with thermoreceptors in the brain as well as in the skin. Pit vipers have special pits on the face that are sensitive to infra-red radiation and are shaped so as to give the animal a directional temperature sense.

Birds are thought to have few thermoreceptors in the skin, except on the tongue and bill of some species. In pigeons (Columbidae), there are thermoreceptors in the brain that influence behaviour and plumage adjustment and others in the spinal cord that control shivering and panting. In mammals distinct heat and cold receptors are distributed in the skin. The heat receptors are usually deeper than the cold receptors. There are also receptors deep in the body – in veins, for example – that can initiate shivering even though the temperature at skin and brain receptors is kept constant. Thermoreceptors in the spinal cord influence shivering, panting and blood flow, and these functions are repeated by thermoreceptors in the hypothalamus. In general, the most sophisticated forms of thermoregulation are found in mammals, and the brain receives information from many parts of the body. The integration of this information leads to appropriate activation of various mechanisms of warming and cooling.

12.2 Mechanoreceptors and hearing

Sound results from minute changes in pressure that originate from a vibrating source within a medium such as air or water. The receptors that detect sound are basically mechanoreceptors that show rapid adaptation (recovery) and that are thus sensitive to vibration.

Vibration-sensitive hairs and receptors in the limb joints have been described for many arthropods. Thus, blowflies have receptors in some joints of the antennae, called Johnson's organs. These organs can follow movements up to 500 hertz. In mosquitoes, the same kind of organ signals direction of sound. The backswimmer *Notonecta* locates prey by detecting waves on

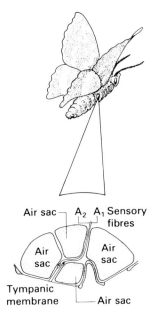

Air sac ⌐ A₂ A₁ Sensory
fibres

Air
sac

Air
sac

Tympanic
membrane └── Air sac

Fig. 12.4 The ear of a
noctuid moth. Vibrations of
the tympanic membrane are
detected by the A₁ and A₂
sensory neurons.

the surface of the water. The receptors are in the legs and are maximally sensitive to vibrations between 100 and 150 hertz. Web-spinning spiders can discriminate live from dead prey by vibrations of the web. Such vertebrates have mechanoreceptors capable of picking up vibrations of the substrate. These have been found in the skin of snakes and the leg joints of cats and ducks (Prosser, 1973).

The auditory systems of animals, though diverse, have certain features in common. For example, there is a peripheral device for converting sound pressure to vibratory motion. Sensory receptors convert this motion into nerve impulses that can be decoded by the CNS. One of the simplest types of peripheral auditory system is found in noctuid moths. There are two ears, each composed simply of a tympanic membrane on the side of the thorax, and two receptor cells that are embedded in a strand of connective tissue, as illustrated in Figure 12.4. This remarkably simple ear enables the moths to hear the ultrasonic cries of hunting bats. By means of an elegant series of experiments, Kenneth Roeder (1963, 1970) showed how this is achieved.

One receptor, called the A_1 cell, is sensitive to low intensity sounds and responds to cries from bats that are about 30 metres away, too far for the bat to detect the moth. The frequency of impulses from the A_1 cell is proportional to the loudness of the sound, so the moth can tell whether or not the bat is approaching. By comparing the time of arrival and intensity of the stimulus at the two ears, the moth can determine the direction of approach. The difference occurs because the moth's body shields the sound from one ear more than the other. The relative altitude of the bat can also be determined. When the bat is higher than the moth, then the sound reaching the moth's ears will be interrupted intermittently by the beating of the moth's wings, but this will not happen when the bat is below the moth.

The A_1 cells give the moth early warning of an approaching bat and may enable it to fly away from the bat before it is detected. By heading directly away from the bat, the moth presents itself as the smallest possible target because its wings are edge-on rather than broadside-on to the bat. It can do this simply by turning, equalizing the sound reaching the bat's two ears. If, however, the bat detects the moth, the moth cannot escape simply by outflying the bat because the bat is a much faster flier. Instead, the moth employs evasive action when the bat comes within two or three metres (Figure 12.5).

The A_2 cell only produces nerve impulses when the sound is loud. It starts responding when the bat is nearby, and its impulses probably cause disruption of the flight control mechanisms of the CNS. The moth consequently flies erratically and drops toward the ground. By means of such evasive action, moths have been observed to escape just as the bats come within striking distance. Thus, Roeder's work on hearing in noctuid moths is a beautiful illustration not only of the workings of a simple ear but also of the fact that an animal's sensory apparatus is often finely tuned to the circumstances of its ecology.

There are many properties of sound to which an animal might respond. As sound travels through a medium, particles move back and forth, creating oscillating pressure waves. The extent or amplitude of the waves determines

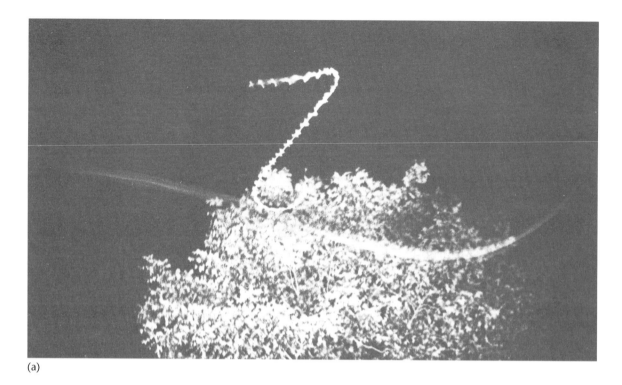

(a)

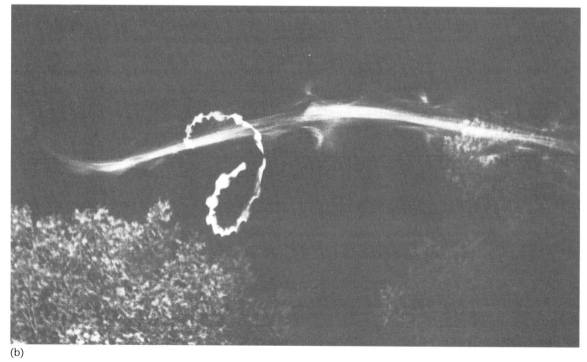

(b)

Fig. 12.5 (a) Streak photograph of a moth escaping from a red bat; (b) streak photograph of a red bat catching a moth (*Photograph: Frederick Webster*).

the intensity or loudness of the perceived sound. The speed of transmission of sound depends upon the density of the medium through which it travels and is independent of the intensity of the sound. In air, sound travels at about 340 metres per second. It travels faster in hot air than in cold air. In water, sound travels about four times as fast as in air.

If we picture sound as a waveform, as in Figure 12.6, the distance between successive peaks, called the **period**, is inversely related to the **frequency**, or number of peaks in a unit of time. Sound frequency is measured in hertz or cycles per second. The simplest sound is a pure tone, which has a single frequency that is perceived subjectively as pitch. Natural sounds are seldom pure tones and are made up of a number of frequencies mixed together. When a complex sound is analysed into its component frequencies, the result is called a **sound spectrum**. A hearing organ may be responsive to a wide range of frequencies, an example being the tympanal organ of a locust that responds to frequencies between 1000 and 100 000 hertz. When a sound

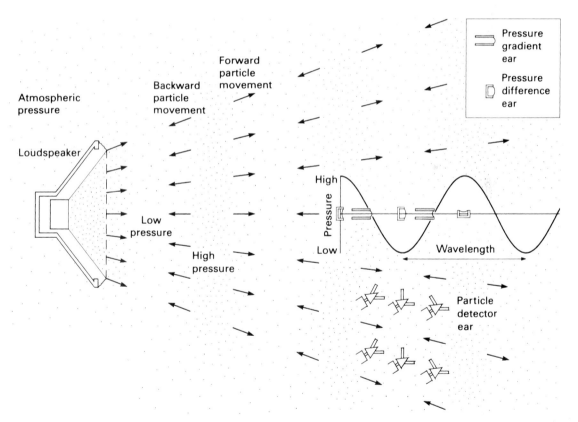

Fig. 12.6 Different types of ear respond to different aspects of sound. Particle-detector ears, found in bees, mosquitoes and some fish, are stimulated by air molecules moving from high to low pressure areas. Pressure difference ears, found in mammals, and some birds, fish and insects, have a sealed chamber that provides a reference pressure, and a membrane that is deformed by changes in the ambient pressure. Pressure gradient ears, found in reptiles, amphibians, and some birds, fish and insects, measure the difference in pressure between the two ends of a tube by means of a membrane in the tube. They are maximally responsive when aligned along the axis of sound transmission (After Gould 1982).

receptor responds to a narrow band of frequencies, it is said to be 'sharply tuned'.

The antennal (Johnson's organ) receptor response of the male mosquito *Aedes aegypti* responds to sounds between 150 and 550 hertz, which corresponds to the wing tone of the female. The wing tone of the male is of a higher frequency and is not detected by the Johnson's organ (Haskell, 1961).

In general, vertebrates' ears are more versatile than those of invertebrates. Thus, few, if any, invertebrates can discriminate between two frequencies unless they happen to have two different kinds of receptor that are tuned to different frequencies (Haskell, 1961). For the vertebrate ear, however, such a discrimination presents no problem. This is partly due to the structure of the ear and partly to the analysing role of the CNS.

The human ear (Figure 12.7), like most mammalian ears, is divided into three parts: the outer ear, the middle ear, and the inner ear. The **outer ear** consists of the **pinna** (ear flap) and the **auditory canal** that terminates in the **tympanic membrane** (eardrum). The **middle ear** consists of a chamber on the inside of the tympanic membrane. This chamber is connected to the **pharynx** (mouth) by the **Eustachian tube**. This arrangement makes possible the equalization of air pressure between the outer and middle ear. Passengers in aeroplanes may experience pain in the ears when they change altitude quickly, as in takeoff or landing. This pain is due to unequal pressure

Fig. 12.7 Diagrams of the human ear, showing the general layout (above), a cross-section of the cochlea (below left), and details of the organ of Corti (below right).

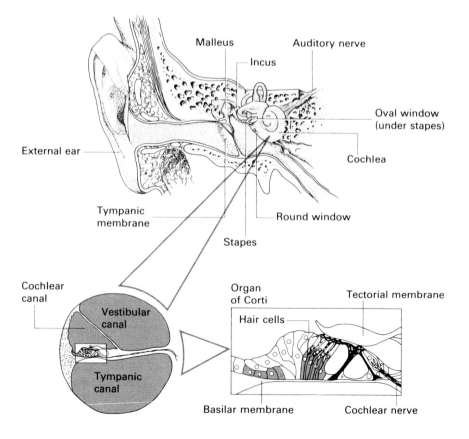

on either side of the tympanic membrane. Passage of air through the Eustachian tube equalizes the pressure, and this is facilitated by chewing, swallowing or yawning. The middle ear contains three small bones – the **malleus** (hammer), the **incus** (anvil) and the **stapes** (stirrup) – that connect the tympanic membrane to another membrane, called the **oval window**, in the wall of the middle-ear chamber.

The **inner ear** is a labyrinth of interconnected fluid-filled chambers and canals. It has two distinct parts: the **vestibular apparatus**, which is concerned with the sense of equilibrium and the **cochlea**, which is a coiled tube that is the organ of hearing. Inside the cochlea are three canals: the **vestibular canal**, which begins at the oval window; the **tympanic canal**, which begins at the round window and connects with the vestibular canal at its other end, and the **cochlear canal**, which lies between the other two canals. The boundary between the cochlear canal and the tympanic canal is formed by the **basilar membrane**, which carries the **organ of Corti**. The organ of Corti carries rows of receptor cells with sensory hairs at their apexes. These hairs project into a gelatinous **tectorial membrane**. Dendrites of sensory neurons terminate on the surfaces of the hair cells, and when vibrations of the basilar membrane cause the sensory hairs to move, deformations of the hairs give rise to generator potentials in the hair cells, and these stimulate the sensory nerves.

Vibrations in the air are caught by the pinna and pass down the auditory canal, causing the tympanic membrane to vibrate at the same frequency. These vibrations are transmitted across the cavity of the middle ear by the three ear bones. The bones are arranged as a lever system that diminishes the amplitude of the vibrations but that increases their force. Moreover, the vibrations of the large tympanic membrane are transmitted to the much smaller oval window. The result is that the sound pressures at the tympanic membrane are magnified by about 22 times at the oval window. This arrangement enhances the detection of faint sounds.

The movement of the membranes of the oval window causes a corresponding movement of the fluid in the cochlea. When the open window bows inward fluid is pushed from the vestibular canal to the tympanic canal, causing the round window to bow outward and to release the pressure in the cochlea. The movement is then reversed during a complete cycle of vibration. The movements of the fluid in the cochlea occur at the same frequency as those of the outside air. As the cochlear fluid vibrates, a travelling wave is set up in the basilar membrane, and as it moves up and down, it deforms the hair cells as they push against the tectorial membrane. As the hair cells are bent they stimulate the sensory neurons.

The point in the cochlear tube with the maximum amplitude of membrane displacement varies with the frequency of the sound stimulus. As early as 1867, Helmholtz had postulated correctly, on anatomical grounds, that the high frequency waves are focused near the base of the cochlea and that low frequency waves have their maximal effect near the apex. Our current understanding of the operation of the cochlea is due to the work of a communications engineer called George von Bekesy (1952, 1960), who was awarded the Nobel Prize for his work. He watched events inside the cochlea by removing the fluid and replacing it with a suspension of coal and powdered aluminium. By reflecting flashes of intense light from this suspension, he was able to observe events like the travelling wave of the basilar membrane. He found that the membrane is stiffer at the base, favouring high frequency vibrations, and less stiff near the apex, favouring low frequencies. Thus, particular frequencies perturb particular parts of the basilar membrane, and each part stimulates different receptors in the organ of Corti.

The nerve fibres of these receptors synapse in the **spiral nucleus**, and the secondary neurons make up cranial nerve VIII. Each signals a particular frequency of sound to the cochlear nucleus of the brain.

The structure of the ear is not the same in all vertebrates. For example, fish and cetaceans (dolphins and whales) have no outer ear, and fish also lack a middle ear with eardrum and ossicles. Since the tissues of fish have about the same density as water, vibrations arriving at the head can pass directly to the inner ear. Some fish, however, have another mechanism that serves a function analogous to that of the inner ear. This is the gas-filled swimbladder that may have a bony connection to the inner ear that provides considerable improvement in hearing ability. In amphibians and reptiles, the tympanic membrane is the outermost part of the ear, but in birds there is an external canal (**auditory meatus**) that leads from the body surface to the eardrum. In birds, a rod-shaped bone, the **columella**, extends to the inner surface of the tympanic membrane and joins the stapes. In amphibians and reptiles, these bones form part of the jaw, though they have some role in hearing in some species.

The lateral line organs of fish and aquatic amphibians are sensitive to vibrations including low frequency sound. They are made up of modified hair sensilla that respond to the flow of water in the lateral line canal or on the body surface.

12.3 Vision

Vision is based upon detection of electromagnetic radiation. The electromagnetic spectrum covers a wide span, of which the visual spectrum is a very small proportion, as illustrated in Figure 12.8.

The energy content of electromagnetic radiation is inversely proportional to the wavelength. The long wavelengths contain too little energy to activate the photochemical reactions that form the basis of photoreception. The short wavelengths contain so much energy that they damage living tissue. Most of the shortwave radiation from the sun is absorbed in the ozone layer of the atmosphere, and it is doubtful that life could have evolved on earth if this had not been the case. All photobiological responses are confined to a narrow band of the spectrum between these two extremes.

Photoreceptor cells contain a pigment that is bleached by the action of light. The bleaching involves changes in the shape of the pigment molecules, and unlike the bleaching with which we are familiar in everyday life, it is reversible. The bleaching process leads to electrical changes in the receptor membrane that are not fully understood (Prosser, 1973).

Fig. 12.8 The electromagnetic spectrum (above) wavelengths in metres, with the visible portion enlarged (below).

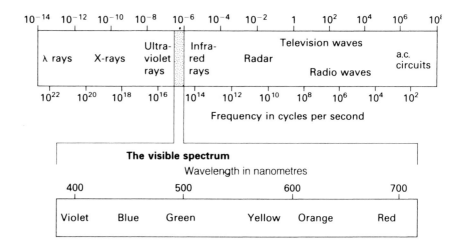

Although photoreceptor cells may be scattered over the body surface, as in the earthworm (*Lumbricus*), they are usually gathered together in clusters. The most primitive type of eye is made up of a cluster of receptors arranged at the bottom of a depression or pit in the skin. Such an eye can give a rough indication of the direction of incident light. As a result of shadows cast by the sides of the pit, light coming from the side will illuminate one part of the pit while the rest remains relatively dark. These differences in illumination can be detected by an array of photoreceptors at the base of the pit, forming a rudimentary retina. The pinhole eye of the mollusc *Nautilus* (Figure 12.9) is a development of the pit eye, the rim having grown inward and the photoreceptor layer forming a retina. The pinhole eye works exactly like a pinhole camera; that is, light from each point reaches only a very small area of the retina so that an inverted image is formed.

The evolutionary development of the eye can be traced among living molluscs, as illustrated in Figure 12.9. From the pinhole eye of *Nautilus*, the eye evolves a protective layer, presumably to exclude dirt. Inside the eye a primitive lens is formed, as in the snail *Helix*. This type of eye is also found in spiders. There are some variations like the eye of the scallop *Pecten* that has an inverted retina and a mirror-like tapetum (see Chapter 13). The eye of the cuttlefish *Sepia* (see Figure 12.9) is very similar to that of vertebrates. The shape of the lens can be changed by ciliary muscles, and there is an iris diaphragm capable of regulating the amount of light falling on the retina.

The eyes of vertebrates, of which the human eye is a good example, conform to a single basic plan, although some ecological adaptation occurs, as we see in Chapter 13. Figure 12.10 shows a horizontal section through the human eye. It is surrounded by a tough coat, the **sclera**, that is transparent at the front of the eye where it is called the **cornea**. Just inside the sclera is a black lining, called the **choroid**, that reduces both transmission of light through the sides of the eye and reflections within the eye. Inside the choroid is the light-sensitive **retina**, which we discuss in detail later. At the front of the eye there is no choroid or retina. A large **lens** divides the eye into **anterior** and **posterior chambers**, which are filled with **aqueous** and

Fig. 12.9 Representative eyes of molluscs. (a) The pinhole eye of the marine mollusc *Nautilus*, (b) the lens-filled eye of the land snail *Helix*, (c) the vertebrate-like eye of the squid *Sepia*, (d) the inverted eye of the scallop *Pecten*.

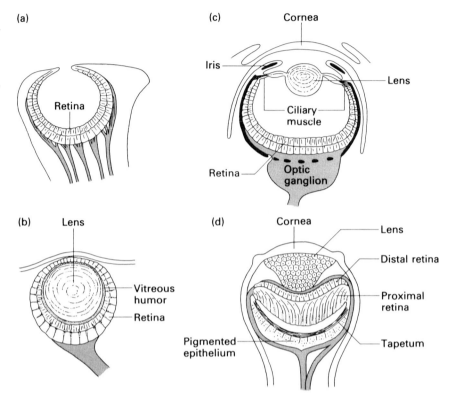

Fig. 12.10 Section through the human eye.

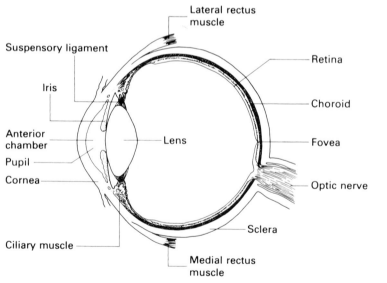

vitreous fluids respectively. In front of the lens is the **iris**, a muscular diaphragm with an aperture called the **pupil**. The iris controls the size of the pupil and the ultimate amount of light entering the eye. The lens is circumscribed by the **ciliary muscle**, which changes the shape of the lens. When the muscle contracts, the lens becomes more convex, bringing close objects into focus on the retina. When the ciliary muscle relaxes, the lens becomes more flattened, bringing more distant objects into focus.

The vertebrate's retina, unlike that of cephalopods like *Sepia*, is inverted. The photoreceptors lie next to the choroid, and light reaches them after passing through layers of neurons, mostly **ganglion cells** and **bipolar cells**. The ganglion cells are adjacent to the vitreous fluid, and their axons pass over the inner surface of the retina to the **optic disc** or blind spot, where they form the **optic nerve** and leave the eye. The bipolar cells are neurons that link the ganglion cells to the photoreceptors, as indicated in Figure 12.11.

The photoreceptors are of two types, called **rods** and **cones**. The rods are more elongated than the cones. They are very sensitive to low levels of illumination and have only one type of photopigment, called **rhodopsin**. Rod vision is therefore colourless. It also provides poor definition (visual acuity) because many rods are connected to a single ganglion cell. Consequently, a single fibre in the optic nerve receives information from

Fig. 12.11 Organization of the primate retina (After Dowling and Boycott, 1966).

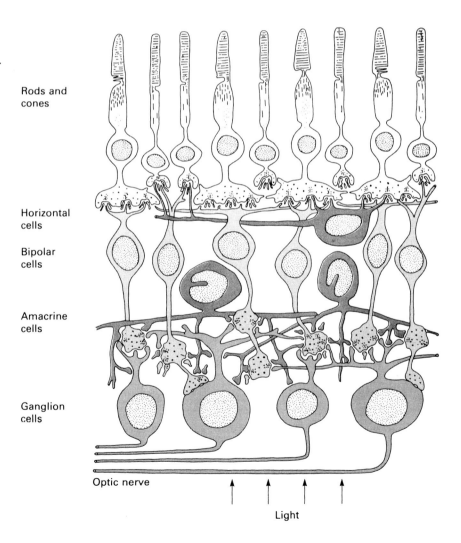

Rods and cones

Horizontal cells

Bipolar cells

Amacrine cells

Ganglion cells

Optic nerve

Light

many rods, an arrangement that increases sensitivity at the expense of acuity. Rods are predominant in nocturnal species where sensitivity is at a premium.

The cones are maximally sensitive to high levels of illumination and provide sharp vision. Only a few cones report to each ganglion cell. Cones may be of more than one type, the photopigments absorbing in different parts of the spectrum. Cones thus provide the basis for colour vision. Cones are most sensitive to the wavelengths that are most strongly absorbed by their photopigments. Vision is **monochromatic** when there is only one active photopigment. This is the case with twilight vision in humans when only the rods are operating (Figure 12.12).

Dichromatic vision can occur when two photopigments are active, as in the grey squirrel (*Sciurus carolinensis*) (Figure 12.12). Each wavelength stimulates the two types of cone in a specific ratio, in accordance with their relative sensitivities in that part of the spectrum. Provided that the brain can recognize these ratios, it should be possible for the animal to discriminate wavelength from intensity. Note, however, that particular ratios occur in more than one part of the spectrum and that some wavelengths therefore will be confused. This also occurs in certain forms of colour blindness in humans. The wavelength that stimulates both types of cone equally (where the absorption curves cross) is indistinguishable from white light and is known as the 'neutral point' of the spectrum. Its existence has been demonstrated behaviourally in the grey squirrel (Muntz, 1974).

Confusion is reduced in visual systems with three receptors, or **trichromatic vision** (see Figure 12.12), which is found in many species including humans. Some confusions still occur, and it is possible, for example, to give the impression of any particular colour by various mixtures of three mono-

Fig. 12.12 Typical receptor mechanism for different types of colour vision (From *The Oxford Companion to Animal Behaviour*, 1981).

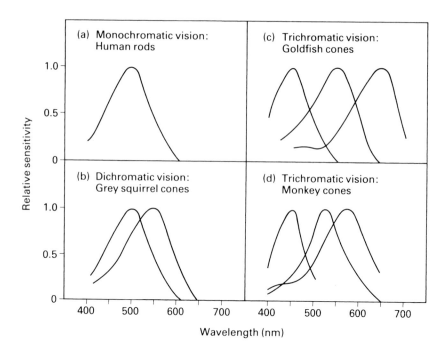

Fig. 12.13 Example of the combination of pigment and oil droplet in the cone of a bird (From *The Oxford Companion to Animal Behaviour*, 1981).

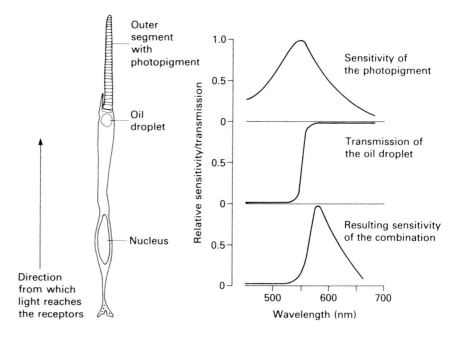

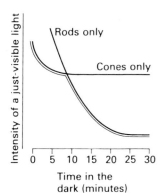

Fig. 12.14 Graph showing how the intensity of a just-visible brief light decreases as we adapt to darkness (double line). The steep single line shows what happens when there are only rods in the retina. The shallow single line shows what happens when only cones are illuminated (i.e. when the light is restricted to the fovea) (From *The Oxford Companion to Animal Behaviour*, 1981).

chromatic lights, suitably adjusted in intensity and saturation. The visual perception of colour photography and colour television would be impossible if this were not the case.

More than three receptor types are found in many birds and reptiles. In addition to various photopigments, the cones of these animals often contain coloured oil droplets that act as filters and combine with the photopigment to determine the spectral sensitivity of the receptor. The oil droplets usually are not distributed evenly over the retina but are concentrated in certain areas, as illustrated in Figure 12.13.

In 1825, the Czech physiologist Jan Purkinje noted that red flowers seem brighter than blue flowers in the daytime but that as twilight falls, they appear to fade before the blue flowers do. This change in the spectral sensitivity of the eye, called the **Purkinje shift**, was shown by Schultz in 1866 to be due to a change from cone to rod vision during **dark adaptation**. The change in sensitivity during dark adaptation can be measured, in humans, by determining the threshold of just-visible light at intervals after entering a dark room. During dark adaptation, this threshold falls progressively, as illustrated in Figure 12.14. The kink in the curve is due to the change from cone to rod vision. The contribution of the cones can be determined by shining a very small light on the **fovea**, which has no rods. The contribution of the rods can be seen in individuals, called rod monochromats, who have no cones in the retina, a rare condition. As we can see from Figure 12.14, the rods are much more sensitive to light than the cones. However, they contain only the photopigment rhodopsin, the sensitivity maximum of which is toward the blue end of the spectrum. This is why blue objects appear to be brighter than objects of other colours during twilight vision.

The range of light intensity over which the vertebrate eye can operate is enormous; there is a billion-fold change in sensitivity. This is achieved by various mechanisms, which vary from species to species. In many fish, amphibians, reptiles and birds, choroid pigment migrates between the outer segments of the receptors during high illumination and retracts as light levels fall. The outer segments of the cones may also move in these animals. In some fish and amphibians, the outer segments of the rods also move in the opposite direction. The level of light reaching the retina can be controlled by contraction of the pupil. This reflex is well developed in eels and flatfish, nocturnal reptiles, birds and mammals (Prosser, 1973).

To produce a sharp image on the retina, the light passing through the eye must be refracted so as to focus at the retina. The light entering the eye is refracted at the cornea and at the lens. In the human eye the cornea provides about twice the refraction of the lens.

The problem is that the cornea is a fixed distance from the retina, and thus, some accommodation is required if objects at various distances are to be brought into focus. The necessary adjustment is provided by the lens. In fish the lens is nearly spherical, with a high refractive index and short focal length. This is necessary because the refractive index of water is about the same as that of the cornea, so that no refraction occurs at the surface of the eye. The lens has a fixed shape, and accommodation is achieved by altering the distance between the lens and the retina. In land vertebrates, accommodation is controlled by the ciliary muscles that alter the shape of the lens. When focusing on nearby objects, the lens is rounded, and when focusing on distant objects it is flattened, as illustrated in Figure 12.15. Animals that live both in and out of water cannot see well in both media. The eyes of the frog, crocodile and hippopotamus are placed high in the head so that the animal can see over the surface of the water while keeping the body submerged. In the so-called four-eyed fish (*Anableps anableps*), each eye is divided into two by a band of skin. The upper portion of the eye protrudes out of the water while the lower portion remains submerged. A single lens in each eye is oval and is shaped in such a way that the lower eye can focus underwater while the upper eye can focus on objects above the surface of the water.

Fig. 12.15 Focusing the eye. The lens is flattened for viewing distant objects (a), and is more rounded for viewing nearby objects (b).

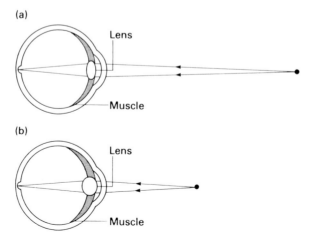

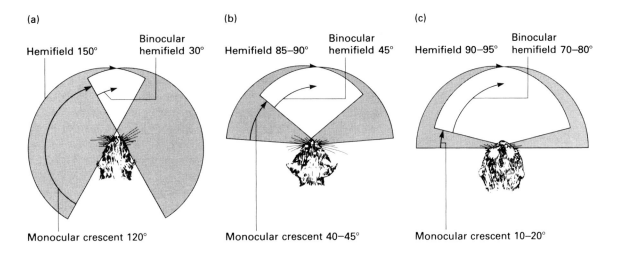

(a)

Hemifield 150°

Binocular hemifield 30°

Monocular crescent 120°

(b)

Hemifield 85–90°

Binocular hemifield 45°

Monocular crescent 40–45°

(c)

Hemifield 90–95°

Binocular hemifield 70–80°

Monocular crescent 10–20°

Fig. 12.16 Visual fields in (a) the squirrel, (b) the cat, and (c) the night monkey (After Kaas *et al.*, 1972).

The eyes have a field of view that is largely dependent upon their position in the head. In vertebrates, the field of each eye is about 170 degrees (Duke-Elder, 1958). Species vary in the degree of overlap of the fields of view of the two eyes. In general, predatory animals have a large overlap at the front of the head and a blind area behind, whereas prey animals have little overlap and a smaller blind area, as illustrated in Figure 12.16.

Where the two visual fields overlap, **binocular vision** is possible. The advantage of binocular vision is that it permits more accurate depth perception and judgement of distance than monocular vision. This is an important attribute for animals that rely on such information in capturing their prey. The advantage of a wide field of view is that movements can be detected readily, even if they occur behind the animal. This is obviously important for animals that need to be vigilant for approaching predators.

During visual examination, when acuity is important, the image is brought into focus on the fovea. In order for both eyes to focus, there must be some convergence of the two lines of sight. The nearer the object under scrutiny, the greater the necessary convergence. The direction of the two lines of sight is adjusted, via the **extraocular muscles**, until the two retinal images coincide and a single image is registered by the brain. If, at the same time, the degree of convergence of the two eyes is noted by the brain, then information about the distance of the object is available. It is not possible, however, for the retinal images of nearby objects to coincide exactly. Depending upon the distance between the eyes, there will be a difference between the positions of the two images. This retinal disparity also provides important information about the distance of objects. The judgement of distance and depth is a complex matter, involving many cues in addition to those provided by convergence and retinal disparity.

Although the basic layout of the eye is the same for all vertebrates, there are a number of variations and specializations, some of which are discussed in Chapter 14. Particularly relevant to this discussion, however, is the variation in foveae. Humans have a single fovea that is placed centrally and that is circular in shape. The human fovea contains only cones, and it is the part of the retina that provides greatest acuity of vision. In the cheetah and many birds, the fovea is elongated in the horizontal plane. This band type of fovea seems to be associated with animals that live in open country or that fly over the sea (Meyer, 1977; Hughes, 1977). Arboreal mammals such as the cat and squirrel have a disc-shaped fovea, as do nocturnal mammals such as the hedgehog and mouse. For such animals, the vertical direction is likely to be as important as the horizontal (Hughes, 1977).

Most birds have some kind of fovea, and about half of those examined have more than one (Meyer, 1977). Many birds have one fovea that serves binocular vision and another that serves the lateral field of view. The adaptive significance of the various arrangements of foveae in birds is not understood fully. Their main functions are probably connected with the complex visual tasks connected with flight, especially those involved in taking prey on the wing and in landing (Lythgoe, 1979).

12.4 Sensory judgements

Psychologists have traditionally made a distinction between sensation and perception. **Sensations** are the basic data of the senses, the raw material from which knowledge is gained. Red and blue are examples of colour sensations. **Perception** is a process of interpretation of sensory information in the light of experience and of unconscious inference. For example, in looking at a landscape we use colour as a guide to judging the distance of hills and other features of the terrain. This is well known to landscape painters who typically use blue shades to give an impression of distance.

For more than a hundred years, psychologists have studied perception in humans intensively. This work has been done mainly in the laboratory and usually by asking subjects to respond verbally to various perceptual situations. The study of sensation and perception in animals is hindered by the fact that they cannot speak to us and that we have difficulty imagining how experimental situations are interpreted by animals. Nevertheless, many of the methods developed by human psychologists have application in the study of animal behaviour.

Gustav Fechner (1801–1887) believed that sensations cannot be measured directly because they are experiences that are entirely private. The magnitude of a physical stimulus can be measured, and a person can be asked to rate the magnitude of the received sensation. The problem is that if two people give different ratings to the same range of stimuli, we cannot tell whether they received different sensations or whether they merely rated them differently. Fechner saw that, although sensations cannot be compared to physical stimuli, they can be compared to each other. A person can compare two sensations and judge whether they are the same or different. One way of doing this is to alter progressively the intensity of a stimulus presented to a subject in an experiment until the subject reports that he or she can detect a change. The difference between the original intensity and the repeatedly changed intensity is called the **just-noticeable difference**, a physical quantity that can be measured over a wide range of physical stimulus intensity. For example, a subject might be asked to compare two weights repeatedly. One would be a standard weight – say, 100 grams – and the other would be taken from a series of comparison weights. The subject would report on whether the latter was heavier, equal to or lighter than the former. From repeated tests of this type, the experimenter would gain an estimate of the uncertainty interval within which the subject was not sure whether or not the test weight was different from the standard. From this testing, the experimenter would obtain an amount called the **difference threshold**, which is the extent to which the stimulus intensity has to be altered to produce an altered sensation.

The German physiologist E.H. Weber (1834) proposed that the difference threshold is a constant ratio of the standard stimulus. In other words, if a change in weight from 100 to 110 grams were the minimum detectable difference, then Weber's hypothesis suggests that the difference between 10 and 11 grams, also a ten per cent change, should be just as detectable as well. If I is the magnitude of the standard stimulus and I is the increment in stimulus intensity required to produce just-noticeable difference, then:

$I/I = $ constant

This is known as **Weber's Law**, and the fraction I/I is called the **Weber fraction**.

Fechner (1860) believed that Weber's Law was the key to the measurement of subjective experience. He and his co-workers carried out numerous experiments to test the law. In general, the law holds true for the normal range of stimulus intensity but tends to break down at the two extremes of the intensity range. Fechner was interested particularly in the relationship between the physical intensity of the stimulus and the subjective magnitude of the resulting sensation. He proposed the following formulation, which has become known as **Fechner's Law**:

$S = \text{k} \log I$

where S is the subjective magnitude, I is the stimulus intensity, and k is a constant. An interesting result of this approach is that different sensory modalities have different typical Weber fractions, as illustrated in Table 12.2. The smaller the fraction, the keener the sense. Thus, the discriminating power of vision is the greatest, and that of taste is the least of the human senses tested.

Table 12.2 Representative (middle-range) values for the Weber fraction for the different senses

Sensory modality	Weber fraction ($\Delta I/I$)
Vision (brightness, white light)	1/60
Kinesthesis (lifted weights)	1/50
Pain (thermally aroused on skin)	1/30
Audition (tone of middle pitch and moderate loudness)	1/10
Pressure (cutaneous pressure)	1/7
Smell (odour of India rubber)	1/4
Taste (table salt)	1/3

This entire line of thought and research is known as **psychophysics**. Various alternatives to Weber's and Fechner's Laws have been proposed, and some researchers claim that a **power law** offers a more accurate interpretation of the data (for a review, see Kling and Riggs, 1971). However, there have been more fundamental objections to the classical psychophysical approach.

The classical psychophysical experiment requires the subject to respond in a yes–no manner. Its interpretation assumes that there is a fixed criterion for saying 'yes I do' or 'no I do not' detect a change; the subject is assumed to have no response bias. The early investigators were aware that the subject's attitude could alter the judgments made in an experiment. They attempted to eliminate the problem by using only carefully trained subjects or by introducing an occasional catch trial and making allowance for false response. A fundamental improvement was obtained by the introduction of methods derived from the theory of signal detectability, originally developed to cope with problems in radio and telephone communications (Swets *et al.*, 1961).

The theory of signal detectability, usually called **signal-detection theory**, assumes no fixed yes–no criterion. It maintains that whether or not a subject says yes depends upon the effect of the stimulus in relation to the prevailing variation (noise) in the same stimulus dimension, the expectation of the observer and the potential consequences of the observer's decision. For example, suppose a subject is required to say whether or not a test sound is louder than a standard sound. The four possible outcomes are illustrated in Table 12.3. Suppose we decide to reward the subject with a payment of 25 cents for each correctly detected loudness difference (hit), but we impose a fine of 25c for failure to detect a difference (miss). We also pay 10c for each correctly identified non-difference (correct negative) and impose a fine of 5c for claiming a difference when there was none (false alarm). The consequences of the subject's responses can now be represented as a payoff matrix, as shown in Table 12.4. Suppose the subject works for 100 trials, on half of which there is a genuine loudness difference and on the other half there is no difference. By consistently saying yes the subject would earn $12.50 by

Table 12.3 The four possible outcomes of a stimulus detection experiment

		Response Yes	No
Stimulus	On	Hit	Miss
	Off	False alarm	Correct rejection

Table 12.4 Payoff matrix. The dollars earned and lost as a result of responding yes and no when the stimulus is on or off in 100 trials

		Response Yes	No
Stimulus	On	$12.50	−$12.20
	Off	−$2.50	$5.00

being correct 50 times and would be fined $2.50 for being incorrect 50 times. By consistently saying no the subject would earn $5 by being correct 50 times and would be fined $12.50 for the 50 incorrect responses. Thus, by consistently responding positively the subject would gain $10, and by consistently responding negatively the subject would lose $7.50. It is not surprising that, with this particular payoff matrix, the subject would be biased toward saying yes.

Another form of bias is generated by the subject's expectations. For example, Linker *et al.* (1964) found that in a taste experiment, the tendency to respond positively depended partly upon how often the stimulus was presented. When the stimulus was presented on 90 per cent of occasions, they obtained a high proportion of hits and a high proportion of false alarms. When the stimulus was presented on only ten per cent of occasions, they obtained low proportions of hits and false alarms.

Signal-detection theory assumes that there is no such thing as a zero stimulus. The subject is assumed to know that sensory events are contaminated by background noise. The task is not, therefore, simply to say whether or not the stimulus is present but to discriminate between noise plus signal and noise alone. Figure 12.17 illustrates the manner in which this is done. The decision criterion is set by the subject's response bias, which in turn is determined by the payoff matrix. Thus, one way of testing the theory is to hold the stimulus situation constant and to manipulate the payoff. Another kind

Fig. 12.17 The basis of signal detection theory: (a) shows the relative frequency of sensations of various magnitudes when noise alone is present; (b) shows a similar normal frequency distribution for signal plus noise; (c) shows that when (a) and (b) are combined, the boundary between 'no' and 'yes' responses depends upon the position of the decision criterion.

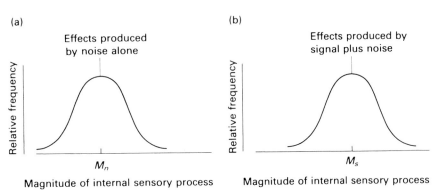

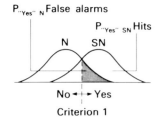

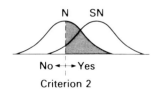

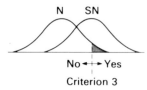

Fig. 12.18 Hypothetical effects of different decision criteria in signal detection theory. The probability of 'yes' responses is given by the area to the right of the criterion, and the probability of false alarms by the shaded areas. It is assumed that stimulus and noise remain unaltered.

of test is to maintain a constant response bias and to alter the stimulus properties. The effect of this kind of experiment on response bias is fairly reliable. When the subject is more conservative, there is a reduction in both the false alarms and hits, and vice versa (see Figure 12.18). When the stimulus magnitude is increased, the number of false alarms diminishes and the number of hits increases.

Psychophysical methods have been applied to a number of aspects of animal behaviour. The first study of this type seems to be that of Donald Blough (1955), who investigated dark adaptation in pigeons. Blough trained pigeons to peck at two response keys just below an illuminated window. The only light in the apparatus came from this source. The birds were trained to peck key A when the stimulus was visible to them and to peck key B when it was not visible. Pecks at key A automatically resulted in dimming of the stimulus, while pecks at key B increased the brightness of the stimulus. Once the bird has been trained, it is then tested in the following way.

Initially, the stimulus window is brightly lit and the bird pecks only at key A. Each peck makes the stimulus less bright, but the bird continues to peck key A until the stimulus becomes so dim that it is below the pigeon's absolute threshold. At this point the bird begins to peck key B, which causes the stimulus to become brighter. When the stimulus again becomes visible to the bird, it returns to key A. Over the course of an hour, the bird alternately pecks keys A and B and the stimulus brightness fluctuates in the region of the bird's absolute threshold. The effect of this procedure is to trace out the pigeon's **dark adaptation curve**. At first the bird is light adapted, but as the stimulus becomes dimmer, it starts to be dark adapted. Thus, the pigeon's threshold progressively changes in the manner illustrated in Figure 12.19. Notice that this dark adaptation curve exhibits the characteristic kink that occurs at the change from cone to rod vision (see Figure 12.14).

12.5 Stimulus filtering

Anatomical and physiological investigations of sense organs and of associated parts of the nervous system can yield valuable information about an animal's sensory capabilities. However, they do not on their own provide conclusive evidence of what an animal does or does not perceive, and behavioural confirmation is usually desirable. Moreover, the demonstration that certain sensory information is available to the CNS does not tell us how it is used.

Much more information is potentially available to an animal than it could possibly register and respond to. In some way the animal has to be selective, to respond to those events that are important and to ignore others. The term for this phenomenon is **stimulus filtering**, the idea being that, at various stages in the causal chain between stimulus and response, certain stimuli are filtered out and do not influence the animal's behaviour.

In a sense, some filtering is inherent in the limited capabilities of the sense organs. For example, the human ear does not respond to sounds with a frequency higher than 20 kilohertz. The human eye filters out the infra-red and

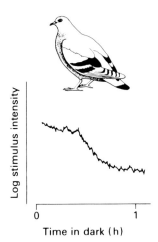

Log stimulus intensity

Time in dark (h)

Fig. 12.19 Dark adaptation curve of a pigeon obtained by a psychophysical method (After Blough, 1955).

ultra-violet parts of the spectrum, although these are detectable by some other animals (see Chapter 13). An interesting example comes from work on the tree frog *Eleutherodactylus coqui*. This frog is named after the distinctive **co-qui** call of the male, which serves to attract females and to repel males. The ear membrane is tuned differently in males and females (Navins and Capranica, 1976). The males hear only the co note, and the females hear only the qui part of the call. Similarly, insect receptors are often highly specialized, responding only to a narrow range of stimulation, as in the case of olfaction in the silkworm moth (*Bombyx mori*).

Once a stimulus has been detected, it may be automatically categorized in a way that filters out other aspects of the stimulus. For example, Jerry Lettvin *et al.* (1959) showed that the photoreceptors of the frog's retina are connected together to form a receptor field, as illustrated in Figure 12.20. Some of these, which they termed bug detectors, respond preferentially to small, dark, moving objects. When feeding, the frogs respond more to such objects than to other stimuli. Some kind of selective responsiveness is occurring here, though it now appears that retinal mechanisms are not exclusively responsible for prey detection (see Chapter 13.3).

Selective responsiveness is common among animals. An animal responding in a particular situation utilizes only part of the potential information available. For example, David Lack (1943) observed that male European robins (*Erithacus rubecula*) often attack other red-breasted robins that trespass on their territory. They will also attack a stuffed robin placed in the territory but only if it has a red breast. It appears that the red breast is a powerful stimulus for labelling another robin as an intruder. Indeed, Lack showed that a territory owner would vigorously attack a bunch of red feathers, just as if it were an intruding robin. No one doubts that a robin is capable of distinguishing between a bird and a bunch of feathers, but in the territorial defence situation robins appear blind to all attributes except the red breast. This type of stimulus is called a **sign stimulus**.

Sign stimuli may change with changes in the animal's internal state. Thus, herring gulls (*Larus argentatus*) will steal and eat another gull's eggs. The raiding gull recognizes the eggs by their shape. For the incubating gull, however, the size and coloration of the egg are most important in retrieving eggs that have rolled from the nest, and shape is relatively unimportant. Thus, the

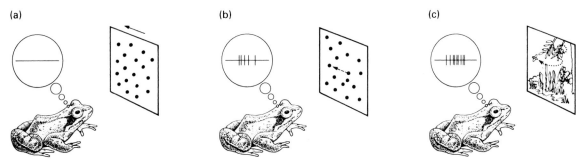

(a)   (b)   (c)

Fig. 12.20 The bug detectors of a frog are unresponsive to movement of a field of dots (a) but are responsive to movement of a single dot (b), especially if the movement is irregular (c).

birds will readily retrieve spherical or cylindrical objects of the same size and coloration as a normal egg. Once the bird has settled down, however, shape becomes important again, and if the eggs do not have rounded edges, the birds will not settle down to incubate (Baerends and Drent, 1970). Thus, the eggs provide three different sets of sign stimuli appropriate to three different kinds of behaviour.

Another interesting example comes from the work of Gordon Burghardt and his co-workers on the feeding behaviour of garter snakes (*Thamnophis*) and water snakes (*Natrix*). These aquatic snakes live in rivers and ponds in the United States and feed on small fish and worms. They detect their prey by taste and smell and have chemoreceptors located in paired pits in the roof of the mouth, called Jacobson's organ. The snake flicks out its tongue, picks up chemicals from the prey and then places the tip of the tongue in the taste pits. Burghardt tested various species of garter snake with cotton swabs containing solutions from extracts of prey such as fish, frogs, salamanders and worms. He found that the different species had strong preferences for the type of prey on which they would normally feed in nature. These preferences were present in newborn snakes that had never fed, and they could not be changed by manipulation of the mother's diet or by force feeding the young snakes on artificial food (Burghardt, 1970). The preferences do change, however, when the snakes switch to another prey. Thus, snakes with a natural preference for minnows would prefer goldfish once they had become used to feeding on them (Burghardt, 1975). Apparently the chemical characteristics of certain prey act as sign stimuli toward which the snake has a genetically determined bias which is modifiable as a result of experience.

While some sign stimuli may be the result of peripheral filtering, it is evident that most must be due to events within the CNS. The early ethologists postulated the existence of an **innate releasing mechanism** (IRM), which they thought was responsible for the recognition of sign stimuli. Nowadays, most ethologists agree that some kind of central filtering mechanism is responsible for the fact that many species show preferential responsiveness to sign stimuli (see Chapter 13.3).

Ethological research has also drawn attention to a type of stimulus filtering that is transitory compared with the relatively permanent recognition of sign stimuli. This is the **searching image** concept first proposed by Jakob von Uexkull (1934). We are all aware of the perceptual phenomenon of suddenly seeing something that we previously overlooked. When we look at a photograph of camouflaged insects we may not see any at first, but then we suddenly may see one. Thereafter, the insects seem easy to pick out. We have developed a searching image for the insect.

A number of studies demonstate the existence of such searching images in animals. Harvey Croze (1970), for example, trained carrion crows (*Corvus corone*) to search for bait hidden under mussel shells of various colours that were widely distributed over the ground. The crows had to turn over each shell to see if there was any food underneath. They tended to concentrate their search on one colour at a time, overlooking shells of other colours even though they had experience of gaining food under all types of shells. A single sample, handed by the experimenter, was sometimes enough to trigger the search for a particular colour of shell. The crows behaved as if they had a searching image for a particular colour of shell that lasted for a certain amount of time but that could be readily switched to another colour.

A most thorough study of searching images is that of Marian Dawkins (1971a, b) on domestic chicks in a laboratory situation. The chicks were fed for three weeks on either orange- or green-dyed grains of rice on a white background. Then they were tested for their ability to detect grains on a background of a different colour (conspicuous grains) and on a background of the same colour (cryptic grains), as illustrated in Figure 12.21. Dawkins found that although the chicks at first were unable to detect the cryptic grains, they showed a marked improvement with experience. She also found that the experience of eating conspicuous grains decreased the chicks' ability to see cryptic food. It appears, therefore, that the ability to detect cryptic food cannot be due simply to improvement with experience.

In further experiments, Dawkins tested chicks that already were feeding on particular sample grains (see Figure 12.22). She found that when the sample grains were conspicuous, the chicks were more responsive to test grains that were distinguishable by colour. When the sample grains were cryptic, the chicks responded more to test grains distinguishable by non-colour cues such as texture and shape. These results can be interpreted in terms of selective attention. Thus, when the sample grains were conspicuous by virtue of their colour, the chicks were attending primarily to colour and consquently found it easier to detect test grains on the basis of colour. When the sample grains were cryptic, they were exactly the same colour as the background, and the chicks attended to non-colour cues. As a result, they found it easier to detect test grains on the basis of cues other than colour.

An animal may be able to discriminate between two stimuli, but it may not normally do so until it has learned to tell the difference. Such learning is

Fig. 12.21 Rice presented to chickens on various backgrounds: (a) green rice on a green background, (b) green rice on an orange background, (c) orange rice on an orange background, (d) orange rice on a green background (From Marian Dawkins, 1971a).

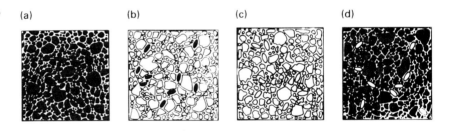

(a) (b) (c) (d)

Fig. 12.22 Choice between two cryptic orange grains (top left and bottom right of test square) and two conspicuous green grains, amid conspicuous orange sample grains (left) and amid cryptic orange sample grains (right) (From Marian Dawkins, 1971b).

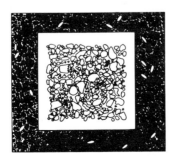 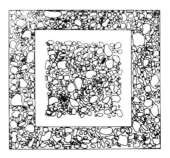

called **discrimination learning**. Appropriate stimuli for investigating discrimination learning in rats would be a black rectangle versus a white rectangle. It is usual to present these on a grey background, and the experiment would incorporate all the normal controls discussed here. Instead of limiting the animal to a single cue, which is essential when investigating sensory capabilities, we might study how the availability of two ways of solving the problem influences the learning process. Thus, the rectangles could be of different sizes or of different orientations.

Let us consider the case of brightness and orientation differences, so that if the black rectangle is presented vertically, the white one is horizontal. Now we have to decide which stimulus is to be rewarded and which is not. Here we run into the problem of confounding the rat's preferences and learning ability. Thus, it may be that rats find it easier to learn about brightness than about orientation but that they also prefer black to white. If we reward the black stimulus, the rat already would be biased toward the correct choice and would learn the problem more easily than if we rewarded the white stimulus. Thus, the rat's initial preferences could easily confound the investigation of its learning abilities. One way of overcoming this difficulty is to divide the twenty rats into two groups of ten, one rewarded for choosing black and the other rewarded for white. However, the same problem arises with respect to the orientation cue, so each group should be divided again into halves, one of which is rewarded for choosing the vertical rectangle and the other for choosing the horizontal rectangle. Table 12.5 shows that, with respect to the configuration of the rewarded stimulus, the experiment has a balanced design.

To solve this type of learning problem, rats have to learn which stimulus to approach to obtain reward and which to avoid. However, reward could be associated with many characteristics of the stimuli. Thus, the rats could learn to associate reward with the shape, size, position, orientation, colour or brightness of the stimuli. In the present case, the rats could never solve the problem on the basis of shape, size, position or colour because the rewarded and unrewarded stimuli do not differ in these respects, whereas they do differ in brightness and orientation. But how are the rats to discover which stimulus characteristics to associate with reward?

Stuart Sutherland and Nick Mackintosh (1971) have shown that many animals have to learn two things in solving problems of this type: (1) which aspects of the stimuli to attend to, and (2) which of the two manifestations of that aspect is rewarded. For example, a rat that was rewarded on a trial in which it had attended to the brightness of the stimuli would be more likely

Table 12.5 Design of discrimination experiment

Group	Number of rats	Orientation of stimulus	Colour of stimulus
1	5	H	B
2	5	V	B
3	5	H	W
4	5	V	W

B = black; W = white; H = horizontal; V = vertical.

to attend to brightness in the future; and if the rat happened to choose the black stimuli, it would be more likely to choose black on those future trials during which it attended to brightness. Rats that learned to attend to brightness or orientation or both could learn successfully to solve the problem outlined here, whereas rats that attended to other aspects of the stimuli, such as size or position, would be able to obtain only 50 per cent reward. Therefore, the rats attending to brightness or orientation or both will always be more rewarded.

To determine which aspects of the stimuli the rats have in fact learned to attend to, unrewarded **transfer tests** are given. During transfer tests, the rats are presented with stimuli differing in orientation or brightness, but not both. Thus, on half the trials the stimuli are black or white rectangles differing only in orientation, and on the other half the stimuli are horizontal or vertical rectangles differing only in brightness. Rats that have learned to attend to orientation only are able to solve the first problem but not the second, while those that had learned to attend to brightness only are able to solve the second but not the first problem. Rats that have learned to attend to both brightness and orientation are able to solve both types of transfer test.

Overall, the evidence strongly suggests that selective attention occurs in many animal species. In many cases the alternative cues are embodied in the same objects, and this means that the selective attention must be controlled centrally and not simply due to the orientation of the head or some other peripheral phenomenon. Selective attention is a type of stimulus filtering in which aspects of the stimulus situation, although registered by the animal's sense organs, are not registered in a way that can influence the animal's learning or behaviour.

There is little difference in principle between the searching image concept investigated by Dawkins (1971) and the selective attention concept proposed by Sutherland and Mackintosh (1971). Indeed, if the theories of selective attention are correct, we would expect to find circumstantial evidence for it in the behaviour of animals in their natural environment. The ecologist Luke Tinbergen (1960) noticed that great tits (*Parus major*), hunting for prey in pine forests in the Netherlands, did not take a new type of food as soon as it became available. Even if the new prey appears gradually, it is often ignored for several days, and then there is an abrupt change in the predation level. Tinbergen postulates that the new prey was overlooked until the birds had formed an appropriate searching image. The search image concept remains controversial among animal behaviourists (see Guilford and Dawkins (1987) for a review).

12.6 Visual recognition

In animals, as in humans, we may describe sensory abilities in terms of the physical attributes of the stimuli, but this does not tell us about the animal's subjective assessment of the stimulus. It is important for ethologists to know what importance animals attach to the various aspects of a stimulus situation. The most successful attempts to do this have been those in which the animal is persuaded essentially to reveal its internal standards of comparison among stimuli. As an example, we first consider the work of Gerard Baerends and Jap Kruijt (1973) on egg recognition in herring gulls. These birds lay three eggs in a shallow nest on the ground, as shown in Figure 12.23. Herring gulls will retrieve dummy eggs provided they resemble real eggs in certain respects.

Baerends and Kruijt set out to discover the features by which herring gulls recognize an egg in this type of situation. They removed two eggs from the nest and placed two dummy eggs on the rim of the nest. The dummies were of various sizes and could differ in colour, pattern or shape. The size of a dummy egg was measured in terms of the area of its maximal projection. This is the maximal shadow that can be obtained when the egg is turned around in a parallel beam of light. All the dummies used in these tests were measured and their sizes expressed in terms of a standard size series ranging from 4 to 16. A real herring gull egg would normally be about size 8.

Fig. 12.23 (a) A herring gull nest containing the typical three eggs. (b) A herring gull retrieving an egg into its nest in preference to a dummy block-shaped egg (*Photograph: Gerard Baerends*).

(a)

(b)

When two identical dummy eggs are placed on the rim of the nest, the gull usually has a marked position preference, retrieving the left before the right or vice versa. To overcome this positional bias, a series of tests were run in which position was titrated against size (Baerends and Kruijt, 1973). The procedure is illustrated in Figure 12.24. If the bird tends to respond to the egg on its right in preference to the one on its left when both eggs are the same size, the experimenter places a larger egg on the left such that the bird is equally motivated to respond to either. By this means, the position preference is given a value r that is the ratio of the sizes of the dummies on the nest rim. Once r has been established for a particular bird, it is possible to introduce a test dummy that differs from a real egg in shape or coloration. Figure 12.24 shows that the titration procedure can yield accurate r values for any test dummy.

The results of large numbers of titration tests are given in Figure 12.25. Four types of dummy egg – green-speckled, plain green, plain brown and cylindrical brown-speckled – are compared with standard-size series of dummy eggs with the same brown-speckled coloration as real eggs. Apparently, the gulls prefer a green background colour to the normal brown background. They also prefer eggs with speckles to eggs without speckles. More important, however, is the finding that the difference among eggs of different sizes remains the same when other features like coloration are changed. This means that each feature adds a specific contribution that is

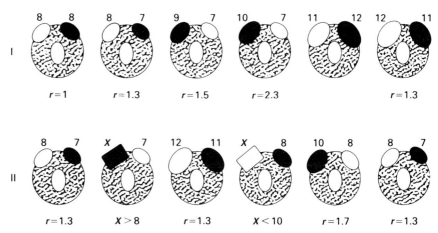

Fig. 12.24 The titration method of determining the value of an egg dummy. The circle represents the nest with one egg in the nest bowl and two dummies on the rim. The numbers 7, 8, 9, 10, 11, 12 refer to the size of the dummies. r is the ratio of the sizes of the dummies on the nest rim. X is the model to be measured. The dummy chosen in each trial is indicated in black.

I. Determination of the value of the gull's position preference. The first (left) shows that the right-hand egg is preferred. The next (to the right) shows that this preference remains when a smaller egg is substituted on the right-hand side. The sequence of tests shows that the value of the position preference lies between $r = 1.3$ and $r = 1.5$.

II. Determination of the value of model X. The control tests show that the position preference remains unchanged. The experimental tests show that the value of X lies between sizes 8 and 10 (After Baerends and Kruijt, 1973).

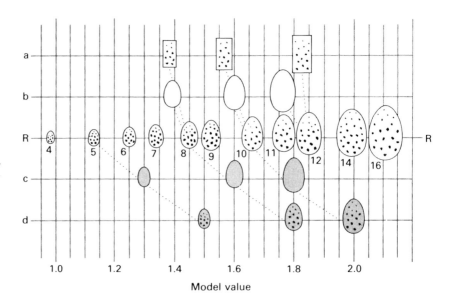

Fig. 12.25 Results of the titration tests. The average values for various dummies are shown in relation to the standard size series R. The dummies used were (a) brown, speckled, block-shaped; (b) brown, unspeckled, egg-shaped; (c) green, unspeckled, egg-shaped; (d) green, speckled, egg-shaped. The numbers 4 to 16 indicate egg size. Equal distances between points on this scale imply equal *r* values (After Baerends and Kruijt, 1973).

independent of the contribution of other features; that is, features such as the size, shape and coloration of the eggs are additive in their effects upon the animal's behaviour. Such a state of affairs had been proposed long ago (Seitz, 1940) and called the **law of heterogeneous summation**. This law states that the independent and heterogeneous features of a stimulus situation are additive in their effects upon behaviour.

Another study that comes to the same general conclusion is that of Walter Heiligenberg and his co-workers on the stimuli eliciting aggression in cichlid fish. Heiligenberg and his co-worker (Leong, 1969) studied the quantitative effect of sign stimuli on the attack readiness of the cichlid fish *Haplochromis burtoni*. Heiligenberg's method was to place an adult male fish in an aquarium together with numerous juvenile fish. The male is free to attack the juvenile fish, but they easily escape and fighting does not occur. A dummy of a rival adult male is presented behind glass at the initiation of each trial. Before the presentation, the attack rate of the resident male toward the juvenile fish is recorded for 15 minutes. During the presentation, usually 30 seconds long, the resident male usually quietly watches the dummy. After the presentation, the attacks on the juvenile fish are recorded for 30 minutes. The rate at which the adult male attacks juvenile fish usually increases markedly just after presentation of a dummy and then wanes to its pre-presentation level. This increment in attack rate is used as an index of the effectiveness of the dummy in increasing the aggression of the resident fish.

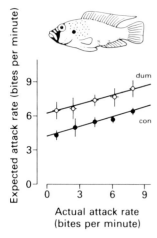

Fig. 12.26 Expected attack rates of male cichlid fish as a function of previous attack rate. These results show that, compared with the control (con) situation, presentation of a dummy rival fish (dum) has an additive effect upon the attack rate (After Leong, 1969).

Because the attack rate fluctuates considerably, presumably reflecting the male's internal state, statistical methods have to be used to compare the observed attack rate with that expected from the immediately preceding observation period. Using this approach, Leong (1969) found that different aspects of the colour pattern of territorial males painted on dummies were additive in their effects upon the attack rate, as illustrated in Figure 12.26. Similarly, Heiligenberg *et al.* (1972) showed that different angular orien-

Box 12.1   Cue strength and egg recognition in the herring gull

All the additive cues involved in egg recognition, summarized in Figure 12.25, can be represented as a graph of egg size versus egg colour pattern index, as shown in (a). Because the isoclines of egg value are straight lines, they can be combined into a single cue strength for egg recognition features and represented along a single dimension, as shown in (b). Other independent features of the egg retrieval situation, like the distance of the egg from the nest, must be represented along a different dimension. Suppose we now measure the gull's tendency to retrieve eggs as a joint function of the stimulus properties of the egg and the distance

from the nest. We would find that certain combinations of these two factors would give rise to the same retrieval tendency. The line joining all points giving rise to a certain motivational tendency is called **motivational isocline** (see Chapter 15).

When independent factors combine multiplicatively, as in the example illustrated in (b), then the motivational isocline is hyperbolic in shape. In general the evidence for multiplicative relationships comes mainly from situations where disparate aspects of the stimulus situation are involved in determining a single response (McFarland and Houston, 1981).

(a)

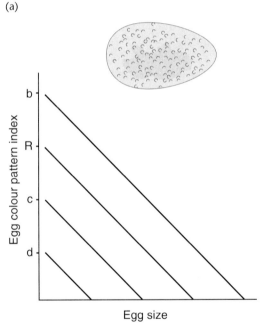

(b)

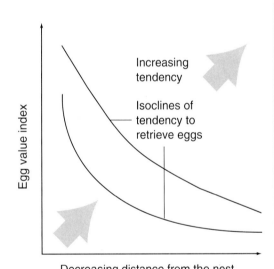

Cue space for egg value. The parallel isoclines indicate that egg colour pattern and egg size are additive in their effects, and could be represented by a single index of egg value

Cue space for egg retrieval. The hyperbolic shape of the isoclines indicates that the egg value index combines multiplicatively with distance of the egg from the nest

tations of the black eye bar of a male *Haplochromis burtoni* were additive in their effects upon attack rate and that the orientation of the whole body also made a contribution, as we can see from Figure 12.27.

These experiments show that the colour markings of a male can increase or reduce the aggressiveness of a rival male in an additive manner, in

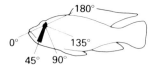

Fig. 12.27 A dummy male cichlid fish fitted with a black 'eye bar' made of metal foil, which can be rotated around the centre of the eye (After Heiligenberg *et al.*, 1972).

agreement with the law of heterogeneous summation. However, Heiligenberg (1976) claims that the body posture has a multiplicative effect upon attack rate. Moreover, Eberhardt Curio (1975), working on the anti-predator behaviour of the pied flycatcher (*Ficedula hypoleuca*), also obtained some evidence of multiplicative relationships among stimulus aspects of models of two natural predators, the red-backed shrike (*Lanius collurio*) and the pygmy owl (*Glaucidium passerinum*). Thus, it appears that the law of heterogeneous summation does not hold universally.

Where stimulus summation operates within a relatively restricted range of features like those pertaining to egg recognition, the additive components can be combined into a single index called the **cue strength** (McFarland and Houston, 1981). In the case of the cichlid fish *Haplochromis burtoni*, for example, we can see that the colour markings on a rival fish are additive in their effect in eliciting aggression in another fish. They combine to indicate the potential threat from the rival. The body posture of the other fish is a separate matter since it signifies the aggressive intention of the rival. The head-down posture is often adopted just prior to attack. Thus, a possible interpretation of Heiligenberg's (1976) study is that a multiplicative relationship exists between the stimuli indicating the aggressive potential of a rival and those indicating his aggressive intention.

In the interpretation of responses to complex stimuli the main difficulty is to interpret the animal's responses in terms of the animal's perception of the stimuli rather than in the experimenter's terms. The problem is that the experimenter is obliged to use an arbitrary scale of measurement without knowing how this corresponds to the subjective evaluation made by the animal. Some ways of overcoming this difficulty have been suggested by Houston and McFarland (1976) and McFarland and Houston (1981).

The vertebrate eye, with its ability to form images, makes visual recognition a distinct possibility. Field studies indicate that many species can recognize breeding sites, mates, neighbours, etc. Laboratory studies show that some species are able to recognize not only ecologically relevant stimuli, such as those with bilateral symmetry (Delius, 1986; Delius and Nowak, 1982), but also some less natural stimuli (Herrnstein, 1985).

Visual recognition requires considerable information processing prowess. Even when the external object, or **scene**, is static, the image on the retina is neither static nor constant. Movement of the image occurs when the animal moves its head, and changes in light conditions, distance and orientation cause considerable variation of the retinal image. If the object is alive, and continually changing its appearance, then the problem of recognition is compounded. Research on human perception shows that there are various perceptual invariance phenomena which contribute to what is called the subjective constancy of objects and scenes. Recently, researchers have started to investigate perceptual constancy in other species, a favourite subject being the pigeon.

Pigeons can be trained to choose one of two stimuli that matches a sample (match-to-sample procedure), or to choose one of two stimuli that differs from a sample (oddity-from-sample procedure). Once they have learned one of these tasks, they can be tested with a series of stimuli that differ slightly from the sample. In this way it is possible to discover which stimuli the pigeons recognize as being the same, and which different. For example, Juan Delius (1986) and his co-workers, working at the University of Bochum, systematically explored orientation, size, contrast, outline and colour invariance of pattern recognition. In the case of testing for size invariance, the pigeon was presented with three two-dimensional stimulus patterns of the type illustrated in Figure 12.28. They were trained with stimuli from the first group. A trial began with the sample shape being displayed on the central key. When the pigeon had pecked this fifteen times, the two comparison shapes were displayed on the side keys. One of these shapes was identical to the sample, the other was different from it. The pigeon was rewarded for pecking the stimulus that was different from the sample. On successive trials this target stimulus appeared on the right or left in a random order, so that the pigeon did not simply learn to peck the left or the right key. The stimuli presented during training were all the same size, being just inscribable inside an imaginary 10 mm diameter circle. Following training the pigeons were given a number of habituation trials in which they were presented with stimuli of varying sizes. The purpose of this was to familiarize the pigeons with these novel stimuli, because pigeons are normally aversive to novelty.

The pigeons were then given a series of unrewarded transfer trials, in

Fig. 12.28 Pigeons were trained with the nine patterns from the first group. The bird obtained a reward if it pecked the stimulus that was different from the central sample pattern. Following training, the pigeons were habituated to stimuli of various sizes. During tests it was found that the birds could pick the odd shape irrespective of size (After Lombardi and Delius, 1988).

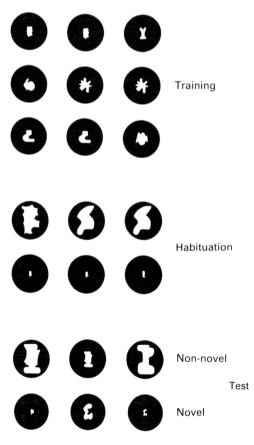

which stimuli of different sizes were used, as illustrated in Figure 12.28. The results of these showed that pigeons could pick out the odd shape irrespective of the size differences. The results of this and other experiments demonstrate that pigeons spontaneously recognize the equivalence or non-equivalence of shapes regardless of size differences between them, without having been specifically taught this skill (Lombardi and Delius, 1988).

The capacity to recognize patterns as being identically shaped even when their sizes are different is a desirable attribute for an image processing system. In humans, this ability is closely connected with size constancy (the object is perceived as being of constant size when viewed from different distances) and distance perception (an object giving a small retinal image is perceived as being further away than the same object giving a large retinal image). Artificial image-processing systems must also cope with these problems, but this has proved to be exceedingly difficult to engineer. Despite considerable effort, the performance of such machines remains rudimentary (Hord, 1982; Reitboeck and Altmann, 1984). Part of the problem is that vast computing power is required.

Similar methods have been used to demonstrate colour and outline invariance. It is particularly impressive that pigeons can recognize shapes even when they are outlined as opposed to solid (see Figure 12.29). This suggests that their encoding of the shape is complex and sophisticated and not based on simple key features. Indeed, it has been shown that pigeons can remember most of 725 black-and-white two-dimensional patterns (illustrated in Figure 12.30) over some seven months. One hundred of these patterns were arbitrarily selected and labelled as positive, while the remainder were negative. They were presented to the pigeon as positive–negative pairs. Responses to the positive shape were rewarded and responses to the negative shape punished. After some 8000 trials the pigeons chose correctly on 82 per cent of the trials (Delius, 1986). Thus, the pigeons' powers of visual recognition appear to be based upon an impressive memory.

Fig. 12.29 In a procedure similar to that shown in Figure 12.28 it was found that pigeons could recognize shapes even when they are outlined as opposed to solid (After Delius, 1986).

Fig. 12.30 Examples of 725 patterns that pigeons can be trained to remember (After von Fersen and Delius, 1989).

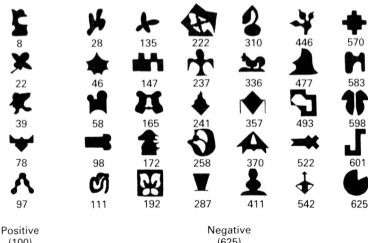

Points to remember

- Sensory processes can be divided into different modalities, each having characteristic sense organs. However, the sensation perceived depends upon the part of the nervous system that is activated and not upon the type of sense organ stimulated (Müller's Law of specific nerve energies).

- Chemoreception and thermoreception both provide information about the situation inside the body as well as about changes in the external environment.

- Hearing is a form of mechanoreception designed to pick up vibrations in air and water. It may be based upon a simple mechanism involving few neurons, as in the special-purpose ears of moths, or it may involve a complex mechanism, as in the sophisticated hearing of humans.

- Vision ranges from simple discrimination of light and dark to the formation and resolution of images. It is widespread in the animal kingdom, but species vary enormously in the type of eye and visual capability.

- Sensory judgements made by humans are not simply a matter of minimal discriminable differences between stimuli but depend upon the ability to separate signal from noise.

- Stimulus filtering, the process of classifying the information that impinges upon the animal, takes place at many levels. It may occur peripherally, as part of the design of sense organs, or it may occur centrally, as a form of selective attention.

- Complex stimuli may be evaluated by animals in different ways according to their internal state. In some cases specific sign stimuli elicit particular responses; in others it is the summation of heterogeneous factors that is important (law of heterogeneous summation).

Further reading

Ewert, J.P. (1980) *Neuroethology*. Springer-Verlag, Heidelberg.

Land, M.F. and Fernald, R.D. (1992) The evolution of eyes. *Ann. Rev. Neurosci.*, **15**, 1–29.

Young, D. (1989) *Nerve Cells and Animal Behaviour*. Cambridge University Press, Cambridge.

Ecology of the senses

13.1 Visual adaptations to unfavourable environments

13.2 Senses that substitute for vision

13.3 Visual recognition of prey and predators

When the animal kingdom is viewed as a whole, we can see evolutionary trends in sensory mechanisms. However, evolution results from natural selection, and natural selection is a form of adaptation to current circumstances. It should not surprise us, therefore, to find that many of the differences among animals are due not so much to ancestry but to their ecological circumstances.

Different sense modalities are best suited to different habitats, and particular modalities may be modified to suit the animal's way of life. This is true not only of the relationship between the animal and its physical environment but also in respect to its social environment. The effectiveness of communication depends upon the relationship between the modality used and the medium through which the information is transmitted.

In the case of the visual modality, the directionality of light permits greater accuracy of stimulus localization and spatial pattern recognition than is possible with any other modality. It is no accident that the predatory animals such as the octopus, hawk and cat have highly developed visual apparatus compared with other molluscs, birds or mammals. The visual modality also permits relatively durable means of communication in the sense that animals may evolve particular surface patterns or may construct artefacts that convey information on a long-term basis. The main disadvantage of vision is that it works well only under certain conditions. Visual signals cannot bypass obstacles in the way that auditory and olfactory signals can. Vision is restricted in dark or murky conditions, although many animals have special adaptations that enable them to overcome this disadvantage to some extent.

Sound has two particular properties that make it an important vehicle for communication. The first is that it is possible to make rapid temporal changes in sound signals. Their transient nature permits rapid exchange of information, and this may be important in a highly mobile species. Sound signals can also be turned off when necessary. For example, the house cricket (*Acheta domestica*) is well camouflaged visually. The male advertises its presence by prolonged bouts of chirping. When it senses danger it simply stops chirping and remains immobile, thus rendering itself difficult to locate by predators. The second important property of sound is that it can be raised in intensity above the background level of the environment. This is not possible with visual signals, except for those few species that emit light. The loud vocalizations of birds and monkeys of the tropical forest are examples of utility of sound in penetrating obstacles and background noise.

Olfaction has some of the advantages of both vision and hearing. Scents can be deposited as a durable message and are used widely in this way by terrestrial insects and animals. Like auditory signals, chemical messages can be transmitted around obstacles and can be raised in intensity above the background level. The sensitivity of olfactory receptors, which we saw in our discussion of the silk moth pheromone bombykol (Chapter 12), makes chemical signals particularly suitable where communication over very long distances is required. The relative durability of chemical signals can sometimes be a disadvantage, and the inadvertent deposition of chemical clues is exploited by many predators. Each modality has its advantages and disadvantages, and many animals make use of all three, deploying them according to circumstances. There are also many other, more specialized senses, which we discuss in the remainder of this chapter.

13.1 Visual adaptations to unfavourable environments

As we saw in Chapter 12 an animal adapted for seeing in bright light is likely to have good acuity, colour vision and movement perception. Such typical diurnal eyes are relatively insensitive to low levels of illumination. Animals adapted for seeing in dim light have greater sensitivity, but this is achieved at the expense of detailed pattern and colour vision.

The need to see in dim light occurs in nocturnal animals and in animals that inhabit deep water and caves. These conditions are not strictly comparable because the spectrum of available light is shifted toward the longer wavelengths at night and toward the shorter wavelengths in deep water. There is no evidence that nocturnal animals have increased sensitivity to long wavelengths (Lythgoe, 1979). However, some do have adaptations that enhance sensitivity to light. Eyes with a large pupil and lens pick up more light than small eyes. Such eyes are found in the opossum, house mouse and lynx (Figure 13.1). In other nocturnal animals such as owls and bush babies, the limitation on lateral expansion of the skull has led to tubular extension of the eye. Tubular eyes are also found in some deep-sea fish (Figure 13.1). The eye of the tawny owl is somewhat more sensitive than a human eye (Martin, 1977), but this is not enough for efficient night hunting. As we shall see owls also rely on other senses.

Many deep-sea animals are adapted specifically to the prevailing light conditions. The visual pigments of deep-sea fish have absorption maximums that coincide with the wavelength of maximum water transparency (Munz, 1958; Denton and Warren, 1957). The deep-diving beaked whale (*Berardius bairdi*) has visual pigments with absorption maximums of shorter wavelength than the shallow-diving grey whale (*Eschrichtius gibbosus*) (McFarland, 1971). Similarly, deep-sea crustacea possess visual pigments of shorter-wavelength maximal absorption than shallow-water crustacea (Goldsmith, 1972).

In addition to the properties of the visual pigments, other specialized fea-

Fig. 13.1 Eyes of nocturnal animals compared with those of a dog which has both day and night vision (After Tansley, 1965).

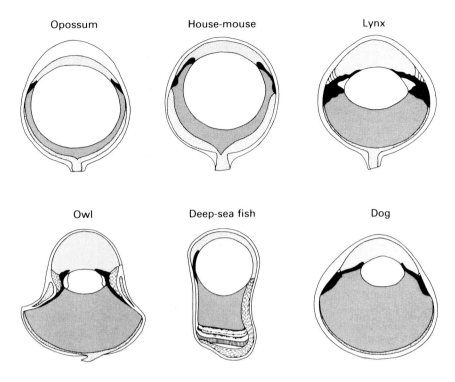

tures enhance vision in dim light. Typically, nocturnal animals have a greater proportion of rods than cones, and some such as the dogfish (*Scyliorhinus*) and bush baby (*Galago*) have few or no cones. Many rods may be connected to the same bipolar cell, thus increasing the sensitivity to light at the expense of acuity (see Chapter 12). Many nocturnal vertebrates have light-reflecting bodies, called **tapeta**, near the photoreceptors. These cause the eyes to glow when they are caught in a beam of light. Light that has passed through a photoreceptor without being absorbed is reflected back, thus increasing the probability of absorption. In some fish, the tapeta can be covered by migrating pigment granules. In the spur dogfish (*Squalus acanthias*), for example, the dark-adapted tapetum reflects 88 per cent of the incident light, but after light adaptation, when the tapetum's reflections have been screened out, only 2.5 per cent of the incident light is reflected (Nicol, 1965). Animals that live in very dim light, as we have seen, may adapt in various ways. Some animals live in turbid media, in which visual contrast is reduced as a result of light scattering by the suspended particles. For such animals, little can be done to rectify the situation (Lythgoe, 1979).

There comes a point at which visual conditions are so difficult that vision has to be abandoned as a primary sense. Most animals that live in caves, the very deep sea or turbid water have rudimentary eyes that are the result of regression. For example, the cave salamanders (*Typhlotriton* and *Proteus*) have eyes at the larval stage but no eyes as adults. If these salamanders are reared in light, the adult develops normal eyes (Lythgoe, 1979). The blind cave fish (*Astyanax mexicanus*) has eyes in the juvenile form that degenerate in the adult. Sadoglu (1975) made genetic crosses between these fish and a surface-living species of *Astyanax* that has normal vision. He found that genes are responsible for the degenerative condition of the eye in the cave-living species. The cusk eels (Ophidiidae) usually live at greater depths and have regressed eyes. Some species (e.g. *Lucifuga subterranea* and *Stygicola dentatus*) have evolved secondarily to live in caves.

Among the mammals, moles and bats provide the most obvious example of visual degeneration. Moles have very small eyes that are covered with skin in some species. The fruit-eating bats (Megachiroptera) have well-developed vision, but the nocturnal bats (Microchiroptera), particularly those species that catch insects on the wing, have very poor vision. Obviously, they have to rely on other senses to forage for food.

13.2 Senses that substitute for vision

Animals that live in dim light and that have poor vision have to rely on other senses. Thus, the bottom-living cat fish and cusk eels have various sensory barbels with which they probe the substratum. These are supplied with numerous touch receptors and chemoreceptors. These senses, however, cannot substitute for vision in the sense of providing information about the size and position of objects in the environment. Fish with neuromast and lateral line organs that are sensitive to vibration can detect the presence of moving objects and may obtain some information about stationary objects from water movements reflected from them (Schwartz, 1974; Pitcher *et al.*, 1976). The electromagnetic senses and developments in the sense of hearing, however, have provided animals with the best substitutes for vision.

Electromagnetic senses

Many lower animals are able to orient themselves in artificial electric fields, but little is known about the sensory basis of this behaviour. A number of species of fish make use of electrical sensitivity in their normal orientation and communication, and scientists know a considerable amount about their electrosensory systems. Sensitivity to magnetic fields has also been demonstrated in a number of animals. Certain bacteria (Blakemore, 1975), for example, orient toward magnetic North and will respond to a magnet in the laboratory. Examination by electron microscope reveals that these bacteria have chain-like structure containing crystals of magnetite, which have also been found in the abdomen of honey-bees and in the pigeon retina. The bacteria in the northern hemisphere follow the declination of the earth's magnetic field, and this steers them down into the anaerobic mud, their normal habitat. Those in the southern hemisphere have the polarity reversed. Magnetically orientated behaviour has been studied also in bees and pigeons, and some researchers have claimed that humans are sensitive to magnetic fields (Baker, 1981).

Fishes use electricity in three principal ways:

■ The so-called 'strongly electric fish' such as the electric ray (*Torpedo*) and electric eel (*Electrophorus electricus*) produce electric shocks capable of stunning prey but may not possess an electric sense.

■ Electrosensitive fish such as the dogfish (*Scyliorhinus*) and sharks do not produce electricity. Dogfish are capable of detecting prey, even when buried in sand, by the local distortion of the earth's electric field. The sense organs responsible are the ampullae Lorenzini, which are distributed widely over the body surface, especially near the head.

■ The so-called 'weakly electric' fish (Gymnotidae and Mormyridae) generate their own electric fields and are sensitive to electrical changes in the environment.

These fish are usually nocturnal and live in turbid water where vision is not practicable. They have two types of electrosensitive receptor: **ampulla** receptors, which respond to slowly changing electric fields; and **tuberous** receptors, which only respond to rapidly changing fields. Some species possess only one type of receptor; others have both. They generate weak electric fields by means of electric organs, which are modified muscles or neuronal axons. The electrical discharges typically are pulsed at up to 300 pulses per second. Some fish can vary the pulse rate as a means of communication with other fish or as a part of a jamming avoidance response designed to reduce interference from the fields generated by other members of the species. In other words when one fish is subject to electrical interference from another fish it can change its pulse rate to reduce the interference. The electroreceptors are also used to locate objects in the surrounding water by the distortions they cause in the electric field. Some fish, like *Gymnarchus*, can discriminate between good and poor electrical conductors such as a metal rod and a plastic rod, as illustrated in Figure 13.2. A more detailed account of the mechanisms of electroreception in weakly electric fish can be found in Ewert (1980).

Fig. 13.2 The electric field of *Gymnarchus*. The field on the animal's right is undistorted. The field on the left is distorted by a good conductor (black) and by a poor conductor (white). The animal can detect the presence of these two objects by sensing the effects they have on the electric field.

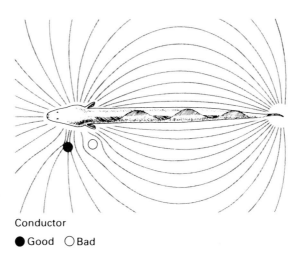

Conductor

● Good ○ Bad

Hearing

Hearing, as a substitute for vision, occurs in a number of species, some of which have developed highly sophisticated and specialized extensions of normal hearing. All these specializations depend upon an ability to locate sound accurately. Comparison of the sounds reaching the two ears is the most important method of sound localization in vertebrates. A person with one ear can turn the head and scan for the direction of maximum sound intensity because the head blocks some of the sound and causes a sound shadow. With two ears, simultaneous comparison is possible, and this is much more rapid and accurate. If the ears are sufficiently far apart, there will be differences in arrival time and phase of the sounds coming from a particular direction. Thus, small animals rely on intensity comparisons alone, whereas humans use all three methods and for a long time were thought to be the species with the greatest ability to locate the source of a sound. However, as a result of the pioneering work of Roger Payne, we now know that the hearing abilities of the barn owl (*Tyto alba*) far exceed those of humans.

The barn owl is a nocturnal hunter. It can locate and capture a freely moving mouse in total darkness. It can even determine the direction of the animal's movement and so aligns its talons with the long axis of the mouse's body. The barn owl is particularly sensitive to the differences in time of arrival of sounds at the two ears. This helps it to determine the azimuth (direction in the horizontal plane) of the sound. Sound intensity differences give further clues to horizontal location. In this respect, the barn owl achieves about the same accuracy as humans, but it is some three times as accurate in determining the elevation (direction in the vertical plane) of the sound source. This accuracy is achieved largely as a result of the owl's facial structure and the positional asymmetry of the ears, which is illustrated in Figure 13.3.

The right ear is directed slightly upward and the left ear slightly downward. The right ear is more sensitive to high frequency (3 to 9 kilohertz) sounds from above and the left ear to high frequency sounds from below the mid-horizontal plane of the owl's head. As the sound source moves downward the high frequency components of the sound become louder in the left ear and softer in the right. When the sound source moves upward the opposite occurs. This gives precise information about the elevation of the sound source because the owl uses low frequency components of the sound to locate sounds in the horizontal plane and high frequency components in the vertical plane. It does not confuse the two types of information, even though both depend upon comparison of the sounds reaching the two ears (Knudsen, 1981).

The most sophisticated substitute for vision is echolocation, in which the animal utters high frequency sounds and detects the presence of objects by the echoes produced. The principle is the same as that used in military sonar. Simple forms of echolocation occur in shrews (*Blarina sorex*) oilbirds (*Steatornis caripensis*) and the Himalayan cave swiftlets (*Collocalia brevirostris*) that roost and nest in dark caves. More advanced forms of echolocation are

Fig. 13.3 Facial structure of owl with some of the surface feathers removed to show the asymmetrical arrangement of the ears (After Knudsen, 1981).

found in dolphins (*Tursiops*) and other marine mammals, but it reaches its pinnacle in the bats (Chiroptera).

The pioneering work of Donald Griffin (1958) has led to a large amount of research into the mechanism of echolocation in bats. We now have a considerable knowledge of the physiology of the bat's vocal and auditory apparatus and of the brain mechanisms involved in echolocation. The student can find more detailed accounts of these in Ewert (1980) and Guthrie (1980).

Echolocating bats emit bursts of ultrasonic sound pulses of short duration (5 to 15 milliseconds) and high frequency (20 to 120 kilohertz). These are beyond the auditory range of humans. The brief pulses permit accurate timing of the echoes so that the bat can determine the distance of the object producing the echo. Natural sounds produced by wind and other animals are usually of low frequency, so by emitting high frequency cries, the bats are unlikely to be subjected to interference. Experiments in the laboratory have shown that artifical sounds above the 20 kilohertz frequency range cause flying bats to become disorientated. Another advantage of high frequencies is that they can be beamed precisely, thus allowing resolution of small objects. Bats produce their ultrasonic cries from a specialized larynx and emit them from the lips, as in the naked-backed bat (*Pteronotus*), or from specially shaped nostrils, as in the horseshoe (*Rhinolophus*) and leaf-nosed (Phyllostomidae) bats.

Bats also have many special adaptations that enable them to detect, time and localize the echoes that result from their ultrasonic vocalizations. Most bats that catch insects on the wing have large external ears, shaped to enhance directional sensitivity. As each loud outgoing sound is produced, the sensitivity is reduced by special muscles in the inner ear. If the outgoing pulses are very short, as in the vespertilionid bats, there is no overlap between the end of the outgoing pulse and the beginning of its echo. Since echoes arrive more quickly from nearby objects, the pulses are shortened progressively as an object is approached so that zero overlap is preserved.

In other bats there is overlap between outgoing pulses and returning echoes, so they require other means of enhancing echo detection. For example, the greater horseshoe bat (*Rhinolophus ferrumequinum*) adjusts the frequency of its cries in such a way that the frequency of the returning echoes is maintained within a narrow range, as illustrated in Figure 13.4. For a bat flying toward an object, the perceived frequency of the echo is always higher than that of the outgoing sound. This is due to the **Doppler effect** that results from the relative motion of the bat and the object. The faster the bat flies, the higher the apparent frequency of the echoes. To compensate for this effect, the bats alter their outgoing frequency in order to maintain the perceived echo frequency as constant as possible. By this means the bat can obtain an estimate of its flying speed and the direction and relative speed of flying prey.

As a substitute for vision, the echolocation abilities of bats are impressive. Laboratory studies show that a bat with a 40 cm wing span can fly in total darkness through a 14×14 cm grid made of nylon threads only 80 µm thick (see Figure 13.5). Bats have also been trained to catch small food particles thrown into the air in complete darkness and to discriminate edible from inedible objects on the basis of small differences in shape (Simmons, 1971). The little brown bat (*Myotis lucifugus*) can capture very small flying insects such as fruit flies and mosquitoes at a rate of two per second.

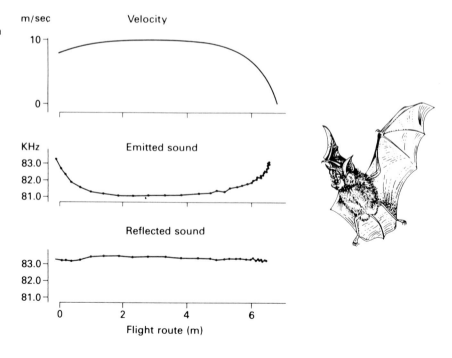

Fig. 13.4 Echolocation of a horseshoe bat approaching a stationary object. As the bat nears the object its velocity declines (top). The bat adjusts the frequency of its cries (middle) in such a way that the frequency of the reflected sound (bottom) remains constant.

Fig. 13.5 Horseshoe bat (*Rhinolophus*) flying through a 14 cm mesh of 80 μm nylon thread. The duration and frequency of echolocation cries is shown on the left.

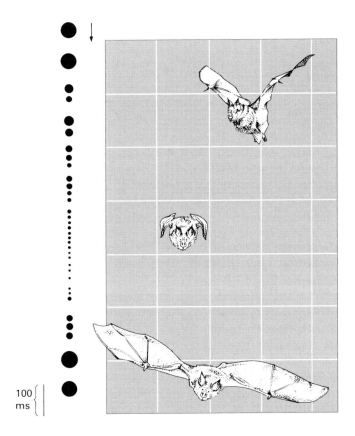

100 ms

13.3   Visual recognition of prey and predators

Most predators encounter a large number of different prey species which they have to discriminate from non-prey. The three most commonly used cues are size, movement and shape.

Predators which have a choice between prey items that differ only in body size usually take the larger. This is the more energy efficient strategy. However, as size increases there usually comes a point beyond which the stimulus is no longer regarded as prey. For example, when the common toad (*Bufo bufo*) is presented with 'prey' stimulus of different sizes, it responds positively to items within a specific size range, but actively avoids larger stimuli, as shown in Figure 13.6. How does the toad judge the size of the prey item? The simplest way is by the size of the retinal image, measured by degrees of visual angle. For an object of constant size, the size of the visual angle changes with its distance from the eye. Nearby objects look larger than far away objects. To select prey of a single size it is necessary for the toad to judge the absolute size of a visual object by taking into account both its retinal size and an estimate of its distance.

Midwife toads (*Alytes obstetricans*) gradually attain size constancy during development (Ewert and Burghagen, 1979). Just after metamorphosis they

Fig. 13.6 Toads respond to small moving squares by orienting towards them, but at a certain size this response wanes and large squares are avoided (After Ewert, 1980).

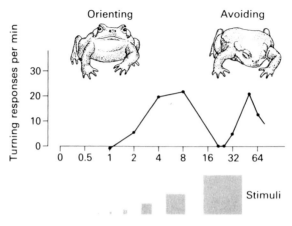

Fig. 13.7 Young toads (above) judge size by the visual angle, but older toads (below) can make size judgements independently of the visual angle (After Ewert, 1980).

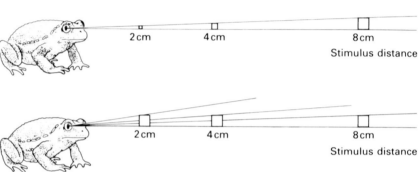

prefer a certain angular size prey dummy, almost regardless of the distance. Six months later they prefer objects of a specific absolute size, regardless of distance (Figure 13.7). In the meantime they have somehow learned to incorporate distance judgements into their size assessments. In some cases there is a bias towards larger prey which is not a result of active preference, but of better detectability. Thus rainbow trout (*Salmo gairdneri*) catch large crustaceans more often than small ones, because the large prey are seen at a greater distance (Ware, 1972).

Movement of the prey-stimulus is essential for prey recognition in some species, such as frogs and toads. The cuttlefish (*Sepia officinalis*) will normally only attack prawns that are moving. But, if a prawn is taken from a cuttlefish that has just captured and paralysed it, then it will immediately be attacked again, even if it is presented motionless (Messenger, 1968). Some predators prefer prey that move in an irregular pattern. Thus dragon-fly larvae prefer prey moving in a zig-zag pattern (Etienne, 1969), and Bluegill sunfish (*Lepomis gibbosus*) attack a wriggling dummy prey-fish rather than one that is smoothly moving (Gandolfi *et al.*, 1968). Sometimes, it is shape in relation to movement that is important. Thus, toads that are presented with a dark moving stripe on a white background most readily attack when the stripe is moving along its own axis, in a worm-like fashion (see Figure 13.8).

Prey recognition by shape is such a complex matter that it is difficult to give any simple rules. In a series of behavioural studies, Mike Robinson (1970)

Fig. 13.8 Toads respond positively to rectangles moved endways, but negatively to rectangle moved broadways (After Ewert, 1980).

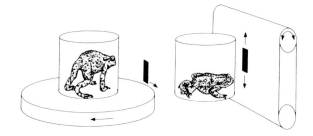

offered insect prey to captive rufous-naped tamarins (*Saguinus geoffroyi*). These are highly insectivorous New World monkeys. Some experiments involved presenting a mantid that was normal, headless, two-headed, etc. The results suggested that identification of the head was important in prey recognition. Experiments using stick-insects as prey showed that the monkeys tended to overlook stick-insects without legs but quickly seized insects with visible legs or small sticks with insect legs attached. Thus the heads and appendages of insects are important cues for prey recognition by tamarins. To counter this the insect prey employ a variety of protective devices that serve to conceal their appendages or disrupt the body outline. In experiments with abstract two-dimensional patterns, Robinson found that bilateral symmetry is one feature of prey that is probably commonly recognized by predators.

Similar principles apply to recognition of predators. For example, a silhouette of a hawk when passed above ducklings or goslings induces fear responses when moved in one direction, but not when moved in the opposite direction, as shown in Figure 13.9. The short neck and long tail is characteristic of a hawk, whereas the long neck and short tail resembles a flying goose (Tinbergen, 1949). Toads avoid snake-like shapes with a raised head (Figure 13.10). A leach walking jerkily along is regarded as prey when its frontal sucker is on the ground, but if it lifts its sucker in the air it is taken for an enemy (Ewert and Traud, 1979).

As we have seen, behavioural studies (reviewed by Ewert, 1980) indicate that toads show prey-catching behaviour toward small elongated objects presented horizontally but not to similar objects presented vertically (see Figure 13.8). The toad normally feeds on insects, larvae, worms, etc. Its prey-catching behaviour involves orientation of the head and body, visual fixation of the prey, snapping at the prey by extension of the neck and tongue, swallowing the prey and wiping the snout with the forelegs (see Figure 13.11). A small moving object is necessary to elicit the prey-catching behaviour. Large moving objects elicit defensive behaviour. Although various purely

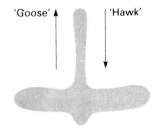

'Goose'   'Hawk'

Fig. 13.9 This silhouette resembles a hawk when moved in one direction and a goose when moved in the other.

(a)   (b)   (c)   (d)

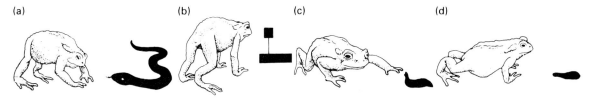

Fig. 13.10 Responses of a toad to simple models of (a) a snake, (b) abstract pattern, (c) leech with raised head, (d) leech with lowered head. The first three are avoided while the last is investigated (After Ewert, 1980).

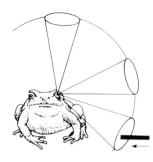

Fig. 13.11 The detection and capture of prey by a toad. From top to bottom: the prey enters the toad's lateral visual field; the toad turns toward the prey, bringing it more toward the centre of the visual field, and then into the binocular field; the toad grasps the prey with its tongue (After Ewert, 1980).

behavioural methods can be used to investigate sensory processes, physiological investigation can also yield valuable information about the functioning of sense organs and the type of information they send to the brain. To discover how the brain makes use of such information, however, a combination of the behavioural and physiological approaches is required. This approach has been used by Jorg-Peter Ewert and his co-workers in their extensive investigations of prey and enemy recognition in toads.

Physiological studies show that some of the prey recognition is carried out at the retinal level. Jerry Lettvin and his co-workers (1959) made electrical recordings from the frog optic nerve while objects were moved across the visual field. They found four main types of response that appear to correspond to the four types of ganglion cell in the retina. These cells were found to be detectors for (1) a stationary boundary, (2) a dark, convex moving object, (3) change of contrast or movement and (4) dimming. The toad appears to have three types of ganglion cell (corresponding to the frog types 2, 3 and 4) that send their axons along the optic nerve to the optic tectum of the brain (Ewert and Hock, 1972). The information provided to the brain thus includes the angular size and velocity of the object, its degree of contrast with the background and the general level of illumination. This information, however, is not sufficient for the toad to discriminate prey from non-prey.

The fibres of the optic nerve can be traced by nerve degeneration studies to the various parts of the brain, including the optic tectum and the thalamus pretectum (see Figure 13.12). The retina of one eye projects topographically to the surface layers of the opposite optic tectum. This projection can be seen as a map in which each area of the visual field corresponds to a particular area of the optic tectum. Electrical stimulation of the optic tectum in the freely moving toad produces an orientation response toward the corresponding part of the visual field, causing the toad to behave as if the appropriate part of the optic tectum had been stimulated by sight of a prey object. When the retinal projection to the thalamic pretectal area is stimulated electrically, the toad responds with avoidance behaviour.

Surgical destruction of the thalamic pretectal region results in animals that react with prey-catching behaviour to any moving object. Destruction of the optic tectum abolishes all reaction to moving stimuli, including the avoidance behaviour. These results led Ewert (1980) to postulate that the projection from the retina to the thalamus pretectum is instrumental in eliciting avoidance behaviour but that some excitatory input from the optic tectum is also required. The projection from the retina to the optic tectum provides a basis for prey-catching responses to all moving stimuli, but responses to large or enemy-like stimuli are inhibited by the thalamus pretectum. Thus, prey catching occurs only to appropriately small stimuli.

Ewert's hypothesis is confirmed by physiological studies in which electrical recordings are made from neurons in the optic tectum and thalamus pretectum in response to stimulation of the retina and of other relevant parts of the brain (Ewert 1980, 1987). Not only is this research of great interest in illustrating the respective roles of the retina and brain in stimulus filtering, but also it is a beautiful example of what can be achieved by a judicious combination of behavioural and physiological methods. From an ethological point

Fig. 13.12 The relationship between the parts of the toad's brain involved in the control of avoiding and orienting toward visually detected objects (After Ewert, 1980).

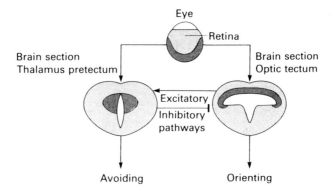

of view, the work of Ewert and his co-workers (Ewert, 1987) provides an excellent example of the neurobiology of the releasing mechanism (IRM). It shows that prey-selective neurons express the outcome of information processing in functional units consisting of interconnected cells. These include units that distinguish the prey from its background, units that select its features, units that discriminate prey from predator and units concerned with prey location.

Points to remember

■ Animals living in dim light often have special visual adaptations, such as tubular shaped eyes, a tapetum and visual pigments designed to maximize sensitivity under the prevailing conditions.

■ Animals that live, or have to operate, under conditions too dark for vision may have degenerate eyes and may have other sensory systems that substitute for vision. These include the echolocation system of bats, the auditory location system of owls and the electro-sensitivity of some fishes.

■ The visual recognition of prey and predators often involves detection of key stimuli (sign stimuli) which enables the animal to make a rapid decision and a rapid response.

Further reading

Ewert, J.P. (1987) Neuroethology of releasing mechanisms: prey-catching in toads. *Behavioral and Brain Sciences*, **10**, 337–405.

Giraldeau, L.-A. (1997) The ecology of information use. In Krebs, J.R. and Davies, N.B. (eds) *Behavioural Ecology*, 4th edn. Blackwell, Oxford.

Gould, J.L. (1980) The case for magnetic-field sensitivity in birds and bees (such as it is). *American Scientist*, **68**, 256–67.

Lythgoe, J.N. (1979) *The Ecology of Vision*. Clarendon, Oxford.

Wehner, R. (1997) Sensory systems and behaviour. In Krebs, J.R. and Davies, N.B. (eds) *Behavioural Ecology*, 4th edn. Blackwell, Oxford.

The animal and the environment

Profile of Claude Bernard

Desert lizard *Aporosaura*
(Photograph: Gideon Louw).

Here we consider the mechanisms used by animals to regulate their relationships with the environment. Chapter 14 begins with the coordination and orientation of the body and progresses from simple orientation to navigation. Chapter 15 considers the animal in relation to its internal environment. The concept of homeostasis is discussed in relation to thermoregulation, feeding and drinking. In Chapter 16 we discuss the ways in which animals are able to adapt their physiology and behaviour to changing environments. The roles of biological clocks in reproductive physiology, hibernation, migration and daily routines are discussed in some detail.

Claude Bernard (1813–1878)

Claude Bernard studied medicine from 1834 to 1843 and then began research at the College de France, as a pupil of Magendie. His main discoveries were made between 1840 and 1859, after which he did little experimental work, instead devoting himself to developing his theories and to writing.

Bernard's major work was the *Introduction à l'Etude de la Médicine Expérimentale*, published in 1865. Bernard had previously published various *Leçons*, but these were usually notes taken by pupils during his lectures and published under his supervision. Bernard published various other contributions culminating in his *Leçons sur le Phénomènes de la Vie Communs aux Animaux et aux Végétaux*, published shortly after his death in 1878.

Prior to Bernard's publications, the body had been regarded as a collection of organs, each with its separate functions. Bernard showed that the various physiological activities are interrelated and that the body should be thought of as a single complex and sophisticated machine.

From 1848 onwards, while working at the College de France, Bernard tested animals on various diets. He discovered that their blood always contained sugar, whether or not the animals had recently eaten. Even dogs fed solely on meat had sugar in the blood, and Bernard showed that this was secreted into the blood from the liver. He demonstrated that the liver stores some of the sugars made available by the digestion of food, and release sugar into the blood when none is available from the diet. Bernard also did important research on digestion, and he documented the regulation of blood supply to the different parts of the body.

Claude Bernard is regarded as the founder of modern experimental physiology. One of his often quoted dictums is: 'Why think when you can experiment. Exhaust experiment and then think'. His experimental work was always directed at a particular hypothesis. Commenting on the 'splendid work of Laplace' and noting that statistics were not much used by biologists, he remarked, 'Of course, I expect that in another hundred years or so, everybody will not only be using but abusing statistics and will rely on this method to salvage work undertaken without the benefit of any working hypothesis'.

Nonetheless, Claude Bernard is remembered mainly for his theoretical concepts. His most famous dictum – 'The stability of the internal environment is the condition for free and independent Life' – was first mentioned in 1859 in one of his *Leçons*, and elaborated in his posthumous work of 1878. Bernard realized that the blood provided an internal milieu for all bodily tissues, and that the state of the blood was influenced partly by internal processes and partly through changes in the external environment. Animals able to maintain the stability of the internal environment had far greater ecological freedom than those at the mercy of environmental fluctuations. The idea of regulation of the internal environment was implicit in Bernard's thinking, though it was not explicitly discussed by scientists until the next century. This basic idea has had profound effects upon modern thinking on physiology and animal behaviour.

Coordination and orientation

In this chapter we look at the animal's spatial relationships with its external environment. In responding to environmental changes, animals have to coordinate and orientate their movements in relation to external stimuli. In this chapter we look first at the subject of coordination and then at orientation. We conclude with a discussion of navigation, the most complex form of spatial orientation.

14.1 Coordination

As we saw in Chapter 11, the effective coordination of movement and loco-motion depends upon the information received by the CNS about the position, tension, etc. of the muscles. This information is provided by internal sense organs including muscle spindles, tendon organs and joint receptors. These sense organs provide information about the relative positions of the limbs or other organs of movement. They also enable animals to make reflex responses to changes induced either by outside agents or by the movement of the animal itself.

Reflex behaviour is the simplest form of reaction to stimulation. Reflexes are normally automatic, involuntary and stereotyped. A sudden change of tension in a muscle may result in automatic postural adjustment. A sudden change in the level of illumination may result in a reflex withdrawal response. Reflexes may be relatively localized, involving a single limb or other parts of the body, as illustrated in Figure 14.1. Sometimes, however, the whole animal may be involved in the reflex response. For example, the startle response of humans requires reflex coordination of numerous muscles. Similarly, the reflex withdrawal responses of invertebrates such as polychaete worms and various molluscs involve the whole body. As we saw in Chapter 11, such reflexes are triggered by giant nerve fibres that carry very fast messages to all the muscles involved so that they contract suddenly and simultaneously.

Although normally automatic, most vertebrate reflexes can be subject to

Fig. 14.1 Diagram of a simple reflex – the knee jerk reflex. A tap just below the knee cap stimulates receptors in the tendon which send messages to the spinal cord. These are relayed to the motor nerves serving the extensor muscles which reflexively contract and cause a kicking action.

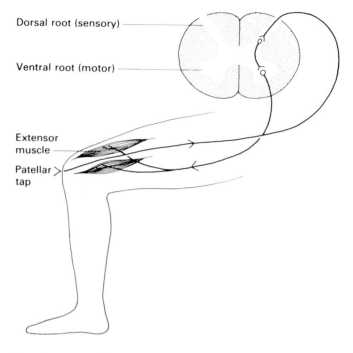

Dorsal root (sensory)

Ventral root (motor)

Extensor muscle

Patellar tap

interference at the synapses that occur within the CNS (see Figure 14.1). In the case of postural reflexes, there is often interference from other incompatible reflex mechanisms. When one reflex utilizes the same muscles as another, the reflexes are incompatible in the sense that both cannot occur simultaneously. Such pairs of reflexes are also neurologically incompatible in that stimulation of one reflex inhibits performance of the other. The inhibition is usually reciprocal so that one activity completely suppresses the other or is completely suppressed by the other. This type of reciprocal inhibition is typical of walking and other types of locomotion. It provides the most elementary form of coordination.

In general, muscular coordination is achieved by two main processes. These are central control and peripheral control. In the case of central control, precise instructions are issued by the brain and are obeyed by all the muscles involved. The coordination of swallowing movements in mammals seems to be of this type (Doty, 1968). Central control is important in the coordination of many skilled movements, which require rapid muscular activity. For example, the cuttlefish *Sepia* catches small crustaceans by means of two extensible tentacles, as illustrated in Figure 14.2. The control of attack falls into two phases. The first is a visually guided system in which the prey is brought into focus binocularly and movements of the prey are followed by movements of the cuttlefish in such a way that visual error is reduced to zero.

Fig. 14.2 A cuttlefish catching a shrimp.

When this dead-reckoning phase is complete, the tentacles are ejected suddenly and the prey is seized. This final phase is so rapid (about 30 milliseconds) that there is no time for the tentacles to be guided visually onto the prey.

The two types of control are illustrated in Figure 14.3. Eye position control, like the strike phase of cuttlefish, is open loop in that it involves no visual feedback. In cuttlefish this can be demonstrated by turning off the lights during tentacle ejection, which does not affect prey capture. However, the prey is missed if it moves during ejection, showing that the cuttlefish is unable to correct its strike using visual feedback (Messenger, 1968). Limb position control, like the targeting phase of cuttlefish, involves a closed loop process by which visual feedback of the target position is compared with the aiming position. Only when these are equal is the visual error reduced to zero.

Locomotor rhythms are often generated centrally. Electrical recording from motor neurons and interneurons in insects has demonstrated central control rhythms for walking (e.g. Pearson and Iles, 1970; Burrows and Horridge, 1974), swimming (e.g. Kennedy, 1976) and flight (e.g. Wilson, 1968) in arthropods. The precise patterns of behaviour, however, are often influenced by reflexes and feedback from the periphery. Thus, removal of proprioceptive input reduces stroke frequency in flying locusts, and wind stimulation increases motor neuron discharge (Wilson, 1968). It appears that the fast walking movements of the cockroach are influenced less by peripheral

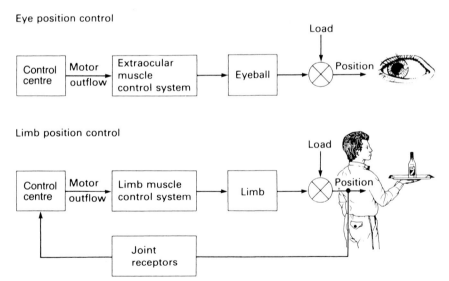

Fig. 14.3 Central versus peripheral control of movement. Movement of the eyeball (above) is under central control. The eyeball is not normally subject to load, so detailed motor commands from the brain can be followed quickly and precisely. If the eyeball is subjected to a load (e.g. from finger pressure), its displacement is not corrected. Peripheral control is important in the movement of limbs (below), because these are often subjected to loads. Displacement of limb position is monitored by sensory receptors and fed back to the central control centre, and the displacement is corrected.

stimuli than the slow walking movements of their relatives the stick insect (*Carausius*) (Wendler, 1966). Peripheral control of coordination is achieved through sense organs in the muscles and other parts of the body that send information to the brain and that thereby influence the instructions issued from the brain to the muscles. Peripheral control usually acts in cooperation with central control. For example, in the coordination of swimming movements in fish, the brain provides rhythmic signals that pass down the spinal cord in waves, coordinating the rhythmic movements of the fins and tail. In the dogfish (*Scyliorhinus*) the rhythm disappears if all nerves leading from the muscles to the brain are cut. If some nerves are left intact, however, the rhythm persists. Thus, it appears that some peripheral feedback is necessary to trigger the centrally produced rhythm (Gray, 1950). In cartilaginous fish (Chondrichthyes), including the dogfish and all other sharks and rays, the fins show little independent rhythmic movement, but in bony fish (Teleostei) the fins can beat at different frequencies under some circumstances. Von Holst (1939, 1973) showed that the rhythms of different fins can influence each other, a feature he called 'relative coordination'. Sometimes the rhythm of one fin attracts and dominates that of another so that they fall into step. In other cases, the amplitudes of fin movements summate so that the movements become smaller when the fins are out of step with each other and larger when they are in step.

In fishes and amphibians, locomotion appears to be mostly under the control of endogenous spinal rhythms. There is no evidence of specialized motor control exerted by the forebrain. Removal of the forebrain in fishes produces no change in posture or locomotion (Bernstein, 1970). In frogs and toads removal causes a general decrease in spontaneous movement, but electrical stimulation of the forebrain has no specific motor effects. The midbrain plays some role in motor control in fishes and amphibians. As we saw in Chapter 13, electrical stimulation of the tectum in toads causes turning of the head and food snapping and swallowing. In higher vertebrates the higher brain is capable of exerting greater control of movement, but the automatic aspects of locomotion are still primarily controlled by the brain stem and spinal cord.

In mammals, the **corticospinal** tract is the most important pathway involved in the voluntary control of movement. It begins in the motor cortex and proceeds through the midbrain and brainstem to the spinal cord. This system, sometimes called the **pyramidal system**, is present in all mammals except the very primitive monotremes (e.g. platypus and spiny anteater). In the marsupial possum (*Trichosaurus*) the pyramidal axons run only to the mid-thorax, where they innervate the forelimbs. The hindlimbs are innervated by an extrapyramidal system.

The **extrapyramidal system** includes all non-reflex motor pathways not included in the corticospinal or pyramidal system. It is thought to be more primitive than the pyramidal system. In animals with little or no cerebral cortex, the basal ganglia of the extrapyramidal system are the most important centres of motor control. They are particularly well developed in birds, which have virtually no cerebral cortex, but a larger striatum than mammals. Thus, it appears that as the cerebral cortex developed, it brought into being a second source of motor coordination that acts through the pyramidal system.

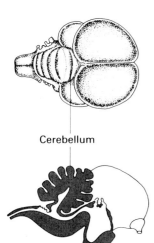

Fig. 14.4 Brain of a pigeon showing the large cerebellum.

In monkeys, neurons in various parts of the corticospinal pathways change their firing patterns during voluntary movements of the eyes, hands and legs (Evarts, 1968). Artificial stimulation of the motor cortex causes responses of single muscles or of single motor units within a muscle. More gross stimulation elicits discrete movements of whole limbs. It is possible to map the surface of the motor cortex according to the parts of the body moved in response to electrical stimulation. A similar map can be obtained for the sensory cortex in that the various parts of the body are represented differentially according to their sensory importance for individuals of the species concerned. Voluntary skilled movements are initiated via the cortical pyramidal neurons, while reflex maintenance of motor responses and of posture is controlled by nearby extrapyramidal neurons. There are estimated to be one million pyramidal neurons in humans (Prosser, 1973).

Another part of the vertebrate brain that is important in coordination is the cerebellum (Figure 14.4). The **cerebellum** is not directly involved in postural reflexes or motor control. It serves as a monitor and coordinator of neural events involved in orientation and balance and in various other refined aspects of motor control. The basic neuronal organization of the cerebellum is similar in all classes of vertebrates, and the cerebellum has undergone less evolutionary change than any other part of the brain. The cerebellum receives information from the senses of vision, hearing, touch, equilibrium and the condition of muscles and joints. It also has connections with the motor areas of the cerebral cortex. In addition to connections through the thalamus to the motor cortex, there are two-way connections to the sensory areas of the cortex. These are important in the maintenance of posture. Upright posture cannot be maintained without visual, vestibular and proprioceptive information. The cerebellum combines visual and vestibular information about equilibrium with the state of contraction of the muscle spindles concerned. It sends appropriate instructions to the muscles, particularly through the gamma efferent fibres (see Chapter 11). The cerebellum thus exerts a modulating and refined control over the muscular contractions involved in the maintenance of posture and in the complex coordinated movements required for locomotion.

14.2 Spatial orientation

The orientation of the whole animal in space may be based on very simple principles but may also involve very complex mechanisms. The simple principles can be seen most easily in certain invertebrate species. Gottfried Fraenkel and Donald Gunn (1940) proposed a system of classification based on the work of earlier writers. This system has provided the basis of more recent discussions and reviews (e.g. Adler, 1971; Kennedy, 1945; Hinde, 1970).

The simplest form of spatial orientation is **kinesis**, in which the animal's response is proportional to the intensity of stimulation but is independent of the spatial properties of the stimulus. For example, common woodlice (*Porcellio scaber*) tend to aggregate in damp places beneath rocks and fallen logs. They move about actively at low humidity levels but are less active at high humidity levels. They consequently spend more time in damp conditions, and their high activity in dry conditions increases the chances that they will discover a damp place. Similar behaviour is shown by the ammocoete larva of the lamprey, which varies its swimming activity in accordance with the light intensity (Jones, 1955).

The type of kinesis in which a relationship exists between the speed of locomotion and the intensity of stimulation is called **orthokinesis**. Another type of kinesis, shown by the flatworm *Dendrocoelum lacteum*, is **klinokinesis**. Here the rate of change of direction increases as the light intensity increases (Ullyott, 1936; Fraenkel and Gunn, 1940; Hinde, 1970).

In a number of types of orientation, usually grouped together as **taxes**, the animal heads directly towards or away from the source of stimulation. For example, when the larva of the housefly (*Musca domestica*) has finished feeding, it seeks out a dark place where it pupates. At this stage it will crawl directly away from a light source, and it is said to show negative phototaxis. The maggot has primitive eyes on its head that are capable of registering changes in light intensity but that are not able to provide information about the direction of the light source. As the maggot crawls it moves its head from side to side, as illustrated in Figure 14.5. When the light on the left is brighter than that on the right, the maggot is less likely to turn its head toward the left. It thus tends to crawl more toward the right, away from the light source. In response to an increase in the level of illumination, the maggot increases its rate of head turning. If an overhead light is turned off every time the maggot turns its head to the right and is turned on every time it turns to the left, then the maggot turns away from the illuminated side, describing a circle toward the right. Thus, although the animal has no directional receptors, it

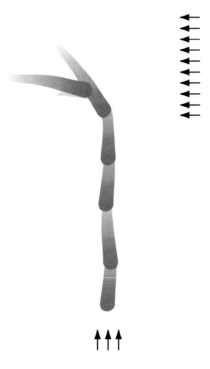

Fig. 14.5 Klinotaxis in a maggot. Notice the side-to-side movement of the head. When the light is moved from below to the side, the maggot turns away (After Mast, 1911).

can perform a directional response. Similar behaviour is shown by the single-eyed organism *Euglena* (Fraenkel and Gunn, 1940).

Orientation by successive comparison of stimulus intensity requires turning movements. Usually it is called **klinotaxis**. Many animals show klinotaxis in response to gradients of chemical stimulation. Simultaneous comparison of the intensity of stimulation received at two or more receptors enables the animal to strike a balance between them. It can then achieve **tropotaxis** that enables it to steer a course directly toward or away from the source of stimulation. For example, the pill woodlouse (*Armadillidium vulgare*), which lives under stones or fallen trees, shows a positive phototaxis after periods of dessication or starvation. With its two compound eyes on the head, the animal is able to move directly toward a light source. When one eye is blacked out, however, it moves in a circle. This shows that the two eyes normally provide a balance of stimulation. When presented with two light sources, the woodlouse often starts off by taking a medium course but then heads toward one light. This happens because equal stimulation of the two eyes can be achieved either by steering between two sources or by moving directly toward one. Deviations from one source tend to be self-correcting, and lateral sources of light tend to be ignored because the eyes are shielded at the back and sides. Deviations from two sources set wide apart, therefore, may result in contact with one being lost.

Eyes that are capable of providing information about the direction of light by virtue of their structure are capable of **telotaxis**. This is a form of directional orientation that does not depend upon simultaneous comparison of the stimulation from two receptors. When there are two sources of stimulation, the animal moves toward one and never in a median direction. This shows that the influence of one of the stimuli is inhibited. An example is depicted in Figure 14.6.

Menotaxis is a form of telotaxis (Hinde, 1970) that involves orientation at an angle to the direction of stimulation. An example is the light compass response shown by homing ants. These animals are guided, in part, by the direction of the sun. Santschi (1911) shielded an ant trail from direct sunlight and used mirrors to reflect sunlight onto the trail from the opposite direction. Ants entering the region illuminated by reflected light turned around and walked in the opposite direction, as shown in Figure 14.7.

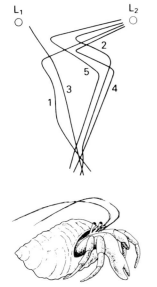

Fig. 14.6 Telotaxis: tracks of hermit crabs presented with two lights. Each part of the track is directed toward one light only (After Fraenkel and Gunn, 1940).

Fig. 14.7 Sun compass orientation in ants. In Santschi's (1911) classical mirror experiment the ant was shaded from the sun and sunlight was reflected onto the ant from the opposite direction. When this was done at positions 1–4 the ant changed direction, as illustrated (After Schöne, 1984).

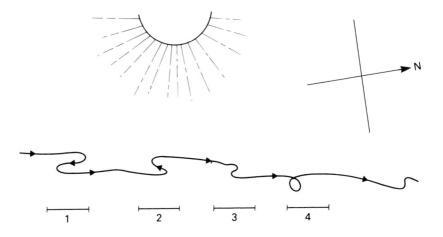

The type of orientation that can be achieved in a given situation depends jointly upon the nature of the external cues and the sensory equipment of the animal. An animal with only one sensor that is sensitive to stimulus strength only is limited to successive measurements of stimulus strength in different localities. If the external cues are inherently directional, then a single receptor that is shielded on one side can provide directional information. Thus, a shielded photoreceptor is useful in this respect, but a shielded chemoreceptor is of no advantage because chemical stimuli are not inherently directional. With two receptors, simultaneous comparison can be used to detect gradients (see Figure 14.8). With many receptors arranged in the form of a **rastor** (a row or mosaic arrangement), more sophisticated types of

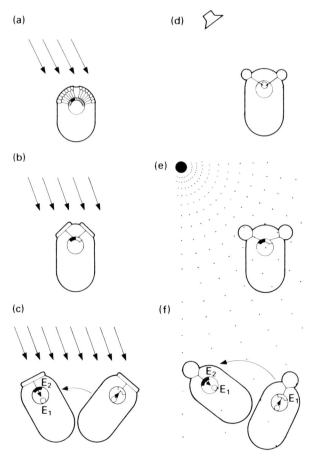

Fig. 14.8 Schematic representation of some of the basic principles of sensory orientation. (a) The direction of stimulation (e.g. light) is registered by a raster of sensory receptors. (b) Direction registered by simultaneous comparison between two receptors. (c) Only one receptor is available and the animal makes successive comparisons by moving its body. (d) Time of arrival of stimulation (e.g. sound waves) is compared by two receptors. (e) A gradient of stimulation (e.g. chemical) is registered by comparison betwen two receptors. (f) A gradient is registered by a single receptor as the animal moves to sample different localities (From *The Oxford Companion to Animal Behaviour*, 1981).

orientation can be achieved (see Figure 14.8). Examples of rastors are the lens eyes of vertebrates and the compound eyes of arthropods (see Chapter 12).

Spatial orientation is often achieved by a combination of methods. For example, some moths are attracted to females as a result of the airborne pheromone released by the female. The scent is windborne, and the flying moth must orientate with respect to the wind. Flying animals normally use visual cues to monitor their progress with respect to the ground. The flight path of the animal is affected by the wind direction, giving a resultant track (Figure 14.9). Experiments with moths show that the track angle changes with the scent concentration. When the scent is absent, the animal flies backward and forward without progressing upwind (i.e. with a track angle of 90°). When scent is detected in the wind, the track angle increases and the animal zigzags upwind. The changes of direction are related to the borders of the scent trail, as illustrated in Figure 14.9. When the scent concentration drops below a certain level, as at the edge of the scent plume, the animal turns in a direction opposite to that of its previous turn. This turning behaviour is not related to wind direction, but it depends upon internal reference, or **idiothetic** information. Thus, the flying moth uses a combination of

Fig. 14.9 Male moth flying upwind in response to pheromone released by a female. The flight path, F, is affected by the direction of the wind, W, giving a resultant track, R. The track angle is the angle between the wind direction, W, and the flight direction, R, with respect to the ground.

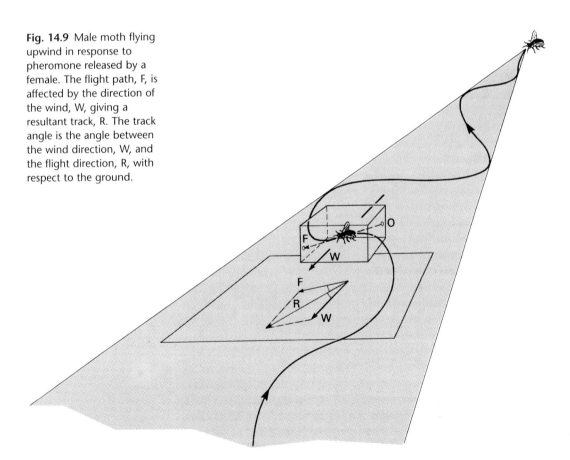

visual, amenotaxic (wind) and idiothetic orientation mechanisms in searching for a mate.

The reafference principle

A sophisticated orientation system must be able to distinguish between stimulation from the outside world and stimulation caused by the animal. In the case of human vision, for example, movement of objects in the outside world causes movement of the image on the retina, which we perceive. Voluntary movement of the eyes, however, also produces movement of the image on the retina, yet we do not perceive such movements. Somehow the brain distinguishes between the movement of the retinal image that is independent of the animal and movement that is due to movement of the eyeball.

Two theories have been proposed to account for this phenomenon: the outflow theory and the inflow theory. The **outflow theory**, originally due to Hermann von Helmholtz (1867), maintains that instructions to the eye muscles to move the eyeball are accompanied by parallel signals to a comparator in the brain. Here they are compared with the incoming visual signals, as shown in Figure 14.10. **Inflow theory**, due to Charles Sherrington (1918), maintains that receptors in the extraocular muscles send messages to the brain comparator whenever the eyes are moved (Figure 14.10). In both theories, the brain comparator assesses the two incoming signals and determines whether the visual signals correspond to the movement that would be expected on the basis of the other signal. If the two signals do not correspond, then some of the movement must have been due to outside causes.

Fig. 14.10 Inflow and outflow theories of eyeball movement control.

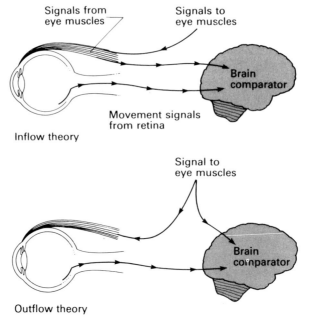

The extraocular muscles contain muscle spindles, and their existence would seem to support the inflow theory of Sherrington (1918). However, it appears that these muscles are not involved in providing a sense of eye position (Howard and Templeton, 1966). Helmholtz 1867) used evidence from mechanical manipulation of the eyeball and apparent motions induced by attempts to move the eye when the extraocular muscles are paralysed to argue that there is no position sense in the eye muscles.

As a matter of common observation, when the eyeball is displaced in its socket by finger pressure, the visual axis is shifted (as can be seen from the double image), and it remains shifted as long as the finger pressure is maintained. The eyeball does not push back against the finger in an attempt to retain its previous position as would be expected if extraocular muscle spindles were involved in the control of eye position (McFarland, 1971). Moreover, when the eyeball is displaced with the finger, movement is perceived, and this would not be expected on the basis of the inflow theory. The muscle spindles should be stimulated however the eyeball is moved, and the brain comparator then should cancel the movement of the image on the retina. The evidence therefore seems to favour the outflow theory.

The outflow theory was extended and generalized by Eric von Holst and Horst Mittelstaedt (1950) (see also von Holst, 1954). According to their **reafference principle**, the brain distinguishes between **exafferent** stimulation (stimulation that results solely from factors outside the animal) and **reafferent** stimulation (that which occurs as a result of the animal's bodily movements). Motor commands not only cause patterns of muscular movement but also produce a neural copy (the 'efference' copy) that corresponds to the sensory input that could be expected on the basis of the animal's behaviour. The brain then makes a comparison between the efference copy and the incoming sensory information (see Figure 14.11). All reafferent information should be cancelled by the efference copy so that the output of the comparator will be zero. Exafferent information will not be cancelled, however, and will be passed on by the comparator to another part of the brain.

Von Holst and Mittelstaedt (1950) showed that the fly *Eristalis*, when placed inside a cylinder painted with vertical stripes, shows a typical **optomotor reflex**, that is, turning in the direction of the stripes when the cylinder is rotated. Such reflexes do not occur when the fly moves of its own

Fig. 14.11 Outline of basic reafference system.

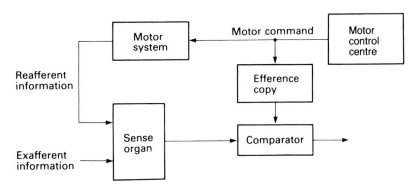

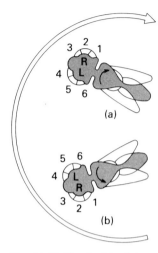

Fig. 14.12 The fly *Eristalis* in a rotating cylinder. L, left eye, R, right eye; (a) head in normal position, (b) head rotated through 180° (After von Holst, 1954).

accord, although the visual stimulation is similar. When the head of the fly is rotated experimentally through 180° (see Figure 14.12), the optomotor reflex is reversed as expected. However, when the fly attempts to move of its own accord, it goes into a spin and its movements appear to be self-exciting. These results can be explained in terms of reafference theory. Normally, the output of the comparator determines bodily movement, and when the fly moves of its own accord, the output is zero and no movement occurs. The optomotor apparatus provides exafferent stimulation that is not cancelled by an efference copy, and the fly responds in a reflex manner. When the head of the fly is reversed, exafferent stimulation has the same effect as before but in the opposite direction. In the case of reafferent stimulation, however, the perceived movement is reversed in sign, and instead of being subtracted from the efference copy, it is added to it. The result is that the comparator now has a magnified output, whereas normally it would be reduced to zero. The more the animal responds, the greater the reafferent stimulation and the greater the amplification of the animal's response. Consequently, the fly tends to spin faster and faster.

The reafference principle is important not only with respect to vision but also in the control of limb position, posture, etc. For example, we can tell the difference between the arm movements involved in shaking the branch of a tree and those, which may be identical, produced while holding passively onto a branch that is being moved by the wind.

14.4 Navigation

The most complex form of spatial orientation is nagivation. Navigation requires not only a compass, or directional sense, but also some kind of map. To illustrate this, we can consider an experiment designed by Robin Baker (1981) to test the possible use of magnetic information by humans. Baker transported a group of students in a small bus from a particular starting point (home) to a particular secret destination. Upon arrival, the students were requested to point in the direction of home. To be able to do this, they would have to have some kind of compass, and Baker was looking for evidence for a magnetic compass. However, such a feat would also require a knowledge of the relative position of the two locations. Even if the students had possessed an accurate magnetic compass, it would not be possible for them to point toward home without some kind of map.

Three types of orientation are important in navigation:

■ **pilotage**, or steering a course using familiar landmarks

■ **compass orientation**, the ability to head in a particular compass direction without reference to landmarks

■ **true navigation**, the ability to orientate toward a goal such as a home or breeding area without the use of landmarks and regardless of its direction

On their long-distance migrations, birds probably use all three types of orientation. Thus, Perdeck (1958, 1967) captured juvenile and adult starlings

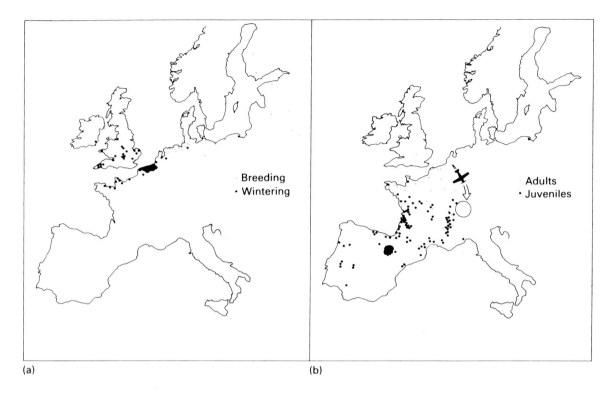

(a)   (b)

Fig. 14.13 (a) Recoveries of breeding and wintering starlings banded in Holland. (b) Recoveries of adult and juvenile starlings flown from Holland to Switzerland, and released there during the autumn migration (After Perdeck, 1958).

as they passed through Holland on their first autumnal migration. They were banded and transported by plane to Switzerland, 750 kilometres southeast of the normal migration route. Normally, the starlings migrate from the breeding grounds in the Baltic to the wintering grounds in Belgium, southern Britain and northern France (see Figure 14.13). After being displaced to Switzerland, the juvenile starlings were recaptured in Spain and southern France, indicating that they maintain the normal southwesterly direction of migration after displacement. Adult starlings, however, were recaptured in their normal wintering grounds. Thus, the adult starlings had corrected for the displacement, whereas the juvenile starlings had maintained the compass orientation that was characteristic of their breeding population. The juveniles of many migrating species reach their winter habitat for the first time on the basis of innate information about the direction and distance of the target area. They are unable to correct for displacement because they lack the map component that is necessary for true navigation (Schmidt-Koenig, 1979).

Animals are known to possess compasses of various types. These are based upon features of the geophysical environment, like the magnetic field of the earth. To be sure that a particular physical feature is used as a compass, it is necessary to show that the animal can detect the phenomenon and that it can use it for orientation under natural conditions. Because of the uncontrolled variability of the outside world, testing for sensory capability is best done in the laboratory. A favourite method used by those interested in navigation is the cardiac-conditioning method illustrated in Figure 14.14. By this

Fig. 14.14 Cardiac-conditioning apparatus for testing the sensitivity of pigeons to various stimuli (polarized light in this example) (After Schmidt-Koenig, 1979).

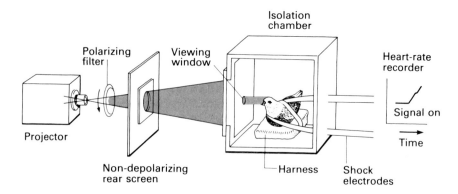

and other methods researchers have shown that birds are sensitive to the following stimuli:

Ambient pressure

Pigeons are sensitive to atmospheric pressure changes in the region of 1 to 10 millimetres of water, equivalent to a change of less than 10 metres in altitude (Kreithen and Keeton, 1974; Delius and Emmerton, 1978). This sensory ability may provide the pigeon with an accurate physiological altimeter.

Infrasound

Sound with a frequency of less than 10 hertz is called infrasound. Humans cannot hear it. Pigeons can hear sounds as low as 0.06 hertz. Infrasound travels a very long distance, and natural sources of infrasound, like surf, could be used by pigeons as a navigational aid (Yodlowski *et al.*, 1977; Kreithen, 1978).

Odour

Birds have long been thought to have a poor sense of smell, but experiments using cardiac-conditioning (e.g. Shumake *et al.*, 1969; Henton *et al.*, 1966) confirm the results from physiological methods, in showing that the olfactory sense in pigeons is sufficiently good to be used in navigation. Similar results have been obtained from other bird species (Schmidt-Koenig, 1979).

That some atmospheric factor is involved in pigeon navigation was established by Wallraff (1966, 1970). Experiments later showed that pigeons deprived of their olfactory sense had severely impaired homing abilities (for reviews see Papi, 1976, 1982). It is now clear that pigeons require an intact olfactory apparatus to navigate to home from unfamiliar distant release sites (Wallraff, 1984), though the exact role of olfaction remains a problem (Wallraff, 1990).

It appears that information relevant to navigation is provided by substances dispersed in the atmosphere. For example, Wallraff and Foa (1981) removed pigeons many kilometres from home in airtight containers, through which ambient air was sucked. The air had to pass through a filter of activated charcoal before reaching the pigeons. Control pigeons were supplied with unfiltered air. In order to prevent the pigeons from smelling natural air upon release, their nasal cavities were anaesthetized with xylocain, which is effective for about 1.5 hours. Thus neither the experimentals or the controls were able to smell while their initial orientation behaviour was being observed. In 14 out of 17 cases of releases conducted in Italy and Germany, at distances of 24–155 km from home, the initial bearings of the controls were more homeward orientated than those of the controls. Similar experiments by Benvenuti and Wallraff (1985) show that pigeons can be systematically misguided by false olfactory information.

Magnetic compass

Scientists had long thought that the energy involved in geomagnetic effects would be too low to be detected by animals. This is now known to be incorrect, and responses to magnetic fields have been shown to occur in many species. Although there were early indications of magnetic sensitivity in birds (e.g. Merkel and Wiltschko, 1965), the possibility was long subject to doubt because of successive failures to demonstrate magnetic sensitivity in cardiac-conditioning experiments. However, positive results have been obtained in pigeons that were free moving rather than strapped down (Bookman, 1978). Miniature magnets have been found in bacteria, honeybees and pigeons (Walcott and Walcott, 1982). Although these bees and pigeons are known to be sensitive to magnetic fields, it is not known if, or how, magnetic information is made available to the nervous system. There is evidence that the internal clock of bees is influenced by magnetic phenomena (Gould, 1980). In birds, it appears that it is the declination of the earth's magnetic field that is important in providing a directional sense, and it would seem that this must be used in combination with visual cues (Wallraff, 1978).

Sun compass

That birds can use the sun as a compass was discovered by Gustav Kramer (1951). He trained starlings in a circular cage to look for food in a certain compass direction. All visual landmarks were excluded, and only the sun and sky were visible. The birds could maintain a particular compass direction throughout the day, showing that they were able to compensate for the movement of the sun.

The starling's internal clock can be reset experimentally (Hoffman, 1954). This is done by confining the bird in a light-proof room and exposing it to artificial photoperiods. Klaus Schmidt-Koenig (1958, 1960, 1961), in experiments with homing pigeons, tested the effects of shifting the clock six hours forward, or six hours backward, or by twelve hours. The results, illustrated in Figure 14.15, show that the bird's orientation in the testing apparatus is exactly that expected if it were using the sun as a compass. Apparently, pigeons have to learn to interpret the sun's movement individually by observation of its whole daily path (Wiltschko and Wiltschko, 1980).

Fig. 14.15 Results of clock-shift experiments with homing pigeons. The animals were trained in a circular cage to look for food in a particular compass direction. They were tested by counting the pecks of food dishes around the periphery (each dot indicates one peck). CET is Central European Time. The arrows indicate the direction expected under the conditions of the experiment. The top row shows experiments, the bottom row shows simultaneous controls (After Schmidt-Koenig, 1960).

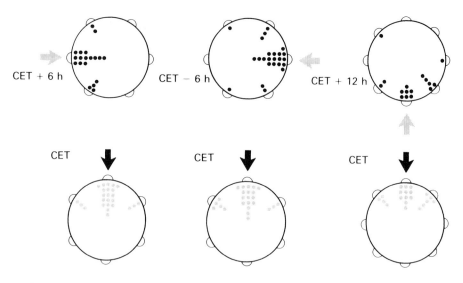

The pigeon's sun compass has been shown to be sufficiently accurate for navigational purposes, provided the birds make corrective measurements at intervals during their journey. However, the compass direction of the sun, in relation to local time, is primarily of use in determining longitude. It is the sun's altitude (or azimuth) that alters with changes in the latitude of the observer. There is some evidence, from operant conditioning experiments, that pigeons can make fairly accurate measurements of changes in azimuth. There is some evidence that pigeons estimate the sun's altitude by measuring shadows rather than the sun directly (McDonald, 1972, 1973). Shadows may magnify the sun's movement by a factor of six, but it is not known whether pigeons are able to make use of this information out of doors. Andy Whiten (1972) trained pigeons to relate sun altitude to the northerly or southerly direction of home. His experiments show that pigeons can associate sun altitude with north–south direction. They do not demonstrate that such abilities are used in navigation.

Some workers, notably Geoffrey Matthews (1955, 1968) and Colin Pennycuick (1960), have attempted to account for the navigation in terms of the sun's movement. In theory, there is sufficient information to determine longitude (from the time-compensated compass direction of the sun) and latitude (from the sun's azimuth in relation to time) and so to construct the equivalent of a map. However, such theories require very considerable visual acuity and accuracy of measurement of sun motion in relation to time. Most scientists doubt that pigeons are capable of such accuracy (Schmidt-Koenig, 1979). Moreover, the ability of a flying pigeon to determine the altitude of the sun has never been demonstrated, and this is an essential ingredient of sun navigation theories. Interestingly, the direction of sunset has been shown to influence orientation, both at twilight, and during the subsequent night (Moore, 1980).

Star compass

If songbirds are caged during the period when they normally would be migrating, they show a typical directional migratory restlessness, and at night

Fig. 14.16 Experimental cage for measuring migratory restlessness. When the bird attempts to leave the cage, it produces inky footprints on the blotting paper lining the funnel. Some examples of records are shown below (After Schmidt-Koenig, 1979).

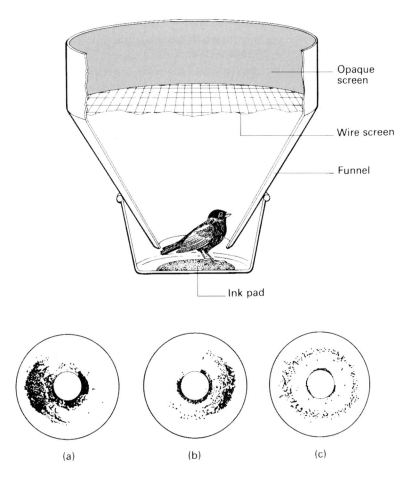

(a)   (b)   (c)

this directionality is related to the stars (Sauer and Sauer, 1955). The birds were orientated under a natural starry sky and under a planetarium sky. The techniques for recording migratory restlessness were later improved (Emlen and Emlen, 1966), using the apparatus illustrated in Figure 14.16. Steve Emlen (1967) demonstrated that the directionality of indigo buntings (*Passerina cyanea*) was the same under a natural sky and a stationary planetarium sky. The birds would follow a shift in the planetarium sky, and Emlen (1972) discovered that their orientation was linked to the rotation of the sky rather than specific constellations.

To a terrestrial observer, the stars appear to fall on the inside of a spherical surface, called the **celestial sphere**. At any given time, the stars form a definite pattern on the celestial sphere. This pattern moves in a manner that gives the impression that the celestial sphere is spinning. In reality, the earth is spinning about its polar axis. The point in the northern sky about which the celestial sphere seems to rotate is called the **north celestial pole**. The star Polaris lies very close to this pole, and for practical purposes this star can be used to determine the observer's northerly direction.

Emlen (1972) showed that buntings that were hand raised without any sight of the sky were unable to orientate during migration. Birds that were exposed routinely to a planetarium sky rotating around Polaris, during the period between fledging and the autumn migration, showed the normal southerly orientation. However, northerly orientation was shown by birds that were exposed to an experimental planetarium sky that was made to rotate around Betelgeuse, a star in the constellation of Orion, which appears in the southern sky for an observer in northern latitudes. Apparently the newly fledged birds learned which part of the sky rotated least and orientated away from that part during their migratory restlessness. An alternative possibility, that is perhaps more appropriate to birds freely flying at night, is that celestial patterns are related to the earth's magnetic field (Wiltschko and Wiltschko, 1975, 1976).

Cardiac-conditioning methods have also been used to investigate the perception of the night sky by birds. There is evidence that mallards (*Anas platyrhynchos*) learn to recognize particular star patterns (Wallraff, 1969). Although star patterns provide potential map-like information, the evidence so far indicates that the stars are used purely as a compass (Schmidt-Koenig, 1979; Wallraff, 1984).

Polarized light

In normal unpolarized light, the wave-like vibration occurs equally in all planes. In polarized light, the vibrations are greater in one plane. When unpolarized sunlight is scattered by atmospheric molecules, polarization occurs. The polarization is greatest for the light scattered at an angle of 90° to the rays of the sun. This means that there will be a pattern of polarization in the sky, which will change according to the position of the sun.

Karl von Frisch (1967) discovered that the orientation of honey-bees, as indicated by their dance (see Chapter 23), depended upon the position of the sun even when the sun was obscured by clouds. He showed that it was necessary for the bees to see only a small portion of sky and that the essential information was confined to the ultraviolet portion of the spectrum. When ultraviolet light from the sun was passed through a polarizing filter, von Frisch found that the orientation of the bee dance changed in accordance with the angle of polarization.

Cardiac-conditioning experiments (see Figure 14.14) clearly show that pigeons are able to perceive rotations in the plane of polarization of light (Kreithen and Keeton, 1974; Delius *et al.*, 1976), but it is not known how they interpret this information. Although the pattern of polarization of the sky can provide an indication of the sun's position even when the sky is overcast by clouds, it appears that pigeons do not use their sun compass under such conditions (Schmidt-Koenig, 1979). Experiments by Able (1982) open up the possibility that the pattern of polarization at twilight is used as a compass cue by night-flying migrants.

Landmarks

Orientation by landmarks may play an important part in navigation, particularly as the destination is approached. A classic example of this comes from

Arthur Hasler's (1960) studies of migrating salmon. Pacific salmon (*Oncorhynchus*) hatch in streams of the western United States and Canada. After the smolt stage of development, they swim downstream to the Pacific Ocean. After two or three years at sea, they become sexually mature and migrate back to the exact stream where they were spawned. The journey to the coast is probably accomplished by means of a sun compass. Once at the coast, however, they have to select the correct river and the correct tributary stream within the river system. After many years of research, Hasler and his co-workers discovered that during the smolting period the fish become imprinted upon the olfactory characteristics of their native stream. During their return journey they are able to discriminate between the water coming from their native stream and that from other tributaries. They are essentially recognizing landmarks, but because the scent of their birthplace flows all along the migratory route, it is not necessary for the young salmon to remember landmarks all along the outward journey. In fact, if salmon are transferred to another stream after the period of imprinting, they return to their native tributary and not to the stream down which they passed on their outward migration.

Various experiments have suggested that allowing pigeons to view familiar visual landmarks before release improves their homing performance. Chappell and Guilford (1997) found that some types of visual landmark provided by experimenters seemed to lack salience for pigeons. Birds allowed to see only two-dimensional landmarks evidently used the sun compass in their initial orientation, because clock-shift trials threw the pigeons off course even though the target was clearly indicated by the landmarks. However, when they introduced a striped cylinder in addition to the two-dimensional landmarks the pigeons orientated correctly even when clock-shifted. Thus the three-dimensional visual cue acted as a beacon, indicating the direction of the target. The experimenters suggest that parallactic changes in the appearance of the cue as the pigeon moves around it are important in providing a directional frame of reference.

Landmarks are particularly important in insect navigation, especially in correcting cumulative errors during path integration (Wehner, 1997). In bees (see Chapter 23) and ants, landmarks are identified by means of a snapshot-matching mechanism, which enables the animal to identify a string of landmarks when it is proceeding along a familiar route (see also Chapter 25).

Combination of factors

Pigeons appear to have a number of possible navigational aids. They most probably use some of these in combination, and others as alternatives. Magnets or Helmholtz coils, attached to pigeons' heads during flight, affect the birds' magnetic compass under overcast conditions, but have little influence under sunny conditions (Keeton, 1971; Walcott, 1977; Walcott & Green, 1974). Wallraff *et al.* (1986) investigated the possibility that magnetic factors might be involved in the process of site localization by means of olfactory cues. Their approach was based on the finding that the initial orientation of pigeons can be influenced by changing either the composition of the air breathed by the birds, or the magnetic field to which they are subjected, for a period before their release. So they decided to test whether the pre-release magnetic treatment affects birds that are prevented from smelling natural odours. What they found was that pigeons that could smell natural air during magnetic treatment prior to release performed poorly in homing tests compared with birds that could smell natural air and had no magnetic treatment. However, birds that received magnetic treatment when they could not smell natural air performed as well as the controls that received no magnetic treatment. The magnetic treatment was ineffective both for birds that were allowed to smell only the air of their own container (sealed from the natural air) and those that could not smell at all. Thus the magnetic treatment had no effect, not because the birds were unable to smell, but because they were unable to smell some particular substances in the local air. One possible explanation is that the magnetic pretreatment disrupts navigation because it provides false information by interfering with the olfactory system. Another possibility is that the magnetic and olfactory cues might be components of an integrated system. It seems likely that there are many types of interaction among the possible navigational cues, and this is a promising area of future research. It has been suggested, for example, that the pigeon 'map' mechanism is based upon olfactory cues, and that this interacts with 'compass' mechanisms based upon other cues (Wallraff, 1988). New methods of investigation are being used, including computer simulation (Wallraff, 1989). Clock-shift effects can also give clues to the nature of interactions between navigational mechanisms (Chappell, 1997). In cases where a pigeon does not alter its orientation as much as would be expected after clock-shift, it may be that there is a distorting effect of visual landmarks, or an interaction between the sun compass mechanism and the magnetic field mechanism.

Points to remember

■ Coordination of movement involves both central control and peripheral control. Peripheral control is largely preprogrammed, but central control may involve feedback (closed-loop control) or it may be predictive (open-look control).

■ The spatial orientation of simple animals is based upon kineses and taxes which vary in type in accordance with the animal's sensory capabilities. Navigation, a more complex form of orientation, requires the equivalent of both map and compass.

■ A wide variety of sensory modalities are known to be involved in navigation, but the way the information is coordinated is not well understood.

Further reading

Papi, F. (1992) (ed) *Animal Homing*. Chapman and Hall, London.

Schmidt-Koenig, K. (1979) *Avian Orientation and Navigation*. Academic Press, London.

Schöne, H. (1984) *Spatial Orientation*. Princeton University Press, Princeton, NJ.

Wiltschko, R. and Wiltschko, W. (1995) *Magnetic Orientation in Animals*. Springer-Verlag, Berlin.

Homeostasis and behaviour

The notion of physiological stability was implicit in Claude Bernard's (1859) concept of the internal environment. He observed that the level of sugar in the blood remained constant, even when the animal was deprived of food or had just consumed meat or food containing sugar. He postulated that some process of regulation and control must be designed to maintain the constancy of the internal environment. He also realized that an animal that is able to regulate its internal environment, in the face of fluctuations in the external environment, has greater freedom to exploit a variety of potential habitats.

Animals can be roughly divided into conformers, which allow their internal environment to be influenced by external factors, and regulators which maintain their internal environment in a state that is largely independent of external conditions. The process by which regulators control their internal state comes under the general heading of homeostasis.

15.1   Homeostasis

The term homeostasis was used first by the American physiologist Walter Cannon (1932) who wrote the following:

'The coordinated physiological processes which maintain most of the steady states in the organism are so complex and so peculiar to living beings – involving, as they may, the brain and nerves, the heart, lungs, kidneys and spleen, all working cooperatively – that I have suggested a special designation for these states, homeostasis'.

Cannon envisaged a situation in which sensory processes, monitoring the internal state of the body, initiated appropriate action whenever the internal state deviated from a preset, or optimal, state. For example, when human body temperature rises above 37 degrees, cooling mechanisms such as flushing and sweating are brought into action. When the temperature falls below the optimal level, warming mechanisms like shivering come into play. By employing a number of such finely tuned mechanisms, humans are able to achieve a precise thermoregulation and thermal homeostasis.

Although regulatory mechanisms of the types envisaged by Cannon now are known to be widespread in the animal kingdom and to involve a great variety of physiological processes, they are not the only types of processes involved in the control of the internal environment. For example, until recently, it was thought that increased drinking by animals in response to high environmental temperature was a result of dehydration arising from the increased water losses involved in cooling responses such as sweating and panting. Such a view is entirely consistent with a theory of the maintenance of homeostasis by regulation, as outlined previously. Evaporation of water is necessary to maintain thermal homeostasis in a hot environment, and this upsets the fluid balance of the body, the restoration of which requires increased drinking. However, we now know that in species such as the rat and pigeon, drinking occurs as a direct response to the temperature change in **anticipation** of any change in fluid balance arising from thermo-regulation and not in response to it as the regulatory theory would require (Budgell, 1970a,b). In other words, the animals drink in order to have water available for thermoregulation.

The role played by behaviour in the control of the internal environment varies considerably with species and with circumstances. Drinking behaviour, for instance, is essential for the maintenance of homeostasis in many species, the physiological mechanisms of water conservation being unable to prevent lethal dehydration after prolonged deprivation. However, some species such as the mongolian gerbil (*Meriones unguiculatus*) and the parakeet or budgeri-gar (*Melopsittacus undulatus*) (Cade and Dybas, 1962) are able to survive indefinitely without water. Because of the great efficiency of their water con-servation mechanisms, they can live on the water contained in the seeds they eat. Others, like aquatic species, need no special behaviour to obtain the water necessary for the maintenance of homeostasis.

In many aspects of homeostasis, the role of behaviour is normally negli-gible. However, experiments involving surgical interference with the normal physiological mechanisms responsible for homeostasis show that animals often have the capacity for appropriate behaviour, even though it is not used in their normal lives. For example, the work of Curt Richter (1943) shows that if the thermal homeostasis of rats is interfered with by the removal of the thyroid gland, they respond by building warmer nests and by engaging in other forms of behavioural thermoregulation, when given the opportunity by the experimenter. Similarly, removal of the adrenal gland, which is involved in salt balance, causes rats to shift their preferences toward saltier food and water.

In other cases, there may be no special organs involved. Thus, rats fed a vit-amin deficient diet are able to select food containing the required vitamins even though they cannot taste the presence of the vitamins in the food. As we see in Chapter 19, they are able to learn which food made them feel better (Revusky and Garcia, 1970; Rozin and Kalat, 1971). Clearly, the mechanisms by which internal stability is maintained are many and various, and it is incorrect to think of homeostasis as necessarily implying a simple regulatory feedback process. Moreover, it is wrong to think that homeostasis simply implies constancy of the internal environment. Many intermediates exist between conformers and regulators, and one species may be able to regulate one body function but not others. This is clearly illustrated by reference to thermoregulation in animals.

15.2 Thermoregulation

Most animals have an optimum body temperature around which they function most efficiently. Below this temperature their metabolism progressively slows down, muscular activity diminishes and the animal may become torpid. Above the optimum temperature metabolic rate rapidly increases, and this may be expensive to maintain. Moreover, there is an upper limit to the temperature at which bodily processes remain viable. For most species this limit seems to be in the region of 47 °C.

Most animals are able to influence their own body temperature to some extent either by employing specialized physiological mechanisms or by appropriate behaviour. In both cases it is necessary for the animal to be able to detect the environmental temperature, its own body temperature or both. This is done by means of various sensory processes that are usually grouped together under the heading of **thermoreception** (see Chapter 12).

The metabolic reactions of the body produce heat continuously, and the more active the animal, the greater the rate of heat production. In a cold environment, however, the heat production from normal metabolism and activity may not be sufficient to counteract the cold, and the animal may produce heat by increasing metabolic rate and taking action to prevent the loss of body heat. Many invertebrates are cold blooded in the sense that their body temperature tends to conform with that of the environment. Because the rate of metabolic reactions is determined by the temperature at which they occur, such animals are forced to reduce their activity when the body temperature falls. However, some invertebrates such as the common woodlouse (*Porcellio scaber*) and millipedes (Myriapoda) are stimulated into extra activity by falling temperatures and thus are able to maintain a body temperature that is higher than that of the environment. In warm-blooded animals, the rate of heat production can be raised by increasing muscular activity, as in shivering, and by direct effects of thyroid hormones on metabolic rate. Food intake can also serve to increase heat production because heat is released during digestion. Many animals increase food intake in response to cold.

Animals that gain heat primarily from external sources like sunlight are called **exothermic**, while those that gain heat primarily from internal processes are called **endothermic**. Exothermic animals sometimes enhance the warming effects of sunlight by colour change. Dark objects absorb more radiant heat than light objects, and some species of lizard are able to change their colour in accordance with their thermal requirements. For example, the desert iguana (*Dipsosaurus dorsalis*) has dark coloration in the early morning when its body temperature is low. It becomes progressively paler as its temperature rises, reaching the halfway stage of colour change at about 40 °C. Similarly, some turtles expose their black feet to sunlight to increase the rate of heat gain.

The metabolic reactions of the body produce heat continuously, and animals can become overheated easily, especially when they are active. Overheating can also occur in especially hot environments or when heat dissipation is impaired. Because the lethal body temperature is not much above the normal body temperature of many animals, cooling mechanisms have to be especially rapid and effective.

There are four main ways of losing heat from the body: conduction, convection, radiation and evaporation.

- **Conduction** is the transfer of heat through solids and liquids, between regions differing in temperature. It can occur within the body tissues or between the body and an external object like the ground. Heat conduction can be reduced by insulation provided by layers of fat within the body and by layers of air trapped in hair or feathers at the body surface.

- **Convection** is the transport of heat in a fluid medium. Heat loss occurs as a result of the circulation of warm blood from the interior of the body to the cooler surface tissues. The control of blood flow to the periphery thus provides an important means of temperature regulation.

- **Radiation** is a form of heat transfer that does not depend upon material mediation but that can occur in a vacuum. Heat loss by radiation is roughly proportional to the temperature difference between the animal and its environment. Animals both gain and lose heat by radiation. An animal's colour makes little difference to its heat loss by radiation, but it does affect heat gain, as we saw earlier. Thus, black animals gain more heat by absorption of radiation than white ones that have greater reflection.

- **Evaporation** of water from moist areas of the animal's body surface involves heat loss and is an important means of cooling in many species.

Animals are able to control the amount of heat lost in these four ways to a variable extent, depending upon the species. To some extent they can control the rate at which the body loses heat by conduction by altering their insulation. Long-term regulation can be effected by increasing deposition of fat and hair growth in winter and by decreasing these in summer. Short-term changes can be achieved by raising or lowering hair and feathers (see Figure 15.1) and so controlling the amount of trapped air, and by altering body posture in relation to the prevailing weather. Surface insulation is influenced by wind and wetness. Wet hair increases heat loss by conduction. When a mammal lies or sits on the ground, the hair is compressed and holds less air. Heat then is conducted into the ground, particularly when the ground is wet. Cows are frequently observed to sit down before rain comes, in order to rest without losing as much heat as they would once the ground was wet.

Some animals can enhance radiant heat losses by behavioural means. For example, certain fiddler crabs (*Uca*), ground squirrels (*Citellus*) and other bur-

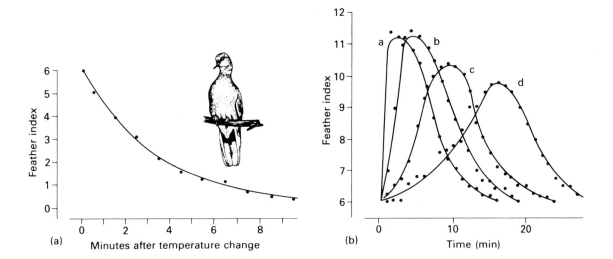

(a)

Minutes after temperature change

(b)

Time (min)

Fig. 15.1 Feather movement in doves (*Streptopelia risoria*) in response to a sudden increase in environmental temperature (a), and in response to drinking 10 ml cold water (b). The feather index is a measure of feather position averaged over the whole body (see Fig. 22.10). The curves a, b, c, d are the responses to drinking at different rates (After McFarland and Budgell, 1970).

rowing animals make sorties between their cool burrows and the warm external environment. In this way they can cool off by radiation if they become overheated. The effective surface of the body can be reduced by curling up or by huddling with other members of the same species, thus lowering the amount of heat lost by radiation. Evaporative water loss through the skin is uncontrolled in amphibians, reptiles and birds, but in mammals it is regulated by the sweat glands. These glands are present in all higher mammals except rodents and lagomorphs (rabbits). In humans, the sweat glands on the palms are controlled emotionally while those on the rest of the body normally respond to thermal control. Sweating is controlled by thermoreceptors in the brain and not by those in the skin. Thus, humans usually sweat during exercise but not necessarily when sitting by a hot fire. Rats and some other animals enhance evaporative cooling by moistening the body surface with saliva or by wetting themselves with water as elephants do.

Respiratory evaporation is controlled to some extent in most animals. Crocodiles, snakes and some lizards gape widely when hot. The desert iguana (*Dipsosaurus*) pants like a dog. Because animals lose so much water in respiratory evaporation, they tend to use it only in emergencies. Birds and mammals pant only when their body temperature approaches the lethal temperature. Flying birds generate a lot of heat and rely primarily upon respiratory evaporation to dissipate it. The camel does not pant at all but relies on radiant cooling during the night. Camels do not store water to a greater extent than other species and cannot afford to expend water in keeping cool (see Figure 15.2).

Ectotherms are able to regulate their body temperature only to a limited extent. Amphibians can keep cool by evaporation of water from the body surface. The leopard frog *Rana pipiens* can maintain a body temperature of 36.8 °C at an environmental temperature of 50 °C. Reptiles have a more limited ability to cool themselves and tend to avoid very hot conditions. The Namib desert lizard *Aporosaura anchietae* burrows into the sand when the midday temperature climbs above 40 °C (see Figure 15.3).

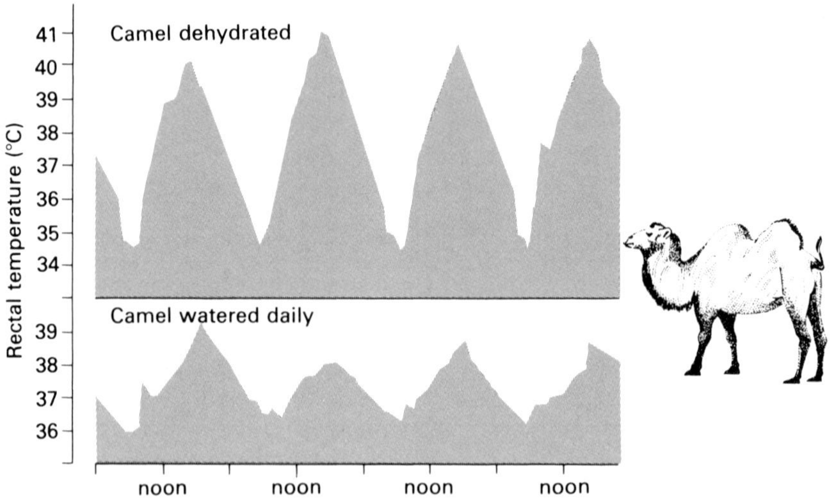

Fig. 15.2 When dehydrated, camels allow their body temperature to rise during the day, so that they do not have to lose water in thermoregulating. During the cold desert night the camel loses the excess heat that it has stored during the day (After Schmidt-Nielsen, 1964).

Fig. 15.3 The lizard *Aporosaura* lives in sand dunes, which have marked fluctuations in surface temperature. The lizard remains under the surface when the surface temperature is too cold, or too hot. This usually means that it has one period of activity on the surface in the morning, and another in the afternoon. Active periods are shown as shaded blocks (After Louw and Holm, 1972) (*Photograph: Gideon Louw*).

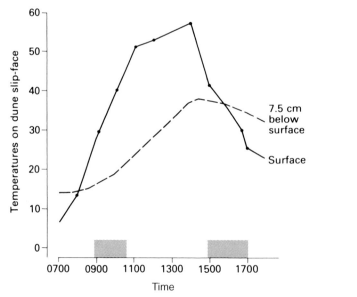

True thermal homeostasis is found in birds and mammals, which are endotherms, able to maintain a constant body temperature despite fluctuations in environmental temperature. Their high metabolic rate provides an internal source of heat, and their insulated body surface prevents uncontrolled dissipation of this heat. Birds and mammals maintain a body temperature that is usually higher than that of their surroundings. The brain receives information about the temperature of the body and is able to exercise control over the mechanisms of warming and cooling. When the brain temperature becomes too high, the cooling mechanisms are activated, and if it becomes too cool, then heat losses are reduced and warming mechanisms are called upon. This feedback principle is the same as that found in a thermostatically controlled electric heater (see Figure 15.4).

Fig. 15.4 The feedback principle involved in a simple thermostatic electric heater. When the temperature reaches a certain preset value, the switch opens, the heater goes off, and the temperature falls. When it has fallen below the preset value, the heater is turned on again.

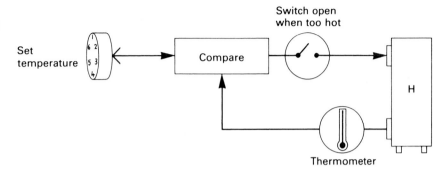

Fine control of body temperature occurs in humans, who have an early warning system consisting of numerous thermoreceptors in the skin. On the basis of this type of information, people are able to take anticipatory action and so forestall any undue fluctuations in body temperature. Controlled temperature changes do occur in birds and mammals, often on a diurnal basis. The average human body temperature is 36.7 °C in the early morning and 37.5 °C in the late afternoon. Many endotherms tolerate some internal temperature fluctuation, probably as a means of energy conservation.

15.3 Water balance

All animals require water to maintain their metabolic processes. Animals continuously lose water by a variety of means including excretion and evaporation from the body surface. In vertebrates, the water of the body is located in distinct compartments, as illustrated in Figure 15.5. The losses occur from the vascular compartment. For example, the lungs are a source of water loss by evaporation. The lungs have a rich supply of blood vessels, and gas exchange takes place across thin membranes that are moistened by the plasma in which the gases dissolve. Exhaled air nearly always contains more

Fig. 15.5 Distribution of water between the cellular and extracellular compartments of the body. The extracellular compartment is made up of the vascular and interstitial spaces.

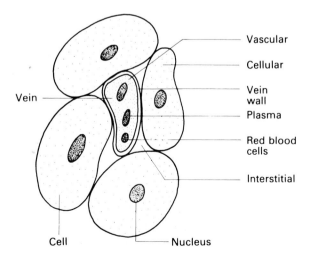

water than inhaled air, and this water comes from the vascular compartment. In many animals, respiratory evaporation is an important means of cooling. As mentioned earlier, birds rely heavily on respiratory evaporation as a means of cooling during flight. Thermoregulation, which involves sweating or spreading saliva over the body surface, also necessitates a certain amount of water loss.

The kidney filters the plasma of the blood, removes waste products and excretes them in urine. Thus, some water loss due to excretion is inevitable, but we shall see that some savings are also possible here. Water is also lost in defecation, but many vertebrates are able to reabsorb water from the small intestine and, thus, excrete very dry faeces in times of water shortage. Loss of water from the vascular compartment results in some redistribution among all the compartments of the body, but the lost water must be replaced eventually.

Most animals obtain water by drinking, but amphibians absorb water through the skin, as do some insects. Some animals, especially those that have infrequent access to water, are able to drink enormous quantities. Whereas a very thirsty man can drink one litre of water in a minute, and possibly three litres in ten minutes, a 352-kilogram male camel has been observed to drink 104 litres in one go. After water enters the mouth, it passes through the oesophagus into the stomach and then enters the intestine. From here it may enter the blood by osmosis. If the concentration of salts in the intestinal fluid is less than that of the blood, then water will pass into the blood. However, if the salt concentration is higher in the intestine, as it may be after ingestion of salty food or water, then water may pass the other way, and there may be temporary dehydration of the vascular compartment.

Loss of water from the vascular compartment results in an increase in the concentration of salts in the extracellular compartment. This induces some redistribution of water between the cellular and extracellular compartment, with the result that the body cells undergo some dehydration and shrinkage. These changes are detected by specialized cells in the brain, called **osmo- receptors**. Opinions differ as to whether the osmoreceptors are sensitive to changes in the extracellular fluid or whether they measure their own shrink- age during dehydration (see Toates, 1980, for a review of this problem). The osmoreceptors are known to occur in the hypothalamic region of the brain, and their stimulation has two main effects. First, it increases the tendency to seek water to drink, and second, it activates various water conservation mechanisms.

There are osmoreceptors in the hypothalamus which control the release of antidiuretic hormone from the pituitary gland, which lies just below the hypothalamus (see Figure 15.6). The presence of antidiuretic hormone in the bloodstream leads to a decrease in the amount and an increase in the con- centration of urine excreted by the kidneys. Damage to the pituitary gland or associated areas of the hypothalamus results in diabetes insipidus, the symptoms of which include excessive urination and consequent thirst. Antidiuretic hormone is thus an important factor in water conservation.

Other conservation mechanisms include increased reabsorption of water in the small intestine so that less is lost in the faeces, and a reduction of food

Fig. 15.6 Diagram of the human pituitary gland. The hypothalamus (H) receives its blood supply from the internal carotid arteries (ICA), which also supply the anterior and posterior pituitary. Neurosecretory neurons (N) shown originating in the hypothalamus are responsible for the release of hormones into the venous system. V is the third ventricle of the brain (After Wilson, 1979).

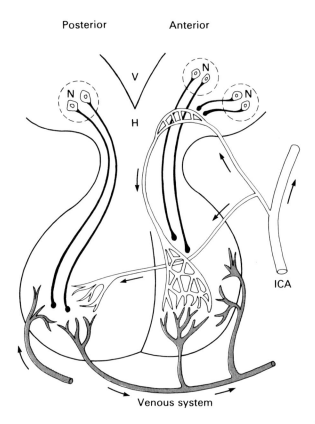

Posterior   Anterior

Venous system

intake. Because the waste products of digestion and food metabolism have to be excreted, some water loss is inevitable, but this can be reduced by eating less. Laboratory studies show that water losses in doves that are deprived of food are only about a quarter of their normal level (McFarland and Wright, 1969). There is also some evidence that doves can control their respiratory water losses (Wright and McFarland, 1969). Water loss resulting from thermoregulation can sometimes be reduced through behaviour such as seeking a cool place and reducing heat production due to exercise and food consumption. When a camel is short of water, it allows its body temperature to rise and stores heat in the fatty tissues of its hump during the daytime. During the cold desert night this heat is dissipated by radiation without any loss of water. Contrary to popular belief, camels do not store water in their humps, although metabolism of the fatty tissue would, of course, release some water (Schmidt-Nielson, 1964).

Animals may become dehydrated as a result not only of intracellular dehydration but also of reduction in the volume of the extracellular fluid. Haemorrhage and other forms of blood loss do not alter the animal's osmotic balance, but the lost fluid has to be replaced. Animals have various mechanisms for detecting such a loss. The hormone **renin** is produced by the kidney in response to a reduction in renal blood flow. Renin is released into the blood where it stimulates production of another hormone, called **angiotensin**. This hormone has two main effects:

1. It acts on the vascular system to maintain normal blood pressure and circulation.

2. It is a powerful thirst stimulus, and animals injected with very small amounts of angiotensin will stop whatever they are doing and seek water (Fitzsimons, 1976).

The main components of the complicated mechanisms involved in extracellular thirst are illustrated in Figure 15.7.

Maintenance of water balance is bound up intimately with thermoregulation and with eating. Many animals greatly reduce their food intake when dehydrated, and this is an important means of water conservation because food intake normally necessitates considerable water loss in the excretion of waste products as we saw earlier. The main water conservation mechanisms are indicated in Figure 15.8. It is important to realize that any method of water conservation disadvantages the animal in some way. It may interfere with normal thermoregulation or it may reduce the intake of energy. The different mechanisms are given different emphasis in different species, in accordance with the animal's normal ecological circumstances. Thus, the camel sacrifices a stable body temperature to conserve water,

Fig. 15.7 Extracellular thirst. Hypovolaemia (loss of extracellular volume) monitored by blood-pressure receptors, causes secretion of renin by the JG (juxta-glomerular) cells. This is converted into angiotensin, which promotes drinking (After Fitzsimons, 1971).

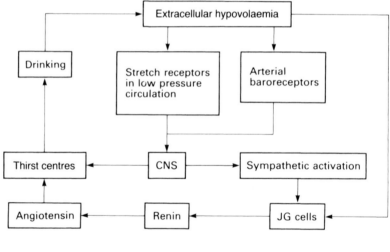

Fig. 15.8 Water conservation mechanisms operate when drinking is not possible. As thirst increases, antidiuretic hormones from the pituitary gland promote water retention in the kidneys, so that less water is lost in urine. Another important means of water conservation is reduction in food intake, because food digestion and excretion normally involve water loss.

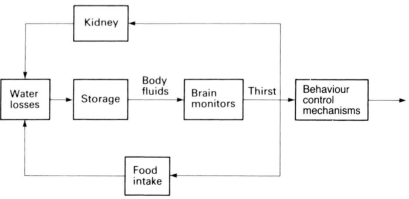

whereas a pigeon sacrifices food intake and maintains a relatively stable body temperature (see Figure 15.2).

At first sight, the mechanisms that maintain water balance would seem to be relatively straightforward examples of homeostasis. The animal detects displacements from the normal amounts (volume) and concentration (osmosity) of extracellular water and takes action to remedy the situation, either by drinking or by reducing the rate of loss of water through various conservation mechanisms. However, the situation is not quite so simple because of interactions with other systems and because drinking often occurs when the animal is not dehydrated.

15.4   Energy and nutrients

Animals require food that can be digested to provide energy and certain specific nutrients and vitamins to provide materials for growth and repair of tissues and to combat parasites and disease organisms.

The cells of the body obtain their energy primarily in the form of glucose dissolved in the extracellular fluid. In the process of metabolism, energy is made available to the cell, and water, carbon dioxide and heat are by-products. The glucose in the extracellular fluid may come directly from the process of digestion or may be released by the liver from its store of glycogen.

The cells of the body can also obtain energy from metabolism of fatty acids, with the exception of the cells of the nervous system, which can utilize only glucose. It is important for the glucose level of the blood to be maintained within fairly narrow limits in order to ensure an adequate supply of energy for the cells of the nervous system. The availability of glucose to cells is controlled by the hormone **insulin**. The cells of the nervous system can absorb glucose in the absence of insulin, but other cells require insulin to transport glucose across the cell wall. During times of glucose shortage, as in fasting, the level of insulin in the blood falls to such an extent that the blood glucose is effectively available only to the cells of the nervous system. The other cells have to depend upon metabolism of fatty acids to obtain energy. During fasting, glucose is derived from the body reserves. These reserves include glycogen stored in liver and muscle, fat stored in various parts of the body and, in the last resort, the protein of muscle and other tissues. Fat is broken down to glycerol and fatty acids, and the glycerol is converted to glucose in the liver. Protein is broken down into amino acids, and these are metabolized by the liver to produce some glucose. The sources of energy available during fasting are summarized in Figure 15.9.

The other major source of energy is food. Species differ greatly in the amount and type of food they require. Small animals that have a high metabolic rate, like songbirds (Passeriformes), quickly suffer an energy shortage when they are unable to obtain food. To maintain its body weight and normal activity, the great tit (*Parus major*) must feed every few minutes. At night when feeding is impossible, some birds become torpid, lower their body temperature and conserve energy. Other animals are able to draw upon energy reserves and may live for long periods without food. For example,

Fig. 15.9 The main sources of energy available during fasting (After Toates, 1980).

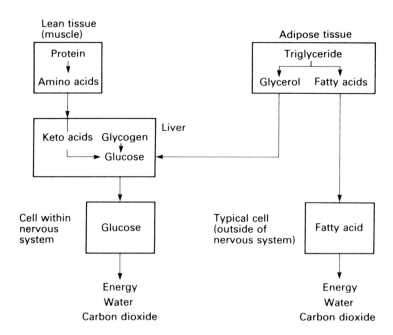

incubating jungle fowl (*Gallus*) do not eat for many days and eat little even if food is placed next to the nest (Sherry *et al.*, 1980). Animals with low metabolic rates at low temperatures, such as fishes, reptiles and hibernating mammals, may not need to eat for weeks.

Food that enters the mouth may be stored immediately, as in the crop of pigeons and the cheek pouches of hamsters, or it may enter the stomach and intestine and undergo the process of digestion. Enzymes in the digestive tract break foodstuffs down into their basic components. Thus, fat molecules are split into glycerol and fatty acids by the action of lipase, and the enzymes trypsin and chymotrypsin split specific amino acid bonds in proteins. The processes of digestion and the enzymes involved vary considerably from one animal species to another. Some animals like the flatworm *Planaria* do not have a complete digestive tract, whereas others like vertebrate herbivores have very complex digestive systems that enable them to cope with plant materials like cellulose that other animals are unable to digest. The products of digestion are absorbed into the bloodstream, partly by diffusion and partly by active transport across the wall of the intestine.

The process of digestion and the pattern of eating are often intimately related. Many animals vary their food intake according to the nutritive value of the products of digestion. A variety of mechanisms are involved in this type of regulation. The simplest mechanism involves direct detection of the substance, as is thought to be the case with sodium. A large and fairly constant quantity of sodium in the body fluids is vital. Sodium is involved in many physiological processes of fundamental importance, including the propagation of nerve impulses (see Chapter 11). Sodium is available in nature in the form of sodium chloride (common salt), but its distribution is patchy. It is not surprising, therefore, that animals should develop a specific appetite for sodium. Animals can detect sodium in the diet in two main ways.

First, salt is a primary aspect of taste in most vertebrates (see Chapter 12). Second, sodium has profound effects on the body fluid compartments, as discussed earlier, and sodium depletion results in secretion of the hormone aldosterone from the pituitary gland. This hormone causes sodium to be reabsorbed from the urine passing through the kidney.

Sodium appetite appears to be innate, but many animals are adept at learning and remembering the location of sources of sodium. For example, rats can be given a choice of pure water or salty water as a reward in a maze. Rats deprived of water developed a preference for the place where water was available in the maze. When these rats later are depleted of sodium but not of water, they immediately switch to the salty water, showing that they remembered the location of sodium, even though it was sampled at a time when they were thirsty and averse to salty water (Krieckhaus, 1970).

Animals are not able to detect many essential vitamins and minerals directly either by taste or by their levels in the blood. Nevertheless, deficient animals develop strong preferences for foods containing the missing substances. For many years, such specific hungers posed something of a problem for scientists attempting to explain how the animals knew which food contained the beneficial ingredient. There were also reports in the medical literature of children eating coal and other unusual substances. These habits were linked to dietary deficiencies for particular essential minerals like cobalt, but how the children learned what to eat remains unexplained.

Thiamine deficient rats show an immediate marked preference for a novel food, even when the new food is thiamine deficient and when the old food has a thiamine supplement (Rodgers and Rozin, 1966). The preference is short lived. If consumption of a novel food is followed by recovery from dietary deficiency, however, then the rat rapidly learns to prefer the novel food. Such rapid learning on the basis of the physiological consequences of ingestion enables the rat to exploit new sources of food, and, thus, to find out which contains the required ingredients. Detailed study of the feeding behaviour of rats shows that they normally tend to avoid novel foods or to sample them a little bit at a time. The intervals between meals are sufficiently long to enable the animal to assess the consequences of sampling new foods. Scientists have also found that, when a rat eats, some of the food is passed rapidly through the stomach into the intestine, bypassing food from a previous meal that is still in the stomach (see Figure 15.10). In this way a rat can test the qualities of the food it has eaten most recently (Wiepkema *et al.*, 1972).

Fig. 15.10 Rat stomachs showing distribution of blue food (dark) eaten after normal food (pale) (After Wiepkema *et al.*, 1972).

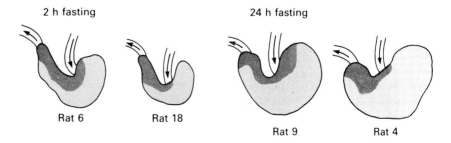

Paul Rozin and James Kalat (1971) and Rozin (1976) point out that a vitamin deficient diet is like a slowly acting poison. Rozin (1967) noted that rats on a thiamine deficient diet were reluctant to eat their familiar food but that they ate avidly from a new deficient food. The aversion to the familiar food persisted even after the rats had recovered from the deficiency. Rats that become sick after eating a poisoned food also show an aversion to such food and show a more than normal interest in novel foods. Thus, there is a close affinity between the rat's responses to a deficient diet and its behaviour toward a toxic diet. This is discussed in Chapter 18.

An animal's dietary requirements are many and various. In the rat they include water, nine essential amino acids, some fatty acids, at least ten vitamins and thirteen minerals (Rozin, 1976). Despite a large amount of scientific work, the question of what physiological factors initiate eating remains something of a mystery. Walter Cannon (1932) thought that stomach contractions and other peripheral factors were responsible. However, when the nerves from the stomach are cut, or when the stomach is removed for medical reasons, human eating behaviour is largely unaffected. Several theories have proposed that receptors within the brain are sensitive to the presence of nutrients in the blood. Substances such as glucose, amino acids or fats might be used as indices of dietary requirements. The level of glucose in the blood increases during digestion, but it also changes in other circumstances such as in autonomic arousal and in anticipation of meals. Jean Mayer (1955) proposed that the brain responds to the **utilization** of glucose rather than to its **availability**, and others have proposed similar mechanisms (Toates, 1980). The suggestion is supported by the results of recent research, but the evidence remains inconclusive.

Some scientists doubt that direct monitoring of the nutrient state of the body is of prime importance in hunger. Although the fact that animals are able to respond to dietary excesses or deficits is suggestive of direct monitoring and regulation, there are alternative possibilities. Animals suffering from a particular deficiency tend to sample a wider range of food and to learn quickly to select appropriate foods. The evidence suggests that such learning is based upon general sensations of sickness and health rather than upon detection of specific deficiencies (Rozin and Kalat, 1971; Rozin, 1976). Curt Richter (1943, 1955) showed that rats left to their own devices with a large variety of purified nutrients from which to choose select a balanced diet from this cafeteria.

How does the rat know which food to select at a particular time? The answer is that for some constituents of the diet, such as water, sodium and possibly glucose, there is different monitoring by taste and by sensory processes in the brain that detect the presence of each of these substances in the blood. For other dietary elements, particularly the vitamins and minerals, there is no direct monitoring. The animal learns what to eat in order to avoid sickness. How this learning occurs we discuss in Chapter 18. In the case of amino acids and fats, the situation remains unclear. Direct monitoring and homeostatic control of these substances have been proposed (see Booth, 1978, for details) but the evidence remains weak.

15.5 Stress

Stress is often characterized as a threat to homeostasis (Chrousos *et al.*, 1988). However some aspects of stress have little to do with the internal environment, especially those arising from some aspect of social behaviour. Both external and internal stimuli can cause stress, and such stimuli are usually called **stressors**. There is a wide variety of stressors, ranging from haemorrhage to signs that a predator is in the vicinity, and stressors can really only be identified by their effects upon the animal's physiology. Selye (1973) referred to the syndrome of physiological changes induced by stressors as the **general adaptation syndrome**. This syndrome consists of specific physiological changes, especially increased secretion of corticosteroids, that result from a diverse range of stimuli (the stressors) and have important effects upon physiological equilibria.

Toates (1995) argues that, given the variety of causes of stress and the variety of stress responses, it is useful to characterize stress as a particular kind of aversive motivational state (see below). Whether it is possible to identify features of the state of stress that are common to all species is doubtful. In particular, there are likely to be large differences between invertebrates and vertebrates.

Behavioural and physiological adaptation to natural stressors has been studied (see Davenport, 1985, for a review) in invertebrates. On the whole it appears that invertebrates have specific automatic responses to single natural stressors, and lack a general mechanism that could be implicated in the response to a variety of stressors. Thus Eisemann *et al.* (1984) and Bateson (1991) conclude that insects do not experience pain. However, some caution is appropriate here, since aversive stimuli can induce a similar range of neuromodulators in both invertebrates and vertebrates (Kavaliers, 1989).

Stress in vertebrates is closely associated with hormonal function (see Chapter 11 for a general review of vertebrate hormones). Animals faced with aversive situations undergo a fundamental reorganization of their physiology. There are two main systems involved, the autonomic nervous system (see Chapter 11) and the **hypothalamic–pituitary–adrenocortical system** (see Box 15.1). This system is largely responsible for the release of adrenocorticotrophic hormone (ACTH).

ACTH exerts its effect upon the adrenal cortex, promoting the synthesis and secretion of corticosteroids. These are important in the regulation of carbohydrate metabolism, and have other metabolic effects which are thought to protect the body against overactivity of its natural defences (Munck *et al.*, 1984; Toates, 1995). Levels of corticosteroids are closely regulated by means of negative feedback to the hypothalamus and pituitary, as shown in Figure 15.11, but this does not mean that the level is held constant. Indeed there is a marked diurnal rhythm in many mammals, including humans (see Figure 15.12).

Corticosteroids are also known to be implicated in the activities of the adrenal medulla, especially the regulation of adrenaline output. Thus the autonomic nervous system and the HPA system are closely linked, and syn-

Box 15.1 The hypothalamic—pituitary–adrenocortical system

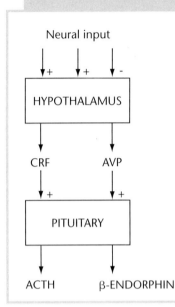

Neural input from various parts of the brain is integrated in the hypothalamus, which acts as a control centre for hormone release. In the paraventricular nuclei of the hypothalamus are the cell bodies of neurons that synthesize and release the corticotrophin-releasing factor (CRF). These neurons secrete their contents into the hypothalamic–hypophyseal portal vessels, which carry them to the anterior pituitary gland. The neurons of the paraventricular region of the hypothalamus are known to be activated by stressors.

The paraventricular nuclei also contain neurons that synthesize and release other neuromodulators, such as argenine vasopressin (AVP), which is transported to the anterior pituitary via the hypothalamic–hypophyseal portal system. Both CRF and AVP have a synergistic effect upon the release of adrenocorticotrophic hormone (ACTH), and β-endorphin, by the anterior pituitary gland.

Fig. 15.11 Negative feedback effects of corticosteroids at both the hypothalmic and pituitary levels. CRF = corticotrophin-releasing factor; ACTH = adrenocorticotropic hormone. (From Toates, 1995).

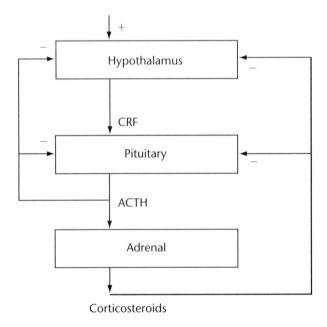

ergistic in their actions. Thus an animal subject to external or internal stressors is likely to have raised levels of adrenaline, noradrenaline, glucocorticoids, thyroxine and growth hormone (in some species). Levels of insulin, testosterone and oestrogen are likely to be depressed (Toates, 1995). Overall, activation of the sympathetic branch of the autonomic nervous system and of the HPA leads to an alteration of hormone balance, which puts

Fig. 15.12 Circadian rhythm of adrenocorticotrophic hormone and corticosterone in young male rats. (From Toates, 1995a, after Dallman *et al.*, 1987)

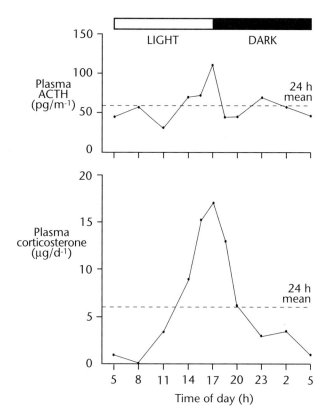

the animals in good physiological condition to take appropriate behavioural action, such as fight or flight (see also Chapter 28).

De Boer *et al.* (1988) subjected rats to a loud noise three times in succession. They found an immediate rise in the plasma concentrations of adrenaline and noradrenaline, followed by a decline to normal within about 20 minutes. There was a delayed rise and fall in corticosterone. The reactions to the first noise were the most marked, and those to the third noise the smallest. Thus the animals habituated to the sound stimulus that was not followed by any aversive consequences. Repeated exposure to foot shock, results in hormonal reactions that do not habituate (Ratner *et al.*, 1989).

The consequences of aversive stimulation can vary according to the circumstances. If there is a predictable relationship between a cue and a stressor, then the animal is likely to learn to anticipate the stressor (see Chapter 17). For example, Burchfield *et al.* (1980) exposed rats to a cold environment for ten minutes per day. After three months of such treatment there was a clear rise in corticosteroid secretion prior to each daily test session. Anticipation of food is also associated with a rise in corticosteroid levels, so it may be that anticipation of any possible future need to take action triggers corticosteroid release (Toates, 1995).

Circumstances in which the animal has some control over aversive stimulation have received a fair amount of study. Hanson *et al.* (1976) looked at plasma levels of cortisol in rhesus monkeys that were able to terminate a loud noise, compared with monkeys that were exposed to the same noise but could do nothing about it. They found that the monkeys that had control over the noise had normal levels of cortisol, while those with no control had elevated levels. Making an effective response to alleviate aversive situations is sometimes called **coping**. Swenson and Vogel (1983) yoked rats in pairs, where one could terminate an electric shock by manipulating a pole, while the other could not. Both copers and non-copers showed an increase in plasma corticosterone levels, but the increase was longer lasting in the non-copers. Similar effects were found for plasma adrenaline levels and hypothalamic hormone levels. Thus having control over the stressor is associated with lesser signs of stress (as measured hormonally) compared with not having control.

Other circumstances that have been found to induce hormonal signs of stress include mother–infant separation, exposure to novelty, conspecific fighting and social subordination (see Toates, 1995, for a review). The variety of situations and circumstances that lead to activation of the HPA axis has led Toates to ask the inverse question – under what conditions is the HPA not activated? He concludes that activation does not occur when the animal is in a familiar situation, having a tried and tested coping strategy available for dealing with any anticipated change in that situation, and where action taken delivers results that come up to expectations.

15.6 Motivational state

Homeostatic requirements place behavioural demands upon animals as we have seen. At any given moment an animal must assess its total internal state, incorporate its knowledge of likely future demands and of the new demands to be created by any given course of action and then must choose what to do next.

The traditional view of motivation is built upon a single feedback principle. A change in the animal's internal state is sensed by the brain and leads to a build-up of drive to perform the appropriate behaviour. The drive gives rise to appetitive and consummatory behaviour. **Appetitive** behaviour involves a search for suitable external stimuli; when these are encountered, **consummatory** activity, like eating or drinking, takes place. The consequences of the consummatory behaviour reduce the drive, either directly or by diminishing the internal or external stimuli that led to the drive. The consummatory behaviour then ceases. For example, dehydration of the body tissue is sensed by the brain and results in a build-up of thirst drive. The drive induces the animal to search for water (appetitive behaviour), and when it finds water it drinks (consummatory behaviour). The consequences of drinking may reduce thirst directly via short-term satiation mechanisms such as the sensation of water in the mouth or loading effects of water in the gut (Rolls and Rolls, 1982). The short-term satiation mechanisms are complemented, or by-passed in some species (Rolls and Rolls, 1982), by absorption of water into the bloodstream, thus alleviating the dehydration that gave rise to the thirst drive. The animal stops drinking either as a result of the short-term satiation mechanisms or because its state of water balance no longer gives rise to thirst.

Drive

The term **drive** was introduced by Robert Woodworth (1918) as an alternative to William McDougall's (1908) concept of **instinct** (see Chapter 20). Woodworth distinguished between the **energizing** (drive) and the **directing** aspects of motivation. Primary drives resulted from tissue needs, and other (**secondary**) drives were derived from learned habits. Similar notions of drive were developed by the early ethologists. Thus, Konrad Lorenz (1950) outlines the ethological view in terms of three successive processes:

1. Accumulation of action-specific energy giving rise to appetitive behaviour.
2. Appetitive behaviour striving for and attaining the stimulus situation activating the innate releasing mechanism.
3. Setting off of the releasing mechanism and discharge of endogenous activity in a consummatory action.

Lorenz postulated 'that some sort of energy, specific to one definite activity, is stored up while this activity remains quiescent, and is consumed in its discharge'.

The basic idea, of drive as an urge to perform particular activities, was common to ethologists and to various schools of animal psychology in the United States. During the fifty years since its introduction, the drive concept has been subjected to numerous quasi-philosophical discussions and analyses. Controversy hinged around questions such as whether drives were inherently purposive (Thorpe, 1956; Peters, 1958), whether drives were general or specific (see Hinde, 1970; Bolles, 1970) and whether drive could be said to energize behaviour (Bolles, 1970; McFarland, 1971). In recent years, a tendency to drop the concept of drive has arisen for reasons we discuss below.

The classical view of hunger and thirst as homeostatic drives implies that feeding and drinking are initiated as a result of monitored changes in the animal's physiological state. Feeding and drinking are said to be under negative feedback control because food and water intake serve to diminish the deviations in physiological state that initiated the behaviour. However, in addition to occurring in response to physiological changes, feeding and drinking often occur in anticipation of such changes. Many animals have distinctive meal patterns that are repeated day after day under constant conditions. Just as humans experience hunger at habitual meal times, so the feeding tendency of animals can be governed by time of day. In conditions in which the environment changes little from day to day, animals quickly establish a daily routine of activities and eat at particular times even when food is continuously available. Physiological processes may become attuned to the routine. In humans, for example, the liver may cease to mobilize glycogen just before a meal is due. This leads to a fall in blood sugar level in anticipation of the increase that will occur when the meal is digested. Experiments have shown that such physiological adjustments can undergo conditioning in relation to the time of day.

We have seen that the long-term consequences of food intake tend to lead to increased thirst. Instead of drinking as a result of such thirst, many animals drink in advance and thus anticipate the dehydrating effects of food ingestion (Fitzsimons and Le Magnen, 1969). Similarly, we have seen that thermoregulation often involves water loss, but instead of drinking in response to thermally induced dehydration, some animals drink in advance and thus have water available for thermoregulation. For example, Phil Budgell (1970a) found that doves (*Streptopelia risoria*) deprived of water for two days at different temperatures drank the same amount when offered water in a test chamber at a temperature of 20 °C. However, doves deprived for two days at the same temperature (20 °C) drank different amounts when tested at different temperatures (see Figure 15.13). Similar results have been obtained with rats (Budgell, 1970b), showing that these animals drink in direct response to environmental temperature changes before any thermally induced dehydration occurs. The term **feedforward** (McFarland, 1971; Toates, 1980) is used for situations in which the feedback consequences of behaviour are anticipated and appropriate action is taken to forestall deviations in physiological state.

There may not always be a one-to-one relationship between a particular activity and its physiological consequences. In some cases, like behavioural thermoregulation, a particular activity may have consequences that are measurable along a single dimension, like a change in body temperature. More commonly, however, behaviour has consequences that affect a number of different aspects of the animal's state. For example, food intake will alter an animal's physiological state in a number of respects, depending upon the constitution of the ingested food. The consequences of such behaviour are said to be ambivalent.

In order to account for homeostasis, in both its physiological and behavioural aspects, it is necessary to go beyond simple negative feedback concepts. The maintenance of physiological state within narrow limits may be achieved by a combination of negative feedback, feedforward and adaptive control. The interactions among the physiological processes that give rise to behaviour are complex. In the case of an apparently simple activity – drinking, for instance – several different factors can contribute to its initiation, mainten-

Fig. 15.13 Effects of room temperature on drinking in doves. Left: the effect of temperature during water deprivation upon subsequent drinking. Right: the effects of temperature during drinking (Data from Budgell, 1970a).

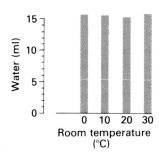
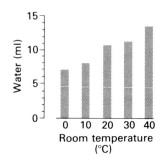

ance and termination (Rolls and Rolls, 1982). In the case of feeding, the situation is much more complex (Booth, 1978). To handle this type of complexity, behavioural scientists have adopted the techniques of control systems theory used by engineers to describe and analyse complicated machines. By making quantitative models of the various components of a system, like the drinking control system, it is possible to carry out a computer simulation of the system as a whole. Such a simulation can be used to make quantitative predictions that can be tested in experiments. The results of the experiments then can be used to update the hypotheses, and in this way increasingly refined models can be developed. Control systems theory has been applied to a variety of aspects of behaviour, including feeding, drinking, thermo-regulation and sexual behaviour (McFarland, 1971, 1974; Booth, 1978; Toates, 1980).

Compared with the precision and rigour of control systems theory, the concept of drive is vague and confusing. Moreover, the new approach has uncovered conceptual problems that effectively mean that the concept of drive is no longer a viable entity. Two of these problems were pinpointed by Robert Hinde (1959, 1960), but the theoretical concepts necessary to handle the issues were not then current in the field of animal behaviour. One mis-conception (Hinde, 1959) is that drives can be considered as unitary vari-ables. When we think of hunger drive, we imagine a quantity that is measurable along a single scale. Thus, we think of an animal as having low or high hunger. However, as we have seen there are many different aspects of hunger. An animal may have a specific hunger for, say, salt or thiamine. It may have a protein deficiency or a shortage of readily available energy. These different aspects of hunger can influence the animal's feeding behaviour, and it is misleading to think of hunger as a unitary variable. An alternative for-mulation (McFarland and Sibly, 1972) represents hunger and other so-called drives in terms of vectors. This approach paved the way for the state-space representation of motivational systems.

The state–space approach

At any particular time, the animal is in a specific physiological state, which is monitored by the brain. The behaviour we see is determined by the brain in accordance with this monitored state in combination with the animal's per-ception of environmental stimuli. The combined physiological and perceptual state, as represented in the brain, is called the motivational state of the animal. It includes factors relevant to incipient activities, as well as to the animal's current behaviour. It is thus fundamentally different from the old concept of drive which it has come to replace.

An animal's physiological state can be represented as a point in a physio-logical space. The axes of the space are the important physiological variables such as temperature, and the space is bounded by the tolerance limits for these variables.

Similarly, the animal's motivational state can be represented by a point in a motivational space. The axes of the space are the important motivational stimuli such as the degree of thirst or the strength of some external stimulus.

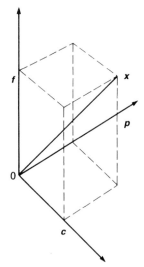

Fig. 15.14 Hunger represented as a multidimensional vector quantity. *f* = fat, *p* = protein, *c* = carbohydrate, *0* = origin, *x* = hunger state (After McFarland and Sibly, 1972).

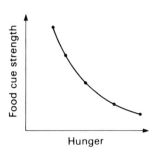

Fig. 15.15 A motivational isocline relating hunger and strength of food cues. The line joins those states (points) that lead to the same strength of feeding tendency.

The state–space concept is important not only in describing motivational systems but also in providing a link between mechanistic and design aspects of decision-making.

One advantage of the state–space representation is that the components of non-unitary aspects of motivation, like hunger, can be portrayed readily, as is shown in Figure 15.14. Another advantage is that the combined effects of internal and external stimuli can be represented in a simple way. In Chapter 12, we saw how quantitative measures of the motivational effectiveness of external stimuli can be combined into a **cue strength** index, and that the relationships among different stimuli can be represented by the motivational isocline. For example, the tendency to seek water that arises from a high degree of thirst and a low strength of cues indicating the availability of water could be the same as the tendency due to the combination of a low degree thirst and a high cue strength for water availability. The points representing these two different motivational states lie on the same motivational isocline, as illustrated in Figure 15.15.

The shape of a motivational isocline gives an indication of the way in which the different factors combine to produce a given tendency. For example, the internal and external stimuli controlling the courtship of the male guppy (*Lebistes reticulatus*) seem to combine multiplicatively (see Figure 15.16), although as we saw in Chapter 12, we have to be careful about our scales of measurement before coming to any firm conclusion.

A major difference between the state–space approach to motivation and the traditional drive concept is that the modern approach makes no assumptions about the ways in which different motivational factors combine or about the relationship between motivation and behaviour. The early psychologists and ethologists tended to assume that motivational factors combined in a particular way. A common assumption was that internal and external factors combine multiplicatively so that the tendency to perform an activity would be zero if internal drive were high and if there were no relevant external cues, and would be zero also if drive were negligible but if the external cues were strong. While this may be true in some cases, the modern approach is to regard this question as an entirely empirical matter, not subject to any particular doctrine.

In the traditional view, behaviour is driven from within, and there is a direct causal relationship between the strength of drive and the properties of the ensuing behaviour. In the modern view, the relationship between motivational state and behaviour is not direct. Although particular combinations of factors can give rise to a given tendency, there is not necessarily a direct relationship between the tendency and the observed behaviour. For example, an animal may have a tendency to feed, which is stronger than any other tendency, but in the interests of some long-term strategy, it may postpone feeding; or an animal may feed even though its feeding tendency is not the strongest tendency. These possibilities have been suggested by observations on animal behaviour. To evaluate them is a complicated task, and many questions currently are unresolved. In this chapter we consider only some aspects of the problem.

In general, the focus of research on this topic has shifted away from

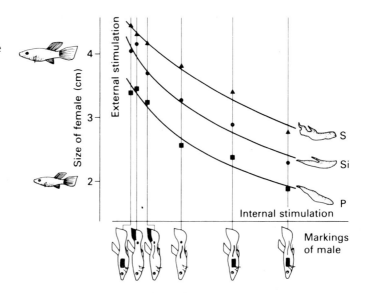

Fig. 15.16 The influence of the strength of external stimulation (measured by the size of the female) and the internal state (measured by the colour pattern of the male) in determining the courtship behaviour of male guppies. Each curve (isocline) represents the combination of external stimulus and internal state that produces the sigmoid courtship postures of increasing intensity P, Si, S (After Baerends *et al.*, 1955).

attempts to establish directly the causal chain linking motivation and behaviour. This is partly because of conceptual difficulties connected with the calibration and measurement of causal factors (McFarland, 1976; Houston and McFarland, 1976; McFarland and Houston, 1981) and partly because of a realization that animals are probably more mentally complicated than hitherto had been thought. The focus of most current investigation is on the consequences of behaviour rather than its immediate causes. What determines the consequences of an animal's behaviour? How do the consequences influence the behaviour of the animal? To what extent is an animal able to take account of the probable consequences of future behaviour? To what extent is it able to assess the costs and benefits of different courses of action?

The consequences of an animal's behaviour are the result of interaction between behaviour and environment. The consequences of foraging behaviour, for instance, depend partly upon the foraging strategy used and partly upon the availability and accessibility of food. Thus, a bird foraging for insects may search in appropriate or inappropriate places, while the insects may be abundant or scarce at the particular time. Similarly, the consequences of courtship display, in terms of the reaction of the partner, depend partly upon the type and intensity of the display and partly upon the motivation and behaviour of the other animal.

The consequences of an animal's behaviour influence the motivational state of the animal in a number of different ways. Two important consequences of foraging are the expenditure of energy and the intake of food. Energy and other physiological commodities like water are expended in all behaviour to a degree that depends upon the level of activity, the weather, etc. Such expenditure must be taken into account because it influences the animal's state. Food intake may alter the animal's state in a number of ways. The presence of food in the mouth may increase appetite temporarily through a positive feedback mechanism (Wiepkema, 1971); it may have a (negative feedback) satiating effect; or it may lead to the rejection of the food as a result of its unpalatability. The presence of food in the gut may have a short-term satiating effect, and it may influence other physiological factors such as temperature and water balance. The food in the gut is digested, and nutrients are absorbed into the bloodstream. The changes in the blood that are consequent upon feeding are complex and may influence many aspects of the animal's physiological state. The point to remember is that all aspects of behaviour have consequences of equivalent complexity. These may include effects upon other animals and upon features of the external environment, like a nest. While it is relatively easy to pinpoint the consequences of feeding because of the vast amount of research that this aspect of behaviour has attracted, we should not forget that the effects of behaviour upon other animals, and upon the environment, influence the motivational state of the animal in a manner equivalent to the consequences of feeding (McFarland and Houston, 1981).

The consequences of behaviour can be represented in a motivational state–space. The motivational state of the animal is portrayed as a point in the space. As a consequence of the animal's behaviour, the state changes, and so the point describes a trajectory in the space, as illustrated in Figure 15.17. On a two-dimensional page it is possible to portray the trajectory in only a single plane of the space, but we can imagine that the changing state describes an equivalent trajectory in a multidimensional space. The advantage of this type of representation is that it enables us to portray very complex changes in state in a relatively simple manner. In the case of an animal drinking cold water, for example, we can see that decisions to change behaviour depend partly upon the trajectory and partly upon the position of the relevant motivational isoclines, as shown in Figure 15.17.

Fig. 15.17 The consequence of behaviour described as a trajectory in motivational space. Left: the consequences of drinking cold water (a short-term reduction in body temperature and a long-term reduction in thirst). Right: where the trajectory cuts a motivational isocline, changes in behaviour can be expected.

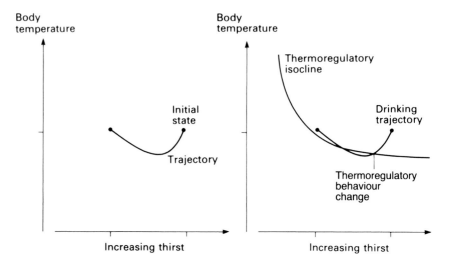

Points to remember

- Homeostasis, the maintenance of an internal steady state, is important for animals that live in a variable environment. It is not, as used to be thought, achieved primarily through negative feedback mechanisms, but by a combination of feedback, feedforward and adaptive control.

- Most animals have some form of thermoregulation. In ectotherms the primary source of heat is outside the body, while in endotherms it is generated by metabolic processes.

- All animals have some obligatory water loss for which they must compensate. Many animals drink more water than is necessary for homeostasis and are able to cut down on their water losses when water is not available. This may, however, involve curtailment of other aspects of behaviour.

- All animals require both energy and nutrients, and many are able to learn how to obtain essential substances, even though they cannot directly detect them in their food.

- The idea of specific drives responsible for different aspects of behaviour has been superseded by the notion of motivational state. This is influenced both by external factors and by the consequences of the animal's own behaviour.

Further reading

Toates, F.M. (1986) *Motivational Systems*. Cambridge University Press, Cambridge.

Toates, F.M. (1995) *Stress*. J. Wiley & Sons, Chichester.

Physiology and behaviour in changing environments

In the last chapter we looked at mechanisms whereby animals maintain their internal environment at steady state through physiological and behavioural adjustment. In this chapter we consider the steps animals must take to deal with the physiological demands of maintaining survival and reproduction in variable environments.

16.1   Tolerance

Life very probably originated in the sea (Whitfield, 1976; Grogham, 1976). Compared with other biomes, those of the ocean environment are relatively stable. Fluctuations in physical characteristics such as temperature, oxygen tension, etc. are small so that the internal environment of many marine invertebrates is subject to little disturbance. Such animals are usually **conformers** in that their bodily condition tends to conform with that of the environment, and they cannot live in variable conditions. For example, the salinity of the body fluids of many marine invertebrates is identical to that of sea water (Baldwin, 1948; Barrington, 1968). Other animals, known as **regulators**, maintain their bodily functions in a condition relatively independent of environmental fluctuations. Such an ability was a prerequisite for the invasion of fresh water and of dry land.

Each species has a characteristic ability to tolerate extreme values of environmental factors, such as temperature, humidity, etc. For example, many marine invertebrate animals are affected by variation in the salinity of the water because their body fluids normally have much the same salt concentration as sea water and their body tissues are adapted to function efficiently in such a medium. If they become immersed in a less saline medium, water enters their tissues as a result of osmosis. In a more saline medium, the opposite process occurs. Animals that cannot control the passage of water into their bodies are limited in their range of living conditions by the tolerable degree of salinity.

An animal's habitat is restricted by the animal's tolerance range. In estuaries, for example, different species of the amphipod *Gammarus* are restricted to different parts of the estuary due to their differing salinity tolerance ranges. As illustrated in Figure 16.1, *Gammarus locusta* has a high tolerance of salt water and is found at the mouths of estuaries. *Gammarus zaddachi* has moderate tolerance of salt water and is usually found in the stretch of river between eight and twelve miles from the sea. *Gammarus pulex* is a true freshwater species and does not occur at all in parts of the river showing any influence of the tide or salt water. In this case each species can tolerate only a restricted range of salinity, but each has been selected towards a salinity spectrum uninhabited by any other member of the genus.

When measures of survival, or fitness, are plotted against important environmental variables, bell-shaped curves like those illustrated in Figure 16.2 usually result. Only a few animals can survive the extreme high or low values and must cluster around some central range. Such curves show not only the limits and range of tolerance of the species but also the optimum

Fig. 16.1 The distribution along a river of three closely related species of the amphipod crustacean *Gammarus*, relative to the concentration of salt water. The degree of freshness of the water is indicated by the degree of shading (From *The Oxford Companion to Animal Behaviour*, 1981).

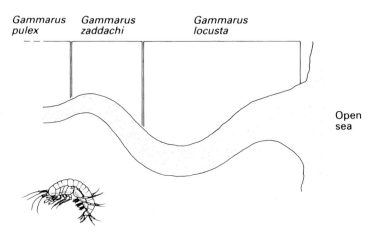

Fig. 16.2 Tolerance of three species of marine wood-boring isopods to different constant chlorinities (After Reish and Hetherington, 1969).

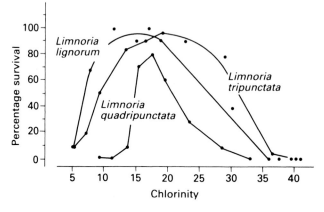

Fig. 16.3 Diagram of population abundant along an ecological gradient (From *The Oxford Companion to Animal Behaviour*, 1981).

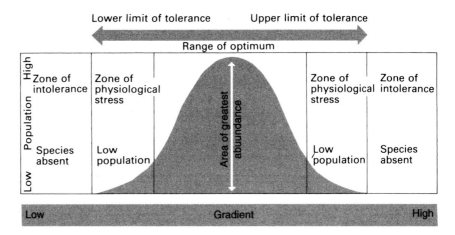

values of the environmental variables. These aspects of tolerance are illustrated in Figure 16.3. Similar considerations apply to the tolerance of animals along all sorts of environmental gradients.

In the case of conformers, tolerance will often be directly reflected in the animal's physiological state. Thus, if an environmental temperature of 40 °C is lethal, it is because an internal termperature near 40 °C causes biochemical disruption. However, the limits of tolerance for a given factor are influenced by the values of other environmental variables. Environmental factors in combination may kill an animal at intensity levels that if taken separately, would not be lethal. For example, for the American lobster *Homarus americanus*, a temperature of 32 °C is lethal when the salinity is 30 parts per thousand and when the oxygen content is 6.5 milligrams of oxygen per litre. If the oxygen content is decreased to 2.9 milligrams of oxygen per litre, the thermal limit is lowered to 29 °C. In one study lobsters were subjected to 27 combinations of temperature, salinity and oxygen content (McLeese, 1956). From the results, he was able to construct the three-dimensional diagram illustrated in Figure 16.4. This diagram shows how the interaction of factors affects the tolerance limits of the species.

Adaptations at the tolerance limits, usually called **resistance adaptations**, are greatly influenced by exposure time, the rate of change in environmental factors and the history of the individual. The physiological mechanisms involved in resistance adaptation require a certain amount of time to adjust to a changing situation. If the environmental change is very sudden, death may occur, whereas if the same change takes place more gradually, it might result in survival. The processes involved in resistance adaptation vary from very rapid regulation to slow acclimatization. Thus, very gradual changes in the environment may enable the individual to acclimatize to new conditions.

16.2 Acclimatization

Acclimatization is a form of physiological adaptation by which an animal is able to alter its tolerance of environmental factors. Usually, the term

Fig. 16.4 Three-dimensional representation of the boundary of lethal conditions for the American lobster *Homarus americanus* for various combinations of temperature, salinity and oxygen (After McLeese, 1956).

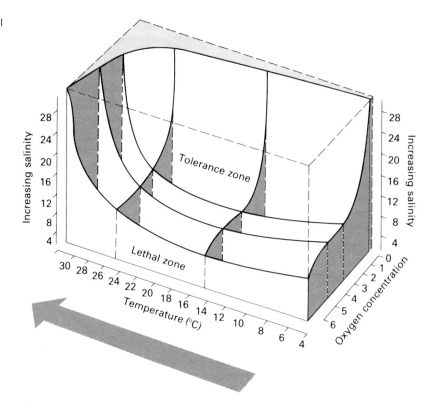

acclimation is used in laboratory studies in which the adaptation occurs to a single factor like temperature. The term **acclimatization** is reserved for the complex of adaptive processes that occur under natural conditions.

Acclimatization often occurs in response to seasonal changes in climate. Thus, seasonal shifts in the upper lethal temperature of freshwater fish are often directly correlated with changes in the temperature of the habitat (Fry and Hochachka, 1970). Because behavioural adjustments can make acclimatization unnecessary and vice versa, there is a wide variety of combinations actually used by animals.

The temperature preferences of the fishes often are related to their state of acclimatization. Many fish species seem to have highly developed behavioural thermoregulation and respond to a temperature gradient by selecting a particular temperature (Fry and Hochachka, 1970). Individual goldfish (*Carassius*) can be trained to maintain the temperature of their aquarium water by actuating a valve to introduce cold water as the temperature rises (Rozin and Mayer, 1961). Such fish maintain the aquarium temperature at about 34 °C with considerable precision.

To a large degree, the phenomenon of temperature selection explains the distribution of fish in nature, and fish usually prefer the temperature to which they are acclimatized (Fry and Hochachka, 1970). Such an arrangement makes biological sense. If slow changes in physiological state due to acclimatization were not followed by corresponding changes in behavioural preferences, then acclimatory processes could be opposed by behavioural mechanisms. For example, acclimatization to cold might be counteracted by a behavioural tendency to select a warmer climate. If a short-lived opportunity to select a warm environment were exploited to the full, then much of the work done in establishing acclimatization to cold would be undone. The obvious alternative is for animals to prefer conditions to which they are acclimatized. The picture is complicated, however, by the phenomenon of **anticipatory adjustment**, by which the animal responds to some aspects of the environment or to its own biological clock and is thereby able to predispose itself to climatic changes.

Acclimatization is normally regarded as a relatively slow process, compared to the rapid physiological adjustments that many animals are able to make in response to sudden changes in environmental conditions. In reality, however, there will usually be a spectrum of adaptive processes, ranging from rapid physiological reflexes to slow acclimatization. For example, when a person is transported suddenly from low to high altitude, the oxygen level of the blood drops because of the rarefied air. This drop is initially counteracted by an increased rate of breathing. However, this type of physiological response involves high energy expenditure. Less costly, though slower, forms of physiological adaptation (see Figure 16.5) then supplant the rapid physiological response, and these in turn are replaced by more long-term forms of acclimatization, like the manufacture of more red blood cells. Even this, however, is not without cost, since energy is required to manufacture the extra cells, and their presence in the blood increases the viscosity and the work that the heart has to do in pumping the blood around the body. When the person is returned to low altitude, therefore, all the adaptive processes occur in reverse, as illustrated in Figure 16.5. This type of reversibility is characteristic of physiological (as opposed to genetic) adaptation.

As we can see from Figure 16.5, acclimatization and regulation together form a spectrum of adaptive processes, ranging from homeostatic regulation involving quick-acting processes that restore physiological imbalance within a day, to acclimatization which may take days, or weeks, to attain physio-

Fig. 16.5 Acclimatization to altitude. Adaptive changes in a man breathing rarefied air for four days, followed by six days at sea level. V = lung ventilation, E = serum erythropoietin, H = rate of haemoglobin synthesis, R = fraction of red blood cells (After Adolph, 1972).

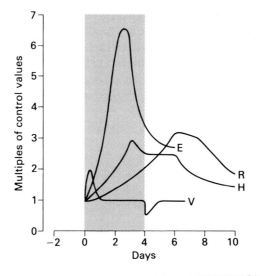

Fig. 16.6 Vectorially additive processes in physiological regulation. The acclimatization vector **a** sums with the regulation vector **p** to give the resultant **r**, which opposed the drift in physiological state **d**. The resultant may be made up of a large contribution from **p** and a small contribution from **a**, as in (a), or a small contribution from **p** and a large contribution from **a**, as in (b) (After Sibly and McFarland, 1974).

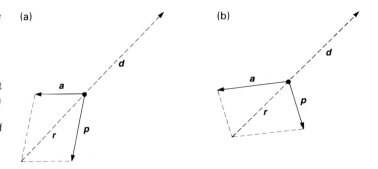

logical steady state. However, the processes of acclimatization are complementary to those of regulation. In acclimatization to altitude, for example, the rapid breathing required at first is gradually slowed as true physiological acclimatization takes place. This means that the effects of regulation and acclimatization upon physiological state sum like vectors, together achieving a desired state (such as a given rate of oxygen transport), with the contributions of each changing as acclimatization proceeds. In cases where there is also a contribution from behaviour, the result will be the vector sum of three processes.

When vectorally additive processes occur at very different rates, as illustrated in Figure 16.6, the state of the slow processes can be regarded as the 'goal' of the faster processes (Sibly and McFarland, 1974). The state of acclimatization can now be regarded as the optimal point of physiological regulation, and the same argument applies in discussing behaviour. For example, if a man is transported from a cold to a hot climate, he has various possible alternative means of adjustment. He can carry on with his normal behaviour, exposing himself to the sun and relying on sweating and other physiological responses to maintain a normal body temperature. After a few weeks he could acclimatize to the heat and would sweat to a lesser extent. Alternatively, he could alter his behavioural routine and seek the shade. He would then rely less on physiological adjustments like sweating, but he would take longer to acclimatize to the new conditions. Thus, as is often the case, the physiological and behavioural solutions to the immediate problem are alternatives. Their effects are complementary and therfore vectorally additive (Sibly and McFarland, 1974; McFarland and Houston, 1981).

As a consequence of this arrangement, we can expect that the effect of acclimatization upon behaviour is to alter the goal of behaviour. There is some evidence that such alteration does occur. For example, the golden hamster (*Mesocricetus auratus*) prepares for hibernation when the environmental temperature drops below about 15 °C. This preparation involves a number of physiological changes, amounting to a form of acclimatization to the cold, and makes hibernation physiologically possible when other conditions are favourable. These conditions include the availability of nest material and sufficient food for the animal to set aside to store. Temperature preference tests, conducted in the laboratory, show that the hamsters develop a marked preference for cold environmental temperatures during the prehibernation period of acclimatization. They prefer an 8°C environment to either a 19°C or a 24 °C environment. Following arousal from a period of hibernation, the situation is reversed, and the hamsters actively prefer the warmer environments (Gumma *et al.*, 1967). As we saw in the case of temperature preferences in fishes, it makes biological sense for behavioural preferences to be linked to the state of acclimatization.

Acclimatory changes can occur as a direct response to environmental changes, but they are also influenced by circannual rhythms. For example, the marine polyp *Campanularia flexuosa* has an annual cycle of growth, development and longevity. This rhythm persists under constant laboratory conditions. Seasonal changes in various metabolic factors are known to occur in mammals including man (Reinberg, 1974) and these are thought to be due to endogenous circannual rhythms (Senturia and Johansson, 1974). Such rhythms would enable the animal to adapt physiologically in anticipation of seasonal environmental changes.

16.3 Biological clocks

Animals commonly confront environmental changes that are cyclical in nature such as days, tides and seasons. In many species animals maintain some sort of internal rhythm, or clock, to predict the onset of the periodic changes and to prepare for them.

There are three basic ways of synchronizing physiology and behaviour to cyclic changes in the environment:

1. There may be a direct response to various changes in external (**exogenous**) geophysical stimuli.

2. There may be an internal (**endogenous**) rhythm that programmes the animal's behaviour in synchrony with the exogenous temporal period, particularly a 24-hour period or a 365-day period.

3. The synchronization mechanisms may be a combination of (1) and (2).

An animal could use many features of the external environment to gain information about the passage of time. The apparent motions of the sun, moon and stars, as seen by a terrestrial observer, provide indications of the time of day, the season of the year, etc. Many animals are known to make use of this type of information. Thus, honey-bees native to Brazil use the sun as a compass in foraging. They can be trained to search for food in a particular compass direction. When they are moved from one locality to another, they continue to forage in the same compass direction, irrespective of the time of day. Thus, bees native to Brazil are capable of compensating for the anticlockwise motion of the sun. However, bees from the northern hemisphere, transported to Brazil, are initially unable to make the appropriate compensation. In the northern hemisphere the sun appears to move in a clockwise direction, and the bees have to learn to adjust to the changed conditions in Brazil (Lindauer, 1960; Saunders, 1976). Similar abilities are well documented in fishes (Hasler and Schwassmann, 1960) and in birds (Schmidt-Koenig, 1979). There is also evidence that some animals respond to the motions of the moon (Papi, 1960) and stars (Schmidt-Koenig, 1979). In addition, it is possible that animals can obtain time cues from factors like changes in temperature, barometric pressure and magnetic phenomena.

The term **circadian** (*circa* means 'about', *dies* means 'day') is used to describe endogenous rhythms that usually fall short of a 24-hour periodicity. The term **circannual** is used to refer to the endogenous rhythm, which is usually less than 365 days. Many animals maintain rhythmic activity when isolated in the laboratory. This suggests that they possess an endogenous clock. However, it is always possible that the animals are responding to some exogenous factor undetected by the experimenter, and it is necessary to establish suitable criteria for judging whether a clock is truly endogenous. There are various ways in which this can be done. First, the rhythm may exhibit a frequency that is not exactly synchronous with any known environmental periodic factor such as light, temperature or other geophysical variables (Weihaupt, 1964). Second, the period of the endogenous rhythm usually deviates from the natural rhythm when studied under constant laboratory conditions. Third, the rhythm may persist when the animal is moved from one part of the world to another. Only when some such criteria apply can we conclude that the rhythm is endogenous.

Since endogenous rhythms tend to gradually deviate from that of the exogenous cycle (such as day–night light or temperature changes), an organism must have a means of synchronizing its endogenous rhythm with the cycle of external events. Jurgen Aschoff (1960) coined the term **zeitgeber** (meaning time-giver) for the environmental agent that entrains the organism's behaviour to the external environment. For example, when lizards (*Lacerta sicula*) are hatched in an incubator with temperature and light regimens designed to simulate a 16-hour or a 36-hour day, they develop normally and exhibit normal circadian rhythms of activity when tested under constant conditions. Thus, the circadian rhythm of these animals is endogenous and does not depend upon individual experience of the day–night cycle. A 24-hour temperature cycle with an amplitude of 0.6 °C is sufficient to act as a zeitgeber and to entrain the lizards' activity rhythm.

In general, it is apparent that many species of animals, ranging from the single-celled (Sweeney, 1969) to the complex multicellular animals (Aschoff, 1965; Bunning, 1967; Pengelley, 1974), have evolved a time sense that combines endogenous clocks and exogenous entrainment.

16.4 Reproductive behaviour and physiology

Seasonal climatic changes in temperature and precipitation exert powerful influences upon the reproductive success of many species. Among birds, the availability of food for the young appears to be the main determinant of breeding success (Lack, 1968). Birds at middle and high latitudes have a seasonal reproductive pattern in which the eggs are usually laid during the spring. This pattern enables the young to mature sufficiently to withstand the winter conditions or to endure a long migratory flight. Sandpipers, for example, breed in arctic regions and nest and incubate in spring when the snow is still on the ground. The eggs usually hatch when the snow melts and there is an abundance of insect life to provide food for the young (West and Norton, 1975).

The reproductive physiology of seasonal breeders is geared to the annual cycle of environmental change in such a way that peak abundance of food, or adverse climatic conditions, are anticipated. There are two main ways in which this is done. First, changes in ambient temperature, day length or other environmental cues provide stimuli that induce physiological change at a particular time of year (see Box 16.1). Second, the physiological changes are programmed on a seasonal basis by means of an endogenous circannual clock.

The control of reproductive physiology involves the complex interaction of a number of hormones, as illustrated in Figure 16.7. In most vertebrates the environmental factors stimulate the production of gonadotrophic hormones by the pituitary gland. These hormones stimulate growth and activity in the testes and ovaries, which in turn produce their characteristic sex hormones. At the end of the breeding season, pituitary activity diminishes, the gonads

Box 16.1   Scharrer's representation of the influence of light on the gonads

Scharrer's (1964) representation of the influence of light on the gonads. (a) In transparent animals, light can exert a direct effect on the internal organs. (b) Some opaque animals have a transparent window which admits light to photosensitive parts of the brain, which then stimulates the gonads via hormones released by the pituitary gland. (c) A system in which light acts on the retina which sends nervous messages to the hypothalamus. This part of the brain stimulates the pituitary gland to release gonadotrophic hormones. (d) Diagram of scheme (c) applied to the human brain (After Bligh, 1976).

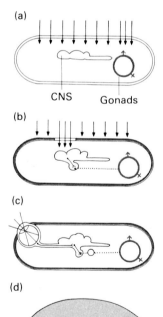

The most regular and predictable seasonal change in the environment is the fluctuation of day length. In transparent organisms, light can have a direct effect on the gonads, bringing them into reproductive condition at the appropriate time (Scharrer, 1964). Some other animals may have a translucent window that admits light to the brain. The photic stimuli are converted to chemical messages by neurosecretory cells. In some mammals the pineal body, on the dorsal surface of the brain, may act as a light transducer (Wurtman et al., 1968). However, the influence of light on mammalian reproduction appears to be primarily via the retina to the hypothalamus, as shown in Scharrer's scheme.

Fig. 16.7 Principal hormonal pathways involved in ovarian function in a mammal. FSH-RF, follicle stimulating hormone (FSH) releasing factor; LH-RF, luteinizing hormone (LH) releasing factor; s.f., small follicle; r.f., ripe follicle; CL, corpus luteum (After Bligh, 1976).

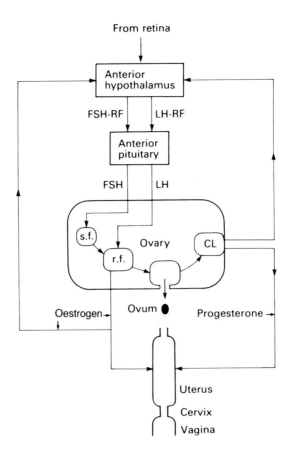

become inactive and reproductive behaviour no longer occurs. In addition to seasonal cycles of sexual activity, many mammals have a much shorter oestrous cycle, or heat. Some, like the red fox (*Vulpes vulpes*), have only one period of heat per year. Others, like domestic dogs, have two, and others have more. Among birds the number of clutches reared in a season may depend upon the food supply (Lack, 1968).

The reproductive pattern of many species is influenced both by photoperiodic factors and by the endogenous circannual clock. Brian Follett (1973) has suggested that the two mechanisms may interact. Species living at high latitudes tend to be most strongly influenced by photoperiodicity. At low (equatorial) latitudes there is less annual photoperiodic change, but an annual reproductive cycle may, nevertheless, be advantageous. Animals occupying arid desert environments, for example, may be dependent upon the occurrence of rainfall to trigger the process of reproduction (Marshall, 1970). Some equatorial species show a marked breeding rhythm that appears to be unconnected with seasonal changes. Several sea birds, including the brown booby (*Sula leucogaster*), the sooty tern (*Sterna fuscata*) and the lesser noddy tern (*Anaus tenuirostris*) breed at intervals of eight–ten months. The progressive shift between such breeding cycles and the annual cycle suggests that no one season is preferable under equatorial conditions. Why then do the birds not breed continuously? Perhaps this would require too much energy and a periodic rest during which moult can occur is the most beneficial arrangement.

The annual reproductive cycle is similar to the process of acclimatization in that it involves slow-acting physiological phenomena that change the animal's physiological state in a fundamental way. Reproductive activities, including territorial defence, courtship, mating and parental care, put extra energetic and physiological burdens on the animal, which have to be catered to by the system as a whole. If the animal is unable to maintain physiological stability while carrying these burdens, then the reproduction must be abandoned for the current year.

16.5 Hibernation

In parts of the world where climate is seasonal, animals may have to adjust to prolonged periods of unfavourable weather. Some animals are able to avoid such conditions by migration, but others are able to survive unfavourable periods by entering a prolonged phase of dormancy, called **aestivation** at high temperatures and hibernation at low temperatures. Some desert rodents, like ground squirrels (*Citellus*), become inactive during the summer. They enter a torpid state in which the body temperature falls and in which a general reduction in physiological activity occurs. This means that energy expenditure is lowered and that the animal can stay alive for long periods without feeding. Water losses are reduced as a result of the lowered food intake and this lowered intake is helped by the habit of retreating into burrows. Some rodents store food in their burrows and maintain high humidity inside by stopping up the entrance (Schmidt-Nielsen, 1964). Aestivation is most common in desert-living species, but it is not confined to them. Thus, many species of European earthworm become dormant during the summer. Each animal hollows out a small chamber deep in the soil and curls up into a ball. The onset of such aestivation is associated with low humidity, and it is possible to prevent it by keeping the worms in a humid atmosphere.

Hibernation occurs in many species in northern latitudes, enabling them to avoid winter conditions that otherwise would make excessive demands on their energy requirements. True hibernation is usually distinguished from partial dormancy, as shown by the European brown bear (*Ursus arctos*) and the American black bear (*Ursus americanus*). During partial dormancy, the bear's body temperature may fall from about 38 °C to about 30 °C, but body temperatures below 15 °C are lethal. True hibernators allow their body temperatures to fall as low as 2 °C. True hibernation is characteristic of small mammals, although similar types of torpor do occur in raccoons and badgers and in some birds. Many species of hummingbird show periods of torpidity during which their body temperature falls to that of the environment, although temperatures below 8 °C are lethal. Torpor usually lasts only a few hours in birds, and seasonal dormancy is known in only a single family of birds, the goatsuckers (Caprimulgidae).

A period of dormancy in response to unfavourable climate also occurs among insects and is given the name **diapause**. Diapause is often associated with a particular stage of life cycle. Insects that avoid freezing by supercooling can survive very cold conditions in diapause, in which the body fluids freeze at temperatures well below 0 °C. The extent of supercooling may depend upon the degree of acclimatization, which progressively alters the chemical constitution of the body fluids. *Bracon cephi*, a Canadian wasp, may lower its freezing point to −46 °C by increasing the concentration of glycerol in the blood.

True hibernation occurs only in small mammals, which cool more readily than large ones due to their proportionately larger body surface. They also warm up more quickly because of their small thermal capacity. Hibernation is characterized by a sleep-like state, with lowered rates of respiration and heartbeat. Hibernating animals often choose a typical sleep site in which to hibernate and adopt a sleeping posture. During hibernation, body temperature is lowered and energy expenditure falls below that of normal sleep. Hibernators often have deposits of a special brown fat that have the primary function of heat production rather than the nutrient energy production that occurs when ordinary fat reserves are mobilized. The brown fat is especially important during periods of arousal from hibernation, when body temperature has to be raised quickly.

In hibernating mammals, short phases with normal body temperature alternate regularly with longer torpid phases. In the alpine marmot (*Marmota marmota*), hibernation occurs in family groups, which huddle together. The short-term cycles of hibernation are often synchronized within groups. In this way they save energy. Torpid marmots are passively warmed by contact with those at normal body temperature, and there is an overall reduction of heat loss because huddling reduces the effective surface area. However, not all members of the group benefit equally. The male of the pair always warms the female, and the parents warm the juveniles.

Some mammals including the golden-mantled ground squirrel (*Citellus lateralis*) and the woodchuck (*Marmota monax*) have a marked circannual rhythm underlying their seasonal hibernation. The ground squirrels are found in western North America at altitudes of 1500 to 3600 metres, ranging from northern British Columbia to southern California. They normally hibernate for three to four months, during which time their body weight falls considerably. After hibernation their food consumption increases rapidly, and their body weight increases up to the onset of hibernation in October. When isolated in the laboratory, the rhythm of hibernation and associated change in body weight may persist for a number of years, even under constant lighting and temperature conditions (see Figure 16.8). If blood serum from a hibernating ground squirrel is injected into a non-hibernating animal, then hibernation is induced (Pengelley and Asmundson, 1974). The evidence strongly suggests that the basic physiological mechanism responsible for controlling hibernation is driven by an endogenous biological clock entrained to the rhythm of external events by a particular external trigger, or zeitgeber. This has been demonstrated clearly in the woodchuck, which is also known to exhibit a circannual rhythm of hibernation. Woodchucks from the eastern

Fig. 16.8 Circannual rhythms in body weight and hibernation periods (shaded portions) of two representative animals (*Citellus lateralis*), maintained under constant laboratory conditions (After Pengelley and Asmundson, 1974).

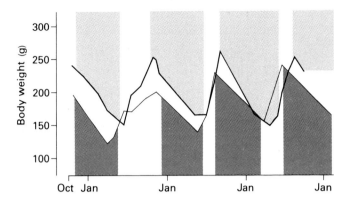

United States were transported to Australia and exposed to the prevailing light and temperature conditions. The woodchucks reversed their normal rhythm within two years, bringing it into line with local conditions, even though the animals were provided freely with food and water and had no need to hibernate.

Hibernation is similar to acclimatization in the sense that it is a long-term change in physiological state that enables the animal to cope with seasonal changes in environmental conditions. The circannual changes that occur in the animal's physiological state can be represented in a physiological space. Both the reproductive and hibernation annual cycles are largely pre-programmed though they are, in reality, part of a single seasonal programme.

16.6 Migration

Migration between two different habitats on a periodic or seasonal basis is shown by many species of animal including butterflies, locusts, salmon, birds, bats and antelopes. This is not the only form of migration, since some species undertake migrations as part of exploration and colonization of new habitats. For example, population explosion among Norway lemmings (*Lemmus lemmus*) leads to dispersal migration that may involve thousands of individuals, usually immature males. Streams of lemmings run down the mountainsides and into the valleys. Many drown in their attempts to cross expanses of water. Lemmings are good swimmers and usually will not enter water unless they can see land on the other side. However, the pressure of numbers on the shore line is sometimes such that individuals are induced to swim out into the open sea.

Colonization of new areas is sometimes attained by progressive migration among succeeding generations as in the European collared dove (*Streptopelia decaocto*), illustrated in Figure 16.9. Exploratory migration is common in young vertebrate animals and may be responsible for successful colonization when ecological conditions are favourable.

Periodic migrations occur in a variety of species in response to changes in environmental conditions. For example, the locust (*Schistocerca gregaria*) inhabits seasonally arid areas. Depending upon their degree of crowding, the insects may develop into one of three adult types. Of these, the *gregaria* indi-

Fig. 16.9 Map showing the
expansion in range of the
collared dove (*Streptopelia
decaocto*) this century (From
*The Oxford Companion to
Animal Behaviour*, 1981).

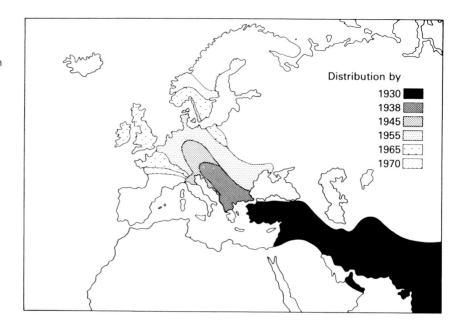

viduals aggregate into dense swarms that migrate downwind into areas of
low barometric pressure where they are most likely to encounter rain. They
fly by day and stop migrating when they encounter wet conditions. Sexual
maturation, copulation and deposition of eggs then occur. Locust swarms
describe seasonal circuits, but the generation time is too short for individuals
to complete the entire circuit.

Some seasonal migrants, in contrast to periodic migrants, initiate their
migration on the basis of a circannual rhythm rather than in response to
changes in environmental conditions (Gwinner, 1986a,b). Studies have been
made of songbird species within the genera *Phylloscopus* (Gwinner, 1971) and
Sylvia (Berthold, 1973) that exhibit a range of migratory habits. It appears
that, in the typical long-distance migrants such as the garden warbler (*Sylvia
borin*), the subalpine warbler (*Sylvia cantillans*) and the willow warbler
(*Phylloscopus trochilus*), there are marked seasonal changes in body weight,
moult, testis size, nocturnal restlessness and food preferences (Figure 16.10).
The European populations of these species winter in Africa and migrate
across the Sahara Desert (Figure 16.11). When they are maintained under
constant laboratory conditions in Europe, having been hand raised from a
few days after hatching, the events that are known to be seasonal in free-
living birds appear as seasonal events in caged members of the same species,
although the period of oscillation is somewhat shorter than the calendar year.

Middle-distance migrants such as the blackcap (*S. atricapilla*) and the chiff-
chaff (*P. collybita*) winter in Europe and Africa and show moderate changes in
body weight and other migratory indexes. These species also show seasonal
changes under constant laboratory conditions. The Sardinian warbler (*S.
melanacephaia*) and the Dartford warbler (*S. undata*) are partial migrants that
winter within the Mediterranean breeding area, while Marmosa's warbler *(S.
sarda balearica)* is resident the year round and endemic on the Balearic Isles

Fig. 16.10 Circannual rhythms of testis length of warblers (*Sylvia borin*) kept under constant photoperiodic conditions for three years. LD is the light–dark ratio (After Berthold, 1974).

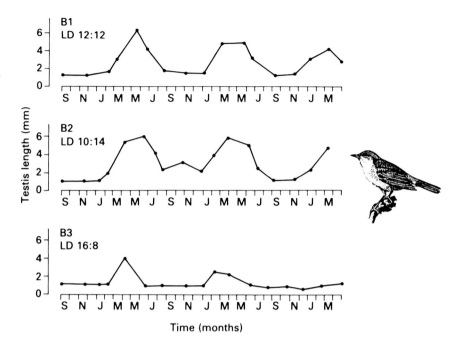

Fig. 16.11 Breeding ranges (dark shade) and winter ranges (pale shade) and their overlap (black) of some migratory warblers (After Schmidt-Koenig, 1979).

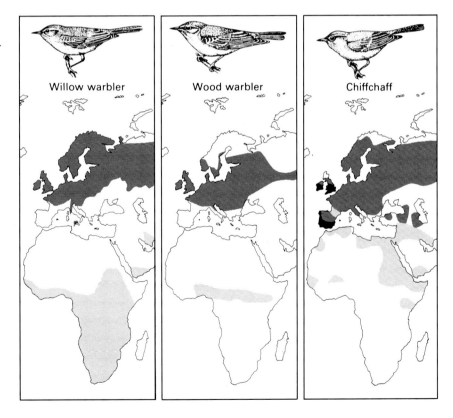

and the Pithyuses in the Mediterranean sea. These species have a fairly constant body weight throughout the year, and they have a moult of body feathers of long duration, which alternates with periods of nocturnal restlessness. When tested in the laboratory, some evidence of seasonal changes can be obtained, although there are marked individual differences (Berthold, 1974).

It appears that the species of these two genera show seasonal changes in migratory indices, the intensity of which is correlated with the migratory habits of the species and the timing of which suggests that they may be under the partial control of an endogenous circannual rhythm. Similar seasonal rhythms have been documented in several avian species with regard to the physiological processes including moult, fat deposition, migratory restlessness and reproduction (Rutledge, 1974; Berthold, 1984a).

There is also evidence that endogenous factors control not only the onset of migratory activity but also its pattern. Many species have migratory routes that are characteristic of particular breeding populations. For example, white storks (*Ciconia ciconia*) that breed in western Europe fly to their wintering grounds in Africa by a westerly route over Spain and Gibraltar, whereas those that breed in eastern Europe take an easterly route, as illustrated in Figure 16.12. Naive

Fig. 16.12 Migratory routes of the eastern and western European populations of the white stork (*Ciconia ciconia*). Dark shading indicates intensity of birds (After Schüz, 1971).

young storks raised in captivity in eastern Europe, but released in western Europe, fly in the southeasterly direction characteristic of birds from eastern Europe (Schuz, 1963). Similar experiments with other species indicate that the direction of migration is influenced genetically. In one study, Biebach (1983) hand raised 46 European robins from a partially migrant population in south Germany. He found that 80 per cent were migrants and 20 per cent non-migrants, as shown by measurements of migratory restlessness. From these birds he formed five migrant/migrant pairs, one non-migrant/non-migrant pair, and four migrant/non-migrant pairs. The proportion of migrants in the offspring of migrant/migrant pairs was 89 per cent, whereas that in the off-spring of pairs with at least one non-migrant was 53 per cent. This and other studies suggest that the migratory disposition is controlled by a genetic polymorphism (Berthold, 1984b).

To reach its winter quarters, a bird must travel not only in the right direction but also the correct distance. Eberhard Gwinner (1972) obtained evidence on the distances travelled on the first migration by juvenile warblers that were not accompanied by adults. The evidence suggests that the birds have an endogenous clock that induces as many hours of flight as are necessary to take the birds along each leg of their migratory journey. There is a good correlation between the number of hours of migratory restlessness of caged birds and the distance normally travelled during migration (e.g. Berthold, 1973, 1984a). The distance travelled by migrating birds that is equivalent to an hour of restlessness can be estimated by comparing the behaviour of caged and migrating birds from the same population (Gwinner, 1972). From knowledge of the speed of flight of migrating birds, Gwinner calculated the distance that his caged birds would have covered had they been going in the right direction. His calculations gave results close to those obtained by observation of freely migrating birds from the same population. Garden warblers (*Sylvia borin*), tested in cages during their first autumnal migration, headed to the southwest during August and September and shifted to south-southeast during October, November and December (Gwinner and Wiltschko, 1978). These results agree with the timetable of natural outdoor migration, as illustrated in Figure 16.13. Thus the circannual programme might be organized on a species or population basis in such a way for the bird to produce the correct hours of migratory activity to reach its specific wintering area. This has become known as the vector navigation hypothesis (Gwinner, 1986a, b). It could explain how young birds which migrate separately from adults find their way to the wintering areas during their first autumn migration. Laboratory experiments indicate that the endogenous programme controlling autumn migration is inflexible and subject to little environmental modification (Gwinner, 1986a). However, it may be that some species are affected by specific environmental conditions pertaining to the ecology of the migratory route. In a study of spotted flycatchers (*Muscicapa striata*) resting during migration at an oasis in the Sahara Desert, Biebach (1985) found that the stopover time of a bird was related to its fat reserves upon arrival. Fatter birds stayed for a shorter time than leaner birds. During their rest at the oasis the birds could increase their fat stores by feeding, and this suggests that migratory activity may be inhibited by the combination of having low fat

Fig. 16.13 Directional preferences obtained during laboratory tests of migratory restlessness in garden warblers. The direction of the arrows indicates the directional preference obtained on the dates shown. The positions of the arrows on the map indicate the places that migrating warblers would have reached (by the dates shown) during their migration to the wintering grounds (shaded area) in southern Africa (After Gwinner and Wiltschko, 1978).

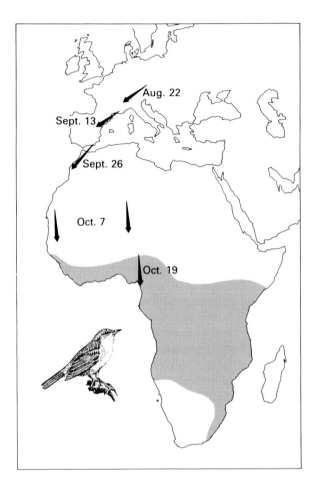

reserves and the opportunity to feed. This hypothesis was confirmed by laboratory experiment. Caged spotted flycatchers were deprived of food in the middle of their migratory season once they reached a bodyweight of 20 g, indicating considerable fat reserves. The birds showed large amounts of migratory restlessness both before and during the food deprivation. When their bodyweight had dropped to about 13.5 g, they were re-fed on a rationed basis, so that their weight increased at a rate similar to that of wild birds preparing for migration across the Sahara Desert. During thus period of weight gain, migratory restlessness was reduced almost to zero, and was resumed only when the birds reached a weight of about 17 g.

Thus it appears that specific factors, relevant to the possibility of stopping at oases while migrating over the desert, influence the migratory pattern in this species. Biebach *et al.* (1986) also showed that the choice of stopover place, and the stopover period, of some passerine migrants is influenced by their fat reserves. Those with sufficient reserves to cross the Sahara stopped for only one day and rested in the shade, while those with insufficient reserves stopped for up to three weeks during which they gained fat. These authors proposed that the birds use a simple set of decision rules.

In general, it appears that endogenous circannual factors are the primary cause of the onset of migration in many bird species, that the pattern of migration is strongly influenced by genetically determined circannual pro-grammes, but that local factors may also exert an influence under particular circumstances.

<table>
<tr><td>16.7</td></tr>
</table>

16.7 Lunar and tidal rhythms

As seen from the earth, the moon moves along a path similar to that of the sun. The moon rises 50 minutes later each day, and it is therefore seen some-times in the daytime and at other times only at night. The lunar cycle of 29.5 days is known to influence a variety of aspects of animal behaviour. In the palolo worm (*Eunice*) of the Pacific Ocean reproductive activity occurs only during the neap tides of the last quarter moon in October and November. As illustrated in Figure 16.14, the posterior part of the worm containing the gen-ital organs becomes detached from the anterior part and this swims to the surface of the water where the eggs and sperms are shed (Korringa, 1947). The worms provide abundant food for sharks and other fish, but by syn-chronizing their reproductive activity the worms are able to swamp these predators and a proportion of the gametes are always able to survive. Worms isolated in the laboratory produce their gametes at the correct time, showing that this activity is probably controlled by an endogenous clock.

Lunar rhythms are also known in terrestial animals. Thus Jamaican fruit bats have a pattern of feeding which involves leaving the day roost in the evenings and feeding throughout the dark nights during the period of the new moon. During the full moon, however, they depart from their day roost at the normal time in the evening, but return there when the full moon is high. They follow this pattern even when the full moon is obscured by clouds. This suggests that they make use of an endogenous lunar clock to time their feeding behaviour.

Fig. 16.14 The mating routine of the palolo worm (After Cloudsley-Thompson, 1980).

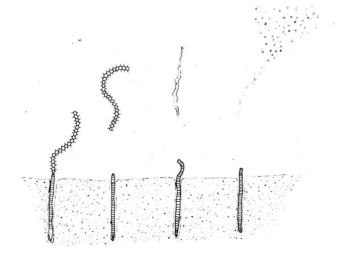

The tides result from changes in the combined gravitational pull of the sun and moon. The tidal cycle is thus repeated twice during the lunar cycle. Many marine animals show rhythms of behaviour that coincide with the tidal cycle and have been shown to be governed by an endogenous clock. For example, the shore crab (*Carcinus maenas*) has a daily rhythm of activity which is superimposed upon a tidal rhythm. The tidal rhythm ensures that activity is synchronized with high water. This rhythm persists under constant laboratory conditions for about a week before fading away. It can be restored by cooling the crab to near freezing point for about six hours. This clock-shock appears to restart the tidal clock, perhaps because it resembles the experience of a beach-stranded crab that is eventually reached by a high tide. In one experiment, shore crabs were raised in the laboratory from eggs to adulthood under a day–night regime without tidal influences. These crabs showed only a circadian rhythm of activity. However, a tidal rhythm appeared after the crabs were given a single cold-shock treatment. It seems that the endogenous tidal clock had been dormant until started by the cold-shock (Palmer, 1973).

Similar phenomena occur in the tidal rhythms of other crabs. Fiddler crabs (*Uca*) show tidal rhythms of activity that may persist for up to five weeks under constant laboratory conditions. The crabs emerge from their burrows at low tide and actively forage, court, etc. As the tide floods, they retreat back into their burrows. In many crab species the drop in temperature brought about by the flood tides acts as a zeitgeber and sets the phase of the activity rhythm (Palmer, 1973).

Superposition of circadian and tidal rhythms can enable an animal to adapt to the irregular changes of high and low tide that occur in some parts of the world. On the coast of California, for example, a period of 13.80 hours between high tides is followed by one of 10.43 hours, as illustrated in Figure 16.15. The intertidal crustacean *Synchelidium* has a pattern of swimming

Fig. 16.15 Changes in tide height at La Jolla, California (above), compared with the swimming activity of *Synchelidium* under laboratory conditions (From *The Oxford Companion to Animal Behaviour*, 1981).

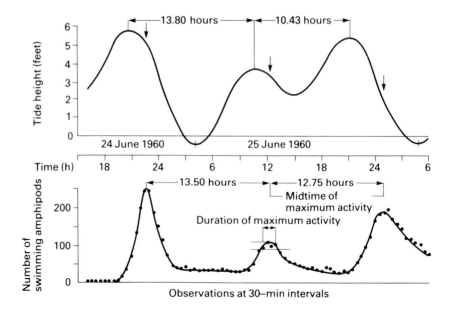

activity which closely follows this tidal pattern. This swimming pattern continues for several days under constant laboratory conditions.

16.8 Circadian rhythms and daily routines

Most animals experience changes in environmental conditions between night and day, which affect animals both directly and indirectly. Thus there may be changes in food availability, and in numbers of predators, which are brought about by changes in temperature, light intensity, etc.

In adjusting to the differences between night and day the animal adopts a daily routine which is made up of many different aspects of behaviour fitted together to form a pattern that tends to be repeated day after day. When a single activity is studied it tends to follow a typical daily rhythm. The daily routines of animals have been subjected to relatively little study (Daan, 1981), but daily rhythms have received considerable attention from those primarily interested in circadian clocks (Rusak, 1981).

The most important daily changes in the external environment are those of light intensity and temperature. Animals specialized for daytime vision may be disadvantaged at night, because they are vulnerable to predators, or because they cannot forage efficiently. In cold climates it can benefit small mammals to be active at night when temperatures are low. Their period of greatest heat production then occurs during the coldest part of the 24-hour cycle, the activity being harnessed as a means of thermoregulation. Small birds, on the other hand, save energy on cold nights by becoming inactive and allowing their body temperature to fall. In hot climates it is advantageous for small mammals to be nocturnal and so avoid the heat of the day.

It is not surprising that rhythms of rest and activity are widespread in the animal kingdom. When it is disadvantageous to be active at night, the best policy is to sit tight in a safe place and save as much energy as possible. This has been suggested as one of the prime functions of sleep (Meddis, 1975). Nocturnal species may hide during the day if they are likely to be preyed upon. If they are themselves nocturnal predators, they may remain hidden and inactive during the day to avoid scaring their prey. Thus the daily rhythms of the physical environment make some activities advantageous at one time and disadvantageous at another. Much depends upon the overall ecology of the species concerned.

An animal which is adapted to its environment will settle into a daily routine designed to maximize the survival value of its various activities. This is partly a matter of making the most of opportunities. For example, the European kestrel is a diurnal predator which specializes in preying upon small mammals. The bird relies on vision to detect and catch its prey and hunts most successfully when the light is good. Field studies reveal that kestrels catch prey throughout the day but do not always immediately eat what they catch. They tend to cache surplus prey in randomly chosen locations in their hunting area. The caching occurs throughout the day (see Figure 16.16) and the cached items are typically retrieved at dusk. This routine enables the kestrel to make the most of the available prey during the day,

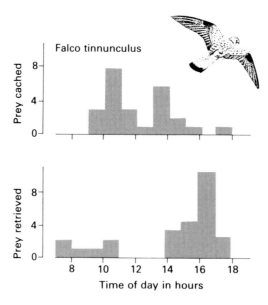

without spending too much time eating the prey. Moreover, if the bird ate all its surplus prey immediately it would become unnecessarily heavy and its hunting efficiency would probably decline.

In addition to the daily routines typical of their own species, individual animals may develop their own daily habits. Thus kestrels that have found food at a particular time and place tend to repeat the same search pattern the next day, as shown in Figure 16.17. Such a strategy is appropriate in circumstances where the prey also has its own typical daily routine.

Fig. 16.17 Movements of a kestrel observed in The Netherlands on 12, 13, 14 and 16 August 1977. The numbers along the flight paths refer to the time of day. Land area is shaded; sea is white (After Daan, 1981).

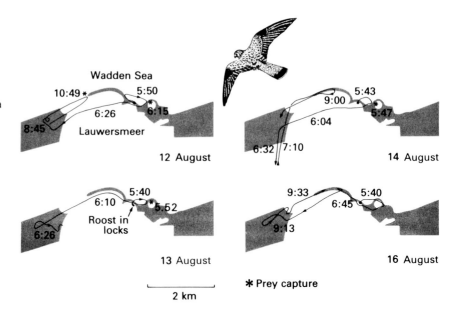

Points to remember

- The range of environmental conditions that an animal can tolerate is limited. It is characteristic of the species but can be somewhat altered by acclimitization.

- Acclimatization involves slow-acting physiological changes which have profound effects on the animal's homeostatic and motivational processes.

- Most animals possess endogenous biological clocks which are entrained to rhythmic aspects of the environment. The clocks may be circannual, lunar, tidal or circadian.

- Many aspects of reproductive physiology and behaviour occur on a circannual basis, as do hibernation and migration. These all involve slow-acting physiological changes, somewhat akin to acclimatization.

- Lunar and tidal rhythms may occur in marine animals in addition to circannual and circadian rhythms.

- Circadian rhythms enable animals to adjust to the day–night cycle in an anticipatory manner and to develop daily routines which make the most of the prevailing opportunities.

Further reading

Berthold, P. (1996) *Control of Bird Migration*. Chapman & Hall, London.

Dingle, H. (1996) *Migration. The Biology of Life on the Move*. Oxford University Press, New York.

Gwinner, E. (ed.) (1990) *Bird Migration*. Springer-Verlag, Berlin.

Kerlinger, P. (1995) *How Birds Migrate*. Stackpole Books, Mechanicsburg, PA.

Animal learning

Profile of Ivan Pavlov

A product of trial-and-error learning: Morgan's dog, Tony, operating the latch of a gate (From Morgan, C.L., *Habit and Instinct*, 1896).

This section is devoted to the psychology of animal learning. Chapter 17 provides an outline of classical and instrumental conditioning from the traditional psychological viewpoint. Chapter 18 introduces evolutionary aspects and compares the biological and psychological perspectives. Chapter 19 discusses various issues that suggest cognitive explanations of animal learning. This chapter is a prelude to the discussion of the mentality of animals that occurs in Part 3.

Ivan Pavlov (1849–1936)

Ivan Pavlov was born in the Russian town of Ryszan, where he attended the religious school and seminary. He entered St Petersburg University in 1870 and graduated in natural science in 1875. After obtaining his doctorate at the Military Medical Academy in 1883, he studied in Germany. He became Professor of Pharmacology at the Military Medical Academy in 1890, and Professor of Physiology in 1895. He received the Nobel Prize for Medicine in 1904 for his work on the physiology of digestion. In 1925 Pavlov resigned his professorship in protest against the expulsion of sons of priest from the Academy. Himself the son of a priest, he nevertheless was not expelled. Up to the end of his life Pavlov continued to conduct research at various laboratories in the Soviet Union.

Pavlov's early work was on the physiology of circulation, especially the mechanisms that regulate blood pressure. He discovered that the vagus nerve controls blood pressure, and investigated nervous control of the rhythm and depth of the heartbeat. In 1879 Pavlov began work on the physiology of digestion, which culminated in his book *The Work of the Digestive Glands*, published in 1879. He investigated the mechanisms involved in the secretion of various digestive glands, and came to the conclusions that they were controlled exclusively by nervous mechanisms. It is now known that control by hormones also occurs. It is this work that led to the award of the Nobel Prize in 1904. During the course of his work on the physiology of digestion, Pavlov noticed that salivation could be induced by the sight of food, or other stimuli that normally preceded feeding. This led him to the discovery of the conditional reflex, now regarded as a fundamental aspect of learning.

From 1902 until his death, Pavlov concentrated his researches on the phenomena of conditioning. He is responsible for many of the basic concepts still current in this field of study, and can be regarded as the founder of the experimental study of animal learning.

In a typical experiment, Pavlov showed that if the presentation of food to a dog was repeatedly accompanied by the sound of a bell, then the dog would come to respond to the bell as if it were food. Pavlov measured the dog's salivary response to paired presentations of food and bell and then measured salivation in response to presentation of the bell alone. He regarded the salivation not the food as an unconditional response, and the subsequent salivation to the bell alone as a conditional response, because it is conditional upon prior pairing between food and bell.

Pavlov's aim was to discover the universal laws of learning and to account for these in terms of brain mechanisms. He suggested that the cells of the central nervous system changed structurally and chemically during conditioning. While this notion is not far removed from the modern view, many of Pavlov's ideas about the role of the cortex in learning were premature. Pavlov's major work, translated into English as *Conditioned Reflexes: an Investigation of the Physiological Activity of the Cerebral Cortex* (1927), had a major impact on the development of psychology.

Conditioning and learning

Ivan Pavlov's work on conditioning became known in the West at a time when considerable interest already existed in mechanistic explanations of behaviour, especially in reflexes. The attempts by Jacques Loeb (1859–1924) to account for animal behaviour in terms of simple tropisms and taxes (see Section 15.2) had considerable influence in Germany and the United States. In physiology, Sir Charles Sherrington published his *Integrative Action of the Nervous System* in 1906. He showed how simple reflexes could combine to produce coordinated behaviour.

In psychology, John B. Watson (1913) launched the behaviourist school of psychology, which was to become very influential during the first part of the century. The behaviourists would allow only observable stimuli, muscular movements and glandular secretions to enter into explanations of behaviour. In order to account for complex behaviour in stimulus–response terms, they postulated covert or implicit stimulus–response relationships. Watson (1907) had already proposed that somesthetic stimuli aroused by the animal's movements served such a purpose. The unobservable processes mediating stimulus and response were said to be comprised of incipient movements and movement-produced stimuli. Thus, Watson (1914) postulated that human thought processes were implicit speech (i.e. talking to oneself), one tongue twitch providing the stimulus for the next response in the chain.

Pavlov gave a lecture in Madrid in 1903, and he delivered the Huxley lecture in London in 1906, of which a report was printed in *Science*. A review of Pavlov's work by Yerkes and Morgulis was published in 1909, and Watson published another in 1916. A translation of Pavlov's book *Conditioned Reflexes* was published in 1927. There was a strong climate of opinion in favour of a purely mechanistic and objective science of behaviour. Pavlov's work added fuel to the extreme environmentalist approach espoused by Watson in his behaviourist approach to psychology. Watson (1916) realized that reflex conditioning might serve as a paradigm for learning in general. Behaviourists such as Watson and, later, B.F. Skinner, claimed that all animal and human behaviour could be accounted for in terms of conditioning. Pavlov's work endowed behaviourism with a certain amount of physiological respectability, and the psychology of animal learning developed to dominate psychology in the United States up to the end of the 1950s.

17.1 The classical-conditioning paradigm

In his original conditioning experiments, Pavlov restrained a hungry dog in a harness (see Figure 17.1) and presented small portions of food at regular intervals. When he signalled the delivery of food by preceding it with an external stimulus like the sound of a bell, the behaviour of the dog toward the stimulus gradually changed. The animal began by orientating toward the bell, licking its lips and salivating. When Pavlov recorded the salivation systematically by placing a small tube in the salivary duct and collecting the saliva, he found that the amount of saliva collected increased as the animal experienced more pairings between the sound of the bell and food presentation. It appeared that the dog had learned to associate the bell with the food.

Pavlov referred to the bell as the **conditional stimulus** (CS) and to the food as the **unconditional stimulus** (UCS). Salivation in response to presentation of food was called the **unconditional response** (UCR), while salivation in response to the bell was called the **conditional response** (CR).

Although Pavlov originally used the words 'conditional' and 'unconditional', this terminology was mistranslated at an early stage, and the terms conditioned reflex, conditioned response, unconditioned response and unconditioned reflex became established in the English-language literature. However, it is modern practice to return to Pavlov's original terminology, and we follow his usage in this book. The rationale behind the terminology is that

Fig. 17.1 Pavlov's arrangement for the study of salivary conditioning.

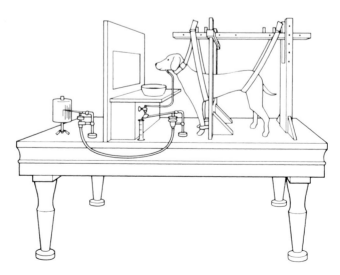

the food unconditionally elicits a set of consummatory responses, one of which is recorded by the experimenter and designated the unconditional response. Conditioning occurs as a result of a contingency, arranged by the experimenter, between the unconditional stimulus (food) and an external stimulus previously unconnected with the feeding situation, like the sound of a bell. After a number of pairings, the bell alone is sufficient to elicit salivation. It is then known as the conditional stimulus because as a result of its training, the dog salivates if and when the stimulus is presented. Similarly, the salivation in response to the bell is known as the conditional response even though it may appear to be the same response as the unconditional response. During the process of conditioning, the presentation of the UCS (food) following the CS (bell) is said to reinforce the conditional reflex of salivation to the CS. The UCS, therefore, is regarded as a **reinforcer**.

Conditioning experiments that employ motivationally beneficial, or positive, reinforcers like the UCS are examples of **positive conditioning**. Conditional reflexes can also be established in experiments that employ negative reinforcers, which are motivationally noxious. For example, the onset of a sound stimulus (tone) preceding the delivery of a puff of air to the eye of a rabbit will come to elicit a blink of the eyelid (strictly the nictitating membrane) as a conditional response. Initially, the air puff (UCS) alone elicits the eye blink (UCR), but after conditioning, the tone (CS) will elicit the eye blink (CR) in the absence of a puff of air. This is an example of negative conditioning.

A reinforcer is characterized not so much by its intrinsic properties as a stimulus but by its motivational significance to the animal. Thus, food acts as a positive reinforcer only if the dog is hungry, and an air puff acts as a negative reinforcer only if it is noxious or unpleasant for the animal. In many cases the reinforcer is innate in the sense that its motivational significance and ability to support conditioning is an integral part of the animal's normal make-up. However, this does not have to be the case, and Pavlov showed that a CS could act as a reinforcer. For example, if a bell is established as a CS by the normal conditioning procedure, it will reliably elicit a CR, like salivation. If a second CS, such as a light, is then paired repeatedly with the bell, in the absence of food, the animal will come to give the CR in response to the light alone, even though food has never been associated directly with the light. This procedure is known as **second-order conditioning**.

Pavlovian, or classical, conditioning is very widespread in the animal kingdom, and it pervades every aspect of life in higher animals, including humans. Pavlov demonstrated that conditioning could occur in monkeys and in mice, and claims have been made for a wide variety of invertebrate animals. In evaluating such claims, however, we must take care to distinguish true classical conditioning from other forms of learning and quasi-learning.

Although classical-conditioning procedures are relatively straightforward, the phenomena they reveal are not so clear-cut and have given rise to considerable discussion in the psychological literature, from the time of Pavlov to the present day. Every student of animal behaviour should be familiar with the basic characteristics of classical conditioning because it is hardly possible to carry out an experiment without some conditioning occurring. It may be simply that the animal becomes conditioned to the time of day at which the experimenter appears or that clandestine conditioning phenomena subtly invalidate the experimenter's conclusions. In any event, as a universal feature of higher animals, conditioning is not only of practical importance but also must be incorporated into any coherent understanding of the way animals work. We now briefly discuss some of the major characteristics of conditioning. A more detailed account is given in the excellent book by Mackintosh (1974).

17.2 Acquisition

We can measure the acquisition of a conditional reflex in various ways. Pavlov used the amount of saliva collected during presentation of the CS, for example. In the case of eyelid conditioning, the probability of occurrence of the response is measured (see Figure 17.2). The rate of acquisition may vary considerably, with species, circumstances and age (see Figure 17.3).

Pavlov held the view that the pairing of CS and UCS leads to the formation of an association between them. The CS becomes a substitute for the UCS and becomes capable of eliciting the responses normally elicited by the UCS. This is usually called the **stimulus substitution theory**. An alternative theory maintains that CRs occur because they are followed by rewards. In other words the CR is reinforced by its consequences. This is usually called the **stimulus–response** theory.

The two theories differ in two main empirical respects. First, on the basis of stimulus substitution theory, we would expect the CR to be very similar to the UCR, whereas it should be somewhat different according to stimulus–response theory. Second, Pavlov maintained that the association occurs between the CS and the UCS, which itself constitutes a reinforcing state of affairs. The stimulus–response theory maintains that learning is dependent upon reinforcing consequences of the CR. Although this topic is controver-

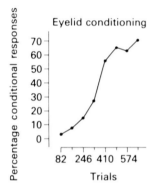

Fig. 17.2 Acquisition of eyeblink response during eyelid conditioning (After Schneidermann *et al.*, 1962).

Fig. 17.3 The effect of age upon eyelid conditioning (After Braun and Geiselhart, 1959).

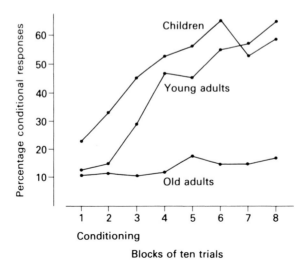

sial, the evidence seems to favour Pavlov's view: the variety of responses elicited by the CS closely matches the variety of responses elicited by the UCS. Apparent exceptions become less convincing on close inspection and attempts to read into an animal's behaviour signs of expectation and preparation tell us more about the experimenter's preconceptions than the subject's behaviour. It is interesting that it was precisely because of the temptation to indulge in such interpretations that Pavlov deliberately ignored the skeletal components of CRs because of the temptation to interpret these as signs of expectation or preparation. To avoid such anthropomorphic interpretations, Pavlov concentrated upon the salivary component. Moreover, as Mackintosh (1974) points out, if responses (such as closure of the nictitating membrane in anticipation of a puff of air, movements of the jaw or licking in anticipation of water, or pecking or salivation in anticipation of food) are relatively unaffected by their consequences when these are explicitly programmed, it is hard to see how they could be established because of their consequences when no such consequences are explicitly programmed. The only alternative is to assume that they are established, as Pavlov argued, because they are elicited by the UCS.

As we shall see, stimulus substitution theory has far-reaching implications for other types of learning and for our understanding of the modifiability of animal behaviour in general.

17.3 Extinction and habituation

We have seen that presentation of a UCS increases the strength of the CR. Pavlov discovered that withholding such reinforcement led to the gradual disappearance of the CR. The process by which learned behaviour patterns cease to be performed when they are no longer appropriate is called **extinction**.

In a classical-conditioning experiment, the dog learns that the bell (CS) signals the presentation of food. Its salivation (CR) is therefore an appropriate response to make in anticipation of the availability of food. If food no longer is made available, then the dog should no longer treat the bell as a signal for food. This is precisely what happens; omission of food results in a decline in the salivary response to the CS. The behaviour of the animal then appears to be the same as it was before conditioning took place. Another example of extinction is illustrated in Figure 17.4.

If, after extinction, the CS is again paired with the reinforcer, the CR reappears much more rapidly than it did during the original conditioning. This suggests that the process of extinction does not abolish the original learning but that it somehow suppresses it. Further evidence for this conclusion comes from the phemomenon of **spontaneous recovery**, by which a response that has been extinguished recovers its strength with rest. For example, Pavlov (1927) reported an experiment in which the number of drops of saliva secreted to the CS was reduced from ten to three in a series of seven extinction trials. The **latency** (time delay) of the response also increased from three to thirteen seconds. After a rest period of 23 minutes, the salivation on the first trial with the CS alone amounted to six drops with a latency of five seconds.

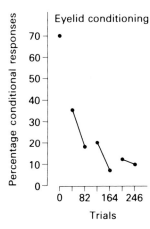

Fig. 17.4 Extinction of eyelid conditional responses in rabbits (After Schneidermann et al., 1962).

Pavlov argued that the decline in CR observed during extinction must be due to an accumulation of internal inhibition. He showed that if an extraneous novel stimulus was presented at the same time as the CS in a conditioned trial, the CR was disrupted. According to Pavlov (1927): 'The appearance of any new stimulus immediately invokes the investigatory reflex and the animal fixes all its appropriate receptor organs upon the source of disturbance. ... The investigatory reflex is excited and the conditioned reflex is in consequence inhibited.' This phenomenon is called **external inhibition**. If the extraneous stimulus is presented during the course of extinction, the CR increases in strength on that trial. This type of effect, called **Pavlovian disinhibition**, provides further evidence of the inhibitory nature of extinction. Unlike external inhibition, it is not due to competition between reflexes but is thought to be caused by an increase in arousal. It is a widespread phenomenon and can be shown to occur when a CR is declining, for whatever reason, including habituation.

The evidence suggests that, during extinction, the animal learns that the CS is no longer followed by reinforcement. The CS is now associated with the absence of a reinforcer, and the CR is inhibited accordingly. As we can see in Chapter 24, the idea that animals can learn that certain stimuli predict no consequences plays an important part in modern learning theory. Scientists have found that an important requirement for the development of inhibition is that the CS should be paired with non-reinforcement in a context in which stimuli already are associated with reinforcement. In the animal's ordinary life, numerous stimuli are not associated with reinforcement, but the animal ignores these and learns nothing about them. Only when the animal is surprised by the absence of reinforcement does it learn that particular stimuli signal non-reinforcement (Mackintosh, 1974).

Repeated applications of stimulus often result in decreased responsiveness. This phenomenon, called **habituation**, is a form of non-associative learning that has some similarities to extinction. For example, the escape response of fish to a shadow passing overhead diminishes progressively if the stimulus is repeated every few minutes, until the fish cease to react at all. Similarly, the toad's (*Bufo bufo*) orientation response toward potential prey progressively declines if non-edible prey-like objects are presented repeatedly. It is well known to fruit growers that scarecrows erected to deter birds are effective for a short time, but the birds soon become habituated to them. Attempts to scare birds from airfields by broadcasting alarm calls have run into similar problems of habituation.

Habituated responses show spontaneous recovery when stimulation is withheld. If habituation and subsequent recovery of the response are repeated a number of times, then the habituation becomes progressively more rapid. In these respects, habituation is similar to extinction. If a novel stimulus is presented during the process of habituation, then there is an increase in responsiveness, as illustrated in Figure 17.5. This dishabituation is thought to be due to changes in the animal's level of arousal and is very similar to Pavlovian disinhibition.

Habituation is usually regarded as a form of learning and can be distinguished experimentally from the waning of a response due to sensory adap-

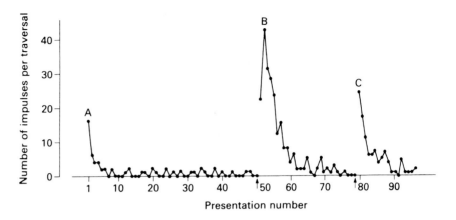

Fig. 17.5 Habituation and stimulus induced dishabituation in a neuron responding to movement stimuli. A is the original habituation, B and C are dishabituations induced by extraneous stimuli (arrows) (After Rowell and Horn, 1968).

tation or to fatigue. Habituation is similar to extinction in that the animal learns to inhibit responses that are not followed by reinforcing consequences. Both extinction and habituation show spontaneous recovery and disinhibition by extraneous stimuli. Extinction differs from habituation in occurring in relation to previously learned responses, whereas the responses that typically show habituation are innate responses that have not been established through any process of conditioning.

17.4 Generalization

When an animal has learned a particular response to a particular stimulus, it may also show the response to other similar stimuli. Thus, Pavlov (1927) noted that: 'If a tactile stimulation of a definite circumscribed area of skin is made into a conditional stimulus, tactile stimulation of other skin areas will also elicit some conditional reaction, the effect diminishing with increasing distance of these areas from the one for which the conditioned reflex was originally established.' This type of phenomenon is called **stimulus generalization**. Pavlov assumed that it was due to a spreading wave of excitation that travelled across the cerebral cortex from the focus of the CS. This explanation, however, does not accord with modern views on the neuronal structure of the brain (Thompson, 1965).

Current explanations of stimulus generalization concentrate on the stimuli involved. A stimulus used in a conditioning experiment consists of a collection of separate elements. For example, a tone has a given frequency, intensity and duration. These dimensions of the stimulus may become conditioned during acquisition of a CR. A new stimulus that has some elements in common with the CS may be capable of eliciting the CR to some extent. For example, while a tone of 1000 hertz is discriminated readily from one of 300 hertz by humans, pigeons that are trained to respond to the former generalize to the latter. However, the tones have other qualities in common, especially the fact that they are unlike natural sounds in being characterized by a single frequency. It is perhaps not surprising, therefore, that pigeons treat them as similar. If a pigeon is rewarded in the presence of a 300 hertz

Fig. 17.6 Generalization gradients for individual pigeons responding to tones. The training tone had a frequency of 1000 hertz (After Jenkins and Harrison, 1958).

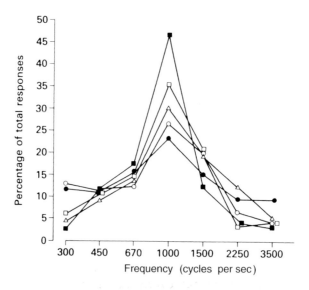

Fig. 17.6 Generalization gradients for individual pigeons responding to tones. The training tone had a frequency of 1000 hertz (After Jenkins and Harrison, 1958).

tone, then it has less tendency to generalize to tones of frequencies other than 1000 hertz. The generalization gradient has been sharpened by discrimination training, as illustrated in Figure 17.6. The discrimination training reduces the number of elements that are associated consistently with reward. The pigeon is encouraged to pay attention to the frequency of the tone by the fact that the other dimensions of the sound are not associated consistently with reward. This is called the common elements theory of generalization.

Stimulus generalization has important implications for many aspects of animal learning. For example, during extinction of a CR, omission of reinforcement changes the stimulus environment in which the conditioning was established. There is some evidence that the reduction in responsiveness seen during extinction is due in part to a generalization decrement. Similarly, a response habituated to one stimulus will show generalization to another similar stimulus. To some extent, the animal treats the new stimulus as if it had been presented previously. The degree of habituation to a new stimulus depends upon how similar it is to the stimulus to which the animal previously has been habituated. If the stimulus is both novel and unfamiliar, the dishabituation will occur. Thus, generalization tends to counteract the effects of novel stimuli on habituated responses.

Pavlov held the view that discriminations could be learned when some aspects of the CS are reinforced while other aspects are not reinforced. Initially, all aspects of the CS elicit a CR, but if the experimenter reinforces some aspects and not others, then we may designate these CS+ and CS−, respectively. Discriminations may be found among aspects of a single physical dimension, such as auditory stimuli of different frequencies or lights of differing brightness. Discrimination may also be formed between combinations of qualitatively different stimuli, called **compound stimuli** by Pavlov (1927). For instance, the CS+ might be the combination of a tone and a tactile stimulus, and the CS− might be the tactile stimulus alone. Once the animal has learned the discrimination, the tactile stimulus alone will no longer elicit a CR. As mentioned, every conditioning experiment involves compound stimuli in the sense that the CS inevitably is presented in a particular stimulus situation and can be distinguished from background stimuli only by the process of discrimination learning on the part of the animal. During the interval between trials, some stimuli that the animal associates with reinforcement will be present during the early part of training, prior to the development of discrimination between CS+ and CS−. Therefore, we might expect to find that dogs salivate during the inter-trial interval of a classical-conditioning experiment, at least during the early stages. Sheffield (1965) reported that his dogs did indeed salivate during the inter-trial interval to an extent that declines progressively during training.

17.5   Instrumental learning

While research on classical conditioning was initiated in Russia the principles of **instrumental conditioning** were discovered and developed in the United States. However, the writings of Conway Lloyd Morgan (1852–1936), of the University of Bristol, United Kingdom, seem to have provided the initial impetus.

Morgan was critical of much of the contemporary research in animal psychology on the grounds of its poor methodology and sloppy reasoning. In his *Introduction to Comparative Psychology* (1894), he enunciated his famous canon:

'In no case may we interpret an action as the outcome of the exercise of a higher psychical faculty, if it can be interpreted as the outcome of one which stands lower in the psychological scale.'

Later (1900) he added the following rider:

'To this it may be added – lest the range of the principle be misunderstood – that the canon by no means excludes the interpretation of a particular act as the outcome of the higher mental processes if we already have independent evidence of their occurrence in the agent.'

Morgan had considerable influence upon the development of behaviourism, particulary upon John Watson and Edward Thorndike. In 1896, he delivered the Lowell lectures at Harvard University, where he provided the stimulus for Thorndike's pioneering research on animal intelligence. Morgan recounted how his dog, Toby, had learned to open the latch on the garden gate by putting its head through the railings. Thorndike started his research soon after Morgan's visit and devised ways of repeating this observation under controlled conditions in the laboratory.

Thorndike became a very important figure in US psychology, and for half a century his theories dominated both animal and educational psychology. An eminent contemporary wrote:

'The psychology of animal learning – not to mention that of child learning – has been and still is primarily a matter of agreeing or disagreeing with Thorndike, or trying in minor ways to improve upon him. Gestalt psychologists – all of us here in America – seem to have taken Thorndike, overtly or covertly, as our starting point.' (Tolman, 1938).

Thorndike carried out a series of experiments in which cats were required to press a latch or pull a string to open a door and escape from a box to obtain food outside. The boxes were constructed with vertical slats so that the food was visible to the cat (see Figure 17.7). A hungry cat, when first placed in the box, shows a number of activities including reaching through the slats toward the food and scratching at objects within the box. Eventually the cat accidentally hits the release mechanism and escapes from the box. On subsequent trials, the cat's activity becomes concentrated progressively in the region of the release mechanism, and other activities gradually cease. Finally, the cat is able to perform the correct behaviour as soon as it is placed in the box.

Thorndike (1898) designated this type of learning as 'trial, error, and accidental success'. It is nowadays called **instrumental learning**, the correct response being 'instrumental' in providing access to reward. This type of learning had been known to circus trainers for centuries, but Thorndike first studied it systematically and developed a coherent theory of learning based upon his observations.

To explain the change in the animal's behaviour observed during learning experiments, Thorndike (1913) proposed his **law of effect**. This stated that a response followed by a rewarding or satisfying state of affairs would increase in probability of occurrence, while a response followed by an aversive or annoying consequence would decrease in probability of occurrence. Thus, the success of instrumental learning is attributed to the fact that learned behaviour can be modified directly by its consequences. Thorndike (1911) assumed that a reinforcer increases the probability of the response upon which it is contingent because it strengthens the learned connection between the response and the prevailing stimulus situation. This became known as the 'stimulus–response theory of learning', and versions of this

Fig. 17.7 A cat in one of Thorndike's puzzle boxes.

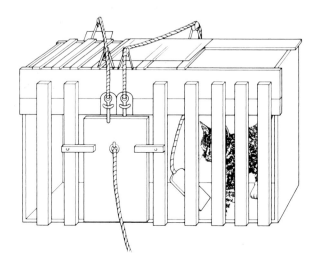

theory were predominant for many years. While recognizing the validity of the law of effect as an empirical statement, most present-day psychologists doubt that behaviour is modified by its consequences in the direct way that Thorndike and his followers supposed. To understand this, we first have to consider the nature of reinforcement.

17.6 Reinforcement

There is a fundamental difference between the way a classical-conditioning experiment is conducted normally and the procedure used in an instrumetal-learning experiment. In classical-conditioning experiments, a contingency is arranged between the CS (e.g. a bell) and the UCS (e.g. food), and the reinforcer (food) is applied regardless of the behaviour of the animal. In instrumental-learning experiments, the reinforcement (e.g. food) is contingent upon some particular behaviour of the animal (e.g. pressing a latch). Thus, in classical-conditioning experiments a contingency is arranged between a stimulus and an outcome, while in instrumental-learning experiments the contingency is arranged between a response and an outcome (Mackintosh, 1974). However, these differences do not necessarily imply that different kinds of learning are involved in the two types of experiments. They do suggest, however, that different processes of reinforcement are involved.

We have seen that, in Pavlov's view, the occurrence of a reinforcing stimulus in a particular context will cause the responses elicited by that reinforcer to occur in anticipation of the delivery of the reinforcer. There have been some remarkable recent demonstrations of this effect, as we shall see. It is evident, however, that reinforcement is not always necessary for the occurrence of learned associations among stimuli. This can be seen most clearly by examining the phenomenon called **sensory preconditioning**. The procedure is to present two conditional stimuli (CS and CS) together for a number of trials before introducing the UCS. Thus, joint presentations of CS and CS are followed by pairing of CS with the UCS. In the final test phase, the strength of the CR in response to the CS is measured.

The first clear demonstration of sensory preconditioning is that of Brogden (1939), who gave dogs 200 trials on which a light and a buzzer were presented simultaneously. One of these stimuli was then paired with electric shock to the paw to establish leg flexion as a CR. Test trials to the other CS produced an average of 9.4 CRs compared with only 0.5 CRs obtained from control experiments that did not involve prior pairing of buzzer and light. More recent experiments show that better results can be obtained if fewer pretraining trials are given and if the two CSs are presented not simultaneously but within a few seconds of each other (Mackintosh, 1974).

The results of sensory-preconditioning experiments clearly show that pairing two neutral stimuli is sufficient to establish some learned association between them. It appears that if the stimuli are presented too often, some habituation occurs and learning does not improve. Quite apart from the fact that these results cannot be explained satisfactorily by a stimulus–response theory of learning, it is apparent that reinforcement is not necessary for associations to occur between two neutral stimuli. Pavlovian reinforcement is, therefore, not a necessary condition for the formation of associations, but it does facilitate their formation and render them resistant to habituation. As we see later, animals are quite capable of learning that certain stimuli are irrelevant to their current motivational requirements, and it is not, therefore, surprising that the associations formed between two neutral stimuli should be temporary.

We now turn to the question of **instrumental reinforcement**. Thorndike's law of effect became the mainstay of the behaviourist's approach to animal learning. An extreme position was reached by Harvard behaviourist Burrhus Frederic Skinner who turned the law of effect into a definition of reinforcement. For Skinner, a reinforcer was any event that, if made contingent upon some aspect of behaviour, would cause the behaviour to increase in frequency. Skinner (1938) also assumed that any reinforcer could strengthen any response in the presence of any stimulus, provided that the stimulus could be sensed by the animal and that the response was within its capacity. Thus, the response and reinforcer were regarded as being essentially arbitrary, and this was a widespread view among learning theorists up to the end of the 1950s.

A reinforcer that encourages an animal to approach the stimuli that it associates with it is generally called a 'positive reinforcer'. If the reinforcer leads the animal to avoid the situation in future, then it is a 'negative reinforcer'.

Animals may learn to fear certain situations as a result of experiencing pain or stress there. They subsequently may show avoidance behaviour when they next encounter the situation. Similarly, situations in which the animal encounters natural fear-inducing stimuli may induce fear by association, without the animal experiencing pain. Repeated presentation of such stimuli, however, may lead to habituation so that the animal no longer avoids them.

As we have seen, aversive stimuli can be used in classical-conditioning experiments. One of the first studies of this type was by Bekhterev (1913) who applied a mild electric shock (UCS) to the forepaw of a dog, following an auditory stimulus like a tone (CS). The dog initially flexed its leg in response to the shock (UCR), and after a number of paired presentations of CS and UCR, it flexed its leg in response to the tone (CR), in the absence of shock. This type of classically conditioned defence reaction remained a paradigm case of avoidance learning for a number of years.

For Thorndike (1913), the effects of punishment were symmetrical to those of rewards. The positive law of effect maintained that responses that were followed by a satisfactory state of affairs increased in probability, while the negative law stated that responses followed by an annoying state of affairs decreased in probability. However, Thorndike (1932), on the basis of several experiments, eventually concluded that the law of effect did not apply to the effects of punishment. In this view, he was followed by Skinner (1938, 1953) and Estes (1944). The consensus seemed to be that punishment was not effective in weakening stimulus–response connections, though it sometimes had the effect of temporarily suppressing the punished behaviour. More recent evidence suggests that punishment can be effective in modifying behaviour when it is made contingent upon a particular response (Church, 1963, 1969).

17.7 Operant behaviour

Skinner (1937) introduced a distinction between operant behaviour and respondent behaviour. **Operant behaviour** he defined as a spontaneous action without any obvious stimulus. **Respondent behaviour** included all behaviour performed in response to an identifiable stimulus. Skinner believed that all operant behaviour was modified and effectively controlled by the prevailing reinforcement contingencies. The idea that animal behaviour could be manipulated entirely by appropriate schedules of reinforcement represented the extreme behaviourist view (Skinner, 1938). Skinner's behaviourist philosophy brought about a revolution in experimental technique that has persisted to this day.

In place of the trial-by-trial procedure characteristic of classical conditioning and experiments using puzzle boxes or mazes, Skinner devised the free-operant procedure in which the animal is allowed to indulge freely in various activities, while the experimenter attempts to manipulate the consequences. Rats and pigeons were chosen most frequently for this type of experiment, although many other animals, including humans, were also used. Operant conditioning consists essentially of training an animal to perform a task to obtain a reward. A rat may be required to press a bar, or a pigeon to peck an illuminated disc, called a key. The typical method of training is called **shaping**.

Let us consider the training of a pigeon to peck a key to obtain food reward. A hungry pigeon is placed in a small cage, equipped with a mechanism for delivering grain and a key at head height, as shown in Figure 17.8. This type of apparatus is nowadays called a 'Skinner box'. Delivery of food is normally signalled by a small light that illuminates the grain. Pigeons soon learn to associate the switching on of the light with the delivery of food and approach the food mechanism and eat the grain whenever the light comes on. The next stage of shaping is to make food delivery contingent upon some aspect of the animal's behaviour. It is usual to require the pigeon to peck the key, but Skinner claimed that any response could be shaped and that pigeons could be taught to preen or turn in small circles to obtain reward.

Fig. 17.8 A pigeon pecking a key in a Skinner box.

Key pecking may be shaped by limiting rewards to movements that become progressively more similar to a peck at the key. Thus, when it has learned to approach a key for reward, the pigeon is rewarded only if it stands upright with its head near the key. At this stage, the pigeon usually pecks at the key of its own accord, but it can be encouraged by temporarily gluing a grain of wheat to the key. When the pigeon pecks the key, it closes a sensitive switch in an electronic circuit that causes food to be delivered automatically. From this point on, the pigeon is rewarded only when it pecks the key and the manual control of reward is no longer required. The animal is now ready for use in an experiment.

Many types of experiment utilize this operant-conditioning procedure. For example, discrimination learning can be studied by rewarding the animals for responding only when a certain colour or pattern is presented, or by allowing the animal to choose between two keys that are differentiated visually. The technique has proved particularly useful for studying the effect of different types of pattern of reward. Thus, instead of rewarding a pigeon for every peck, it can be rewarded for every *n*th peck, so that there is a fixed ratio between number of pecks and number of rewards. This procedure is called a **fixed ratio reward schedule**. Other commonly used schedules include **variable ratio**, **fixed interval** and **variable interval**. On an interval schedule, reward is given at intervals of time specified by the experimenter. The animal is rewarded for the first response made after a given interval has elapsed. Different schedules of rewards have been found to have different effects on the animal's performance. For example, a variable interval schedule produces a very uniform rate of responding and provides a good benchmark against which to test the effects upon behaviour of various factors like reward size.

Skinner is regarded as having treated behaviourism as a philosophy of the science of behaviour, not the science itself. His approach was operationist and his psychology was anti-theoretical. Although Skinner maintains that all behaviour is produced by reinforcement, he recognized (1975) that 'natural selection is responsible for the fact that men respond to stimuli, act upon the environment and change their behaviour under contingencies of reinforcement'. Similarly, 'the fact that operant conditioning, like all physiological processes, is a product of natural selection throws light on the question of what kind of consequences are reinforcing and why'.

Skinner's approach depends upon the efficacy of reinforcement in modifying behaviour. His claim that any activity can be modified is illustrated by the various games that pigeons can be trained to play. Thus, Skinner (1953) reports that 'the pigeon was to send a wooden ball down a miniature alley towards a set of toy pins by swiping the ball with a sharp sideward movement of the beak. The result amazed us. . . . The spectacle so impressed Keller Breland that he gave up a promising career in psychology and went into the commercial production of behaviour'.

Ironically, it was Breland and Breland (1961) who first cast doubt upon the assumption that any activity could be modified by reinforcement. They found that, in attempting to train animals to perform various tricks, some activities appeared to be resistant to modification by reinforcement. For example, they attempted to train a pig to insert a coin into a piggy bank. The pig would pick up a wooden token, but instead of dropping it into the container, it would repeatedly drop it on the floor, 'root it, drop it again, root it along the way, pick it up, toss it in the air, drop it, root it some more, and so on' (Breland and Breland, 1961). Similarly, they encountered chickens that insisted on scratching at the ground when they were supposed to stand on a platform for 10–12 seconds to receive a food reward. Subsequently, there have been numerous reports of this type. Thus, Sevenster (1968, 1973) successfully trained male three-spined sticklebacks (*Gasterosteus aculeatus*) to swim through a small ring to obtain access to a female. He was unable, however, to train males to bite a glass rod to gain access to a female because the male insisted on directing courtship behaviour toward the rod. These and other studies are reviewed by Shettleworth (1972).

Breland and Breland (1961) interpret their findings as evidence of instinctive drift by which 'learned behaviour drifts towards instinctive behaviour' whenever the animal has strong instinctive behaviour that is similar to the conditioned response. They point out that their results violate Skinner's (1938) principle of least effort, by which animals are supposed to obtain their rewards in the quickest and most comfortable manner. In all cases, reward is delayed considerably by the misbehaviour of the animals in their studies. The general question of the relationship between instinct and learning is the subject of Chapter 20. In Chapter 18 we discuss further examples of this type and try to assess the extent to which they can be accounted for by a universal theory of learning.

Points to remember

- Classical, or Pavlovian, conditioning leads to the establishment of an association between a previously meaningful stimulus (the unconditional stimulus) and a previously neutral stimulus (the conditional stimulus) such that the animal comes to respond to the latter (the conditional response) in the way that it previously responded (with the unconditional response) to the former.

- The process of establishing the association, called acquisition, usually takes a number of trials during which some reinforcement is given.

- A decline in responsiveness can occur as a result of habituation or extinction. Habituation results from repeated presentation of a stimulus without immediate consequences. Extinction results from repeated presentation of a conditional stimulus without reinforcement.

- Generalization occurs when an animal responds to a stimulus which is similar to a conditioned stimulus.

- Instrumental learning appears to differ from classical conditioning because the animal is required to make a response before receiving reinforcement. However, it is not possible to conduct an instrumental-learning experiment without also setting up the conditions for classical conditioning. It is possible that reinforcement is not really instrumental in the formation of learned responses.

Further reading

Breland, K. and Breland, M. (1961) The misbehaviour of organisms. *Am. Psychol.* **16**, 661–664.

Mackintosh, N.J. (1983) *Conditioning and Associative Learning*. Clarendon, Oxford.

CHAPTER

Biological aspects of learning

The mechanisms of learning in animals are complex partly because simple forms of learning like habituation coexist with more complicated aspects. In people, for example, we can find examples of habituation, conditional reflexes, instrumental learning, reasoning and cognition. While other animals may not be so adept at learning by experience and reasoning, many undoubtedly possess some of these abilities. However, animal species vary considerably in their ecological circumstances and evolutionary history. Should we not expect natural selection to have tailored each animal's learning abilities to suit its ecological niche? Many biologists are inclined to take this view. Psychologists tend to assume that some learning mechanisms are common to many species. Is this view justified? In this chapter we attempt to answer this question.

18.1   Evolutionary aspects of learning

When an animal learns, its behavioural repertoire is changed forever. Although an animal may extinguish or forget behaviour it has learned, the animal can never return to its previous state. As we saw in Chapter 18, most apparent unlearning is learned inhibition of previous learning. Learning changes the animal's make-up and is, therefore, likely to alter its fitness.

The processes of learning have long been subject to natural selection. Therefore, we can expect the consequences of learning to be adaptive and to increase the animal's fitness. The idea that learning is adaptive has long been recognized by both psychologists (e.g. Thorndike, 1911; Seligman, 1970) and ethologists (e.g. Lorenz, 1965). However, for many years there was little systematic attempt to look at learning in terms of evolutionary theory (Shettleworth, 1984).

If the environment never changed, animals would gain no benefit by learning. A set of simple rules designed to produce appropriate behaviour in a given situation could be established by natural selection and incorporated as a permanent feature of the animal's psychological make-up. Some features of the environment do not change, and we usually find that animals respond to these in a stereotyped innate manner. For example, gravity is a universal feature of the environment, and the anti-gravity reflexes tend to be consistent and stereotyped (Mittelstaedt, 1964; Delius and Vollrath, 1973).

Some features of the environment change on a cyclic basis, either diurnal, lunar or seasonal. Here again, there is no real need for the animal to learn to adapt to such changes. As we saw in Chapter 16, fundamental aspects of the animal's physiological and motivational make-up can change on a periodic basis, as a result of acclimatization or as dictated by a biological clock. In general, environmental changes that are predictable from an evolutionary point of view can be handled on the basis of preprogrammed changes in the animal's make-up (McFarland and Houston, 1981).

Change throughout an animal's life history may be predictable also from an evolutionary point of view. In some cases, the animal's behaviour may change by maturation, without learning being involved. In other cases, the animal's behaviour may change as a result of learning, but the learning is preprogrammed in the sense that it occurs at a particular stage of the animal's life, more or less independently of the particular consequences of the animal's behaviour. For example, human children have an immense facility for language learning between the ages of two and seven. They learn whatever language they hear around them. Similarly, some birds have distinct sensitive periods of song learning, during which they normally learn by hearing their fathers or other conspecifics (see Chapter 3). Such learning is preprogrammed in the sense that it occurs during a particular period of the life history, relatively independently of the individual's circumstances. A child in one environment may learn German, another English. A bird in one environment may learn one type of song, while in another environment the song it learns may be different.

In Chapter 20, we see how young animals may become imprinted upon certain features of the environment at particular stages of development. In this way they may learn about certain characteristics of their parents, siblings or habitat. Such imprinting may influence the future habitat selection and courtship and social preferences of the individual animal. Why do such animals have to learn the characteristics of their habitat, parents or future mates when other species show no such imprinting? This is a difficult question to answer. In the case of habitat selection, imprinting may enable animals to choose habitats similar to those of their kin and thus to enhance their inclusive fitness.

Pat Bateson (1979) has suggested that some young birds may need to discriminate between the parent that cares for them and other members of the species because parents discriminate between their own and other young and may attack young that are not their own. He also suggests that imprinting enables an animal to learn about certain characteristics of close kin so that it subsequently can choose a mate that is slightly different and so attain a balance between inbreeding and outbreeding. Whenever the problems facing the juvenile animal differ slightly from generation to generation, there may be an advantage for individuals that are predisposed to learn certain types of things like the nature of the habitat, of close relatives and of food sources. This type of learning is a form of **contingent maturation** in the sense that the course of development is contingent upon certain experiences.

Unpredictable environmental changes that occur within an individual animal's lifetime cannot be anticipated by preprogrammed forms of learning. The individual must rely upon its own resources in adapting to such changes. The ability to modify behaviour appropriately in the face of unpredictable environmental changes usually is taken as a sign of intelligence, but we should not forget that much can be explained in terms of conventional learning mechanisms.

Historically, learning was investigated in artificial laboratory conditions, without much regard for the animal's special requirements and preoccupations. As we saw at the end of Chapter 17, however, doubts about the efficacy of reinforcement first came from within psychology.

A common reaction to the discovery that reinforcement is not always effective in altering behaviour in the way the experimenter intended has been to deny that any general laws of learning exist. Instead, the learning abilities of different species are said to be adapted specifically to the ecological constraints typical of the animal's normal way of life (e.g. Rozin and Kalat, 1972; Shettleworth, 1972). This is obviously an attractive idea for ethologists and evolutionary biologists, but does it mean that we necessarily have to abandon all attempts to lay down general principles of learning? The issue is controversial, and before coming to any conclusion, we should look again at the role of reinforcement in instrumental learning.

18.2 Constraints on learning

In an instrumental-learning situation, the animal is reinforced for performing some act. The reinforcement must occur in some stimulus situation. Particular features of the stimulus situation will be associated regularly with reinforcement, thus providing the essential conditions for classical conditioning. The reinforcer used in an instrumental-learning experiment is also a potential classical UCS that will automatically elicit an innate set of responses. In chickens, these responses include scratching for food, and in pigs, rooting. These responses will become classically conditioned to the stimuli that regularly accompany food (Breland and Breland, 1961) (Figure 18.1).

Fig. 18.1 Examples of tricks taught to animals by Keller Breland, using operant conditioning methods.
(a) The rabbit plays up and down the keyboard several times to obtain a reward.
(b) The duck must collect a number of rings before obtaining a reward.

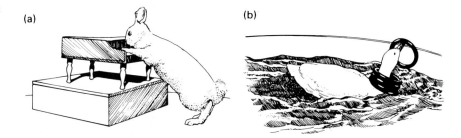

What, then, of the arbitrary responses selected for reinforcement by Skinner and his followers? Moore (1973) has pointed out that in almost every case the responses are not really arbitrary but part of the repertoire of instinctive behaviour that normally is associated with the reward. Thus, the swiping movement employed by the pigeon in Skinner's bowling alley, mentioned previously, is a normal part of pigeon feeding behaviour used to flick loose earth aside and so uncover seeds. If this interpretation is correct, then we might expect that pigeons pecking a key for food reward would do so in a manner different from that employed when they are working for water reward. Examination of a film of pigeons during key-pecking behaviour shows that this is in fact the case (see Figure 18.2). When students were asked to distinguish between filmed food and water test sessions, without knowing which reinforcer was being used, they were able to judge with 87 per cent accuracy (Moore, 1973). Birds pecking for food strike the key with an open beak using sharp, vigorous pecks. When pecking for water, the bill remains closed and contact with the key is more sustained. Water pecking is often accompanied by the pumping (swallowing) movements typical of pigeon drinking behaviour (unlike other birds, pigeons drink by sucking up the water).

Further evidence comes from the phenomenon of **autoshaping**. Brown and Jenkins (1968) discovered that, if pigeons were exposed to repeated pairings of key illumination and food presentation, they would begin to peck the key without any training or shaping of the type described before. The key-pecking behaviour typical of innumerable operant-conditioning experiments was established not as a result of instrumental reinforcement but by a straightforward Pavlovian procedure. The lighted key was paired with grain, and the pigeons pecked the key as if it were grain. The point can be demonstrated even more clearly if the key illumination is paired with a reinforcement to which pecking is not the pigeons' natural response. This can be done by offering the pigeons a sexual reward. Paired male and female pigeons are housed in adjacent chambers separated by a sliding door. Once a day, the stimulus light is turned on and the sliding door is removed so the male can begin his courtship display. Within five to ten trials, the males begin to make

Fig. 18.2 Pecking for food and water rewards by pigeons: (a) shows pecking for water and (b) shows pecking for food in an autoshaping (*Photograph: Courtesy of Bruce Moore*).

(a)

(b)

conditioned responses toward the stimulus light. The pigeons direct their courtship toward the light, thus behaving toward it as if it were a female (Moore, 1973).

To prevent the pigeon from obtaining any instrumental reinforcement during the autoshaping experiment, it is possible to arrange matters so that food reward is withheld whenever the pigeon pecks the key. Under these conditions, autoshaped pigeons persist in pecking; such is the strength of the Pavlovian conditioning (Williams and Williams, 1969). These and other autoshaping procedures have established beyond doubt that complex behaviour patterns such as feeding, drinking and courtship can be established by means of classical conditioning. They also suggest that much of the so-called operant behaviour established by instrumental-conditioning techniques like shaping is in fact the result of classical conditioning. This does not mean, however, that the law of effect necessarily has to be abandoned.

18.3 Learning to avoid enemies

Avoidance reactions are a form of defensive behaviour by means of which animals minimize their exposure to situations that appear to be dangerous. Fear-provoking stimuli may be sign stimuli (see Chapter 20) that are responded to without any prior experience, or they may be stimuli to which a fear or avoidance response has become attached through the process of conditioning. Natural stimuli that induce innate avoidance behaviour include those associated with predators, such as the avoidance of hawk-like silhouettes by young birds, and with poisonous plants and animals such as snakes, fungi, etc. Innate avoidance behaviour varies considerably from species to species. It includes freezing posture, especially in animals that normally are well camouflaged; running-for-cover thigmotaxic behaviour, which involves contact with objects and avoidance of open areas; and warning signals, designed either to deter an attack or to warn other animals. Figures 18.3 and 18.4 provide some examples.

In its natural habitat, an animal that learned how to escape from predators

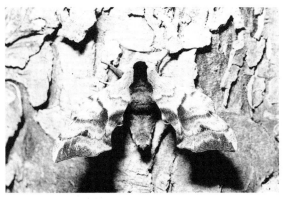

Fig. 18.3 Warning display of a hawkmoth disturbed by a potential predator. The sudden appearance of the eyespots often scares away small birds (*Photographs: P. Ward*).

Fig. 18.4 An example of a warning device.

by means of a trial-and-error procedure probably would survive for a few trials only. Animals tend to have innate defence reactions that are essentially reflex in nature though modifiable by learning. Robert Bolles (1970) pointed out that, in an avoidance-learning experiment, the experimenter selects, either deliberately or by virtue of the design of the apparatus, the avoidance response that will be effective. This selected response may or may not coincide with the animal's innate response. The degree of compatibility between the innate and to-be-learned behaviour greatly influences the ease with which the avoidance response is learned.

For a number of years, psychologists had reported considerable variation in the facility with which animals learn avoidance responses (Mackintosh, 1974). Thus, rats learn to run from one box to another to avoid shock within about five trials but require hundreds of trials to learn to press a lever to avoid shock. Pigeons have great difficulty in learning to peck a key to avoid shock but have less difficulty in learning to depress a treadle. Another complicating factor is that innate avoidance responses may vary in accordance with the nature of the situation. Thus, in response to shock, rats tend to run when escape is possible and to freeze when escape is not possible.

If the situation includes a target like another member of the species, then shock is likely to induce aggressive behaviour (Logan and Boice, 1969). If the situation requires the animal to approach the stimulus that is associated with shock, then the training of avoidance responses is very difficult. Thus, if a rat is required to press a lever directly below the light that serves as a warning signal for shock, avoidance learning is difficult to establish. If the warning light is located far away from the lever, then conditioning is much easier (Biederman *et al.*, 1964). Grossen and Kelley (1972) observed that exposure to shock in an open-field apparatus increased the amount of time rats spent in close contact with the wall. They required the rats to jump onto a ledge to avoid shock and found that the rats learned this response more rapidly when the ledge was next to the wall than when it was in the centre of the apparatus.

In terms of conventional learning theory, innate defence reactions are UCRs. A Pavlovian analysis of avoidance learning involves two cardinal propositions (Mackintosh, 1974). First, successful avoidance responses are related closely to the responses normally elicited by the aversive stimulus. As we have seen, avoidance learning is much more rapid if the response required by the experimenter is compatible with the animal's natural reaction to the punishing stimulus. We will shortly see this principle illustrated in another context.

The second proposition of the Pavlovian view of avoidance learning is that the establishment of an avoidance response should not be directly dependent upon the avoidance contingency as such. It is the presentation of a negative reinforcer that strengthens the response elicited by that reinforcer. Omission of reinforcement, whether or not a result of avoidance behaviour, should not improve the process of classical conditioning. At this point, however, it is worth noting that omission of reinforcement sometimes has no effect on avoidance learning, while it sometimes has profound effects (Mackintosh, 1974). For example, Woodard and Bitterman (1973) showed that goldfish

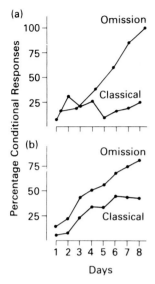

Fig. 18.5 Comparisons of classical and omission schedules of reinforcement: (a) wheel running in guinea pigs, (b) leg flexion conditioning in dogs ((a) After Brogden *et al.*, 1938 and (b) After Wahlsten and Cole, 1972).

would swim from one compartment to another when this resulted in the omission of shock, but that they would swim just as readily when shocks were unavoidable. Conversely, omission of shock is an effective reinforcer to induce guinea pigs to run in a wheel (Brogden *et al.*, 1938) or rats to run from one compartment of a box to another (Miller, 1948). We saw that this procedure is superior to the classical-conditioning procedure (see Figure 18.5).

Mackintosh (1974) points out that the distinction between classical and instrumental analysis of avoidance learning may hinge on the question of whether or not external stimuli are established as a single shock or for safety, and whether or not the responses to these stimuli provide the animal with adequate feedback. In cases where the animal avoids shock by jumping out of the box or by running away from a noxious stimulus, the avoidance response itself is readily distinguishable from the animal's other behaviour, and it also transports the animal into a different external stimulus situation. However, for a goldfish that is swimming continually, swimming to a particular place (to avoid shock) is not a very distinctive response. For a rat that is motionless, however, running to a place where shocks are not received is a distinctive response with readily apparent consequences. It should not be surprising, therefore, that the rat finds this type of avoidance response easier to learn than pressing a bar.

Rats that can press a lever to avoid shock and gain access to another compartment, where shock was never delivered, learn to press more quickly than rats that could only press the lever but not leave the first chamber (Masterson, 1970). Pressing a lever leaves the animal in much the same external stimulus situation as before, whereas escape to another compartment removes the animal from the stimulus associated with shock. Avoidance learning is much more effective in a one-way shuttle box than in a two-way shuttle box, presumably because the two-way box lacks a place associated with safety.

Hudson (1950) studied one-trial avoidance learning in rats. The animals received a single electric shock to the mouth while feeding in their home cage from a distinctive food-holder. Four minutes after the shock the food-holder was removed from the cage. Learning was assessed by later replacing the food-holder in the cage and recording the rat's reaction. A common observation was that, on the first occasion that the food-holder was replaced, the rat pushed bedding material onto and over the food-holder, as if to bury it. Thus after a single trial, the rats exhibited **conditional defensive burying** behaviour.

Subsequent work has shown that this phenomenon occurs in a number of species, in response to a variety of noxious stimuli (see Pinel and Wilkie, 1983, for a review). Thus rats attempt to bury objects associated with a conditional taste aversion (see below), and coyotes, after eating mutton containing lithium chloride, sometimes bury it. Ground squirrels show the response to snakes, in a natural situation (Owings and Coss, 1977) as well as in the laboratory (Coss and Owings, 1978).

Pinel *et al.* (1980) compared the burying response of rats in cages differing in size. They found that the larger the cage the less time was spent performing the burying response (Figure 18.6). Thus the burying response still occurs when the rat is free to escape, though to a lesser extent.

Fig. 18.6 Mean duration of burying by separate groups of shock and no-shock treated rats, tested in chambers of different size (from Pinel and Wilke, 1983).

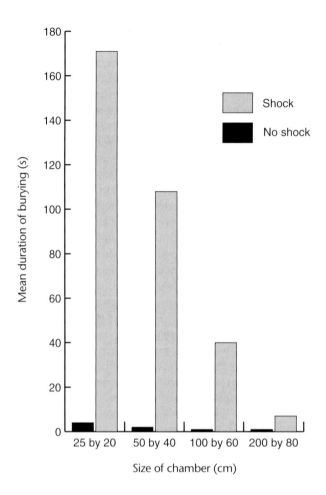

18.4 Learning to avoid sickness

We have seen that interoceptive cues can act as Pavlovian conditional stimuli (Chapter 17) and that learning plays an important role in homeostasis (Chapter 15), especially in the regulation of food intake. Evidence for the importance of learning was available at an early stage of research into specific hungers. Classic studies by Harris *et al.* (1933) and Richter *et al.* (1937) clearly demonstrated that thiamine deficient rats developed a preference for foods containing thiamine. Scott and Verney (1947) offered distinctively flavoured thiamine supplemented food and an unflavoured thiamine deficient food to thiamine deficient rats. After a preference had developed for the flavoured food, the flavour was switched to the other food, but the rats continued to prefer the flavoured food, even though that food was not the one deficient in thiamine. This result suggested that there was no specific recognition of thiamine.

The main problem in recognizing that learning was involved in the development of specific hungers was the long delay between the ingestive behaviour and its reinforcing consequences. In the case of sodium appetite, the sodium deficient rat can learn readily which food contains salt because it can taste and recognize salt and so obtain rapid reinforcement for selecting that particular food. In the case of a vitamin deficiency, however, the animal cannot detect directly the presence of the vitamin in the food. It has an opportunity to learn about the properties of the food only after some of it has been digested and there has been an appropriate alteration in the animal's physiological state. Although it was suggested (e.g. Scott and Verney, 1947) that the animals learned to associate the stimulus characteristics of the food with the subsequent beneficial feelings of recovery from sickness, this view was not acceptable to most psychologists at the time, for two main reasons. First, a generalized feeling of well-being is a rather vague concept, with overtones of hedonism (Young, 1961, p. 179). Second, the delay between the ingestive behaviour and its consequences is very much greater than learning theorists thought possible. Attempts were made to bypass the problem by postulating some mediation process like prolonged aftertaste resulting from feeding. However, such possibilities were disproved.

John Garcia and his co-workers (1955) fed rats a harmless substance and then produced a physiological after-effect by an independent means. After gamma radiation the rats developed an aversion to previously preferred saccharin-flavoured water. Garcia *et al.* (1961) pointed out the importance of this type of effect for the psychology of animal learning and cited nine confirmatory studies. It was another ten years, however, before the full implications of the phenomenon were appreciated and before researchers realized that specific hungers and poison avoidance were different aspects of the same phenomenon.

For centuries, people have attempted to eradicate local rat populations by various means, including poisoning. A wide variety of poisons have been tried (Chitty, 1954) with limited success. A perennial problem has been the rat's proverbial bait shyness (Rzoska, 1953). As we saw in Chapter 15 rats tend to avoid unfamiliar foods and to sample them tentatively. A rat that takes a small amount of poisoned food and survives will never touch the same type of food again. The following account of the laboratory equivalent of this ocurrence is given by Revusky and Garcia (1970):

'An animal is made to consume a flavoured substance, such as saccharin solution, and is later subjected to toxic after-effects produced by such independent means as injection of poison or X-irradiation. After it has recovered from the toxicosis the animal will avoid consuming the flavoured substance. The animal behaves as though it thinks that consumption of the substance made it sick. This specific aversion will not develop if toxicosis occurs in the absence of previous consumption, or if consumption occurs without being followed by toxicosis. It differs from the types of learning usually investigated in that it can occur after a single pairing even when the interval between ingestion and toxicosis is a number of hours.'

Fig. 18.7 A rat feeding in its natural environment.

Little is known about the nature of the internal cues by which animals assess their relative sickness or well-being. Treisman (1977) has suggested that the nausea and vomiting typical of motion sickness in humans, which arise from abnormal visual–vestibular stimulation (Howard, 1982), occur because one of the first indications that an animal has that it has been poisoned is a disturbance in the vestibular system. The vomiting, therefore, serves to expel any poison from the system, and the ensuing depression serves to keep the animal quiet while it recovers. Thus, Treisman suggests that the vestibular system, which is particularly sensitive and susceptible to penetration by small molecules, has evolved a secondary function as a poison detector, but this hypothesis has not been confirmed.

Sickness can be regarded as a state function, that is an index of the animal's overall physiological state. It can be argued that it must be the rate of change of this function that the animal associates with the feeding stimuli (McFarland, 1973). This line of argument leads McFarland (1973) to claim:

'It is impossible for an animal both to learn to prefer flavoured substances on the basis of improved physiological state, and to avoid them on the opposite basis. The two processes will simply cancel each other out. For example, an animal that learns to avoid a flavour because of subsequent sickness must also learn to approach the same flavour when it recovers from the sickness.'

Paul Rozin and James Kalat (1971) note that much of the evidence that appears to demonstrate positive preferences can be reinterpreted in terms of learned aversions. The problem is that many of the experiments designed to test for preferences (e.g. Garcia et al., 1967; Campbell, 1969; Zahorik and Maier, 1969) do not give the rats a fair choice. Zahorik and Maier gave a choice among a taste associated with recovery from dietary deficiency, a taste associated with deficiency and a novel taste. However, they did not offer the rats food with a familiar safe taste. Thus, it is possible that the apparent preference for the food associated with recovery was simply a contrast to the fact that the rats were aversive to the other foods on account of the novelty of one and the association of the other with dietary deficiency.

A number of studies have attempted to demonstrate a clear positive preference (e.g. Revusky, 1967; Zahorik et al., 1974), but the question remains controversial (McFarland, 1973; Rozin, 1976). The situation is complicated because rats have a natural aversion to novel foods (see Chapter 19), but this is modified in deficient animals. Rodgers and Rozin (1966), for instance, found that thiamine deficient rats show an immediate marked preference for new foods, even when the new food is thiamine deficient and the old food has a thiamine supplement. (The preference reverses within a few days.) Rozin (1968) found that a rat suffering from effects of poisoning or from a dietary deficiency, when faced with a choice among a familiar safe food, a familiar aversive food and a new food, shows a preference for the familiar safe food. It seems (Rozin and Kalat, 1971) that the rat learns to avoid the deficient food when it is the only food available. It tends to become anorexic, refusing to eat. As a result of this learned aversion, the rat shows an immediate preference for a novel food, thus enabling it to learn about its consequences.

Rozin (1976) has suggested that rats categorize food into four classes: novel, familiar-safe, familiar-dangerous and familiar-beneficial. The last category remains a subject of controversy. In any event, it is usually agreed that rats 'are strongly biased toward learning effectively and rapidly what makes them sick, and rather poor at learning what makes them well' (Rozin, 1976). We now turn to the question of how this learning occurs.

18.5 Stimulus relevance

How does an animal learn to associate eating a particular food with the delayed physiological consequences? During the interval between eating a poisoned food and suffering toxic effects, there will usually be numerous events including meals with which the toxicosis could be associated. The problem can be met in part by appealing to the principle of **stimulus relevance** (Capretta, 1961). According to this principle, the associated strength of a cue with some consequences (reinforcer) depends partly upon the nature of the consequences. Similar notions are those of **preparedness** (Seligman, 1970) and **belongingness** (Garcia and Koelling, 1966). Revusky and Garcia (1970) take the view that:

'The relevance principle responsible for association of delayed physiological consequences with flavours is that flavour has a high associative strength relative to physiological consequence, while an exteroceptive stimulus has low associative strength (at least in the mammal). If the consequence is an event which normally emanates from the environment, such as shock or receipt of a pellet food, the converse is true.'

Various experiments seem to substantiate this view. For example, Garcia and Koelling (1966) labelled food with light, sound and taste cues simultaneously and paired them with either electric shock or poisoning. Different groups of rats were presented with different pairs of stimulus and consequence. They found that the rats would associate the taste of food with sickness but not with electric shock. They would associate the visual and auditory stimuli with shock but not with sickness. Similarly, Garcia *et al.* (1968) used two sizes of food pellets, coated with flour or powdered sugar or uncoated. The design of this experiment is illustrated in Table 18.1. Rats ate pellets differing either in size or in flavour. They were punished with electric shock or with irradiation. The results indicate that eating pellets that were distinguishable by flavour was suppressed markedly by sickness but not by shock. Conversely, eating pellets that differed in size was suppressed by shock but not by sickness.

Table 18.1 Plan of the experiment by Garcia *et al.* (1968) and the results expected on the basis of stimulus relevance

Variable	Punishment	Expected result
Size of pellet	Shock	Aversion
Size of pellet	Toxicosis	No aversion
Flavour	Toxicosis	Aversion
Flavour	Shock	No aversion

Some scientists (e.g. Garcia *et al.*, 1970; Rozin and Kalat, 1971) claim that the rat's ability to associate flavours selectively with sickness over long intervals represents a specialized learning system that does not obey the conventional law of learning. Others (e.g. Mackintosh, 1974; Revusky, 1977), however, maintain that almost every property of conventional laboratory conditioning is also present in the learning of taste aversions.

Birds can readily learn aversions to the sight of food (Brower, 1969). Wilcoxon *et al.* (1971) found that Japanese quail, in contrast to rats, learn poison-based aversions more rapidly to the colour than to the taste of drinking water. Moore and Capretta (1968) obtained similar results with chickens. Foree and LoLordo (1973) found that if pigeons were trained to press a treadle in the presence of a combined visual and auditory stimulus (tone), the relative importance of the visual and auditory cues depends upon the nature of the reinforcement. Pigeons that were responding for food would not respond to the tone presented alone, but would respond to the light alone. Conversely, pigeons responding to avoid electric shock responded in the presence of the tone, but not of the light. Thus, it appears that the cues to which the animal normally attends while feeding are those that are effective in learned aversions to the physiological consequences of feeding. Rats normally attend to olfactory and taste cues, while birds normally attend to the visual characteristics of their food. In responding to electric shock, rats attend to visual stimuli and pigeons to auditory stimuli.

Some evidence indicates that animals perceive electric shock as an external stimulus, similar to an attack from an opponent or predator (Ulrich and Azrin, 1962). For a bird, auditory stimuli, like alarm calls, may be the most effective in situations where there is an external threat. Mackintosh (1974) points out that, during their lifetime, rats may learn that changes in visual or auditory stimulation are uncorrelated with changes in their internal state, whereas changes in taste stimuli do predict such changes. The tendency of rats, as adults, to associate flavour rather than visual or auditory stimuli with poisoning may result from the rats' ability to learn about, and generalize from, the correlations to which they are exposed during their lifetime.

The long delay that can occur between stimulus and reinforcement in taste aversion learning poses a more serious problem for conventional learning theory. In the conventional learning experiment, the reinforcer starts to lose its effectiveness after only a few seconds of delay. An association between flavour and sickness can occur even if there is an interval of several hours between the two (Revusky and Garcia, 1970; Rozin and Kalat, 1971). It has been suggested that the flavour may persist for a long time after ingestion of the food, but a number of studies rule out this possibility. Nachman (1970), for example, showed that rats could associate the temperature of their drinking water with subsequent delayed sickness, a factor that could not possibly persist for more than a few seconds. Moreover, as we have seen, birds are able to associate the colour of water with subsequent sickness. Rozin (1969) found that rats could learn aversions to particular concentrations of substances, and it is difficult to see how different concentrations could remain discriminably different half an hour after ingestion. Several studies show that exposure to a second flavour during the interval between ingestion of a novel test substance and subsequent induced sickness does not abolish the association between the test substance and sickness. If the second flavour is similar, it will not become associated with sickness and will not interfere with the aversion to the test substance (Revusky, 1971).

Normally, when a rat is poisoned after exposure to a novel flavour, few, if any, events occur during the interval between ingestion and sickness that would have sufficient stimulus relevance to be associated with the sickness. However, this is not the case with normal exteroceptive conditioning. Revusky (1971) argues that the possible delay of reinforcement is very short in such cases because a longer interval would permit interference from other stimuli, which could become established as signals for the UCS. Long delays are possible because only flavours are associated readily with sickness (in rats), whereas a variety of stimuli can be associated with electric shock. When a novel flavour (high stimulus relevance) is introduced into the interval between ingestion of a test substance and subsequent sickness, then interference does occur (Revusky, 1971; Kalat and Rozin, 1971). Revusky's suggestion goes some way toward reconciling the phenomena of taste aversion learning with conventional learning theory, but difficulties remain. For example, birds tend to associate the visual characteristics of food with subsequent sickness, and there must be numerous visual stimuli that could interfere with the formation of such associations.

18.6 The biological and psychological perspectives

Psychologists are often accused of thinking that all animals are alike and of placing too much reliance on experiments with laboratory rats and pigeons. Many ethologists and some psychologists (e.g. Rozin and Kalat, 1972) argue that the learning capacities of a species are tailored to its particular niche. This view might suggest that there is no general learning process common to many species and that the learning capacities of a given species are an amalgam of specific learning processes. For example, we have seen that the processes involved in learning to avoid noxious food seem different in some respects from the learning processes traditionally studied in laboratories.

There are some types of learning, such as the song learning of birds and imprinting of juvenile animals (see Chapter 20), which are obviously different from the associative learning discussed in Chapter 17. The animal may be evolutionarily preprogrammed to learn certain things at a certain age.

Other aspects of learning, such as learning to avoid noxious foods, have also been seen as learning that is biologically tailored to the animal's way of life. This view is open to dispute, because almost every property of conventional laboratory conditioning can be demonstrated in food-aversion learning (Revusky, 1977). Knowledge of which foods cause illness and which are nutritious, which route is dangerous and which is safe, requires some integration of the predictive relationships among events in the environment. Such knowledge can be gained either through genetic programming or by learning.

Some psychologists working on animal learning have concluded that all types of learning have basic similarities due to the common problem of learning about causal relationships. Learning enables animals to associate cause with effect, and so to predict significant events. Predictive relationships are based upon events in a causal chain that has universal properties. Thus, events do not occur without a cause and do not occur before a cause. We take these relationships for granted, and it is reasonable to suppose that animals do the same. The best predictors of events are the causes of the events, and animals that can detect and learn about such events will be well equipped to respond to an important and universal feature of the natural world.

On the basis of such considerations, some psychologists (e.g. Dickinson, 1980) believe in 'the existence of a basic associative learning mechanism, which is common to a variety of species and designed to detect and store information about causal relationships in the animal's environment'. This does not mean that there are not also ways of learning unique to particular species or that some animals may not have a range of learning capacities. It simply means that most animals have one aspect of learning in common. However, this approach differs from previous psychological approaches to learning because it has a biological perspective. As we see in the next chapter, it is possible to ask what features of causal relationships are likely to be important to animals. This is a question about design, an essentially biological way of thinking. An interesting, and perhaps surprising, result of this approach is that it leads to an essentially cognitive view of simple associative learning.

Points to remember

■ Learning does not necessarily confer an evolutionary advantage. Some learning is evolutionarily preprogrammed as part of normal development, some occurs only in certain circumstances, and only some is innovative.

■ Learning can be constrained by the nature of the relationship between the type of reinforcer and the type of response to be learned. Such constraints can be seen in learning to avoid enemies and in learning to avoid sickness.

■ Constraints on learning also occur as a result of stimulus relevance. In other words, the associative strength of a cue which has a particular consequence depends partly upon the nature of the consequence.

■ The biological view of learning is that the constraints are primarily innate, while the psychological view is that the same laws of learning apply to all animals. In some cases these two views provide different explanations of the same phenomena.

Further reading

Bolles, R.C. (1970) Species-specific defense reactions and avoidance learning. *Psychol. Rev.* **77**, 32–48.

Shettleworth, S.J. (1972) Constraints on learning. *Advances in the Study of Behavior* **4**, 1–68.

Shettleworth, S.J. (1984) Learning and behavioural ecology. In Krebs, J.R. and Davies, N.B. (eds) *Behavioural Ecology*, 2nd edn. Blackwell, Oxford.

Cognitive aspects of learning

Most present-day psychologists recognize that there is a spectrum of learning ability ranging from the simple learning of primitive animals to the cognitive abilities of humans. The problem is to assess the extent to which these abilities occur in particular animal species. We start by considering some aspects of learning that cannot easily be explained in terms of conditioning.

19.1   Hidden aspects of conditioning

Pavlovian conditioning traditionally has been viewed as the most mechanistic and least cognitive aspect of learning. Indeed, as we have seen, a major objective to the behaviourist school of psychology was to account for the behaviour of animals without reference to unobserved processes, whether physiological or cognitive. Thus, classical conditioning has often been used as an explanatory device to account for apparently cognitive aspects of complex behaviour (Rescorla, 1978), but the reverse perspective has not found favour with behaviourists. It has been evident for a long time, however, that some aspects of classical conditioning do not conform easily to the behaviourist view.

When a CS is present together with a UCS, we observe an increase in conditional responses to the CS. How can we know that these responses are a consequence of the experimentally arranged relationship between the CS and the UCS? It is possible that the CS may elicit certain reflexes that are enhanced by exposure to the UCS. In some cases these reflexes may resemble the CR that is to be conditioned. In human eyelid conditioning, for example, a visual CS may reflexly elicit eyelid closure. In conditioning emotional reactions, any novel stimulus may elicit an emotional reflex, like the galvanic skin response (a change in electrical conductivity of the skin due to the action of the sweat glands). Such potentiation of the CR by the CS is called sensitization (Gormezano, 1966). For example, the common octopus (*Octopus vulgaris*) can be trained in a laboratory aquarium to emerge from its home to attack a crab or a neutral stimulus associated with food (see Figure

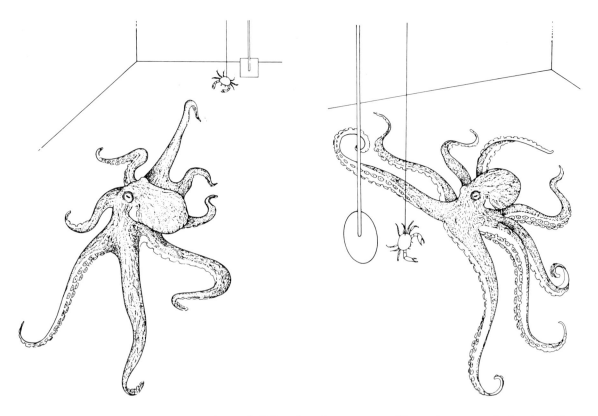

Fig. 19.1 Octopus learning to associate two neutral stimuli with food.

19.1). The probability that the octopus will attack a neutral stimulus, like a white disc suspended on a rod, is increased if the octopus has received food recently and decreased if it has received a mild electric shock recently. The effect occurs even if the octopus is fed or shocked in its home, so it cannot be due to positive or negative reinforcement of attack behaviour.

It is possible that the UCR may come to be elicited by stimuli other than the UCS even though there is no contingent relationship between them. This is usually called **pseudoconditioning** (Grether, 1938). A possible explanation is that there is generalization to stimuli similar to the UCS. It can also happen that exposure to electric shock alters the animal's internal state so that it comes to avoid any external stimulus. This has been shown to occur in polychaete worms (Evans, 1966) and octopus (Young, 1960) and appears to be an adaptive form of behaviour (Wells, 1968).

Experimental psychologists normally take precautions against sensitization and pseudoconditioning (Mackintosh, 1974), but other aspects of conditioning are not readily observable and may be difficult to account for in behaviourist terms. These include interoceptive and temporal conditioning.

Secretions of internal organs can be conditioned readily to external stimuli. They can also be conditioned to internal stimuli such as changes in body temperature, blood sugar, the carbon dioxide content of inhaled air, etc. (Bykov, 1957). Such **interoceptive conditioning** may be no different in principle than ordinary classical conditioning, even though it does not always involve observable behaviour.

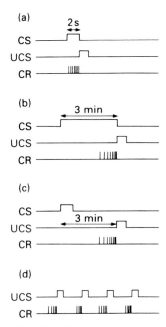

Fig. 19.2 Schematic representation of Pavlovian conditioning: (a) usual procedure with the UCS occurring immediately after the CS; (b) delayed conditioning; (c) trace conditioning; (d) temporal conditioning.

In simple **temporal conditioning** experiments, there is no CS. The UCS is presented at fixed intervals of time, and the CR develops as a correspondingly periodic response (see Figure 19.2). Such experiments were conducted in Pavlov's laboratory, and there have been many variations (Church, 1978; Richelle and Lejeune, 1980). The most natural explanation of temporal conditioning is that the sense of time, based upon some internal clock (see Chapter 16), can act as a stimulus just like any external stimulus. However, time lacks the physical attributes normally associated with a stimulus, and entities like the internal clock are not directly observable. For these reasons the simple explanation has not found favour with the behaviourists, and they have made many attempts to account for temporal conditioning in terms of mediating behaviour. However, it is now normally accepted that animals do have internal clocks and are capable of learning on the basis of temporal cues. There is good evidence that the internal clock advances at a fixed rate and that the animal can read the clock when necessary. There is also evidence that animals can exercise some control over the external clock and can make it stop or run (Roberts, 1981).

In the traditional description of Pavlovian conditioning, a UCS that normally evokes a UCR is arranged to follow a CS that initially fails to evoke the response. After repeated pairings of UCS and CS, the CS comes to evoke the response, which is now called the CR. This description implies the incorporation of a new stimulus into an existing reflex system. An alternative view, anticipated to some extent by the American psychologist Edward Tolman (1932), is that the animal learns about the relationships among events outside its control and that it then generates appropriate behaviour (Rescorla, 1978). Rather than emphasizing the animal's responses to stimuli, the cognitive approach asks whether or not the animal's behaviour changes as a result of exposing it to particular relationships among events. Whether or not the events initially evoke responses is largely irrelevant.

The cognitive approach emphasizes the distinction between learning and performances. As we see in Chapter 24, animals may form associations among events without altering their behaviour at the time. On the one hand, the reflex tradition places Pavlovian conditioning in a compartment of its own, as a particular type of learning. The cognitive approach, on the other hand, sees Pavlovian conditioning as an example of associative learning. Animals can associate many types of stimuli – exteroceptive, interoceptive, those arising from an internal clock and those emanating as feedback from the animal's own behaviour. In some cases the animal may choose to reflect its learning directly in its behaviour, and in some cases it may not.

19.2 Insight learning

The idea that animal learning involves cognitive processes has a long history. It gained clear expression in the work of the Gestalt school of psychologists, who believed that animals gained insight into problems through an innate tendency to perceive the situation as a whole.

A classic series of experiments was carried out by Wolfgang Kohler between 1913 and 1917. Throughout World War I, Kohler was interned on the island of Tenerife. He devoted his energies to studying the chimpanzees at the Anthropoid Station there and reported his work in *The Mentality of Apes*, published in 1925. Kohler's experiments required chimpanzees to use tools to obtain food rewards. For example, in one experiment the chimpanzee was required to use a stick to rake in food from outside its cage, as shown in Figure 19.3. Kohler claimed that his problems differed from those set by Thorndike (see Chapter 17.5) in an important respect. He noted that Thorndike's animals could not develop an understanding of the latch mechanism that opened the cage door because the latch was on the outside of the cage and hidden from view. They had little alternative but to solve the problem by trial and error. In Kohler's experiments, all the ingredients necessary for the solution of the problem are visible.

One of Kohler's chimpanzees was given two bamboo poles, neither of which was long enough to reach the fruit placed outside the cage. However, the poles could be fitted together to make a longer pole. After many unsuccessful attempts to reach the fruit with one of the short poles, the chimpanzee gave up, started playing with the poles, and accidentally joined them together by pushing the narrower pole inside the hollow end of the other. The chimpanzee then jumped up and immediately ran to the bars of the cage to retrieve the fruit with the long pole. Kohler interpreted this as an example of insightful behaviour. In another experiment, fruit was suspended high up in the roof of the chimpanzee's cage. It could be reached by stacking boxes on top of each other, as shown in Figure 19.4. Some chimpanzees learned to solve this problem.

In accounting for the results of his experiments, Kohler claimed that his animals exhibited **insight**, a term used to denote the apprehension of relationships among stimuli or events. Insight learning differs from trial-and-

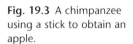

Fig. 19.3 A chimpanzee using a stick to obtain an apple.

Fig. 19.4 One of Kohler's chimps standing on stacked boxes to obtain food suspended from above (From Kohler, 1925).

error learning because it involves the sudden production of a new response. The philosopher Bertrand Russell was amused by the difference between the views of Thorndike and Kohler. He noted:

'All the animals that have been carefully observed … have all displayed the national characteristics of the observer. Animals studied by Americans rush about frantically, with an incredible display of hustle and pep, and at last achieve the desired result by chance. Animals observed by Germans sit still and think, and at last evolve the solution out of their inner consciousness.'

The Gestalt interpretation of animal problem solving has been criticized on various grounds. The experiments are devised to determine whether or not animals behave with insight under conditions that are supposed to require such behaviour. There is no independent evidence that the task requires insight, but if the animal succeeds in the task, it is said to have demonstrated insight. This line of reasoning is not universally acceptable. Another difficulty is that it is very hard to know whether the insightful response is genuinely new. The chimpanzees engage in a considerable amount of irrelevant behaviour, play and abortive attempts to obtain the reward. Is it possible that they arrive at the solution to the problem through a cumulative process of trial and error?

This question was examined by Paul Schiller (1952), who systematically investigated the innate components of the problem-solving behaviour of chimpanzees. In one study, for example, he provided pairs of sticks that would fit together to 48 new chimpanzees without any problem to solve. Of these, 32 fitted the sticks together within an hour, and 19 of the 20 adults in the group fitted them together within five minutes. On the basis of these and other investigations, doubt has been cast on the Gestalt interpretation of problem-solving behaviour (e.g. Chance, 1960). It appears that the previous experience of the animal is very important. Familiarity with sticks and boxes makes an enormous difference to the way in which the problems are tackled. Chimpanzees that are allowed to play with these objects learn about their properties. The ability to manipulate objects in ways that are relevant to problem solving is largely a matter of maturation. Once the animal discovers that it can achieve a particular manipulation, it tends to repeat it over and over again. Some manipulations are simply too difficult for the younger chimps, but once the behaviour becomes established in the animal's repertoire, it can be deployed in a variety of contexts.

The main difference between insight learning and other forms of learning appears to lie in the ability of the more intelligent animals to draw on experience gained in other contexts. However, this does not necessarily mean that insight must be ruled out as an aspect of learning. On the one hand, the problem-solving abilities of animals are difficult to investigate because the human investigator may have little idea as to how the animal sees the situation. It is obviously unfair to set a problem that is beyond the animal's manipulative ability or against its natural inclination. On the other hand, each species is well suited by nature to perform some apparently clever feats, and we must not be misled into thinking that these are evidence for insight or intelligence. Indeed, these words are really only labels for phenomena that still require explanation.

Associative learning

To investigate the hypothesis that animals possess mechanisms to detect and learn about causal relationships, we must specify the nature of such relationships. There are basically two types of causal relationship, and there is little doubt that animals can learn about both (Dickinson, 1980). An event (the cause) can cause another event to happen (the effect) or not to happen (non-effect). The first event need not be an immediate cause of the effect or non-effect, but it may be an identifiable link in a chain of cause and effect. Indeed, the event noticed by the animal may not be part of the causal chain but may merely be a sign that the causal event has occurred. It is the apparent cause that is important to the animal.

That animals can learn about both types of causal relationship can be demonstrated by a simple experiment. Hungry pigeons can be placed in a Skinner box fitted with two illuminated keys and a food delivery mechanism. One group of pigeons is exposed to a light–food (cause–effect) relationship, and another group is exposed to a light–no food (cause–non-effect) relationship. A third group is exposed to the light alone. In the first case, one of the discs is illuminated for ten seconds, at irregular intervals, and food is presented as soon as the light goes off. In the second case, the light and the food are presented the same number of times as in the first case, but care is taken to ensure that food is never presented soon after the light. In the third case, no food is presented. In the first case the light signals the presentation of food, and in the second case it signals the non-presentation of food. Either of the lights can be illuminated on a particular trial, and a record can be kept of whether the bird tends to approach or withdraw from the light. The results of this experiment (see Figure 19.5) show clearly that the pigeons exposed to the light–food relationship tend to approach the light, as might be expected on the basis of normal classical conditioning. However, the pigeons exposed to the light–no food relationship were not indifferent to the light but definitely avoided it (Wasserman *et al.*, 1974). Only pigeons that were exposed to the light alone apparently were indifferent to it (Figure 19.5).

In considering the results of this experiment, we must bear in mind a number of important points. First, we cannot claim that the pigeon learned that the light caused food or no food. What animals learn when exposed to causal relationships can be discussed only in terms of theories about the internal changes that occur as a result of the learning experience (see following discussion). Second, the fact that the pigeons approached the light in one case and withdrew from it in another is interesting but not directly relevant to the matter in hand. Many differences in behaviour caused by the presentations could be used as an indication that learning had taken place. The particular behaviour shown may become important when we come to consider what the animals learned. Third, it makes sense to refer to the light as the apparent cause of the non-occurrence of food only if the food would have been presented had the light not occurred. We cannot expect the pigeon to learn that the light is a cause of the non-occurrence of food if the food never occurred in the situation, as was the case for the third test group. An

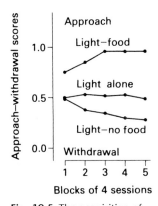

Fig. 19.5 The acquisition of a tendency to either approach or withdraw from a lighted key by pigeons exposed to various correlations between illumination of the key and delivery of food. In the light–food condition illumination of the key was paired with food, and the pigeons developed an approach response. In the light alone condition the pigeons were never presented with food and developed no approach or withdrawal response. In the light–no food condition they were presented with food, but never during, or shortly after, illumination of the key. The pigeons developed a withdrawal response (After Dickinson, 1980).

animal has the opportunity to learn about a cause–non-effect association only if there is some reason to expect the effect to occur (Dickinson, 1980).

Although sometimes we can conclude that a change in behaviour shows that learning has occurred, we cannot assume that nothing is learned in the absence of behavioural change. Anthony Dickinson (1980) calls this the problem of **behavioural silence**. In the experiment described previously, for example, we cannot assume that the pigeons in the light-alone condition learned nothing. Indeed, there is evidence that rats can learn to ignore stimuli that predict no change in the consequences of their behaviour (Mackintosh, 1973; Baker and Mackintosh, 1977). If rats are first exposed to a schedule in which two stimuli occur independently, the rate at which they subsequently learn to associate the two stimuli is retarded compared with animals that had no prior exposure to the stimuli. The implication is that animals can learn that particular stimuli are irrelevant and that the learned irrelevance interferes with subsequent learning about the stimuli.

Other forms of behaviourally silent learning can also occur. Animals can learn that two events are unrelated, either in the sense that an effect is unrelated to a particular cause or that an effect is unrelated to a whole class of causal events. If the class of causal events is the animal's behavioural repertoire, this form of learning is called learned helplessness (Maier and Seligman, 1976), meaning that animals can learn that there is nothing they can do to improve a situation, and such learned helplessness retards future learning in that situation.

For an animal to learn about a simple causal relationship, there must be an overall positive correlation between the two events (Dickinson, 1980). From the animal's viewpoint, there are always a variety of possible causes of an event, apart from that provided by the experimenter. As illustrated in Figure 19.6, there must be a fairly close relationship between two events for one to be seen as the cause of the other. To test the importance of such background, or contextual, cues, Mackintosh (1976) trained rats to press a lever for food and then introduced a variety of stimuli. One group was given a light stimulus on each trial, while the other groups were given a compound stimulus consisting of light plus noise. The noise was soft (50 decibels) for some groups and loud (85 decibels) for others. On all trials the animals were given a mild electric shock just after the presentation of each stimulus. At the end of the experiment, the light was presented alone to all groups to see how much the animals had learned about the light–shock relationship.

The results (Figure 19.7) showed that the presentation of the light suppressed the bar-pressing behaviour to different extents in different groups. The rats that had been given light alone or light plus soft noise showed considerable suppression, but those that had experienced the light in the presence of a loud noise showed much less suppression. Thus, the presence of a powerful second stimulus decreased the amount by which the light was associated with the shock, even though there was perfect correlation between light and shock. This phenomenon is called overshadowing. The degree of overshadowing depends upon the relative salience of the overshadowed and overshadowing stimuli. This is why the soft noise had little overshadowing effect.

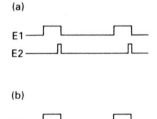

Fig. 19.6 The pattern of events when there are different correlations between E1 and E2. In (a) E2 only occurs during or shortly after E1, and the two events are positively correlated. In (b) E2 is just as likely to occur when E1 is absent as when it is present, and the two events are uncorrelated (After Dickinson, 1980).

Fig. 19.7 The degree to which a light (a) and a noise (b) suppressed lever-pressing for food in rats which had received prior pairing of a shock with the light (L), a weak noise (n), an intense noise (N), a compound of light and weak noise (Ln), or a compound of light and intense noise (LN). High score indicates little suppression (After Mackintosh, 1976).

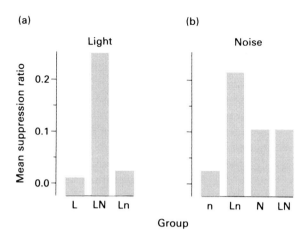

Animals will learn to associate two events only if they are accompanied initially by an unexpected or surprising occurrence (Mackintosh, 1974). In a normal conditioning experiment, the surprise is provided by the reinforcer. Thus, if a stimulus is paired with shock, and if neither the stimulus nor the contextual cues initially predict its occurrence, the shock will be surprising. Suppose, however, that the animal already has experienced shock in the presence of stimulus A, then if both A and B are correlated with shock, the presence of A will block learning about B. This phenomenon, first discovered by Kamin (1969), is known as blocking. An experiment by Rescorla (1971) demonstrates this effect and also shows that the more surprising the reinforcer, the more the animal learns (see Figure 19.8).

In the natural environment, certain types of cause are more likely to produce certain effects than other types of cause. For example, if a cat jumps into an apple tree at the same time as a dog barks and then an apple falls to the ground, we are more likely to think that the cat was responsible than that the dog's bark caused the apple to fall. Both the cat's jumping and the dog's barking bear the same temporal relationship to the fall of the apple, but other aspects of these events lead us to assume that the cat was the cause of the apple's falling. Similarly, we can show that animals are more likely to form associations among certain types of stimuli rather than others. For example, rats readily associate taste with subsequent illness but do not easily learn a tone or light illness association (Domjan and Wilson, 1972; see also Chapter 18). Rats also learn to associate two events when they are in the same sensory modality (Rescorla and Furrow, 1977) and when they are in the same spatial location (Testa, 1975; Rescorla and Cunningham, 1979).

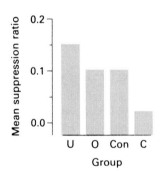

Fig. 19.8 The degree to which stimulus A suppressed lever-pressing for food in rats previously exposed to pairings of an AB compound with shock. The uncorrelated group (U) had received prior pairings of B with shock. The simple overshadowing group (O) was pre-exposed to the shock alone. The control group (Con) had experienced random presentations of B with shock. The correlated group (C) had experienced a negative correlation of B with shock. High score indicates little suppression (After Rescorla, 1971).

Finally, for an animal to form an association between two events, it is usually necessary for the events to occur close together in time. The temporal relationships between two events have been the subject of numerous experiments in animal learning (see Dickinson, 1980), and it appears that learning is most effective when the onset of one event (the cause) occurs shortly before the onset of the other (the effect). However, some evidence supports the view that this relationship is not due to a direct effect of the time interval on the learning process, but that changes in the time interval increase the extent to which the first event is overshadowed by contextual cues (Dickinson, 1980). An implication of this hypothesis is that the extent to which the time interval influences learning should depend upon the relevance of the contextual cues. In particular, we would expect contextual cues to be much less important in the case of taste–illness learning than in the case of tone–shock learning, because the background stimuli provided by the experimental situation are not normally relevant to taste aversion learning.

Sam Revusky (1971) was the first to suggest that lack of overshadowing might be responsible for the very long time intervals between food ingestion and subsequent sickness that can occur in taste aversion learning. He showed that the effective time interval could be reduced by interposing a background of relevant (taste) stimuli.

In summary, we have seen that animals can learn to associate two events if the relationship between them conforms to what we normally call a causal relationship. Thus, animals can learn that one event (the cause) predicts another event (the effect) or that one event predicts that another event (non-effect) will not occur. They can also learn that certain stimuli predict no consequences in a given situation or that a class of stimuli (including the animal's own behaviour) is causally irrelevant. The conditions under which these types of associative learning occur are those that we would expect on the hypothesis that animals are designed to acquire knowledge about the causal relationships in their environment. Thus, the animal must be able to distinguish potential causes from contextual cues, and for this to occur there must be some surprising occurrence that draws the animal's attention to particular events, or the events must be (innately) relevant to particular consequences. If these conditions are not fulfilled, contextual cues may overshadow potential causal events, or learning may be blocked by prior association with a now irrelevant cue. Thus, the conditions under which associative learning occurs are consistent with our commonsense views about the nature of causality. They are not consistent with the traditional view that animal learning is an automaton-like association of stimulus and response. The phenomenon of behaviourally silent learning suggests that some kind of cognitive interpretation of animal learning is required. However, this does not mean that we can jump to conclusions about the cognitive abilities of animals, or about the nature of the animal mind.

19.4 Animal thinking

We have seen that animals can learn to associate two events if the relationship between them conforms to what we normally call a causal relationship. Some of the conditions under which associative learning occurs are not consistent with the traditional view that animal learning is an automaton-like association of stimulus and response (S–R learning). There are two basic alternatives that have a long history, but are still considered important:

1. The view that animals can make associations between different stimuli (S–S learning).

2. The view that animals are designed to acquire knowledge about various relationships in their environment.

First of all, we have to distinguish two types of knowledge: 'knowledge how' and 'knowledge that'.

What do we normally mean by know-how? We mean knowing as a matter of knowing how to *do* something, such as how to swim, how to ride a bicycle. This type of knowledge cannot be transferred from one task to another, and cannot be articulated. Thus knowing how to ride a bicycle involves a type of knowledge that cannot be used for anything except riding a bicycle, and which cannot be transferred to another person by speech or writing.

Riding a bicycle involves a procedure that is implicit, involving no explicit representations that can be transferred to another task or person. This type of knowledge is generally called 'procedural knowledge.'

What do we normally mean by 'knowledge that'? We mean knowing as a matter of having accessible, in different ways, usable information about an object, a person, a place, etc. In studies of humans, this has usually been called 'declarative knowledge', being knowledge that people declare that they possess, or declare to be the case. In this sense, declarative knowledge can only be had by language-using agents.

One can summarize this by saying, that 'knowledge how' is (somehow) in, or part of, the system, in contrast to 'knowledge that' which is available for the system to work on (Pylyshyn, 1984).

Animals do not have language abilities, and so cannot have declarative knowledge. However, it is perfectly possible that some animals may have the non-linguistic equivalent of declarative knowledge, and be able to deploy such knowledge in a variety of tasks. To allow for this possibility Lanz and McFarland (1995) revised the traditional terminology as follows:

■ **Procedural knowledge** is knowing how. It is knowledge that is tied to a procedure. It cannot therefore be used in another procedure, or be accessed by another process. Dickinson (1985) equates procedural knowledge in rats with habits.

■ **Explicit knowledge** involves explicit representations of facts that are (by the definition of explicit representation) accessible to many processes. Explicit knowledge, therefore, involves tokens, which represent 'facts'.

■ **Declarative knowledge** (that people declare) is available only to language-using agents. It is a variety of explicit knowledge that pertains to humans, and possibly to artificial agents in the future. Where Dickinson (1985), and others who have experimented on animals, use the term 'declarative knowledge', we substitute the term explicit knowledge.

The empirical issue in animal behaviour research is whether or not animals possess explicit representations that permit them to manipulate explicit knowledge. An animal that did not have such representations would only have procedural knowledge. It could learn (e.g. by forming S–S associations), but it could not think.

An animal that did have explicit representations that it could manipulate to apply its explicit knowledge to a variety of tasks would be capable of some form of **cognition** (defined as the manipulation of explicit representations by Lanz and McFarland, 1995). Such an animal would be able to think. The distinction between procedural and cognitive systems in animals is illustrated in Box 19.1.

In a procedural system the form of representation, i.e. implicit representation (Lanz and McFarland, 1995), directly reflects the use to which the knowledge will be put. For example, Holland (1977) exposed rats to a tone–food relationship by occasionally presenting an eight-second tone and then delivering food pellets into a receptacle. Holland noticed that the rats developed a tendency to approach the food receptacle during the tone. This observation might suggest that during learning, the procedure of approaching the food receptacle is established, so that the learned information is stored in a form that is related closely to its use. The alternative possibility is that the rat learns that the tone causes the food, and thus establishes an

---

Box 19.1   The procedural and cognitive distinction in rats

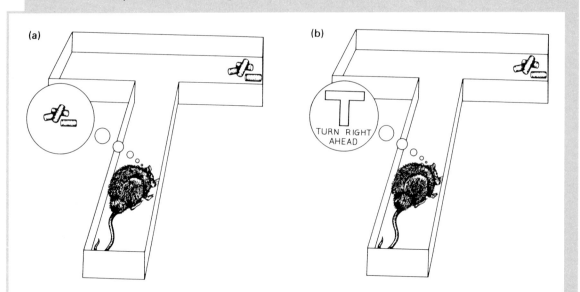

A simple example of explicit (left) and procedural (right) representation. In one case the rat has a mental image of the goal, in the other it follows a simple rule or procedure.

Representation is an issue that is usually considered to be central to the question of animal cognition. Do animals have internal representations, or mental images, or goals to be achieved, of objects they are searching for, or of complex spatial or social situations?

An explicit representation is a representation of knowledge about something. If a rat uses an explicit representation to find food in a familiar maze, it knows that the food is in a particular location. We might imagine that the rat has a mental image of the food, as shown in (a).

A procedural representation is a set of instructions relating to some procedure. Thus, if a rat uses a procedural representation to find food in a maze, we can say that it knows **how** to behave in the maze, but not that the food is present. A possible procedure is illustrated in (b).

explicit representation. The observation that the rat tends to approach the food receptacle during the tone would then have to be accounted for in some other way, because an explicit representation is passive in the sense that it does not control the animal's behaviour. The procedural representation thus provides a more parsimonious explanation of the rat's behaviour.

Suppose, however, that after the tone–food association is established, the rats are exposed to a food–illness relationship to the point where they refuse to eat the food when it is presented. The rats now will have formed two separate associations, tone–food and food–illness. The question is whether they are capable of integrating the two. On the one hand, a procedural account of learning would imply that the rats should not be able to integrate the two procedures, which have no factors in common. An explicit system, on the other hand, provides a basis for integration because both representations have the food term in common. Holland and Straub (1979) showed that rats can integrate information from such associations learned at different times. Rats exposed to a food–illness following tone–food showed a disinclination to approach the food receptacle when the tone was presented again.

There is little doubt that animals can integrate separately formed associations, and this is explained most easily in terms of an explicit system. However, whether or not such findings really require a cognitive explanation is very controversial. It is a topic that we come back to in Chapter 25.

Cognition refers to the mental processes that cannot be observed directly in animals but for which there is, nevertheless, scientific evidence. Suppose we allow hungry pigeons to observe food presentations that are accompanied by the illumination of a small electric light. During the observation period, the pigeons are not allowed to approach the light or the food. Other (control) pigeons observe light and food presentations that are unrelated in time. At the end of the initial observation period, the pigeons are allowed to approach the light and food stimuli. All the pigeons tend to peck at the food delivery mechanism. However, the experimental pigeons also peck at the light, while the control pigeons do not. This result shows that the experimental pigeons must have formed some kind of association between the food and the light, even though their behaviour during the initial phase of the experiment was the same as that of the control pigeons. The question is what kind of association – procedural or cognitive?

Of course, many aspects of cognition exist apart from those involved in associative learning. Some of these seem to be more in the nature of innate skills rather than forms of learning. For example, pigeons are capable of feats of navigation and time perception that are beyond the ability of human beings (see Chapter 14). Pigeons also seem to be extraordinarily good at the formation of natural concepts. They can be trained readily to discriminate between photographs showing water versus non-water, tree versus non-tree or human versus non-human (Herrnstein et al., 1976; Malott and Siddall, 1972; Siegel and Honig, 1970). They can make these discriminations even though the relevant cue (see Chapter 12) is presented in a variety of ways. For example, the pigeon can recognize water in the form of droplets, a turbulent river or a placid lake. Humans can be picked out whether they are clothed or naked, alone or in a crowd, etc.

Other examples of apparent cognition in animals include demonstrations of apparent linguistic abilities in chimpanzees and other species; the ability of monkeys to perform complex serial ordering tasks; the apparent ability of animals to imitate others, etc. These and other examples are discussed in Chapter 26.

Points to remember

- Some aspects of conditioning are not apparent to an observer, and this suggests that cognitive processes may be involved.

- Animals that appear to arrive at the solution to a problem suddenly are sometimes said to have shown insight. However, it is not always clear exactly how this differs from ordinary learning.

- Some aspects of associative learning are said to require cognitive explanation, because it seems that they must involve a mental image of the goal to be achieved. The alternative possibility is that the animals are simply following complex procedural rules.

Further reading

Bekoff, M. and Jamieson, D. (1996) *Readings in Animal Cognition*. MIT Press, Cambridge, MA. & London, England.

Dickinson, A. (1980) *Contemporary Animal Learning Theory*. Cambridge University Press, Cambridge.

Understanding complex behaviour

3.1 Ethology

3.2 Animal competence

3.3 The mentality of animals

In the third part of this book we look at complex behaviour patterns. To understand these we need to think in terms of both the design and mechanisms of behaviour.

The combination of design and mechanism has traditionally been the province of ethology, as distinct from the related scientific disciplines of evolutionary biology and psychology. Evolutionary biologists (including sociobiologists) look at behaviour in terms of its past evolution and do not address problems of present-day causation. Psychologists are primarily concerned with the proximate causes of behaviour and rarely make use of rigorous argument based upon the theory of natural selection. Ethologists, on the other hand, have sought to combine the mechanistic and evolutionary approaches to behaviour, asking not only how behaviour is controlled but how the mechanism evolved and why particular mechanisms appear in particular circumstances.

In this third part of the book we look first at some areas of classical ethology in which this dual approach has been employed. We then look at more recent attempts to combine design and mechanism. Finally, we consider the difficult areas of animal language and cognition, in which it seems that the dual approach is likely to prove useful in the future.

SECTION 3.1

Ethology

A herring gull fitted with a radio transmitter (*Photograph: Robin McCleery*).

In this section we discuss issues from classical ethology which involve consideration of both design and mechanism in behaviour. Chapter 20 deals with the problem of instinct, starting from early ideas and progressing to interactions with learning as exemplified by imprinting. Chapter 21 reviews communication, which still poses problems of causal explanation but which has provided fertile ground for consideration of evolutionary aspects. Chapter 22 is concerned with human behaviour.

Konrad Lorenz (1903–1989) and Niko Tinbergen (1907–1988)

Photograph: Hermann Kacher.

Photograph: B. Tschanz.

Konrad Lorenz and Niko Tinbergen are generally regarded as the founders of modern ethology. Although their approach was anticipated by the work of Charles Whitman (1842–1910) and Wallace Craig (1876–1954) in the USA and Oskar Heinroth (1871–1945) in Germany, their work provided the basis for the future development of ethology and their approach offered an alternative to the then dominant American behaviourism.

Konrad Zacharius Lorenz was born in Austria. He studied medicine in Vienna and also studied comparative anatomy, philosophy, and psychology. He became demonstrator and then lecturer in comparative anatomy and animal psychology. At the same time he studied animal behaviour at his family home in Atlenberg. In 1940 he was appointed Professor of Philosophy at the University of Königsberg, but in 1943 he was drafted into the army medical service. In 1944 he was taken prisoner of war by the Russians. He was released in 1948, became attached to the University of Münster, and then moved to Seewiesen with the founding of the Max Planck Institute for Behavioural Physiology, where he remained until his retirement in 1973.

Nikolaus Tinbergen was born in The Hague, The Netherlands, and studied biology at the University of Leiden. In 1930 he went on an expedition to Greenland, and in 1938 he visited Lorenz at Altenberg. During World War II he was interned in a hostage camp in The Netherlands, afterwards to become Professor of Zoology at the University of Leiden. In 1949 he was invited to become a Lecturer in Zoology at the University of Oxford, where he founded the Animal Behaviour Research Group. He retired in 1974.

In 1973 Konrad Lorenz and Niko Tinbergen, together with Karl von Frisch, were awarded the Nobel Prize for Medicine. Both Lorenz and Tinbergen emphasized the importance of straightforward observation of animal behaviour under natural conditions. Lorenz's approach was somewhat more philosophical and his numerous theories became quite influential. Tinbergen was a gifted field biologist who carried out many elegant experiments in the natural environment. A significant feature of the work of Lorenz and Tinbergen was their attempt to combine evolutionary, or functional, explanations of behaviour with causal, or mechanistic, explanations. For example, in his 1963 paper *On aims and methods in ethology*, Tinbergen posed four questions he thought necessary to a full account of any aspect of animal behaviour. The ethologist, he believed, should aim to answer the questions of the causation, development, survival value, and evolution of any behaviour pattern under study.

This aim is probably the most notable characteristic of ethology. While evolutionary biologists seek functional accounts of behaviour, and psychologists seek explanation in terms of proximate causes or mechanisms, ethologists have followed the example of Lorenz and Tinbergen in retaining interest in, and enthusiasm for, all four approaches to animal behaviour.

CHAPTER

20 Instinct and learning

Classical ethology is an approach to animal behaviour that involved a resurrection of the concept of instinct. The history of the concept of instinct is bound up inextricably with notions of voluntary behaviour and of our responsibility for our actions.

In this chapter we discuss the relationship between instinct and learning. You will find that the dichotomy is not nearly so wide today as it was in the past. To understand how this shift in opinion has taken place, it is helpful to start with an historical sketch.

20.1   The concept of instinct

Early writers regarded instinct as the natural origin of the biologically important motives. Thus, Thomas Aquinas wrote that animal judgement is not free but implanted by nature. Descartes regarded instinct as the source of the forces that govern behaviour, being designed by God in such a way as to make the behaviour adaptable. The associationists appeared to reject all notions of instinct, although Locke did write of 'an uneasiness of the mind for want of some absent good .... God has put into man the uneasiness of hunger and thirst, and other natural desires ... to move and determine their wills for the preservation of themselves and the continuation of the species.'

While the associationists believed that human behaviour is maintained by the knowledge of and desire for the consequences of behaviour, others like Hutcheson argued that instinct produces action prior to any thought of the consequences. Whereas instinct previously had been regarded as the source of motivational forces, Hutcheson made instinct the force itself. This concept of instinct was seized upon by the new rationalists such as Reid, Hamilton and James in 1890 as a convenient vehicle for the non-rational elements of

Fig. 20.1 A robin attacking a dummy. The red breast of the dummy is a sign stimulus (Drawn from a photograph in Sparks, 1982).

behaviour. Thus, human nature was seen as a combination of blind instinct and rational thought.

The idea of instinct as a prime mover was taken up by psychologists such as Freud (1915) and McDougall (1908). Sigmund Freud developed a motivational theory of neuroses and psychoses, emphasizing the irrational forces in human nature. He saw behaviour as the outcome of two basic energies: a life force underlying life-maintaining and life-continuing human activities and a death force underlying aggressive and destructive human activities. Freud thought of the life and death forces as instincts, the energy of which was seen as requiring expression or discharge. McDougall thought of the instincts as irrational and compelling sources of conduct that orientated the organism toward its goals. He postulated a number of instincts, most of which had a corresponding emotion – for example, flight and the emotion of fear, repulsion and the emotion of disgust, curiosity and the emotion of wonder and pugnacity and the emotion of anger.

These various conceptions of instinct were derived from subjective human emotional experiences. This essentially unscientific practice involves difficulties of interpretation, of agreement among different psychologists and of determining the number of instincts that should be allowed or recognized. Darwin (1859) was the first to propose an objective definition of instinct in terms of animal behaviour. He treated instincts as complex reflexes that were made up of units that were compatible with the mechanisms of inheritance. Instincts were thus the product of natural selection and had evolved together with the other aspects of the animal's life. Darwin's concept of instinct is thus similar to that of Descartes, with evolution replacing the role of God.

Darwin laid the foundations of the classical ethological view propounded by Lorenz and Tinbergen. Lorenz (1937) maintained that much of animal behaviour was made up of a number of fixed-action patterns that were characteristic of the species and largely genetically determined. He subsequently postulated that each fixed-action pattern, or instinct, was motivated by **action-specific energy** (Lorenz, 1950). This was likened to a liquid in a reservoir. Each instinct corresponded to a separate reservoir, and when an appropriate releasing stimulus was presented, the liquid was discharged in the form of an instinctive drive that gave rise to the appropriate behaviour. Tinbergen (1951) proposed that the reservoirs, or instinct centres, were arranged in a hierarchy so that the energy responsible for one type of activity, like reproduction, would drive a number of subordinate activities, such as nest-building, courtship and parental behaviour. Lorenz and Tinbergen provided numerous examples of what they regarded as instinctive behaviour patterns.

The classical ethological concept of instinct is currently regarded as unsatisfactory for two main reasons. The first reason is connected with the suggestion that instinctive forces or drives energize certain aspects of behaviour. For reasons that are discussed fully in Chapter 15, motivation is no longer seen in terms of drives, super-reflexes or instinctive urges. The second reason concerns the implication that certain aspects of behaviour are innate in the sense that they develop independently of environmental influences. As we saw in Chapter 3, genetic influences upon behaviour are no longer thought of as being independent of environmental influences. The term **innate** has come to be used for activities that are characteristic of the species, bearing in mind that the early environmental influences are also characteristic of the species in the sense that the circumstances in which the different members of the species are born and raised are often very similar.

The primitive idea of instinctive behaviour was that detailed instructions relating to the performance of the behaviour and the stimuli that elicit the behaviour are encoded in the genes. The ontogeny of the behaviour is fixed in the sense that, within limits, the developmental circumstances make no difference to the form of the behaviour. Instinctive behaviour thus is characteristic of the species and is made up of fixed-action patterns that are released by specific sign stimuli. Instinctive behaviour is adaptive because natural selection acts upon it as it does on other genetically determined traits. There is also a tendency in the early ethological literature to imply that where behaviour is clearly adaptive, it must therefore be instinctive, as opposed to learned behaviour that is not acted on by natural selection.

20.2 The innate releasing mechanism

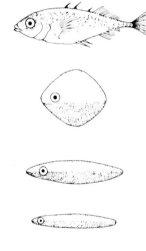

Fig. 20.2 Models of male sticklebacks that elicit attack by territorial males. At the top is a deal male lacking in nuptial colours. The four lower models are very crude, but their red underside provides a sufficient sign stimulus to elicit attack (After Tinbergen, 1951).

The early ethologists (e.g. von Uexkull, 1934; Lorenz, 1935) thought that animals sometimes respond in an instinctive manner to specific, though often complex, stimuli. Such stimuli came to be called **sign stimuli**, an example of which is illustrated in Figure 20.1. A sign stimulus is part of a stimulus configuration, and it may be a relatively simple part. For instance, a male three-spined stickleback has a characteristic red belly when in breeding condition. This is a sign stimulus that elicits aggression in other territorial males. As we can see from Figure 20.2, crude models suffice to elicit aggression provided they have a red underside. In contrast, a freshly killed male stickleback without a red belly is ineffective in provoking attack from other males. Thus, many of the details of the structure and texture of a male stickleback apparently are ignored by other males. The red coloration is much more effective if it is on the underside of the model.

Such a configurational relationship is a common feature of sign stimuli. In the case of sticklebacks it does not seem to be an essential feature, however. Tinbergen (1953) describes how he was studying the behaviour of territorial male sticklebacks in aquariums placed in a window. Whenever a red post office van passed along the road outside the window, the sticklebacks immediately attempted to attack it as if it were a rival male. Recently the BBC was able to repeat these observations while making an historical documentary film for television (Sparks, 1982).

The selective responses to stimuli suggested to early ethologists that there must be some built-in mechanism by which such sign stimuli were recognized. This supposed mechanism came to be called the **innate releasing mechanism** or IRM (Lorenz, 1950; Tinbergen, 1950). There are three important aspects of this concept. First, the mechanism is envisaged as being innate in the sense that both the recognition of the sign stimulus and the resulting response to it are inborn and characteristic of the species. As we saw in Chapter 3, however, the notion of innateness used by early ethologists is somewhat different than that current today. Second, the IRM has the role of releasing the response to the sign stimulus. The implication here is that the IRM holds back the pent up action-specific energy, or drive, until the appropriate sign stimulus is recognized. The energy then is released in the form of appropriate behaviour. So central was this aspect of the IRM that sign stimuli were referred to often as releasers. Third, the response released by the IRM was stereotyped and was part of the animal's innate repertoire of fixed-action patterns. Fixed-action patterns, as originally conceived by Lorenz (1932), were activities with a relatively fixed pattern of coordination, somewhat akin to reflexes. They were innate and typical of the species.

Lorenz drew several distinctions between fixed-action patterns and reflexes. First, fixed-action patterns can be released by a variety of stimuli, whereas reflexes are elicited by specific stimuli. Second, animals are motivated to perform fixed-action patterns, but this is not true of reflexes. Third, fixed-action patterns can appear in the absence of external stimulation and are then called **vacuum activities**. Some of these points would be disputed now. For example, the startle reflex occurs in response to a variety of stimuli. Many fixed-action patterns are now known to be generated in a predictive manner, without feedback control, which accounts for their stereotyped appearance.

A typical example of the IRM concept is provided by Baerends (1950):

'There is important evidence to support the conception that every releasing mechanism has its own sign stimuli. For example, the digger wasp *Ammophila adriaansei*, that catches caterpillars and drags them to its nest as food for its larva, may respond to the perception of a caterpillar in different ways, all depending on which instinct is activated. When it is hunting, a caterpillar is caught and stung; when it is found near the nest opening, just after the wasp has opened the nest, it is drawn in; but when it lies close to the nest when the wasp is filling the nest entrance, it may be used as filling material. Finally, when we put it into the nest shaft when the wasp is digging out the nest, then it brings the caterpillar away, exactly as she would deal with another obstacle – for instance, a piece of plant root. It is, therefore, the same object, that with different conditions of the animal, releases different responses. Still, in every situation the caterpillar is always sending a visual as well as chemical stimuli to the sense organs of the wasp where they will always be transformed into impulses. But then it depends on the instinct activated in the wasp which of these impulses will be intercepted somewhere and which can pass along a still unknown way in the nervous system finally to stimulate the principal motoric centre of the reaction. There are indications that in each case different stimuli are working. When hunting, *Ammophila* very likely become aware of the presence of a caterpillar by its odour, but when it loses the caterpillar during the transport to the nest in the first place optical stimuli are used to find it.'

Here we have the elements of many of the features of stimulus filtering discussed in Chapter 12. In demonstrating that animals are selective in their responses to complex stimuli, the ethologists made a major contribution to our understanding of perception in animals. However, their concept of the

Fig. 20.3 A greylag goose retrieving an egg into its nest (From *The Oxford Companion to Animal Behaviour*, 1981).

IRM is open to a number of criticisms. Hinde (1966), for example, pointed out that the term IRM is often used with the implication that the mechanism is specific to a particular response when the evidence for its existence is based solely on the study of that response. It may be the case that a particular sign stimulus may be relevant to more than one aspect of behaviour. Second, the selectivity in responsiveness need not be confined to stimuli that release responses. It may also occur with respect to those aspects of the stimulus that are important in the orientation of the behaviour. Third, the process of selectivity need not be purely innate but may be influenced by learning. As we saw in Chapter 12, the notion of a sign stimulus is not far removed from that of selective attention, a concept that originated in the study of animal learning.

Some early ethologists were well aware of these problems. Thus, Tinbergen (1951) specifically distinguishes between releasing and directing aspects of the stimulus and illustrates this distinction by reference to the studies of egg-retrieval in the greylag goose (see Figure 20.3), conducted by Lorenz and Tinbergen in the 1930s.

20.3 The discovery of imprinting

One of Lorenz's main contributions to ethology is his work on the development of social relationships, especially the phenomenon of imprinting. In this he was influenced greatly by his observations on jackdaws and by the previous work of Heinroth. Although Heinroth is often given the credit for being the first to describe the phenomenon of imprinting, Spalding had conducted extensive studies on imprinting many years earlier. Between 1872 and 1875, Spalding published some six papers reporting his extensive observations of the hatching of domestic chicks and their behaviour during the first few days of life. This work anticipated much of the early ethological work on instinct and included the observation that chicks only two and three days old would follow any moving object and develop an attachment to it. Spalding died in 1877 at the age of 37, and his work was forgotten until it was discovered and republished by Haldane in 1954. Had Spalding lived longer, he probably would be regarded as the founder of ethology (Thorpe, 1979).

Heinroth published his major papers in 1911 on the ethology of ducks and geese. He made detailed studies of various species and was a pioneer of the comparative method. He observed the behaviour of goslings hatched individually in an incubator and then handled by a human prior to being introduced to a goose family. Although the parent goose regarded each gosling as its own, the goslings showed no inclination to regard the geese as their parents. 'The gosling runs off, piping, and attaches itself to the first human being that happens to come past; it regards the human being as its parent' (Heinroth, 1911). Heinroth explains that to introduce such goslings successfully into a goose family, they must be removed from the incubator and immediately placed in a sack so that they do not catch sight of a human being.

Lorenz (1935), in extending Heinroth's observations, emphasized the difference between imprinting and ordinary learning. He argued that imprinting, unlike ordinary learning, took place at a particular stage of development and was an irreversible process. Lorenz confirmed Heinroth's observations on goslings (see Figure 20.4) and also studied imprinting in mallard ducklings, pigeons, jackdaws and many other birds. He confirmed Heinroth's observation that birds imprinted on humans would often direct their subsequent sexual behaviour toward them. Thus, Lorenz (1935) notes how a Barbary dove (*Streptopelia risoria*) that was imprinted on humans would direct its courtship behaviour toward his hand and would attempt to copulate if the hand was held in a certain position. Lorenz (1935) emphasizes that the behaviour shown as a result of imprinting is innate but that the recognition of the object of imprinting is not innate. He maintains that the young animal becomes imprinted upon whatever moving object is encountered during a particular phase of development and that it subsequently directs its filial, sexual and social behaviour toward this object.

As Heinroth (1911) and Lorenz (1935) observed, the young of many precocial species (see Chapter 3), which can run around soon after birth, show a fairly indiscriminate attachment to moving objects. Newly hatched goslings and ducklings that are separated from their mother will follow a slowly walking person, a crude model duck or even a cardboard box. A lamb will follow the person who feeds it on a bottle, although it may not be hungry. Even when the lamb has been weaned and has joined a flock it will approach and follow its former keeper. Thus, as a juvenile the lamb follows the person as if it were its parent, and as an adult it retains some attachment to the person, illustrating that imprinting can have both long- and short-term aspects.

Although the following response is elicited by a wide range of stimuli, some are more effective than others. Up to a point, the effectiveness of a stimulus increases with its conspicuousness (Bateson, 1964), but if an object is too startling, it elicits fleeing rather than approach. Thus, ducklings approach a human that sways from side to side but flee if the same person moves vigorously. Some species have particular preferences. Domestic chicks, for example, most readily follow blue or orange objects. Mallard ducklings prefer yellow–green objects, and their following response is enhanced if the object emits appropriate sounds. Wood ducks (*Aix sponsa*) nest in holes in trees, and the young are normally called out of the nest by the mother from some distance away. These ducklings will approach a source of intermittent sound in the absence of any visual stimuli (Gottlieb, 1963).

In general, the more an animal follows and becomes familiar with one object, the less it is attracted to others. The following response can be enhanced with food rewards, and in nature it is rewarded by contact with the mother and the warmth she provides.

20.4 Sensitive phases of learning

Lorenz (1935) recognized that imprinting was confined to a particular period of development, and he thought that this was due entirely to endogenous factors, similar to those involved in embryological induction. However, we

Fig. 20.4 Konrad Lorenz followed by goslings (*Photograph: Dmitri Kasterine*).

know now that the period during which imprinting can occur is affected considerably by experience. Ducklings and domestic chicks tend to stay close together, even in the absence of a parent. Guiton (1959) found that chicks kept in groups cease to follow moving objects three days after hatching but that those reared in isolation retain the following response for much longer. He was able to show that socially reared chicks become imprinted upon each other.

If developmental age is measured from the beginning of embryonic development, then the onset of the sensitive period of mallard ducklings is much more marked than if age is measured from hatching (Gottlieb, 1961, 1971). This suggests that posthatching experience is relatively unimportant and that the onset of the sensitive period is due to maturation. More recent studies, however, indicate that posthatching experience is also important (Landsberg, 1976). Changes in ability (Hess, 1959a) and maturation of the visual system and parts of the brain have been suggested (see Bateson, 1979, 1990).

Newly hatched birds of many species do not avoid novel objects initially but tend to approach and explore them. After a few days they become more timid and show signs of fear of unfamiliar objects. The time at which this occurs is influenced by rearing conditions (Bateson, 1966). For the newly hatched bird, nothing is familiar and nothing is strange, but as it becomes familiar with some stimuli, it is able to differentiate others. Bateson (1964) found that chicks avoid moving objects less if they have the same colour pattern as the walls of the pen in which the chicks were reared. This shows that within a few days of hatching the chicks can learn the characteristics of their immediate environment and discriminate them from novel stimuli. They avoid objects they detect as being unfamiliar. In the natural environment the mother and siblings would soon become familiar, and most other objects would be initially unfamiliar. Hess (1959) found that the graph of increase in speed of locomotion with age corresponded closely to the graph showing the onset of the sensitive phase in chicks (see Figure 20.5). He also showed that the graph of fear responses with age (measured as the proportion of birds giving distress calls in a standard situation) corresponded with the end of the sensitive period, as shown in Figure 20.6.

Fig. 20.5 The sensitive period in chicks: scores obtained by chicks of different ages in a laboratory test of the following response (After Hess, 1959a).

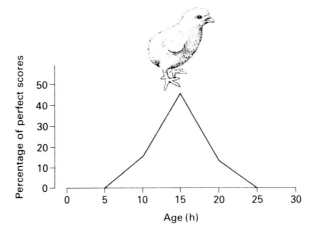

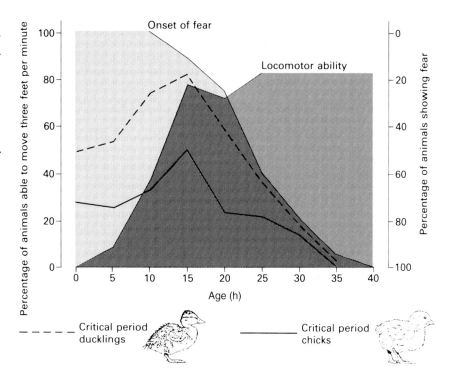

Fig. 20.6 Empirical and theoretical critical periods for chicks and mallard ducklings. The empirical critical periods for chicks and ducklings are shown by continuous and dashed lines, respectively. The pale shaded area indicates the proportion of birds showing fear responses. The darker shaded area shows the proportion of birds passing a locomotion test. The dark area (where the two shades overlap) indicates the theoretical critical period, on the assumption that it is determined by these two factors (After Hess, 1959b).

Hormones are also thought to be important in determining the sensitive phases of imprinting (Landsberg, 1981). In zebra finches there are peaks of testosterone production at 18–22 and 34–38 days of age. The first peak coincides with the age at which the young leave their dome-shaped nest and see their parents in full size and colour for the first time. The second peak occurs shortly after weaning, when the young zebra finches leave their parents and join the flocks of non-breeding birds. Adult levels of testosterone are reached at 70 days of age, when the birds become sexually active (Immelmann, 1985). It is possible that testosterone has an organizational effect on brain development in finches, as has been found in other species.

The evidence suggests that there may be more than one type of sensitive period. Early studies of imprinting show that both filial and subsequent sexual preferences can be affected by early experience. More recent studies (e.g. Schutz, 1965; Vidal, 1976; Gallagher, 1977) suggest that the sensitive phase for sexual imprinting occurs later than that for filial imprinting. For example, Vidal exposed cockerels to a moving model for one of three periods: 0–15 days, 16–30 days or 31–45 days after hatching. Subsequently, they were reared with a female or isolated up to the age of 150 days, at which time they were given a series of choice tests. He found that the chicks that had been exposed to the model from 31 to 45 days showed the strongest sexual preference for the model even though they had shown the least amount of filial behaviour to the model during the period of exposure. The sensitive period for filial imprinting normally would be from one to 30 days. Sensitive periods of development are known to occur also in many other contexts. Thus, starvation has stunting effects on the growth of rats if it occurs early in life

(Dobbing, 1976; Smart, 1977). Handling influences the subsequent behaviour of rats, provided it is done while they are still being reared by the mother (Denenberg, 1962). Sensitive periods occur in the learning of song in birds and language in humans. Many other examples are reviewed by Bateson (1979).

20.5 Long-term aspects of imprinting

As a result of filial imprinting, an attachment develops between offspring and parent or foster parent. This attachment ceases to be important once the juvenile reaches adulthood. However, such early experience may have long-term effects upon subsequent social behaviour. Among dogs, for example, there is a sensitive period from three to ten weeks, during which normal social contacts develop. If a puppy is reared in isolation beyond about fourteen weeks of age, its social behaviour will not develop normally. Like some birds, dogs readily accept humans as social partners, and a puppy will form a lasting relationship with its owner if it is able to form an attachment at the height of the sensitive period. In primates also, contact between mother and infant is of great importance for the development of normal social relationships (Hinde, 1974).

In birds, as noted earlier, imprinting can have profound effects upon subsequent sexual preferences, and this is often called sexual imprinting. Cross-fostering experiments with domestic breeds of fowl, ducks and pigeons can be carried out readily. If individuals of one distinctively coloured breed are reared by parents of a different breed and colour, the offspring usually prefer to mate with birds of their foster parents' colour rather than their own. For example, Warriner et al. (1963) used black and white varieties of domestic pigeon (*Columba livia*). Sixteen previously unmated pigeons were placed together in a large cage and allowed to form pairs. These pigeons had been raised exclusively by black or white pigeons, as shown in Table 20.1. Their courtship and pairing behaviour in the large cage was observed and recorded. This experiment was repeated four times with fresh birds each time, so that 64 pigeons were involved in the experiment as a whole. The results showed that males paired with females of the same colour as the male's foster parents in 26 out of the 32 cases. In five of the six remaining cases, the female paired with a male of the same colour as her foster parents. Thus, in most pairs the

Table 20.1 Characteristics of pigeons used in each replication of the sexual imprinting experiments of Warriner et al. (1963). Sixteen previously unmated pigeons of the sex, colour, and rearing-parent type indicated were placed together in a large cage and allowed to form pairs

Colour of pigeon	Sex	Rearing parents White	Black
White	Male	2	2
	Female	2	2
Black	Male	2	2
	Female	2	2

male's preference predominated over that of the female. The results show a clear effect of early experience on choice of the male pigeons. In the case of females the evidence is inconclusive because the female preference is so often masked by the dominance of the male.

Similar cross-fostering experiments can be done with birds of different, though closely related, species. Experiments have been carried out with ducks and geese, pigeons and doves, jungle fowl and domestic fowl, house and tree sparrows and various species of gulls and finches. The results usually show a sexual attachment to the species of foster parent. For example, if male zebra finches (*Taeniopygia guttata*) are raised by Bengalese finches (*Lonchura striata*), then they court Bengalese finches when adult. Even if given a choice between an enthusiastic female zebra finch and an unenthusiastic female Bengalese finch, a cross-fostered male zebra finch will prefer the female Bengalese finch (Immelmann, 1972). The sexual preference for the foster-parent species is not restricted to particular individuals, but generalizes to all members of the species. Among male zebra finches, a preference for a particular individual may develop after pair formation, but the initial preference is for the species of female that played a parental role. In filial imprinting, by contrast, an attachment to particular aspects of the parent or parent substitute is much more likely to develop.

Schutz (1965, 1971) carried out a series of experiments with various species of duck. He found that males tend to prefer sexual partners that are similar to the female that reared them, whereas females prefer to mate with males of their own species, irrespective of their early experience. Of 34 male mallard ducks raised with other species or domestic varieties, 22 mated with their foster-parent species while twelve mated with their own species. However, among eighteen mallard females raised by another species, all but three mated with their own species. This type of result was obtained with the other duck species studied by Schutz, with the exception of the Chilean teal (*Anas flavirostris*). Of seven Chilean teal females fostered by mallards, all subsequently paired with mallard males. The other species studied by Schutz are sexually dimorphic, the female coloration differing from that of the males (see Figure 20.7). Chilean teal are monomorphic, both sexes having incon-

Fig. 20.7 Sexually dimorphic mallard ducks (left) and monomorphic Chilean teal (right) (*Photograph: F. Schutz*).

spicuous female-type plumage. Ducklings typically are cared for by the female parent and normally would not be greatly exposed to the male colour pattern. Different species of duck are often found together on a lake, and the full-coloured females look more alike than the conspicuous males. Thus, the females of dimorphic species can discriminate easily among males, but the males have a more difficult task. It has been suggested that the males of dimorphic species have to rely more on early experience to learn to identify their own species (Schutz, 1971) as we see later in this chapter.

Sexual imprinting occurs most readily to conspecifics, and less readily to inappropriate species like human beings. In the absence of any alternative, however, reliable sexual imprinting may occur and may be long lasting. Thus, birds that are hand reared can become sexually imprinted on people, and this has been reported for more than 25 species (Immelmann, 1972). Sexual preferences based upon imprinting often persist for a number of years. Thus, mallards fostered by other species of ducks and geese continue to court members of the foster species, even though they obtain little cooperation. Immelmann (1972) cross fostered Bengalese and zebra finches and then isolated them from their foster species for a number of years. Most bred successfully with members of their own species, but when they eventually were given a choice between their own and their foster species, they strongly preferred to court the species they had been raised by years before.

This means that brief contact with foster parents early in life exerts a longer-lasting influence than does social contact of long-term duration during adult life. However, if male zebra finches are foster-raised by Bengalese finches and then mixed with their own species, instead of being kept in isolation, they may transfer their attachment to their own species. Thus for adolescent males, in contrast to adults, it is possible to alter a previously established preference and 're-imprint' the birds on their own species (Immelmann, 1985).

20.6 | Imprinting and learning

Lorenz (1935) believed that imprinting was fundamentally different from other forms of learning, but this is not a popular view today (Bateson, 1966, 1990; Hinde, 1970). Imprinting involves a narrowing of pre-existing preferences, a process that has much in common with other forms of perceptual learning.

Perceptual learning is a rather controversial topic because it is often difficult to determine to what extent animals have to learn to organize their perceptual world and to what extent their recognition of external stimuli is innate. Numerous experiments show that animals deprived of perceptual experience during early life do not discriminate stimuli as well as normal animals (see Hinde, 1970). However, the interpretation of sensory deprivation experiments in terms of perceptual learning is open to criticism on a number of grounds. On the one hand, deprived animals may perform poorly in discrimination tests because they have learned to rely on other senses, because exposure to novel stimuli is emotionally traumatic or because of deterioration of sensory mechanisms.

On the other hand, experiments that provide specific experience with particular stimuli can provide good evidence for perceptual learning. The development of full song in the chaffinch requires both exposure to the song during a particular phase of early life and the opportunity to practise singing it at a later stage (see Chapter 3). The phase during which the bird stores a description of the complete song can be described as perceptual learning. Similarly, the zebra finch that is raised by a Bengalese finch to which it responds sexually after years of separation provides a clear example of perceptual learning.

In a classic study, Gibson reared rats in cages with shapes cut out of metal fixed to the walls. In her tests, the ability of the rats to approach one of two shapes to obtain a food reward was evaluated. The observers found rats that previously had been exposed to one of the shapes learned more quickly than rats that had not been so exposed (Gibson and Walk, 1956; Gibson et al., 1959). Bateson's (1964) finding that chicks imprint more readily to patterns with which they are familiar from their home environment is comparable and illustrates the importance of perceptual learning in imprinting. Recent studies (Bateson, 1990) suggest that at least two types of plastic change seem to be involved: establishing an internal representation of the familiar object, and the pre-emptive capturing by that representation of the systems controlling filial and sexual behaviour.

Imprinting also appears to involve instrumental learning. Ducklings exposed to a toy train one day after hatching can be trained to peck at a pole by running the train past them just after they had pecked (Hoffman et al., 1966). Ducklings exposed to the train at a later stage of development could not be shaped in this way. Bateson and Reese (1969) found that mallard ducklings and chicks would learn to depress a treadle to gain exposure to a rotating light to which they had not been imprinted previously. They found that the birds would learn this response only during the sensitive period for imprinting. Thus, it appears that the chick may learn about its mother through perceptual learning and also that it learns responses that bring it into her presence through instrumental conditioning (see Hinde, 1974). Bateson and Wainwright (1972) showed that as the chick becomes familiar with an imprinting stimulus, it begins to prefer stimuli slightly different from it. They tested chicks in the apparatus shown in Figure 20.8. In this way they were

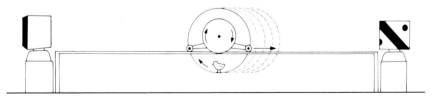

Fig. 20.8 Apparatus used by Bateson and Wainwright (1972) to test chick's preferences among various imprinting stimuli. As the chick attempts to approach one stimulus, it is propelled away from it (by a special arrangement of gears, not shown here). As the chick nears the other stimulus, it reverses its preference and attempts to approach it. The relative preference between two stimuli can be gauged from the positions of the chick when its preference changes.

able to obtain a quantitative measure of the chick's preference when given a choice between familiar and unfamiliar stimuli. The results showed that as familiarity with the initial imprinting stimulus increased, the chicks developed an increasingly strong preference for stimuli differing slightly from the familiar one. They suggested that, in the natural situation, this tendency would have the effect of familiarizing the chick with different aspects of the mother. The ability to recognize a parent from many angles could only develop as the young bird built up a composite picture of its parent's characteristics (Bateson, 1973).

As we saw in Chapter 18, some types of learning can occur in the absence of reinforcement and yet can be modified by reinforcement. Moreover, animals will perform responses that are instrumental in providing access to familiar stimuli such as the song of a conspecific or the sight of a mate. The role of reinforcement in the imprinting situation clearly falls into this same category, and the claim that imprinting is a special form of learning cannot be upheld on these grounds. Imprinting can be classified, on functional grounds, as a form of preprogrammed learning. However, this term also applies to many other aspects of learning such as song learning, place learning and latent learning. The features that once made imprinting seem to be different from other forms of learning have been shown by research to be commonplace, partly as a result of our changed view of learning (Bateson, 1990).

Indeed, it has been suggested (Bischof, 1983) that early sensory stimulation may lead to morphological alterations in particular brain areas, but as morphological plasticity is reduced, at the end of the sensitive phase, external stimulation can produce only biochemical changes in nervous tissue. The greater stability of the morphological features of the brain may explain the relative irreversibility of imprinting. Thus the distinctive features of imprinting may relate more to the form of information storage than to the mode of acquisition of the information (Bateson, 1990; Immelmann, 1984).

20.7 Functional aspects of imprinting

Lorenz (1935) suggested that imprinting was important in species recognition, but later studies indicate that it does not play an essential role. A bird of a species that imprints readily, nevertheless, can respond to conspecifics even if it has no relevant experience (Schutz, 1965; Immelmann, 1969; Gottlieb, 1971). Thus, although imprinting may provide an addition to innate recognition of members of one's own species, it probably also serves other functions.

From an evolutionary viewpoint, it is often important that mating should occur only between members of the same species and that parents should care only for their own offspring. Imprinting tends to occur in species in which attachment to parents, to the family group or to a member of the opposite sex is an important aspect of the social organization. For example, among flocks of goats or sheep there is always a possibility of kids and lambs losing contact with the mother and approaching other females. Shortly after she gives birth, the mother goat labels her offspring by licking them and is sensitive to the smell of her kid for about an hour. During this period, a five-minute contact with any kid is sufficient for it to be accepted as her own. If no such contact occurs, the kid will not be allowed to suckle (Klopfer and Gamble, 1966). Similarly, chicks of colonial gulls may be abandoned by their parents during a disturbance in the colony, and they usually hide under nearby vegetation. When the parents return they call the chicks from hiding, but some chicks may have strayed outside their parents' territory. Many gull chicks develop a specific recognition of their parents' call and vice versa. These contact calls, learned at a particular stage of development, are used after periods of separation and help to maintain the integrity of the family units (Beer, 1970).

The sensitive period for sexual imprinting in ducks, geese and finches corresponds closely to the period of parental care and is more prolonged in geese than in ducks. Normally, the young bird is susceptible to imprinting while it is a member of a family group. The sensitive period usually ends before the juvenile is likely to mix with birds other than its immediate kin. Bateson (1979) suggests that sexual imprinting is more important for recognition of kin than for species recognition. He suggests that sexual imprinting enables an animal to learn the characteristics of its close kin and subsequently to choose a mate that appears slightly different, but not too different, from its parents and siblings. In this way it is possible that the animal could strike a balance between the advantages of inbreeding and outbreeding. The advantages of outbreeding are commonly held to be that it introduces beneficial genetic variety and reduces the potency of lethal recessive genes. Although the validity of these suggestions is disputed (see Maynard Smith, 1978a), the existence of some selection pressure from outbreeding is widely recognized (Bischof, 1975). Similarly, there are thought to be certain advantages in inbreeding, particularly in maintaining the integrity of co-adapted complexes of genes. To strike a balance between these opposing pressures, the animal should choose a mate with a particular degree of relatedness like a first cousin. But how are such kin to be recognized?

Bateson (1979) suggests that in order for a bird to recognize its kin it should delay learning about its siblings until they are old enough for their juvenile characteristics to provide a reliable indication of their adult appearance. As Figure 20.9 shows, the appearance of the mallard duckling, Japanese quail chick and domestic fowl chick changes with age at a rate that is different for each species. Sexual imprinting in the mallard starts at about week four and lasts about a month (Schutz, 1965). This coincides with the time when the duckling begins to take an adult-like appearance. In the quail, sexual imprinting occurs in the first few weeks after hatching (Gallagher, 1977), and by week three the chick's plumage is already fairly adult-like. By contrast, the domestic chick takes much longer to develop adult plumage, and the sensitive period for sexual imprinting appears to be about weeks five and six (Vidal, 1976).

Bateson (1979) suggests that the timing of sexual imprinting is associated with the development of adult plumage, thus giving the juvenile bird a good opportunity to learn about the appearance of its siblings. However, this appears to be true only for precocial birds (Immelmann, 1985). Bateson

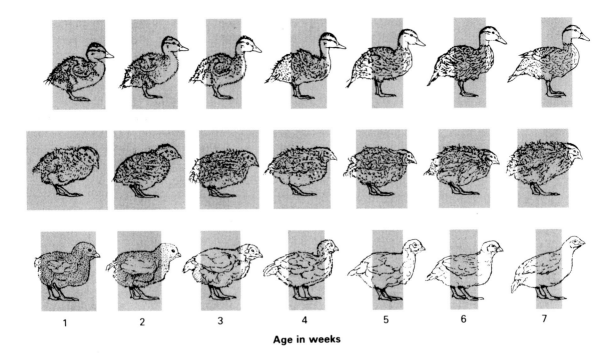

Age in weeks

Fig. 20.9 Changes in external appearance with age in a mallard duckling (above), Japanese quail (middle), and domestic fowl chick (below). The 10 cm scale (shaded) shrinks as the birds increase in age, thus presenting a sibling's eye view of each bird (After Bateson, 1979).

(1980) found experimentally that Japanese quail prefer to mate with birds that differ slightly in plumage colour from their parents. Normal brown males reared in groups preferred to mate with a brown female when given a choice between a brown and a mutant white female. However, when given a choice between a strange brown female and a familiar one with which they were reared, they preferred the former. Similarly, there is some evidence that Bewick's swans (*Cygnus columbianus bewickii*) avoid mating with close kin. These swans have distinctive facial markings (see Figure 20.10), and members of the same family tend to have similar faces. Mated pairs have facial patterns that differ more than would be expected by chance (Bateson *et al.*, 1980), suggesting that the young birds avoid inbreeding by choosing mates with facial markings that differ from the family pattern.

Some researchers have suggested (Westermark, 1891) on the one hand that humans tend to choose mates that are socially, psychologically and physically similar to themselves (Lewis, 1975). On the other hand, there is evidence that satisfactory marriages are not formed between people who spend their early childhood together. Studies of Taiwanese arranged marriages (Wolf, 1966, 1970) and Israeli kibbutzim (Shepher, 1971) indicate a lack of sexual attraction between people who spend their childhood together. In Taiwanese arranged marriages the bride is adopted into the family of the husband as a small child. In many kibbutzim the children are raised from birth in peer groups, consisting only of children of the same age. In both cases there is some social pressure in favour of ultimate marriage, yet is is rarely successful. Although such human relationships are complicated by social convention and taboo, the evidence for some biologically based negative imprinting is considerable.

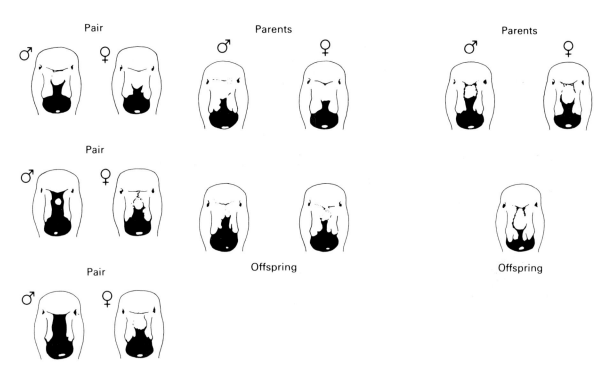

Fig. 20.10 Facial pattern in Bewick's swans. Offspring tend to resemble their parents (right), suggesting that the pattern is inherited. Mates are often different in appearance (above), suggesting that the swans actively outbreed (From Bateson *et al.*, 1980).

Points to remember

- The concept of instinct has changed over the years with the growing realization that all behaviour is the result of genetic and environmental influences.

- The early ethologists thought that the recognition of sign stimuli required a special mechanism, which they called the innate releasing mechanism. This notion has much in common with selective attention.

- The young of many bird species develop an attachment to moving objects soon after hatch-ing. This imprinting process occurs during a particular sensitive period of development.

- Imprinting may have long-term consequences, particularly upon the sexual behaviour of animals.

- Imprinting used to be thought of as a special form of learning, but it has much in common with ordinary conditioning.

- Imprinting may be important in the development of kin recognition and in preventing interbreeding among close relatives.

Further reading

Bateson, P.P.G. (1979) How do sensitive periods arise and what are they for? *Anim. Behav.* **27**, 470–486.

Horn, G. (1985) *Memory, Imprinting and the Brain*. Clarendon Press, Oxford.

Tinbergen, N. (1951, 1965) *The Study of Instinct*. Clarendon Press, Oxford.

CHAPTER

Ritualization and communication

21.1 Ritualization

21.2 Conflict

21.3 Communication effectiveness

21.4 Evolution and communication

21.5 Manipulation

The early ethologists made considerable progress in understanding animal communication. Charles Darwin (1872) had set the stage in emphasizing the role of communication in the emotional expressions of animals. For many years, however, the subject remained curiously neglected. Konrad Lorenz (1932, 1935) used the term releaser for 'those characters exhibited by an individual of a given animal species which activate existing releasing mechanisms in conspecifics and elicit certain chains of instinctive behaviour patterns'. Thus, Lorenz laid the foundations of the classical ethological view of communication, further developed by Niko Tinbergen (1951, 1953).

Various specific features of an animal's morphology may be ritualized (see Chapter 20) and act as sign stimuli to which other members of the species respond in an instinctive manner. In the social context these sign stimuli were often known as social releasers. For example, Tinbergen (1951) describes how the chick of the herring gull is fed by a parent. When the parent arrives at the nest after foraging, it calls the chicks out of hiding. The chick approaches the parent and pecks at the red spot at the tip of the parent's beak. This stimulates the parent to regurgitate food, which it then picks up from the ground and holds at the tip of its bill as illustrated in Figure 21.1.

The stimuli that elicit the begging response of the herring gull were subjected to extensive analysis by Tinbergen and his co-workers (e.g. Tinbergen, 1949; Tinbergen and Perdeck, 1950). Using a series of cardboard models of a gull's head, they measured the responsiveness of chicks in terms of the number of pecks aimed at the model in a given period of time. The models were varied systematically in shape, coloration and patterning, as illustrated in Figure 21.2. These experiments showed that the greatest response was shown to models that were held fairly close to the ground, were moving slightly and had a long, thin protrusion pointing downward. The chicks normally aimed their pecks at the tip of the bill and were most responsive if this was marked with a red spot on a contrasting background. The colour of the bill and of the head were found not to affect the begging response.

Classically, social releasers are characteristic of each species and have

377

evolved through a process of ritualization. Their recognition by means of the innate releasing mechanism is also a species characteristic, and the communication system evolves in such a way that the mechanisms that send and receive the signal are kept in tune with each other. Subsequent research has greatly modified this basic picture, but before we discuss this we should look more closely at the phenomenon of ritualization.

21.1 Ritualization

Many animal activities may become ritualized during the course of evolution so that they come to serve a communication function. Any activity that is already a potential source of information to other animals may become ritualized. Darwin (1872) noted that the protective facial expressions of mammals play a role in communication. The protective reflexes, which include narrowing the eyes, flattening the ears and raising the hair around the neck, serve to protect the sense organs at moments of danger. Such responses give potential information to other animals, who can interpret them as signs of fear or anger. Thus, primitive facial expressions provide good material for the selection of an efficient communication system. The expressions can be made more effective by exaggeration, by accompanying vocalization and by distinctive markings that draw attention to the face or that emphasize a change in facial expression. The absence of hair on parts of the human face draws attention to the main features used in communication, and other examples are illustrated in Figure 21.3.

Darwin noted that signals with opposite meanings are often conveyed by expressions or postures that are opposites. Human facial expressions indicating pleasure and anger use opposing sets of muscles, and the posture of an angry dog is in many ways the opposite of its posture when friendly (see

Fig. 21.2 Models used in experiments on the food-begging behaviour of herring gull chicks. The length of the grey bar indicates the number of pecks delivered to the model. (a) Models differing in bill colour, (b) models differing in the colour of the spot on a yellow bill, (c) models with grey bills and spots differing in contrast (After Tinbergen and Perdeck, 1950).

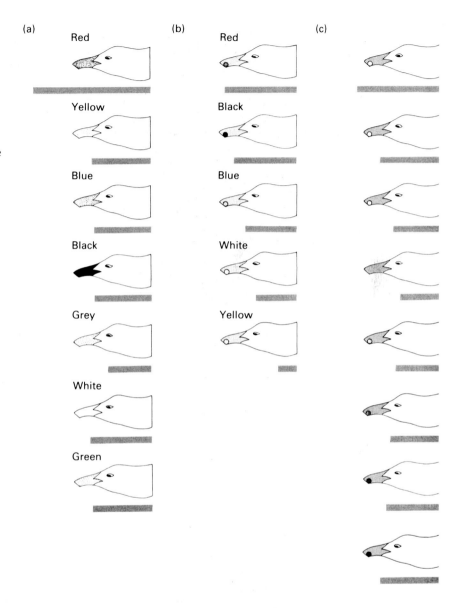

Figure 21.4). Darwin (1872) called this the **principle of antithesis**. Protective responses and their antitheses are thought to have been particularly important in the evolution of facial expressions among primates (Andrew, 1963).

Another aspect of behaviour that is thought to have provided an important origin for ritualized displays is **intention movement**. This is an incomplete behaviour pattern that provides potential information that an animal is about to perform a particular activity. For example, when a bird is about to take off in flight, it first crouches, raises its tail and withdraws its head, as illustrated in Figure 21.5. The crouching may occur a number of times before the bird takes off, or it may not precede flight at all.

Fig. 21.3 Plate from Darwin's book *The Expression of the Emotions in Man and Animals.*

Fig. 21.4 Darwin's principle of antithesis illustrated by the posture of a dog when friendly (left) and angry (right).

Fig. 21.5 Many birds crouch and raise the tail as a prelude to take-off.

(a)

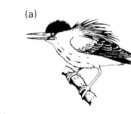

(b)

(c)

Fig. 21.6 Displays of the green heron *Butorides virescens*. (a) Forward threat display, (b) snap display, (c) stretch display (From *The Oxford Companion to Animal Behaviour*, 1981).

The importance of intention movements as signals to other animals can be seen from the study of Michael Davis (1975) on the flight reactions of pigeons. He found that a pigeon, on leaving a flock, does not disturb the others, provided the normal flight-intention movements occur. If a pigeon flies away suddenly, without any prior intention signals, then all the birds fly off. It appears that flight not preceded by intention movements is a type of alarm signal.

A number of ethologists have studied ritualized intention movements (e.g. Daanje, 1950; Tinbergen, 1953; Andrew, 1956c). They occur, for example, in the courtship of the American green heron and the golden-eye duck. The American green heron, studied by Andrew Meyerriecks (1960), nests in dead trees in salt marshes. The males arrive in spring and defend a nest tree against rival males. Intruders are challenged by a forward threat display (Figure 21.6) that is thought to have evolved from intention attack behaviour. The male adopts a horizontal posture, points its beak toward the opponent, erects its feathers and vibrates its tail. Females are attracted by the male advertising call but initially are threatened by the male. As the females persist, the behaviour of the male changes, and his readiness to accept the female is signalled by the snap display (Figure 21.6). The beak is pointed diagonally downward and the mandibles are snapped together. This display is similar to the behaviour in which the male breaks twigs from trees to build its nest, and it may be a ritualized form of displacement nest building.

After accepting the female, the male performs the stretch display, which appears to be a ritualized form of flight intention. It is the antithesis of the forward display in many respects. Whereas the forward display is accompanied by a harsh call and ruffled feathers that increase the bird's apparent size, the stretch display is accompanied by a soft call and sleeked feathers. In the forward display the beak, the bird's main weapon, is directed toward the opponent, thus exhibiting the bright red lining of the mouth. In the stretch display the beak is directed away from the female. Whereas the forward display signifies threat, the stretch display symbolizes appeasement and is followed by mutual displays on the part of male and female, including mutual billing and preening. Copulation occurs shortly after this contact between male and female.

21.2 Conflict

Motivational conflict is a common starting point for ritualization. It occurs when two tendencies compete for dominance in the control of behaviour. Because conflicting tendencies cannot be expressed simultaneously in behaviour, the behaviour seen during conflict is very different from the normal smooth run of activity.

There are three logical main types of conflict, though only one is of practical importance.

- **Approach–approach** conflict occurs when two simultaneous tendencies are directed toward different goals. Although it is possible for an animal to reach a point where the tendencies are equal, such a situation is usually transitory because any departure from the point of balance will result in an increased tendency to approach the other. This instability occurs because of the goal gradient (Figure 21.7) by which the tendency to approach a goal increases with proximity to the goal.

- **Avoidance–avoidance conflict** occurs when two avoidance tendencies occur simultaneously. Since the tendency to avoid objects usually increases with proximity to the object (see Figure 21.7), the animal will tend to move to a position where the avoidance tendencies are equal and then to escape from the situation by moving away at right angles to the line between the two objects. An avoidance–avoidance situation is therefore unstable.

Fig. 21.7 The goal-gradients of approach and avoidance.

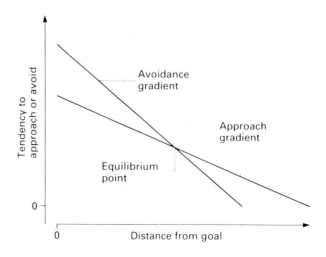

■ **Approach–avoidance** conflict occurs when an animal has simultaneous tendencies to approach and to avoid the same object. As illustrated in Figure 21.7, the gradients of approach and avoidance normally cross in this type of situation – that is, near the object the avoidance tendency is stronger than the approach tendency, while the reverse is true when the animal is far from the object. At some intermediate distance is a point of equilibrium where the tendencies to approach and avoid are the same. Thus, approach–avoidance conflict tends to be stable because the animal moves toward the equilibrium point whichever side it starts from.

Approach–avoidance conflict is very common in animal behaviour. Thus the male three-spined stickleback tends to find itself in such a conflict (between fear of and aggression towards a rival male) when near the boundary of its territory. Courtship also involves approach–avoidance conflict since each animal is initially wary of the other yet sexually attracted. In theory, the animal reaches an impasse at the equilibrium point in a conflict situation because whichever way it moves, its approach or avoidance tendency brings it back to the equilibrium point. In practice, one tendency eventually dominates, either because the animal's fear of the object declines or its tendency to approach increases or because some other motivational factor intervenes. However, animals may be readily observed to oscillate backward and forward during approach–avoidance conflict or to remain stationary in an ambivalent posture. Behaviour that consists of separate components of the conflicting tendencies is called compromise behaviour. Alternatively, the animal may take up an ambivalent posture that compounds elements of the conflicting tendencies (Andrew, 1956a). For example, when a duck is offered bread by a person in a park, it may approach and then stop, craning its neck forward to reach the bread while turning its body away.

Displacement activities are also typical of conflict situations and are thought to occur at the point of equilibrium. Displacement activities were described by a number of early ethologists, (e.g. Huxley, 1914; Makkink, 1936; Kortlandt, 1940; and Tinbergen, 1940). Tinbergen (1952) emphasized the following characteristics of displacement activity:

■ displacement activities are recognizably similar to, or derived from motor patterns that are normal for the species

■ the movements shown appear to be irrelevant, entirely out of context with the behaviour immediately preceding or following them

Figure 21.8 shows some of the classical examples of displacement activities. Tinbergen (1951) noted that displacement activities occurred when 'there is a surplus of motivation, the discharge of which through the normal paths is somehow prevented'. He identified the causes of such prevention as conflict of two strongly activated antagonistic drives and as strong motivation in situations where there is a lack of the external stimulation required to release the relevant consummatory behaviour. A conflict between two antagonistic drives is found in animals fighting at the boundary line between their territories. Thus, male sticklebacks, when meeting at the boundary between their territories, adopt a head-down attitude, as illustrated in Figure 21.9.

Fig. 21.8 Examples of displacement activities. 1, Displacement nesting behaviour during aggressive encounters in herring gulls. 2, Displacement grass-pulling (nest-making) during a territorial dispute. 3, Displacement sleep during a flight between oystercatchers, and 4, between avocets. 5, Displacement sand-digging (nest-making) by male three-spined stickleback during a territorial dispute. 6, Displacement preening during courtship in the sheldrake, and 7, in the gargeney, and 8, in the mandarin, and 9, in the mallard, and 10, in the avocet. 11, Displacement food-catching during courtship in the European blue heron. 12, 13 and 14, Displacement sexual behaviour in the European cormorant during aggressive encounters. 15, Displacement food-begging in the herring gull during courtship. 16, Displacement feeding in domestic cocks during fighting (From Tinbergen, 1951).

Tinbergen (1951) identified this posture as belonging to the digging behaviour that occurs at the start of nest building. Reviews of field studies (e.g. Tinbergen, 1952; Bastock *et al.*, 1953) indicate that displacement activities tend to occur in three types of situation:

■ physical thwarting of appetitive behaviour

■ thwarting of consummatory behaviour by removal of its objective or goal

■ simultaneous activation of incompatible tendencies.

These observations are confirmed by laboratory studies (e.g. McFarland, 1965). The feature that these situations have in common is that the ongoing

Fig. 21.9 Threat posture of the male three-spined stickleback, derived from the sand-digging behaviour that precedes nest-building.

Fig. 21.10 The upright threat posture of a herring gull (After a photograph by Niko Tinbergen).

behaviour is thwarted, either physically or by non-availability of an expected consequence of the behaviour or by an incompatible activity.

The motivational state of an animal in a conflict situation usually is readily apparent. It is therefore ideal material for ritualization. Many displays seem to consist of ritualized aspects of conflict behaviour, and many have been analysed from this point of view. For example, the upright threat posture of the herring gull (Figure 21.10) contains elements of both fear and aggression (Tinbergen, 1959). On the one hand, the sleeked feathers, stretched upright neck and sideways position of the body with respect to the opponent are signs of fear. On the other hand, the downward pointing bill and raised wing carpels (elbows) are signs of aggression.

The interpretation of displays, in terms of their evolutionary origin and present meaning, is fraught with difficulties. Nevertheless, ethologists have been fairly successful in analysing the diversity of behaviour shown in threat and courtship in terms of ambivalence among a relatively small number of behavioural tendencies. The usual approach is to postulate three basic tendencies, one of which, if acting alone, would lead to sexual behaviour, one to attack and one to fleeing. These activities are rarely observed in unadulterated form. The behaviour that is observed is interpreted in terms of a mixture of the three basic tendencies. Thus, a threat posture will usually be due to a particular combination of the tendencies to attack and to flee. This approach has been applied to a wide variety of activities including courtship, nest-relief ceremonies and mobbing behaviour. Recognition that the behaviour involves conflict and analysis in terms of the underlying incompatible tendencies is the essence of the approach, which has come to be known as the **conflict theory of display** (Baerends, 1975). It is typified by the work of Tinbergen (1959, 1962) and his co-workers.

The interpretation of displays in terms of conflict involves four lines of evidence (Tinbergen, 1962; Hinde, 1966).

1. The **situation** often gives some clues to the animal's motivation. Thus, near a territory boundary the animal is likely to be both fearful and aggressive, while in the presence of a potential mate, sexual motivation may be involved as well.

2. The **behaviour that accompanies a display** often provides evidence about the animal's motivational state. There may be obvious conflict behaviour like alternately approaching and retreating from a rival. There may be colour changes that are correlated with the animal's motivation. Thus the colour patterns of male guppies give an indication of the extent of the fish's sexual motivation.

3. The **behaviour immediately preceding or following a display** may be used in evidence in interpreting the display. Martin Moynihan (1955) used this method to assess the relative strengths of the attack and flight tendencies that are involved in various displays of the black-headed gull (see Figure 21.11). However, there are a number of difficulties with this method. First, the behaviour shown immediately after a display may be a reaction to the response that the other animal makes to the display. This difficulty can be circumvented by including in the analysis only that

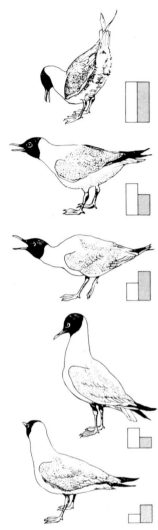

Fig. 21.11 Threat postures of the black-headed gull The different postures are associated with differences in the balance between the tendencies to attack (dark column) and escape (pale column) (After Moynihan, 1955).

behaviour shown when the behaviour of the rival is unchanging or by testing theories about displays by using motionless dummies of rival animals. A second problem is that the behaviour that precedes or follows a display may be due to a complex of motivational factors rather than an expression of a simple tendency. Third, the behaviour may be motivationally unrelated to the display, as in the cases of displacement activities.

4. The **nature of the display**. Each posture can be analysed in terms of its components, such as the angle of the head, limbs, etc. For example, Tinbergen (1959) distinguished between the aggressive threat posture of the herring gull (Figure 21.10) and the anxious threat posture in which the beak is held more horizontally. Comparison of species sometimes can be used to substantiate this type of evidence. For example, the herring gull uses wing beating during fights, and its upright threat posture includes raising of the wing carpels as if in preparation to fight. Skuas do not use wing beating in fights, and their upright threat posture does not involve exposure of the carpels.

While these lines of evidence can often be used to obtain plausible accounts of the motivation of displays, there is sometimes an element of circularity in the procedure. The circumstantial evidence obtained from field observations have been substantiated in some cases by experiment. Robert Hinde (1952) described various threat postures in the great tit (*Parus major*), which he observed in territorial boundary disputes. The postures are summarized in Figure 21.12. Hinde formed the hypothesis that the threat displays were caused by simultaneous arousal of attack and fleeing behaviour. Nick Blurton-Jones (1968) set out to test this hypothesis in experiments in which the tendencies to attack and to flee were manipulated independently.

In the initial part of his study, Blurton-Jones made observations of birds in uncontrolled situations using the methods of analysis outlined earlier. As a result of very thorough analysis, he concluded that the threat displays resulted from the same causal factors as attack and that the factors normally causing fleeing contributed little to the display. He postulated that threat displays occurred when causal factors for attack were prevalent but when attack was prevented from occurring by the tendency to flee or some other factor. A tendency to stay in one place, resulting from a tendency to feed there, was related to the proportion of horizontal displays, as opposed to head-down or head-up displays. Thus, the type of threat display seemed to be influenced by the circumstances that prevented attack.

In the experimental analysis, the tendencies to attack, flee and feed were manipulated independently. Attack provoked by poking at the bird with a pencil was accompanied by the head-down, horizontal, head-up and wings-out postures. Fleeing was evoked by a small lamp bulb and was accompanied by crest raising and feather fluffing. Presentation of food elicited hopping toward, reaching for and taking the food. By presenting stimuli simultaneously, Blurton-Jones was able to create artificial conflicts. Simultaneous presentation of attack and flight stimuli created an attack–flee conflict in which attack was reduced greatly and the head-up and other threat postures

Fig. 21.12 Postures of the great tit (*Parus major*) seen in territorial disputes: (a) is the normal relaxed posture, which grades into a head-down threat display (b–d), or into a head-up threat display (e–g). The horizontal display (h, i) is sometimes accompanied by a wings-out display (j, k) (After Blurton-Jones, 1968).

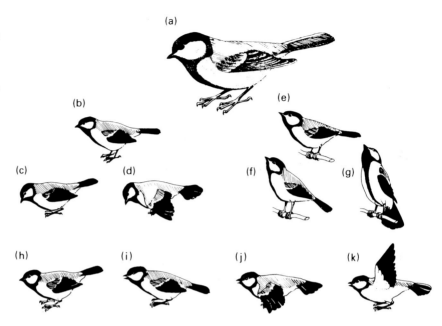

increased. Thus, a prolonged bout of attack behaviour could be changed to threat display by presentation of the flee stimulus outside the cage where the bird could not reach it. This confirmed the conclusion from the observational part of the study that anything that stopped attack in the presence of the attack stimulus would evoke threat. Blurton-Jones reasoned that this result would not be expected if the threat displays were unritualized combinations of components of attack and escape. However, the displays might be ritualized to the extent that they were no longer influenced by the fleeing tendency but only by the attack tendency. It is possible for a display to become completely emancipated from its original causal factors and to acquire causal factors of its own. However, this does not seem to have occurred in the threat displays of the great tit. Both the observational and experimental methods indicate that threat displays have essentially the same motivation as attack behaviour, though the exact form of the display depends upon the circumstances.

21.3 | Communication effectiveness

The science of communication began with the invention of telegraphy in the eighteenth century (for an historical account, see Cherry, 1957), and the view developed by telecommunication engineers persists to this day. In this view, people are seen as 'sources' or 'receivers', and information is transmitted through a medium which has a specifiable 'information capacity', a commodity which can be bought and sold.

In a simple communication system, a **source** encodes and transmits a signal, which is detected by a **receiver**, and decoded into meaningful terms. The problems of interest to the communications engineer are the encoding, transmission, detection and decoding of the signal. Encoding involves transformation of a message from one representation to another by operation of code rules. A message is defined as an ordered selection from an agreed set of signs designed to communicate information. **Signs** are simply constructs by which an organism affects the state of another. The **signal** is the physical embodiment of the message. Detection of the signal is an important part of communication theory, and problems in this field led to the development of signal detection theory (see Chapter 12).

The efficient decoding of the received signal is seen by communication engineers as a problem of devising appropriate code rules. The cost of transmission can then be measured by the number of basic instructions required for decoding. The most economical decoding procedure is one in which each instruction halves the range of n equally likely possibilities (Mackay, 1972). The cost of this is measurable by log n, called the **selective information content**. Where some forms are selected more frequently than others, the average cost is reduced if the more frequent items are encoded in a way that enables them to be selected by fewer steps than the less frequent items. Shannon (1948) showed that the minimum average cost per selection achievable by encoding in this way is given by a weighted mean of the selected information content. Thus Shannon devised a measure of the prior uncertainty, or statistical unexpectedness, of information.

In the natural environment, it is important to distinguish between **transmitted** information and **broadcast** information (Wiley, 1983). An observer measures the first by an increase in predictability in the receiver's behaviour following activity by the source, or sender. The latter is an increase in the predictability of the sender's identity or behaviour after a signal. Thus broadcast information is a measure of the information obtained from a signal by an observer.

The broadcast information depends upon the state of the sender. The mapping from the sender's internal state to its communicative behaviour is an aspect of encoding (Wiley, 1983; Green and Marler, 1979). In other words, it is envisaged that the signal is encoded as a result of a translation of the sender's internal state into behaviour and other aspects of communication (such as pheromone release). By the same token, the receiver translates the signal into a change in its own internal state, a process called decoding. The exact nature of the translation depends upon the signal, the concurrent influence of other external stimuli, and the animal's current internal state.

Errors in reception may occur as a result of poor detection of signals, failure to classify them correctly, and confusion between signals and irrelevant stimuli (see also Chapter 12). Thus errors in detection can arise in two basic ways: missed detections (missing some occurrences of the signal) and false alarms (reacting to stimuli that are not signals). A receiver cannot minimize missed detections and false alarms simultaneously, but must trade off between them.

Problems for the classical communication theory arise in considering the decoding of signals. In decoding, the message is transformed from one representation to another. It is usually assumed that, at this stage, the message becomes meaningful to the recipient. The problem is that much of communication theory applies to ordinary perception, as well as to communication. So how does communication differ from ordinary perception? 'There is here immediately a difficulty of definition. How can we distinguish between communication proper, by the use of spoken language or similar empirical signs, and other forms of causation? (Cherry, 1957; see also Mackay, 1972). Part of the problem is that the recipient is seen in anthropomorphic terms. While this may seem to be satisfactory, to some, for human communication, it is clearly an unsatisfactory way to define communication in animals.

In formal terms, information can be said to have passed from one individual to another when the behaviour of the former becomes more predictable to the latter. But not all such instances of information transfer should be regarded as communication. For example, just before a dead bough falls from a tree there may be a creaking noise. Anyone standing below the tree is warned of the impending danger and can take evasive action. Similarly, just before a cow defecates it lifts its tail. Anyone standing just behind the cow is warned of the impending danger. We would not want to say that the tree is communicating with the animal or person standing underneath it, because the creaking noise is a fortuitous byproduct of the breaking process. It is not designed (by natural selection) as a warning signal. We would not want to say that the cow is communicating with the person standing behind, because the function of raising the tail in these circumstances is not to provide a warning signal, but to promote cleanliness. To count as communication, the behaviour must be designed to influence the behaviour of another animal. This means that natural selection must have acted on the sender to fashion the signal, and on the receiver to detect the signal. As we see below, the design of communication systems by natural selection has led to all sorts of complications not envisaged by the early communications engineers.

The effectiveness of animal signals is influenced by the physical environment, the nature of the receiver and the influence of other signallers. As we saw in Chapter 13, the different sensory modalities are best suited to different habitats (see Table 21.1). Much depends, therefore, upon the physical nature of the habitat. In the case of bird song, for example, there are two fundamental problems of transmission, that of attenuation and that of distortion (Catchpole and Slater, 1995).

Attenuation occurs as a result of spherical spreading of the sound waves sent from the source (Figure 21.13). The further the receiver from the source,

Table 21.1 Advantages of different sensory channels of communication (From Alcock, 1984)

Feature of channel	Type of signal			
	Chemical	Auditory	Visual	Tactile
Range	Long	Long	Medium	Short
Rate of change of signal	Slow	Fast	Fast	Fast
Ability to go past obstacles	Good	Good	Poor	Poor
Locatability	Variable	Medium	High	High
Energetic cost	Low	High	Low	Low

Fig. 21.13 Spherical spreading (From Catchpole and Slater, 1995).

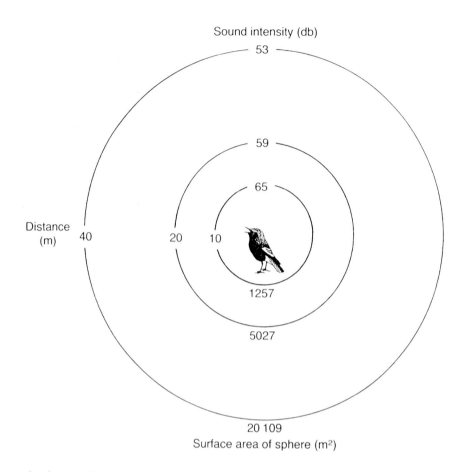

Sound intensity (db)

the fainter the sound will be, because the energy is physically diluted. Pure spherical spreading assumes a source that is isolated in an environment that is homogeneous in all directions. In the real world, sound attenuation will be greater as a result of weather conditions and obstacles to sound transmission. Higher sound frequencies tend to show greater attenuation, because they are more likely to be absorbed by the atmosphere, especially in hot or humid conditions. High frequencies are also more easily scattered by obstacles. Moreover, sound travels faster in warm air, and as the air is usually warmer near the ground, the sound waves tend to be bent upwards causing a sound 'shadow' (Wiley and Richards, 1978).

Low frequency sounds, especially where the source is near the ground, tend to be prone to greater interference effects (due to cancellation and summation of sound waves) than high frequency sounds (Catchpole and Slater, 1995). Thus the highest and lowest frequencies are not generally ideal for communication purposes, and this leads to the idea of a middle frequency 'sound window' (Morton, 1975; Marten and Marler, 1977; Marten *et al.*, 1977).

Sound distortion may result from scattering by foliage and reverberation from rocks or tree trunks. Pure tones are less subject to distortion but suffer greater attenuation through interference. However, if coding rather than sound transmission is important, then frequency coding is less affected by weather factors than is amplitude coding.

In practice, we would expect from the theory that small birds, which cannot produce low frequency sound, should minimize interference by singing from a high position, well clear of obstacles. The review by Catchpole and Slater (1995) suggests that this is often the case. When we look at geographical variation within a species the picture is mixed (Catchpole and Slater, 1995). In those species that are tutored by their parents (see Chapter 3), one might expect dialects that are adapted to the habitat (Hansen, 1979), and this has been found in some cases (see Box 21.1).

In addition to the physical structure of the environment, the effectiveness of communication is also influenced by the nature of the receiver. A signal is effective only so far as it is tailored to the sensory apparatus and **receiver psychology** of the recipient. The receiver psychology embraces the detectability, discriminability and memorability of the signal (Guilford and Dawkins, 1991). In addition, the effectiveness of communication will be influenced by other signallers, particularly where there are closely related

Box 21.1 Habitat variation in great tit song

Hunter and Krebs (1979) recorded great tit song in a number of European countries plus Iran and Morocco. Regardless of geographic variation, they found consistent differences in song structure between two types of habitat. Songs recorded from open habitats, like parkland and open woodland, had a wider range of frequencies, higher maximum frequency, and more rapidly repeated notes than those recorded in dense forest.

(From Krebs and Davies, 1993.)

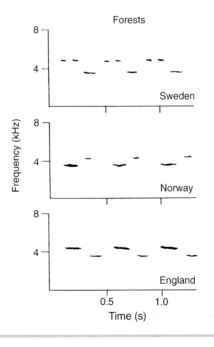

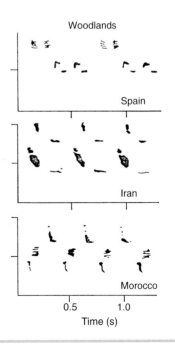

species active in the same season and habitat. To reduce interference, such species should employ distinctly different signals. To some extent there may be conflicting selective pressures in the design of signals that both match receiver psychology and remain distinct from other signallers (see Johnstone, 1997, for a review).

The tendency for closely related species to diverge in characteristics is called **character displacement** (see Chapter 5). Loftus-Hills and Littlejohn (1992) studied two species of toad, *Gastrophryne carolinensis* and *G. olivacea*. These species overlap in Texas and eastern Oklahoma. Recordings of the male advertisement calls showed that the calls of the different species were more dissimilar in habitats where they overlapped than in habitats where they did not overlap.

21.4 Evolution and communication

For the classical ethologists, as we have seen, animal communication was a relatively straightforward matter. Various specific features of an animal's morphology become ritualized during the course of evolution, and act as sign stimuli to which other members of the species respond in an instinctive manner. Thus the communication system evolves in such a way that the mechanisms that send and receive the signal are kept in tune with each other (see 'receiver psychology' above).

In the case of communication among members of the same species, it is clear that the classical ethological view must hold in those cases where the communication is especially elaborate. For example, the ritualized displays of the male green heron (see Figure 21.6) can hardly have evolved without some corresponding evolution of female recognition mechanisms. Similarly, the elaborate fern-like antennae by which the male silkmoth detects the female pheromone (see Figure 12.1) are unlikely to have evolved to serve some other function. Thus, communication systems often involve complex coevolution by actor and reactor.

On the other hand, in considering communication among members of different species, it is hard to understand, in evolutionary terms, how an individual animal can benefit by informing another individual about its true motivational state or about what it is likely to do next. It seems more plausible that such an animal would attempt to gain an advantage by deceiving the recipient animal. Indeed, there are many instances where natural selection has favoured misleading communication among species. This is a particularly common feature of anti-predator devices.

When discovered by a predator, many animals adopt a posture designed to intimidate the predator. In some cases the display is pure bluff. For example, many species of moth and butterfly suddenly expose eye-like spots on their hind wings when they are disturbed while resting. Such eye spots also are found in cuttlefish, toads and caterpillars (see Figure 21.14). Some researchers have shown experimentally (e.g. Blest, 1957; Coppinger, 1969, 1970) that the sudden appearance of bright colours may startle predatory birds, giving the moth a chance to escape. Eye spots, whether permanent or

Fig. 21.14 Hawkmoth (a) in resting posture, (b) exposes its eye spots when disturbed by a predator.

(a)

(b)

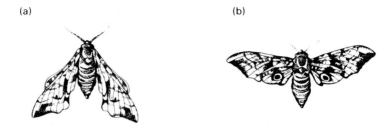

suddenly revealed, also have a deterrent effect, presumably on account of their resemblance to the bird's predators. David Blest (1957) placed dead mealworms on a box and encouraged birds such as chaffinches, buntings and tits to feed on them. When the birds were accustomed to the situation, he tested various eye-like patterns. As soon as the bird alighted on the box, it completed an electric circuit that lit up two patterns each side of the mealworm. Blest found that circular patterns were more effective than crosses in frightening the birds and that the more eye-like the pattern, the more effective it was in eliciting escape behaviour (see Figure 21.15). Blest found that birds soon habituate to the eye spots, and it therefore seems advantageous for insects to keep their eye spots hidden until they are needed.

The display of eye spots is a form of **mimicry** of the sign stimuli used by other species. Many types of display involve mimicry of the markings or of the behaviour of other animals. The saber-toothed blenny mimics the distinctive colour patterning of the cleaner wrasse (see Figure 7.4) and is thus able to deceive large fish into permitting it to approach. Instead of removing parasites like the cleaner wrasse, however, the blenny bites a piece out of the large fish and then escapes. Some snakes mimic the colour patterning and

Fig. 21.15 Three of the models used by Blest (1957) in his experiments on eye-spot patterns. When a bird landed on the apparatus to obtain a mealworm (centre) it activated a switch that lit up the circles, or eye spots, on either side. The birds were least frightened by the model shown at the top, and most by that at the bottom.

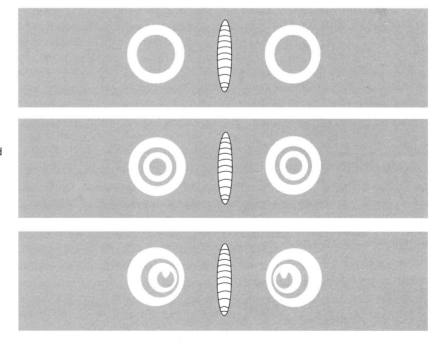

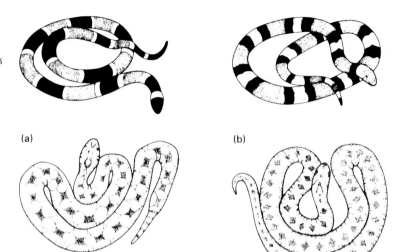

Fig. 21.16 The poisonous coral snake *Micrurus fulvius* (left), and its non-poisonous mimic *Lampropeltis elapsoides* (right) (After Edmunds, 1974).

Fig. 21.17 (a) The swirling display of the poisonous carpet viper (*Echis carinata*), and that (b) of its harmless mimic *Dasypeltis* (After Edmunds, 1974).

warning displays of poisonous species. Thus, the harmless coral snake *Lampropeltis elapsoides* mimics the distinctive bands of red, yellow and black of the poisonous coral snake *Micrurus fulvius* (see Figure 21.16). The African carpet viper (*Echis carinata*) has a warning display in which it folds its body upon itself and produces a rasping or hissing noise by rubbing the folded parts against each other (see Figure 21.17). This display is mimicked by some of the harmless snakes of the genus *Dasypeltis*. Some hole-nesting birds hiss like a snake if disturbed on the nest. Since it is dark in the hole, predatory mammals might be intimidated by the display, even though the bird does not visually resemble a snake (Hinde, 1952). Some hawkmoth caterpillars have markings on the head that closely resemble the head of a snake when inflated, as shown in Figure 1.3. When the caterpillar is disturbed it inflates its snake-head and moves it around. It may even strike at the predator (Wickler, 1968; Edmunds, 1974).

Mimicry is a form of **deceit**. The display of eye spots or snake-like stimuli give protection to the extent that they induce in the predator behaviour appropriate to dangerous stimuli. If the potential prey is not really dangerous, then the predator is deceived. Another example of deceit can be seen in the angler fish (*Lophius piscatorius*) that dangles a worm-like bait on the end of a rodlike appendange (see Figure 21.18). When a prey fish approaches the lure, it is captured by the angler fish. Evolutionary deceit exists in situations in which natural selection favours developments in members of one species that trick members of another species into behaviour that is deleterious. Natural selection will, of course, tend to sharpen the powers of discrimination of the victim, but this can be counteracted by the evolution of more effective mimicry. If the model is sufficiently common in relation to the mimic, it is very difficult for the victim species to avoid being deceived. Thus, because worm-like objects are a common form of prey, the angler fish can easily exploit the prey-recognition system of its victims. To be able to distinguish true prey from the mimic, the victim would have to spend more time

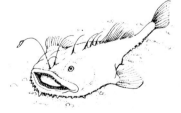

Fig. 21.18 The angler fish *Lophius piscatorius*.

inspecting each prey item, at the expense of feeding efficiency. Provided the dangerous prey are not too common relative to real prey, natural selection will result in a compromise; that is, if the risk of encountering the dangerous prey is low, it can be offset by the benefits of efficient feeding. Nevertheless, it is important to recognize that in cases of communication among species that involve deceit, the forces of natural selection acting on each species tend to work in opposition.

Situations in which communication among species is mutually beneficial usually involve **symbiosis**. In one form of symbiosis, known as **commensalism**, one species benefits from the relationship while the other remains unaffected. For example, the trumpet fish (*Aulostomus*) sometimes joins schools of yellow sturgeon and takes advantage of this camouflage to approach smaller prey fish. It darts out from among the sturgeon and seizes its prey. The sturgeon fish remains unaffected by the association, and there appears to be no communication between the two species. Similarly, cattle egrets (*Bubulcus ibis*) live in close association with cattle and feed on insects disturbed by them. The birds do not remove parasites from the cattle as do tick birds. There is no communication between the cattle and the cattle egrets, and it seems that the cattle do not benefit from the relationship.

In true symbiosis, or **mutualism**, both species benefit and communication usually occurs between them. For example, the honey badger (*Mellivora capensis*) lives in symbiosis with a small bird called the honey guide (*Indicator indicator*). When the bird discovers a hive of wild bees, it searches for a badger and guides it to the hive by means of a special display. Protected by its thick skin, the badger opens the hive with its large claws and feeds on the honeycombs. The bird feeds upon the wax and bee larvae, to which it could not gain access unaided. If the honey guide cannot find a badger, it may attempt to attract people. Local people understand the bird's behaviour and follow it to the hive. It is an unwritten law that the bird be allowed to take the bee larvae.

In the case of communication among members of different species it is easy to see how one animal may deceive or manipulate another. However, Dawkins and Krebs (1978) suggest that the same reasoning applies to communication between two members of the same species. For example, they note that:

'If a dog can cause rivals to flee simply by baring his teeth, selection will favour dogs who exploit this power. Tooth-baring will become ritualised, exaggerated for increased power to frighten, and the lips may be pulled back further than is strictly necessary merely to get them out of the way. Over evolutionary time teeth may get larger, even if this makes them less significant for eating.' [p. 288]

This view may not seem very different from the traditional ethological view. However, Dawkins and Krebs maintain that the recipient of the signal is deceived into responding by the similarity between the ritualized signal and other sign stimuli that the reactor normally would respond to, just as we saw in our discussion of communication among members of different species. This is where they depart from the traditional ethological view, which has always maintained that evolutionary developments in communication must occur in both actor and reactor (e.g. Blest, 1961); that is, during the ritualization at displays, or the evolution of any specialized mode of communication, any change in the signal must be matched by a corresponding change in the signal recognition mechanism.

On the other hand, it is clear that it will not always be the best strategy for an animal 'honestly' to inform another individual about its true motivational state or about what it is likely to do next. A population of honest animals is always open to invasion by dishonest cheaters (see Chapter 7). Blushing in humans is an example of non-verbal communication that is probably ritualized. Blushing is not usually under voluntary control, and it occurs in people who are embarrassed or mildly frightened. It is usually confined to the face and neck, the focal area for communication, and it is the opposite of what we would expect on simple physiological grounds. Normally, mild fear results in sympathetic arousal that causes blood to move away from the skin and other peripheral parts of the body. If blushing is a ritualized activity that informs other people that the actor is embarrassed, it is difficult to see how it has evolved. How does the actor benefit by providing this information? Why are populations of blushers not invaded by non-blushing cheats?

Another interesting example is provided by a study of the Harris sparrow (*Zonotrichia querula*) (Rohwer and Rohwer, 1978). These birds form flocks in winter, and dominant birds exert priority over subordinates in gaining access to food. The dominant males have more black plumage than subordinates, although all males are capable of growing black feathers, which is what they do when they adopt their breeding plumage in the spring. Why the paler males do not simply grow dark plumage and gain the benefits of apparent dominance is something of a puzzle. The Rohwers painted subordinates to look like dominant sparrows, but these birds did not rise in status. Pale birds treated with the hormone testosterone became more aggressive, but their opponents did not retreat during disputes. However, birds that were both painted and treated with testosterone were able to assert their dominance. Thus, it looks as though a combination of dark coloration and dominant behaviour is necessary to attain a dominant status that is recognized by other birds. If a subordinate male attempts to cheat by growing dark feathers, it would not be successful. However, it is difficult to see why a cheat could not also secrete more testosterone. Perhaps high levels of testosterone carry costs that subordinate birds cannot afford.

21.5 Manipulation

Design of communication systems, whether by natural selection, or by man, should be viewed in terms of the ultimate payoff, which in the case of natural systems is fitness (see Chapter 6).

A useful classification of communication can be based upon Hamilton's (1964) classification of social behaviour in general, which is formulated in terms of the design opportunities that present themselves to an evolving communications system. **Mutuality** occurs when both signaller and receiver benefit from the interaction. The dance of the cleaner wrasse, mentioned above, is an example of this. **Deceit** occurs when the signaller's fitness increases at the expense of the receiver's. **Manipulation** by receivers can occur when the receiver obtains information about the signaller, against the interests of the signaller. For example, the courtship display of a male may attract rivals. This form of manipulation has been called **eavesdropping**. The situation in which an animal reduces its own fitness in the process of harming another animal has been called **spite** (Hamilton, 1964). It is important to remember that these terms, borrowed from everyday language, are names of evolutionary strategies (see Chapter 7).

Wiley (1983) points to two important considerations for the success of manipulative strategies. First, misleading signals must occur rarely in relation to normal signals. Otherwise, receivers will adjust their rules for decoding signals. The argument here is similar to that for Batesian mimicry (Edmunds, 1974). For example, many ground nesting birds feign injury when their nest is approached by a predator, such as a fox. The bird moves away from the nest trailing an apparently broken wing, thus luring the fox away from the nest. Should the fox adopt the rule 'stalk the bird', or the alternative, 'look for the nest', when it encounters an apparently crippled plover?

The answer to this question depends upon the balance between the probability of success in stalking the bird and the probability of success in searching for the nest. (Strictly, we take the probability of success when stalking times the amount of food obtained if the stalk is successful, versus the probability of finding the nest times the value of the food in the nest.) All depends upon the proportion of plovers that are really crippled as opposed to feigning. Stalking will be superior to searching when this proportion is high, but if it is low then foxes do better by searching, and the injury-feigning strategy will not pay off for plovers (Wiley, 1983).

The second consideration for the success of manipulative strategies is that natural selection should act on senders to increase the efficiency of transmitting information whenever the sender's fitness is increased by the response. Moreover, selection should also act on receivers to minimize this efficiency whenever the receiver's fitness is reduced. Thus deceit by signallers is likely to be followed by **retaliation** by receivers. Deceit occurs when signallers can take advantage of receiver's rules. Retaliation can arise by **devaluation** of the signal by the receiver, or by increased **discrimination** by receivers.

Devaluation is a strategy evolved in response to bluffing. For example, threat displays often involve bluff in the form of exaggerated body size, or weapons. Such signals deceive the receiver into responding inappropriately in relation to the real threat. Receivers that devalue the threat will have an advantage over receivers that remain deceived. Thus devaluation will spread through the population with the effect that all signallers that do not bluff will be placed at a disadvantage. Natural selection now favours further bluffing and the situation escalates, as pointed out by Dawkins and Krebs (1978). Eventually, the threat signal is likely to become too costly to be worthwhile.

Evolutionary escalation between signaller and receiver is less likely to occur when the receiver adopts other countermeasures such as improved discrimination among signals. By scrutinizing the sender's behaviour, the receiver should be able to focus on details that give a better indication of the true situation than does the exaggerated broadcast signal. In this way the receiver knows when to call the sender's bluff. However, the sender can respond by improving the signal, making it harder for the receiver to discriminate. There is good evidence that such a process has occurred in the evolution of brood parasitism by cuckoos. Other examples come from studies of social dominance in sparrows (Rohwer and Rohwer, 1978) and of male rivalry among elephant seals (Cox and Le Boeuf, 1977; see also Wiley, 1983).

The cuckoo (*Cuculus canorus*) is a brood parasite, laying its eggs in the nests of various species of passerine bird. In Britain, the three main hosts are meadow pipits (*Anthus pratensis*), reed warblers (*Acrocephalus scirpaceus*), and dunnocks (*Prunella modularis*), in moorland, marshland, and farmland, respectively. Individual female cuckoos specialize on one of these hosts, and mimic the shade and colour of the host eggs (see Figure 21.19). The cuckoo lays one egg in the host nest, usually during the host laying period. The cuckoo chick generally hatches first and immediately ejects the other eggs from the nest. It often remains the sole occupant of the nest. The hosts desert

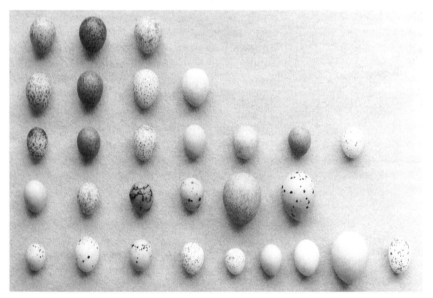

Fig. 21.19 From left to right. *Top row:* Real cuckoo eggs from reed warbler gens, meadow pipit gens and pied wagtail gens. *Second row:* Model cuckoo eggs representing reed warbler gens, meadow pipit gens, pied wagtail gens, redstart gens. *Third row:* Eggs of current favourite British hosts – reed warbler, meadow pipit, pied wagtail, dunnock, robin, sedge warbler, wren. *Fourth row:* Eggs of species which are suitable but rare hosts – redstart, spotted flycatcher, reed bunting, chaffinch, blackbird, song thrush. *Bottom row:* Eggs of species unsuitable as hosts – linnet, greenfinch, bullfinch, great tit, blue tit, pied flycatcher, wheatear, starling, swallow (From Davies and Brooke, 1989).

single eggs, but not single chicks (Davies and Brooke, 1988); consequently the host often pays the cost of raising an alien chick, and obtains no reproductive benefit, if it is parasitized by a cuckoo.

We would expect natural selection to favour the ability of the host to decrease successful parasitism by cuckoos, and to lead to improvements in the parasitic strategies of cuckoos. Thus co-adaptation between brood parasite and host can be expected to lead to ever more refined adaptations and counteradaptations. To investigate these, Nick Davies and Mike Brooke (1989) tested the discrimination abilities of host species by parasitizing the nests experimentally with model cuckoo eggs. They found that species that were suitable as hosts, including those currently used and those rarely used by cuckoos, rejected model eggs that were unlike their own. An exception is the dunnock (currently parasitized but not mimicked by cuckoos), which did not discriminate between its own eggs and model cuckoo eggs unlike its own. Evidence from old literature suggests that many of the species that do discriminate against alien eggs do have a previous history of interaction with cuckoos.

Species with no history of interaction with cuckoos, either because they are unsuitable species, or because of geographical isolation (e.g. meadow pipits and white wagtails in Iceland), show little discrimination against alien eggs. There was no difference in the distinctiveness of egg markings between species that have interacted with cuckoos and those that have not. Thus there is no evidence that host egg patterns evolve in response to cuckoos. None of the four host species tested discriminated against the presence of an alien chick in the nest. It appears that hosts evolve discrimination against odd eggs, but not against odd chicks. Davies and Brooke (1989) come to the conclusion that the present-day picture of egg mimicry by cuckoos and egg discrimination by host species is part of a continuing evolutionary 'arms race'. The cuckoos eventually lose the race with a particular host species, because they cannot completely counter discrimination against their eggs, but they then switch to a new host species and begin a new arms race. It appears that the dunnock is a relatively new host species in this respect.

Finally, we have the possibility of 'deliberate', cognitively- based deceit, which we discuss in Chapter 26. In the case of humans, other primates, and even other mammals, it is sometimes claimed that an individual may intentionally set out to deceive another. However, as we shall see, it is not always easy to draw the line between deceit that comes from individual initiative and that which is an evolutionary manipulative strategy (see also Trivers, 1985).

Points to remember

- The problems of interest in classical communication theory are: (1) the encoding, (2) the transmission of the signal, (3) the detection and (4) the decoding of the signal.

- Transmitted information is a measure of the increase in predictability in the receiver's behaviour following activity by the sender. Broadcast information is a measure of the predictability of the signaller's identity or behaviour after a signal. Thus broadcast information is a measure of the information obtained from a signal by an observer.

- Information can be said to have passed from one individual to another when the behaviour of the former becomes more predictable to the latter. But to count as communication, the information must be designed to influence the behaviour of another animal. This means that natural selection must have acted on the sender to fashion the signal, and on the receiver to detect the signal.

- The design opportunities that present themselves to an evolving communications system can be classified into four types: (1) Mutuality occurs when both signaller and receiver benefit from the interaction. (2) Deceit occurs when the signaller's fitness increases at the expense of the receiver's. (3) Eavesdropping occurs when the receiver obtains information about the signaller, against the interests of the signaller. (4) Spite is the name for the situation in which a signaller reduces its own benefit in the process of harming another agent.

- There are two important considerations for the success of manipulative strategies. First, misleading signals must occur rarely in relation to normal signals. Otherwise, receivers will adjust their rules for decoding signals. The second is that natural selection should act on senders to increase the efficiency of transmitting information whenever the sender's fitness is increased by the response.

- Selection should also act on receivers to minimize this efficiency whenever the receiver's fitness is reduced. Thus deceit by signallers is likely to be followed by retaliation by receivers. Deceit occurs when signallers can take advantage of receiver's rules. Retaliation can arise by devaluation of the signal by the receiver, or by increased discrimination by receivers. Devaluation is a strategy evolved in response to bluffing.

Further reading

Halliday, T.R. and Slater, P.J.B. (1983) *Animal Behaviour: 2, Communication*, Blackwell Scientific Publications, Oxford.

Johnstone, R.A. (1997) The evolution of animal signals. In Krebs, J.R. and Davies, N.B. (eds) *Behavioural Ecology*, 4th edn. Blackwell, Oxford.

Human behaviour

The early ethologists did not hesitate to extrapolate from animals to humans. For example, Lorenz (1943) identified those characteristics of the human infant, such as a relatively large head, protruding forehead, large eyes, etc., that he thought of as sign stimuli acting on the innate releasing mechanism (see Chapter 20) for human parental behaviour. Tinbergen (1951) thought that displacement activities showed that the organization of instincts in man was basically similar to that of animals. A more systematic approach was taken by Eibl-Eibesfeldt (1970) who looked at the development of reflexes in human infants, and the behaviour of children born blind and deaf–blind, and made good use of the comparative approach in his studies of human non-verbal communication.

Human behaviour is also studied in other disciplines. Anthropologists, such as Lee (1979), carry out systematic studies of primitive peoples in their natural environment, and psychologists (e.g. Passingham, 1982) study behavioural differences among primates (including humans) in relation to differences in brain structure and function. Evolutionary biologists address those aspects of human behaviour that are thought to have a genetic and evolutionary basis. These disparate approaches to the study of human behaviour complement each other. In this chapter, and in later parts of this book, we take selected topics in human behaviour and see how the different disciplines contribute to the overall picture.

Human non-verbal communication

Darwin (1872) realized that there was much in common between the facial expressions of humans and those of other animals, especially the primates. Today, scientists widely recognize that the non-verbal aspect of communication in humans invites direct comparison with the displays of animals. However, we must recognize that some aspects of human non-verbal communication are related directly to language. An obvious example is the sign language used by deaf people. However, many simple gestures, like the thumb-up sign, are probably also derived directly from language.

In Chapter 21, we saw that many of the displays and facial expressions of animals may be the result of ritualization. They may be derived evolutionarily from protective responses, intention movements, displacement activities, etc. Do equivalent stereotyped displays occur in humans?

One problem in studying human communication is that it may vary according to the cultural context. For example, a number of ways of signalling assent and denial involve head movements such as nodding and partial rotation (shaking) of the head. However, the meaning attached to these movements varies widely from culture to culture (Leach, 1972). In Greece, and some other cultures 'no' is expressed by a strong backward jerk of the head. Cross-cultural studies do indicate nevertheless that some human facial expressions are universal and appear to have the same meaning in many cultures. An example is the eyebrow flash (see Figure 22.1), in which both eyebrows are raised momentarily. This signal is usually given as a form of greeting at a distance (Eibl-Eibesfeldt, 1970). There are some small differences of usage between cultures. In Europe it is used as a greeting to good friends and relatives; in New Guinea it is used toward strangers; in Japan it is suppressed, being considered indecent. In general, the eyebrow flash is used as a friendly form of greeting or approval, but it may be omitted by people who are reserved or suspicious (Eibl-Eibesfeldt, 1970).

A number of basic expressions such as smiling, laughing and crying are universal. Not only are they found in different cultural groups, but they also occur in people born deaf and blind (Eibl-Eibesfeldt, 1970) and are prominent in young children (Blurton-Jones, 1972). These can be compared directly with the facial expressions of other primates. Analysis of human

Fig. 22.1 Eyebrow flash: an example of facial expression in humans.

facial expressions reveals a number of basic situations that give rise to fairly stereotyped responses (van Hooff, 1972, 1976). These include alertness, which is evident in the relatively fixed gaze and a certain tension of the facial muscles. This expression is similar to that found in other primates, except that in some the ears also may be focused on the centre of attraction. Surprise is characterized by prolonged eyebrow raising, open eye and often an open mouth. This expression does not seem to have a counterpart in non-human primates. Fear is similar in a wide range of primates and is characterized by wide-open eyes and withdrawn lips. Disgust involves wrinkling of the nose and raising the upper lip, screwing up the eyes and turning away the face. These components of the disgust expression are derived from protective responses that serve to exclude the noxious stimuli. In humans the disgust expression is ritualized, but in other primates it is merely a collection of protective responses that has no special role in communication. Sadness is accompanied by arched eyebrows, retracted corners of the mouth turned downward, and outward curling of the lips. Tears may be shed in intense cases. This expression appears to be ritualized in humans, and perhaps in chimpanzees. It elicits comfort behaviour from other members of the group. Anger takes a variety of forms, usually involving withdrawn lips and bared teeth, staring eyes and a frowning expression. Joy is expressed by smiling and laughter (see Figure 22.2), which often are associated with humour and therefore considered to be uniquely human. Some ethologists question this view, however (e.g. van Hooff, 1972; Eibl-Eibesfeldt, 1970).

Fig. 22.2 Open-mouth laughter of a human infant in a tickling game. This type of open-mouth laugh without retracted lips becomes rare in adulthood (From *The Oxford Companion to Animal Behaviour*, 1981).

Some monkeys and apes have a relaxed open-mouth display that is associated with playfulness (see Figure 22.3). It is superficially similar to human laughter, although there are a number of differences, particularly in the accompanying pattern of breathing. In both chimps and humans, the laughing-type display is elicited by an element of surprise in the situation. In chimpanzees and small children, this is of a purely physical nature, while in human adults it may be physical or purely intellectual. In both cases, however, the context is usually a social one. Human smiling is not simply a low intensity form of laughter, though the two are often associated. Smiling also occurs in situations involving mild fear or apprehension, such as social greeting and reassurance. In many primates a silent bared-teeth expression is typical of subordinate animals in acknowledging or appeasing dominant members of the group. In the chimpanzee, however, it is a reassuring and friendly expression, something akin to the human smile.

The facial expressions of humans, and of other mammals, are usually regarded as expressions of emotion, or of affect (Ekman, 1971). However, many other types of non-verbal communication can be classified in various ways. In addition to affect gestures, other classes of non-verbal communication can be distinguished (Ekman and Friesen, 1969). These include adaptors, which serve both a communication and a non-communication function. Examples are grooming movements and intention movements. Emblems are non-verbal acts that have a verbal counterpart. They include the sign languages used by deaf people, obscene gestures and various signalling movements used at a distance, like beckoning. Illustrators are movements that are

Fig. 22.3 Relaxed open-mouth face in a capuchin monkey being tickled by a human (From *The Oxford Companion to Animal Behaviour*, 1981).

used to illustrate points that are also being made verbally. They include gestures of emphasis, pointing, etc. Regulators are gestures that are used to control the flow of conversation between two people. Examples are head nodding, eye-contact movements and various shifts of body posture.

Michael Argyle (1972) proposes a somewhat simpler classification that is based upon the consequences of the behaviour. He recognizes three basic categories:

1. **Managing the immediate social situation**. This involves gestures and postures that convey the person's attitude and emotional state. Thus, a person may indicate feelings of superiority, dislike, sexual desire, etc. without expressing these attitudes in words.

2. **Sustaining verbal communication**. This includes regulators and illustrators and various gestures whose presence or absence may alter the meaning of the words being spoken. Eye contact has been found to be particularly important in this context (see Figure 22.4).

3. **Replacing verbal communication**. Sign languages are found among many different cultures and may be employed whenever verbal communication is difficult or impossible. Communication among hunters, when there was a necessity to keep quiet, was probably one of the first situations in which sign language was found to be useful. A modern equivalent is the signalling used by workers in noisy surroundings. Sophisticated language-based signalling systems are used to communicate at a distance (e.g. semaphore) and with deaf people (e.g. American Sign Language, or Ameslan). Writing and Morse code perhaps should come into this category also.

Fig. 22.4 Direction of gaze (ordinate) at the beginning and ending of long utterances is an important aspect of non-verbal communication between two conversing people. Same gaze direction indicates eye contact. Note that participants are furthest from eye contact at the switch from one speaker to the other (After Kendon, 1967).

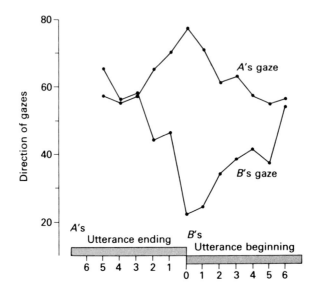

Thus human non-verbal communication can be classified into two basic types:

- communication that is ancillary to the use of language or that is part of language

- communication that is independent of language and that is akin to the mode of communication employed by the majority of animals

22.2 Mate selection and sexual strategy

Darwin (1871) thought that sexual selection was of paramount importance in human evolution but he probably underestimated the importance of natural selection (see Chapter 8). While some of the secondary sexual characteristics of humans, like differences in hair pattern, are attributable to sexual selection, others are probably due to natural selection, such as differences in strength and body size.

Primate sexes may differ in body weight, hair colour, skeletal characteristics, and secondary sexual characteristics. Such sexual dimorphism is more marked among the apes and humans than among the monkeys (Crook, 1972). Some of the differences between the sexes can be attributed to natural selection and to the differing roles of males and females. For example, the differences between the skeletons of male and female humans are primarily due to the greater muscularity of the man and to the fact that the female pelvis is designed to enable her to give birth to an infant with a large head. Among our remote ancestors, women with too small a birth passage or men too weak for the rigours of the hunt would soon be eliminated by natural selection.

From the point of view of evolutionary theory we would expect both men and women to select mating partners that are likely to enhance their reproductive success. Thus a woman might increase her reproductive success by choosing a high status male who can provide material resources that help her raise her offspring. A man is likely to increase his reproductive success by choosing a woman who is receptive, fecund and has characteristics that suggest she will be a successful mother. While the reproductive value of a man might be obvious, that of a woman is generally concealed. Men are obliged to use indirect cues such as physical attractiveness, including features that indicate the health status of the woman. Singh (1993) presents evidence indicating that body fat distribution, as measured by the waist-to-hip ratio, is correlated with good health and high reproductive potential. Her studies show that men prefer female figures with a low waist-to-hip ratio, and judge them to be more attractive and have higher reproductive potential that those with a higher ratio.

Hamilton and Zuk (1982) suggested that good health is the defining feature of female attractiveness in animals. Singh (1994) argues that sexual dimorphism in body fat distribution as measured by waist-to-hip ratio is unique to humans, and is an accurate predictor of risk of various diseases, and of female fecundity, irrespective of overall body weight. In her study, physicians were presented with line drawings of female figures that varied independently in waist-to-hip ratio and overall body weight. They were asked to rate the stimuli for the following attributes: good health, youthfulness, attractiveness, desire for children, likelihood of being a good mother, intelligence and aggression. The results (Figure 22.5) showed that the variables for attractiveness, healthiness, intelligence and youthfulness were located close to each other by both male and female physicians. The variables indicating desire for children, capability for children and good mother were grouped in a different part of the analysis space by both male and female physicians. The aggressiveness variable was located differently, but apart from this, the scores made by male and female physicians were highly correlated. The analysis showed that both utilized the waist-to-hip ratio, within each bodyweight category, to make judgements about healthiness, attractiveness, youthfulness and reproductive capability. These findings are consistent with those reported for non-physician male and female judges.

Although it may be true that strong, or high status, men may be sexually attractive to women and that female body shape is attractive to men, this does not necessarily mean that sexual selection is responsible for these fea-

Fig. 22.5 Multidimensional unfolding analysis showing the location of various personal qualities and stimulus figures following ranking of photographs of women by (a) female physicians, (b) male physicians. Dimension I represents waist-to-hip ratio, and dimension II represents body weight (After Singh, 1994).

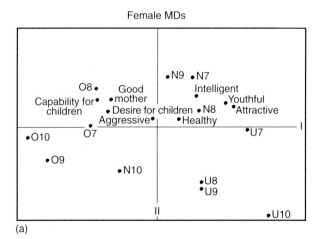

(a)

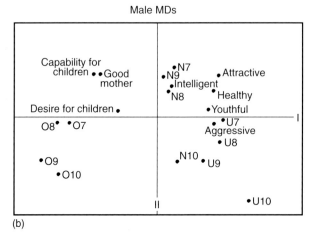

(b)

tures. Even in the total absence of sexual selection, it may be that these male and female features have no relevance to reproductive activity.

The secondary sexual characteristics of humans include the beard (in some races) and other body hair of the man, the change in the male voice that occurs at puberty, and the protruding and rounded breasts of the women. A number of authors (e.g. Goodhart, 1964; Morris, 1967; Wickler, 1967) have suggested that human breasts are the result of sexual selection because they are attractive to men and, by comparison with other primates, appear to be unnecessarily large for their role in providing milk for infants. Other authors, however, point out that the attractiveness of female breasts may be a cultural phenomenon, since it is apparently not a feature of sexual behaviour in all human societies (Crook, 1972).

Imagine a society in which women prefer redhead husbands and in which this preference is determined genetically. The hair colour of men is inherited. Redhead men will always have a choice in the selection of mates and are likely to marry early and to produce more offspring. If the society is not strictly monogamous, there will be many opportunities for sexual selection to exert its effect. In a strictly monogamous society, sexual selection will have a weak effect unless the fertility of redhead men is greater than normal or redhead men marry women who have more children for some other reason (Maynard Smith, 1958).

This analysis suggests that degree of polygamy is an important factor in determining the effectiveness of sexual selection in human society. This is a difficult matter to investigate, but a number of studies have been done. Among the Xavantes and Yanomamas, two of the surviving tribes of South American Indians, the men vary considerably in their reproductive performance (Salzano et al., 1967; Chagnon 1968). In one of the villages studied, a quarter of the total population were the offspring of two head men.

Present-day human mating patterns range from strict monogamy to various forms of polygamy. In some societies, mating follows strict culturally based rules and taboos. In others, there is relative freedom from social control. Some authors take the view that sexual selection has been of the utmost importance in the evolution of early hominid society (Fox, 1972), but others are more cautious (Caspari, 1972). It seems likely that hominids have become progressively more monogamous in response to an increasing requirement for parental care. Among the primates, humans have by far the largest period of development from birth to sexual maturity. It is difficult to see how infant humans could be successfully brought to maturity without close cooperation between the parents.

If our primitive ancestors were polygamous, then sexual selection probably would have been more effective than it is now, and this may account for some of the differences, like lack of body hair, between humans and other primates (Crook, 1972). How, then, are we to account for present-day polygamous societies? Societies are regarded as primitive if they do not use metal tools and practise little or no agriculture, so that their subsistence is derived mainly from hunting and gathering. They are often adapted to life in specialized habitats, such as arctic, desert, or forest conditions. The question arises, therefore, whether societies that we label as primitive in their biological and cultural characteristics are truly primitive or whether these characteristics are special adaptations to particular living conditions.

Sexual selection could be an effective evolutionary force in a polygamous human society, but in a monogamous society few opportunities exist for choosing a new mate. The frequency of strict monogamy among humans is difficult to evaluate, and we must remember that it is reproduction outside marriage, not sexual activity itself that is of evolutionary importance. Anthropological estimates of the proportion of monogamous societies range from 16 per cent to about 50 per cent. However, these estimates usually reflect the customary or official practice and may not indicate the true situation. It is probable that the extent of strict monogamy is rather low. In comparison with other primates, this low extent is what one might expect. If we could be confident in identifying primitive societies, then we could look to see if polygamy is made more prevalent in such societies. As things stand we have no way of doing this.

Only a few mammalian species are monogamous. These include foxes, jackals, beavers, five species of New World monkey and two ape species, the gibbon and the siamang (Passingham, 1982). However, some features of human social relationships seem to indicate that a degree of monogamy is important. In most societies, the father stays with the mother when his children are young. He often has more than one child by a particular woman, and he usually helps to support her and to care for the children. It is not common for the father to desert his offspring as in most polygamous species. A similar situation exists in two other primate species. In the hamadryas baboon (*Papio hamadryas*) of Ethiopia, adult males have small harems that they guard from the advances of other males. They remain with their females for life and do not frequently change the members of the harem as do most other polygamous species. Young males acquire females by kidnapping infants and looking after them in a parental manner. These females are thought to remain with their captor and become members of his harem (Kummer, 1968). Mountain gorillas (*Gorilla gorilla*) live in groups in which fully adult males consort with a number of females. Particular females consort with the same male over many years (Harcourt, 1979).

The reproductive strategy characteristic of humans involves high investment in each of very few offspring. Infants are born one, occasionally two, at a time, at intervals of about two years. Reproductive maturity occurs late, and a woman will give birth to few offspring during her lifetime. These offspring must receive a considerable amount of parental care if the strategy is to be successful. It is difficult for the human mother to raise her children without help. This is a result of the helplessness characteristic of the human infant, compared with those of other primates (Passingham, 1982). Chimpanzee mothers, for example, nurse their infants for several years but retain sufficient freedom to manage unaided. The infant clings to its mother's fur, leaving her free to forage and to keep up with other members of the group. Chimpanzees are polygamous and the males have no parental role. In contrast, the human mother must hold her infant in her arms because it is unable to support itself. Even when able to walk, the child could not keep up with other members of the group.

Studies of the Kalahari !Kung bushmen living a hunter–gatherer existence show that the women are severely encumbered by their children (Lee, 1972). The women provide about two-thirds of the calorie income by foraging for plant foods. Mongongo nuts are their most important plant food source. These are in plentiful supply in the dry season but usually are situated some six miles from suitable campsites. The women make excursions to gather these nuts every few days, taking their children with them. The men take no part in food gathering but confine themselves to hunting. Lee shows that the weight carried by mothers on these excursions increases with the frequency of having babies, not only because there are more mouths to feed but also because the small children have to be carried. Nick Blurton-Jones and Richard Sibly (1978) show that average birth spacing of four years is the optimum under the available conditions. Thus, the women maximize their reproductive success by spacing births widely and by foraging seldom. Bushman women have children more frequently when they do not have to carry them on foraging expeditions.

The helplessness of the human infant is due largely to its immature brain. The human brain is four times as large as would be expected for a primate of equivalent stature. The size of the brain at birth is very large compared to the size of the body (Figure 22.6), but it is nevertheless only partly functional. It takes a baby twice as long as a gorilla or chimpanzee to reach the stage where it can support its body with its limbs. Although the newborn baby has a strong grasp reflex and can even support its own weight (McGraw, 1945), this ability soon disappears. The infant ape can hang onto its mother's hair, using both hands and feet, but the human infant could not do this even if sufficiently strong. The feet are the wrong shape for grasping, and the mother has too little hair.

Under the circumstances, we might expect a woman to exercise caution in choosing a mate, to try to ensure that he will be a good father. However, although female coyness is a feature of human courtship, the young woman does not always have much say in her choice of mate. In many societies

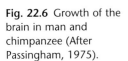

Fig. 22.6 Growth of the brain in man and chimpanzee (After Passingham, 1975).

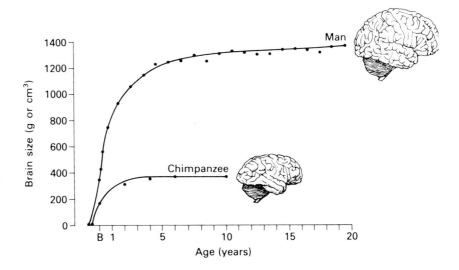

matters are arranged by a woman's parents or choice is limited by a shortage of men of appropriate social status. The method of mate selection varies widely from one human society to another but there is some evidence that females (or their parents) are influenced by material benefits. Among bushmen a man was not eligible for marriage until he had killed a large antelope. Irons (1979), studying the Yomut society, found that the wives of males of the wealthier half of the population have many more descendants than the wives of males from the poorer half (see Figure 22.7). So material benefits would seem to be linked to reproductive success.

The human female has evolved various mechanisms for encouraging male fidelity. In common with some other primates, but unlike other mammals, humans have a menstrual cycle, rather than an oestrous cycle of sexual receptivity. The menstrual cycle is characterized by periodic bleeding due to the sloughing off of the uterine lining. It occurs in all species of ape and some kinds of monkey. Animals with a menstrual cycle are sexually receptive most of the time, in contrast to those with an oestrous cycle which are receptive only at the time of ovulation (Figure 22.8). Continual sexual receptivity helps to maintain the interest of the male partner. Moreover, the time of ovulation is concealed in women, unlike most mammals in which it is advertised. This means that a man must copulate regularly with the same woman to ensure fertilization (Lovejoy, 1981). He must guard the woman against advances from other men if he is to be sure that he is the father of her offspring. The low and uncertain chance that a particular mating will lead to fertilization not only encourages continual male attentiveness but also reduces the benefits of opportunistic copulation with other women. The man has to run the risk of an aggressive encounter with another man for a payoff that is uncertain and unlikely (Halliday, 1980; Smith, 1984).

Although men have a vested interest in the welfare of their children, they might seem to have something to gain from promiscuity, especially if there is a chance that their children by other women will be cared for by other people. Many human societies attach considerable importance to the estab-

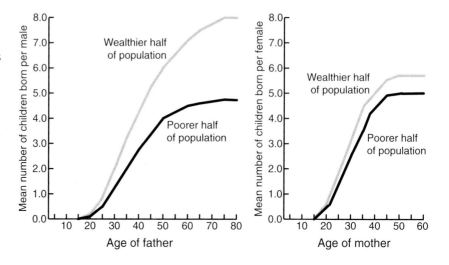

Fig. 22.7 Reproductive success and wealth in a polygynous society, the Yomuts of Turkey. Both males and females in the wealthier half of the population have higher reproductive success than individuals in the poorer half of the Yomut society (From Irons, 1979).

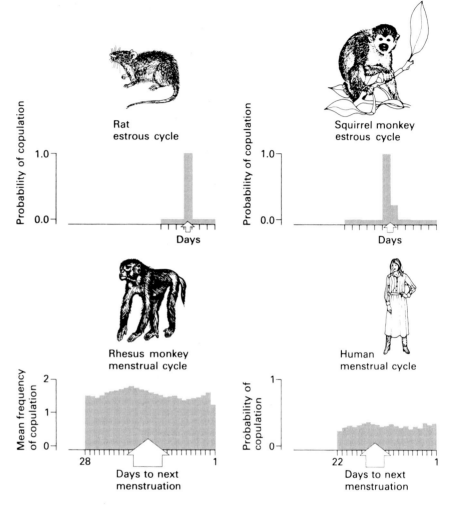

Fig. 22.8 Probability or frequency of copulation at different times during the sexual cycle of the rat, squirrel monkey, rhesus monkey, and man. Arrow indicates ovulation date (After Daly and Wilson, 1978).

lishment of paternity. A family man, who has parental duties, has more to lose than a woman does if she has a child by another man. Her genetic relationship to the child is not in doubt, so her maternal care benefits her own genes. A man has no such certainty, and he usually guards his woman from other men. In some societies virginity is highly prized, especially in brides. In others, adultery is punished and prostitution outlawed. In some societies, where female promiscuity is commonplace, the inheritance of property is through the sister of the man at the head of the household and not through the wife. The children of a man's sister are certain to be genetically related to him, while some of his wife's children may be unrelated (Halliday, 1980). Many of these features of human society have been attributed to sperm competition (see Chapter 8) (Smith, 1984; Baker and Bellis, 1995).

Male sexual strategy in humans involves rivalry in acquiring women and a certain amount of aggressiveness in protecting them from other men. This may account in part for the man's greater size and strength, though it is more likely that this has to do with division of labour within the family

(Passingham, 1982). To raise his children successfully, the man is usually obliged to assist in their care, and it is in his interests to establish their paternity. The alternative male strategy is to exploit sexual opportunities without providing parental care for any ensuing children. In stable societies in which the women are protected, the opportunities may be few, but it is evident that such opportunities are often taken during warfare or other social disturbances.

It is not possible to give an entirely satisfactory account of human sexual strategy from a purely evolutionary viewpoint. Humans have tremendous capacity for innovation, and many of the products of this become culturally established. Consequently, we see variation in sexual practice among different human societies. Nevertheless, it would appear that there are some universals. Statistical studies of urban man show evidence that active mating preferences (assortative mating) occur for physical traits like stature and for psychological traits such as intelligence and musical ability (Parsons, 1967). In a more recent study Buss (1989), and numerous collaborators, obtained data from 10 047 subjects from 37 distinct cultures, embracing wide variations in mating systems (polygamy/monogamy), religion, and race. They found remarkable uniformity; women seeking mates who are mature and affluent and men preferring younger, physically attractive women. There were some cultural differences. Thus in a society where food is in relatively short supply, plumpness in a woman is valued. In some societies virginity in a bride is considered essential, in others it is irrelevant.

22.3 Family relationships and altruism

It is the custom in all human societies that people marry, at least in the biological sense. Whether or not marriage involves some sort of ceremony or formal agreement is a matter that varies from culture to culture. In humans three patterns of association are prevalent (Passingham, 1982). The first is a partnership between a man and a woman which lasts for their lifetime, and excludes sexual relationships outside marriage. Such strict monogamy, though held up as the ideal in many societies, is probably not very common. The second is serial monogamy in which one person is married to several partners in succession. This may occur as a result of separation, divorce, or the death of a partner. The third is polygamy, which usually takes the form of one man having several wives.

Amongst sociologists there is some controversy as to whether the human family is natural or whether it is a purely cultural institution (see Eibl-Eibesfeldt, 1989, pp 184–187, for a discussion). Some sociologists take the view that showing preference for members of one's own family is an obstacle to extending friendship and responsibility to all. Others maintain that men and women only have separate roles within the family because of early upbringing and cultural pressure. One way to investigate such matters is to study early childhood development and socialization, as do developmental psychologists. Another way is to compare different primitive societies that have not been subject to modern ideology, as many anthropologists have done.

Most ethologists and developmental psychologists take the view that there is a special relationship between mother and child that starts at birth and is largely instinctive (Hinde, 1974; Blurton-Jones, 1975; Eibl-Eibesfeldt, 1989). Not only does the human child show special behaviour towards the mother from an early stage, as is the case in many other animals (see Chapter 20), but the mother shows distinctive behaviour towards her child. An example of this is **baby talk**. Mothers speak to their babies in a manner that differs considerably from normal speech, and this trait seems to be culturally universal (Eibl-Eibesfeldt, 1989). Four-month-old infants prefer to listen to baby talk than to normal speech (Fernald, 1985).

The basic repertoire of affectionate behaviour by mothers towards their children is identical across cultures, and much the same is true of fathers (see Eibl-Eibesfeldt, 1989, for a review). Thus there is little doubt that, at the behavioural level, the mother–child–father triad is the nucleus of the human **family** and society. The family is often extended to include grandparents and other relatives. This is a family type that is uncommon in other primates. Chimpanzees, our nearest relatives, live in territorial closed groups comprising several males, females and juveniles. The groups are structured hierarchically, and there are no long-term sexual partnerships. Chimpanzees are promiscuous, but a long-term relationship may persist between mother and offspring.

The human family is bonded together by love and affection, but sexual love within the family is rare. In all cultures **incest** is prohibited by an incest taboo, and for a long time this was thought to be a learned cultural adaptation, lacking any biological basis. Incest has genetically deleterious consequences, and is disruptive of society. Therefore those societies that could ban incest would seem to be better off. Doubts about this sociological theorizing began to emerge in the 1970s (see Bischof, 1975, for a review).

Incest inhibitions occur in a wide variety of animals, including some non-human primates (Bischof, 1975). Moreover, it had been observed long ago (Westermark, 1891; Ellis, 1906) that people who grow up with each other experience little mutual sexual attraction, and often a sexual aversion develops. Wolf (1970) confirmed these observations by comparing two kinds of Chinese marriage arrangement on the island of Formosa. In one kind the couple meet as adults and form a mutual sexual attraction. In the other, called a *sim-pua* marriage, the bride and groom meet in childhood and the bride is adopted by the groom's family and brought up like a sister. The *simpua* marriages are characterized by lack of sexual interest, and fewer children and higher divorce rate than the adult marriages.

Shepher (1971) studied kibbutz children brought up communally in age classes. Girls and boys live together, sharing showers and toilets, and show no embarrassment with the opposite sex up to the age of twelve. After twelve the girls avoided undressing in front of boys and showed the signs of modesty that the kibbutz leaders had sought to eradicate. After puberty children raised together developed friendly sibling-like relationships, but were not interested in each other sexually. Age–class members do not marry. Shepher (1983) investigated 2769 marriages between individuals raised in the kibbutz, and there was no instance of a marriage between people brought up in the same age–class. It appears that there is a sensitive period (see Chapter 20) up to the age of six or seven, during which the child learns which individuals are to be sexually excluded, a process called negative imprinting.

The extended family usually forms part of a group or tribe, the members of which have mutual obligations. The families which comprise a group are genetically related, and **kin selection** (see Chapter 8) must be an important factor in the evolution of group behaviour. Bonding and xenophobia (see below) tend to delimit the group, and both these are likely to be the result of kin selection.

The group is characterized by lasting interpersonal relationships which extend across family borders, and these are largely based upon some form of **reciprocal altruism**. Among the !Kung San of the Kalahari there are no chieftains. Decisions are made collectively and food is shared (see Chapter 24). Individual families are bonded together by a system of exchange interactions (see Box 22.1), which act as a form of social insurance during a time of emergency, but otherwise the families are self-sufficient (Eibl-Eibesfeldt, 1989).

Food obtained by animals is sometimes donated to others, especially kin and potential mates. Among humans, **food sharing** usually takes place according to certain rules. The successful hunter is considered to be the possessor, who is granted the right to divide the food. This system also occurs among chimpanzees (McGrew, 1975). Among the !Kung San meat distributed to the successful hunter's in-laws is often given in bulk to the father-in-law, who in turn redistributes the meat. Mothers give to children, and children often share or swap food items among themselves.

With the invention of tools it became important for harmonious group life

---

Box 22.1   The *hxaro* exchange system of the !Kung San

The highly reciprocal exchange system of the !Kung San was investigated by Wiessner (1977). Each member of the group develops a network of personal relationships based upon mutual obligations of gift giving. These *hxaro* partnerships are very durable, and may be passed on by parents to their children. On average each person has sixteen such partnerships, 18 per cent of which bind him/her to others within his/her own camp, 24 per cent to individuals in neighbouring camps, and 58 per cent to persons in two or three distant camps (up to 200 km). Each individual regards his/her total network of alliances as 'my people'.

The maintenance of an alliance is fairly onerous. The average amount of time spent making gifts for *hxaro* exchange is five working days per partner per year (e.g. the time taken to make or to earn the money to buy a blanket). The partner being visited also incurs costs. For the first two or three days the host is expected to provide for the visitor and his/her family. Thereafter the guests pay for themselves, hunting and gathering food in the locality, but not doing other work. Overall, an individual spends about three months of the year with a *hxaro* partner, over and above the time (5 × 16 days) taken to make the gifts. This is a substantial investment in the system.

Forty-five per cent of the exchange partners are not closely related. Thus the *hxaro* system is a form of social insurance that is only partly based on kin. During a partner's lifetime 55 per cent of the partnerships will be created, the rest being inherited. Thus *hxaro* partnerships are highly valued, long-term reciprocal relationships.

that respect for possession of objects, as well as food, be maintained. Respect of ownership develops early in humans (Bakeman and Brownlee, 1982), and the donation of objects, as gifts, is a culturally universal phenomenon, serving to establish friendly relations, or appease aggression (Eibl-Eibesfeldt, 1989).

Gift exchange fulfils a social rather than an economic function. Usually, there is an obligation to offer a gift modestly, an obligation to receive a gift graciously, and an obligation to reciprocate in one way or another. An obvious example is the gift exchange that occurs in some Western societies at Christmas time.

To some extent altruism seems natural to humans: we routinely help each other in time of danger, share food in times of shortage, and help the sick and wounded. We routinely share tools, and knowledge. We may do these things more often for relatives, but we often help people who we know are not related to us. From an evolutionary viewpoint, although reciprocal altruism can act as a form of social insurance, benefiting the group as a whole, it is open to cheating (see Trivers, 1985, for a discussion). Cheaters accept help but may not reciprocate, or may reciprocate little. Trivers suggests that our ability to calculate may stem from the need to remember who gave what to whom. Thus the human altruistic system is likely to be unstable. It is not surprising, therefore, that gift exchange among primitive peoples is often highly ritualized, as in the *kula* exchange system of the Trobriand islanders of New Guinea (see Eibl-Eibesfeldt, 1989, for a discussion).

22.4 Territorial behaviour and aggression

Xenophobia is a universal human quality (Eibl-Eibesfeldt, 1989). People who know each other are bonded by familiarity, whereas strangers are met with reserve. Human groups are relatively stable bounded entities, whether they occupy a single locality, or are nomadic. Group identity is delineated by dialect, dress and certain behavioural norms. Individuals who deviate from the norm are often subject to aggression.

Where a well-defined human group inhabit a particular patch of land, they come to regard it as their **territory**. As in other animals, territorial boundaries are protected in part by the mere knowledge of the possessors' readiness to defend the territory.

In Chapter 9 we saw that, in some animals, territories are formed only when a region is economically defensible. Dyson-Hudson and Smith (1978) have argued that the same holds for some human tribes, citing examples from North American Indians. However, in Australia tribes inhabiting the vast central arid regions maintain uneconomical territories by relying on mythical totemic ancestors to guard the boundaries. Each group derives its territorial claim from a totemic ancestor who once lived in the area and whose spirit still watches over it (Eibl-Eibesfeldt, 1989). Prominent landmarks associated with the mythical ancestor are sacred sites and only initiated males may visit them. Females and males from other groups are excluded and violators may be punished by death. The watchfulness of the totemic ancestor is said to bring misfortune to those who trespass. Thus the territorial integrity is maintained by a purely cultural means, and trespassers are deterred only if they share the belief-system of the owners.

The !Kung San of the Kalahari also inhabit vast tracts of arid land, and whether or not they are territorial is controversial (see Eibl-Eibesfeldt, 1989). However, the neighbouring !Ko, inhabiting more hospitable land, do seem to be territorial (see Box 22.2).

Territorial defence in animals usually involves **aggression** and threat, as well as sign stimuli marking territory boundaries (see Chapter 21). Aggressive behaviour in humans has been studied from the point of view of many disciplines, including ethology, sociobiology, social and developmental psychology, comparative psychology, and behavioural endocrinology (see Archer, 1988, for a review). In humans, as in many other mammals, aggressive behaviour appears quite early in development. It is a cultural universal (Eibl-Eibesfeldt, 1989). Its manifestation in adult life is strongly regulated by ritual, religion, law, and other cultural norms. In primitive peoples, at the individual level, aggression occurs as a result of disputes over possessions, sexual rivalry, and social dominance. When we consider the functional role of aggression in protection of self and offspring, in sexual rivalry and mate guarding, and in acquisition and defence of resources (see Archer, 1988), it is not surprising that aggression at the individual level in humans is much the same as in other species.

At the group level humans do seem to differ from other species. Although various forms of group defence do occur in animals (Edmunds, 1974), they do not usually involve organized aggression. In humans, **warfare** (armed conflict between groups) is widespread, and has a long history (see Figure 22.9). Even the normally peaceful !Kung San have a history of inter-group aggression (Woodhouse, 1987).

Human dispositions that are conducive to warfare include the following:

- loyalty to the group
- aggressiveness to outsiders that threaten the group
- dominance tendencies among males
- territoriality, regarding territory as a possession
- xenophobia, or fear of strangers

These dispositions, together with the complex social organization of humans, and their ability to plan ahead, might be thought to lead inevitably to inter-group aggression. This tendency is counteracted, however, by strong instinctive and cultural inhibitions against killing fellow human beings (see Eibl-Eibesfeldt, 1989, for a review), and it may be that in primitive man inter-group conflict was no more common than it is in other animals. The invention of weapons, however, totally alters the picture. Weapons enable humans to kill at a distance, and as a tool for killing prey they are invaluable. When deployed against fellow humans, weapons that act at a distance neutralize many of the signs and signals that normally inhibit people from killing each other. Once weapons are incorporated into inter-group aggression, then a new phenomenon materializes, namely warfare.

Box 22.2 The nexus territories of the !Ko

The social organization of the !Ko is based upon the family, the local group or band, and the nexus. A !Ko achieves access to territory through birth, adoption and marriage. Each family has a small patch of land near to their hut. They have no exclusive claim to the land, but their rights are respected by custom so that families can gain sufficient to live on during times of hardship.

A number of families band together and have jurisdiction of a territory with clear boundaries. Each band has a headman who exercises authority over the territory, and visitors must deal with him personally.

Several bands form a **nexus** (or maximal band),

the members of which regard each other as members of a larger alliance. The members of a nexus join together for ritualistic occasions, and maintain common security arrangements. In an emergency, members of a band can request permission to hunt within the territory of another band. Such requests are never refused to nexus members. Nexus territories are exclusive, and members of one nexus do not cooperate in this way with members of another. Nexus territories are usually separated by a strip of no-man's land, but this is not the case for territories within a nexus. The nexus system of the !Ko serves a similar function to the *hxaro* exchange system of the !Kung San (see Box 22.1).

Fig. 22.9 Depiction of warlike activities in South African rock paintings of Bushmen (From Bleek, 1930).

Points to remember

- Human behaviour is studied by anthropologists, ethologists, evolutionary biologists, psychologists, and sociologists.

- The non-verbal aspect of communication in humans invites direct comparison with the displays of animals.

- From the point of view of evolutionary theory we would expect both men and women to select mating partners that are likely to enhance their reproductive success.

- There is controversy as to whether the human family is natural or whether it is a purely cultural phenomenon.

- Where a well-defined human group inhabit a particular patch of land, they come to regard it as their territory. As in other animals, territorial boundaries are protected in part by the mere knowledge of the possessors' readiness to defend the territory, and in part by outright aggression.

Further reading

Eibl-Eibesfeldt, I. (1989) *Human Ethology*. Aldine de Gruyter, New York.

Animal competence

Profile of Karl von Frisch

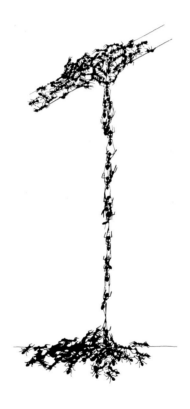

In this section we look at complex behaviour of animals, taking as an initial example the behaviour of honey-bees. This example, which takes up the whole of Chapter 23, shows how complex behaviour can occur in a relatively simple animal. In Chapter 24 we look at decision-making in animals from an evolutionary viewpoint and discover many affinities with consumer economics. In Chapter 25 we look at mechanisms of decision-making, in animals and robots, and see how animal competence can be achieved by simple mechanisms.

Karl von Frisch (1886–1983)

Photograph by courtesy of Maximillian Renner.

Karl von Frisch was born in Vienna and educated at the universities of Munich and Vienna. He became Professor of Zoology at the University of Rostock in 1921, at Breslau in 1923, and at Munich in 1925. As a child, Karl von Frisch was interested in natural history, and while still at school he published some of his observations, including experiments on the light sensitivity of sea anemones. Most of his subsequent research on animal behaviour was concerned with how animals obtain information about their environment. For much of his life von Frisch spent winters in his laboratory, studying fish, and summers at his family home at Brunnwinkl, studying honey-bees. He discovered that fish are capable of colour vision and of discriminating underwater sound waves.

Both these discoveries were contrary to the prevailing scientific opinion, and thus aroused opposition. Von Frisch also discovered that when the skin of a minnow is damaged, a pheromone is released that causes other minnows to flee from the area. In these studies von Frisch's success was based upon careful behavioural observation and a profound understanding of biological function. This was also shown in his work on honey-bees.

It was widely thought that bees were colour blind, but von Frisch reasoned that the colour of flowers must function to attract bees and other insects. He demonstrated that, although bees ignore the wavelength of light when escaping from a box, they are responsive to colour when foraging for food. During these experiments he noticed that a single 'scout' would appear at coloured food dishes set out in the open, but once the scout had departed it was only a short time before many bees arrived. This observation led von Frisch to the discovery of the bee language system.

In 1973 Karl von Frisch shared the Nobel Prize for Medicine with Konrad Lorenz and Niko Tinbergen. Although von Frisch's greatest material contribution was his work on honey-bee communication, he also stands out as a pioneer of the argument from design. Time and again he made important discoveries on the basis of his understanding of biological function.

The complex behaviour of honey-bees

In this chapter we examine the behaviour of a particular species in detail. This species is the honey-bee (*Apis mellifera*). It has been the subject of considerable scientific research, and quite a lot is known about its behaviour. Because the honey-bee is an insect, we tend to think of it as a mere automaton. However, its behaviour is surprisingly complex. By understanding the nature of this complexity we hope to gain some insights into the organization of complex behaviour in general. This will, perhaps, help us to understand the complex behaviour of other species, especially those more closely related to mankind.

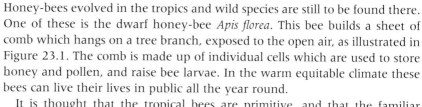

23.1   The honey-bee life cycle

Honey-bees evolved in the tropics and wild species are still to be found there. One of these is the dwarf honey-bee *Apis florea*. This bee builds a sheet of comb which hangs on a tree branch, exposed to the open air, as illustrated in Figure 23.1. The comb is made up of individual cells which are used to store honey and pollen, and raise bee larvae. In the warm equitable climate these bees can live their lives in public all the year round.

It is thought that the tropical bees are primitive, and that the familiar honey-bee (*Apis mellifera*) was able to colonize the temperate zones of the world by evolving the practice of building its comb within hollow trees. This enables them to gain protection from the changeable weather and to withstand very cold temperatures by massing together and generating metabolic heat in the insulated cavity. During the spring, when food becomes abundant, the queen bee lays many thousands of eggs. She is constantly attended by young female 'nurses' which groom and feed her. They are attracted by a special pheromone and as they groom the queen they ingest chemicals which suppress the activity of their ovaries, and prevent them from becoming

Fig. 23.1 Comb of the tropical dwarf honey-bee *Apis florea* (After a photograph in Gould, 1982).

possible rivals to the queen. The queen is capable of laying up to 3000 eggs per day. Each egg hatches two days after deposition and the resulting larva is fed by the workers. It receives food roughly once per minute for an entire week. The queen lays a few eggs in especially large cells prepared by the workers. The larvae in these cells are given special nourishment and develop into fertile queens rather than sterile workers. Of these virgin queens one becomes the new queen of the hive, but before this happens the old queen prepares for departure.

Once the new queen cells have been started, the workers tending the queen stop feeding her. Her egg production rate declines and she loses weight. When the queen has lost enough weight to be capable of flight, she starts to communicate with the virgin queens. She produces a pulsed tone by vibrating her thorax, and any mature virgin queens respond from their cells with a similar pulsed tone of higher frequency. In this way the queen indicates that she is ready to depart, and the virgin queens signal that they are ready to take her place.

The old queen departs from the hive together with about a half of the total population. The departing bees swarm together and form a compact cluster on a nearby tree, where they remain for several days. Scout bees fly off and investigate possible sites for a new hive. They return and perform a dance on the vertical surface of the swarm. This dance provides information about the direction, distance and quality of a new hive site. The dance also stimulates other workers to fly off and inspect the site. When they return they also dance in a way that indicates the quality of the site. A long and vigorous dance indicates an attractive site. Usually more than one possible site is initially reported, and each report recruits further scouts. Differences in the quality of the potential sites become apparent from the number of dancing bees and the vigour of their dances. Eventually the bees from the inferior sites cease to advertise their discoveries and universal agreement is reached. The swarm then moves to the chosen site and begins to build a new hive. The bees prefer sites, such as holes in trees, or in the ground, that are of a suitable size and are protected from climatic extremes. The new hive should not be too far for the queen to fly from her temporary resting place, nor too near the old hive.

When the old queen has left the new ones emerge from their cells. The first to emerge usually kills the ones still in the cells. If two emerge simultaneously they fight until one is killed. A few days later the new queen flies out of the hive to mate with several drones. She then returns to begin her egg-laying career.

The drones are few in number, and spend most of their time in the hive doing nothing. On sunny afternoons they may fly out to visit the traditional mating places in the locality. These are usually about 20–30 metres above ground and are simply air-spaces of about 20 m in diameter. The same localities are used year after year, even though the drones do not survive the winter and the new queens have not visited the mating area before. How the bees find these places is a mystery. The drones chase the arriving queens and the successful one mates with her, and dies soon afterwards. The vast majority of drones never mate and they are evicted from the hive when autumn comes.

In midsummer the bees start preparing for the winter. The colony size increases slowly, the workers devoting some time to raising young, but most of their time and energy is spent in collecting and storing food for the winter. Each worker takes on a series of tasks. Initially she cleans the cells, but as her mandibular glands develop she feeds the queen and her brood. When her wax glands become functional, she helps to cap cells and build the comb. At about three weeks old she begins to forage, and by about six weeks she is dead, usually from wear and tear.

23.2 Foraging by honey-bees

The foraging honey-bee, scouting for food, is faced with a formidable task. She must leave the hive and search for food. She must recognize suitable sources of food. She must then register her whereabouts in relation to the hive and make her way back to it. Upon arrival she must communicate her findings to other workers and persuade them to fly out and collect food from the newly discovered source. As with the bumblebees discussed in Chapter 24, this must all be accomplished in the most economical manner, taking account of the quality of the food and its distance from the hive.

Honey-bees forage primarily upon flowers that produce copious pollen and secrete a sugary nectar. Many plant species have flowers that are designed to attract bees and other insects, which act as vehicles for transporting pollen from one flower to another, thus cross-fertilizing the plants. Some flowers are open only at specific times of day, so the bees have to learn not only which kinds of flowers are producing nectar, but at what time and what place it is available. The first question we have to consider is: 'How do bees recognize suitable flowers?'

When a foraging honey-bee finds a source of food it may be a long way (up to 10 km) from the hive. The discoverer must return to the hive to inform other workers about the find. To do this she must use information about the direction of the hive, even though she may have arrived at her present position by a circuitous route. The second question that we shall consider is: 'How do bees navigate?'.

Upon arrival at the hive the forager must communicate to the other workers the direction and distance of the food source, and must give some indication of the quality of the food. To do this she must attract the attention of the other workers, who may already be engaged in other tasks, or may have already received messages from other foragers. Our third question is: 'How do bees communicate?'

23.3 Flower recognition by bees

In 1912 Karl von Frisch started his experiments on honey-bees. Contrary to the prevailing view, he reasoned that honey-bees were likely to have colour vision. Why else would flowers be so colourful? He found that the bees would quickly learn to visit a dish of sugar solution placed near the hive.

Fig. 23.2 Flowers that
appear to us to be white in
daylight (a) may have
distinct patterns (called
honey guides) when viewed
in ultraviolet light (b). Bees
are sensitive to ultraviolet
light and respond to the
honey guides.

(a)

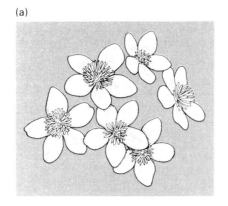

(b)

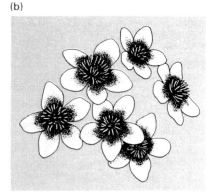

They would collect some of the food, and take it to the hive. Von Frisch then placed the dish on a coloured piece of paper. After a number of further visits by the bees, he set out many pieces of paper, some coloured and some various shades of grey. The bees searched for the food only on papers of the appropriate colour, thus demonstrating that they could distinguish the colour from the shades of grey. Von Frisch also discovered that the bees were unable to tell red from grey, but that they could distinguish between grey paper made by different manufacturers. Further investigation showed that some sheets of paper reflected more ultraviolet light than others and that the bees were sensitive to this.

We now know that honey-bees have well developed colour vision that differs from human vision in being insensitive to red, but extends into the ultraviolet where humans are totally insensitive. Von Frisch discovered that many flowers have well developed markings, called honey guides, some of which are visible only in ultraviolet light (see Figure 23.2). These markings are thus normally invisible to humans but visible to bees.

When honey-bees are foraging they readily alight upon coloured flower-like shapes. They can be trained to alight upon particular shapes to obtain food, but they prefer to alight on shapes with a broken outline and a radial pattern. Experiments with honey guides indicate that small details of flower pattern can influence the behaviour of honey-bees (Manning, 1956). In one group of orchids, the bee orchids, each species mimics the odour and appearance of a different species of bee. The male polinates the flower while attempting to copulate with it (Baerends, 1950).

23.4 | Honey-bee navigation

When a bee flies out from the hive in search of a new food source, it takes a circuitous route as it visits various possible feeding localities. It flies home by a direct route without retracing its outward path. It has been suggested that the bee keeps track of each leg of its outward path by measuring the distance in terms of the energy expended and the direction by reference to the angle with respect to landmarks, and to the sun.

By requiring bees to walk to a source of food along a gallery (Figure

Fig. 23.3 Martin Lindauer's experiment to investigate the energy expelled by bees during the flight from the hive to a food source. In (a) the bee has a long return course, while in (b) it has a short one. In both cases the outward course is the same, and when the bees returned to the hive their dance indicated this.

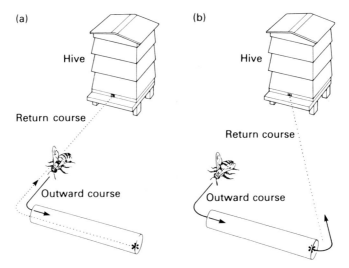

23.3), Martin Lindauer (1963) was able to manipulate the distance of the return flight. He found that the bees correctly gauged the energy expended in the outward flight after both long and short return flights. Von Frisch showed that bees fitted with 55 mg lead weights, or with tinfoil flaps that increase drag, overestimate distances because of the extra energy they expend.

In investigating the importance of landmarks, von Frisch and Lindauer (1954) trained bees to obtain a honey reward by flying in a particular direction. In one experiment the flight path was along the north–south edge of a pine forest. When the bees had been trained, they were tested near the edge of a similar forest, the edge of which ran east–west (see Figure 23.4). Most bees followed the line of the plantation, only a few taking the correct southerly course. Landmarks are most effective when they are linear and lead directly to the food. Trees in the middle of a field may be ignored (see Figure 23.5), even though they would appear to be useful landmarks.

Once a scouting bee finds a food source it flies directly home. A simple experiment shows that they use the sun as a compass (see Chapter 14). Bees are trained to a feeder, and then the position of the feeder is moved while some of the bees are feeding. When they depart for home, these bees fly in the direction that would have been correct if the feeder had not been moved. If the bees are trapped in the feeder for a period of time long enough for the sun to move appreciably, they still fly in the correct direction. This suggests that they have a time-compensated sun compass (see Figure 23.6). Other work (see Box 23.1) indicates that insects can learn about the sun's course (the solar ephemeris) even though they may have experience of only part of it. How they might do this is discussed in Chapter 25.

Bees may often find themselves in a situation in which there are no suitable landmarks and the sun is not visible because it is behind clouds. Under such circumstances they can still fly in the homeward direction, so they are obviously capable of using some other navigational cues. As we have seen

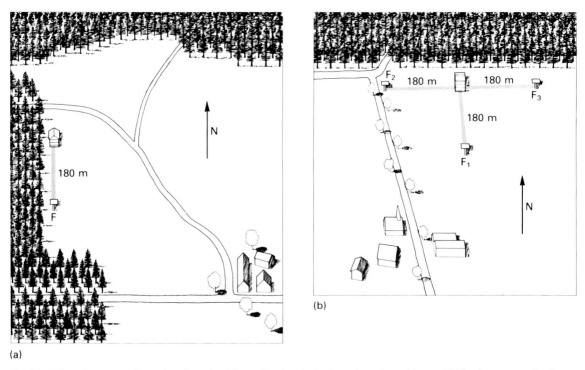

(a)

(b)

Fig. 23.4 Experiments to determine the role of linear landmarks in the orientation of bees. (a) The bees are trained to visit a food table F along the north–south edge of a pine plantation. (b) They are then tested near an east–west edge, where they were given a choice of three food tables F_1, F_2 and F_3. Most bees chose to fly east–west, along the edge of the wood, even though they had been trained to fly north–south (After Lindauer, 1961).

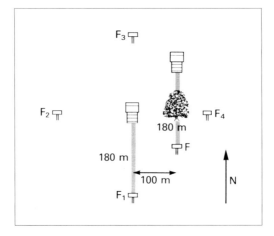

Fig. 23.5 Bees trained to fly past the tree to the food table F, ignored the tree and flew due south when the hive was placed west of the tree (After Lindauer, 1961).

(Chapter 14), bees are sensitive to the plane of polarization of sunlight in the ultraviolet region of the spectrum.

Von Frisch made an octagonal filter from eight pieces of triangular polaroid, as shown in Figure 23.7. When he looked through it at different parts of the sky he saw different patterns of brightness even when the sun

Fig. 23.6 When bees are trapped at the feeder for a few hours, and the feeder is moved during this time, they fly in the correct direction when released. This shows that they can compensate for the change in the sun's position with time.

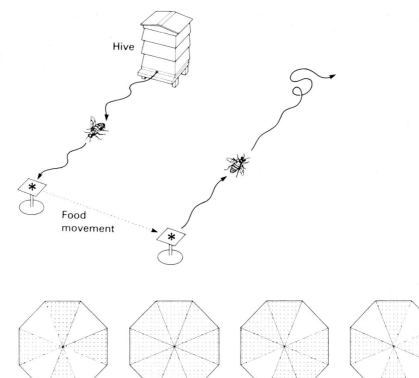

Fig. 23.7 Four different brightness patterns seen at the same time of day when different parts of the sky were viewed through an octagonal filter made of eight pieces of polaroid. Depth of shading indicates brightness (From *The Oxford Companion to Animal Behaviour*, 1981).

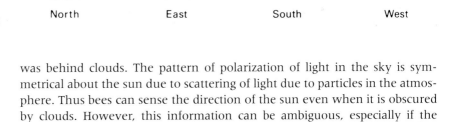

was behind clouds. The pattern of polarization of light in the sky is symmetrical about the sun due to scattering of light due to particles in the atmosphere. Thus bees can sense the direction of the sun even when it is obscured by clouds. However, this information can be ambiguous, especially if the animal has a restricted view of the sky, as primitive bees have in the African forest.

Only a small patch of sky has to be visible to the bee, but there will usually be two patches that look the same, and are positioned symmetrically with respect to the sun. The bees overcome this problem by assuming that the patch they see is the one to the right of the sun. This convention will sometimes lead to errors, but if all bees consistently use the same convention, then the errors will tend to cancel each other. The problem can also be overcome by knowing where the sun ought to be at each time of day. Like many other insects, bees have an endogenous clock which is entrained by a zeitgeber (see Chapter 16). It has been suggested that, since worker bees spend much of their time in the darkness of the hive, the zeitgeber is not provided by the time of sunrise or sunset, as it is in some animals. Instead it may be that daily changes in the earth's magnetic field provide the necessary information. Bees are known to be sensitive to magnetic fields, and during magnetic storms their time sense is disrupted (Gould, 1980).

Box 23.1 Insects can learn about the sun's course

Lindauer (1959) reared bees in an incubator, allowing them to see the sun only in the afternoon. During a daily flight period he trained them to fly south to find a feeding station. After training he moved the experiment to a new location, so that the bees could not rely on local landmarks and could only rely on the sun as a navigational aid. He tested the bees in the morning, at a time when they had never seen the sun. He found that the bees flew predominantly to the south, which suggests that they had some information about the most likely morning position of the sun. Subsequent work has shown that a variety of insects can estimate the sun's position at a time when they have never experienced it. Thus bees (Lindauer, 1957; Dyer, 1985) and desert ants (Wehner, 1982) can estimate the sun's position at night.

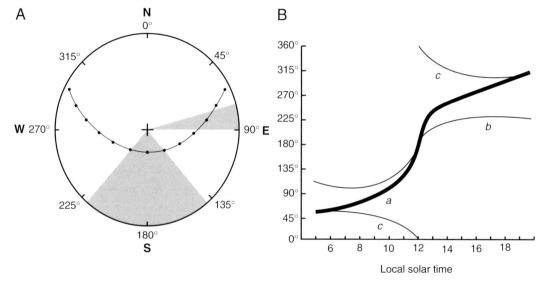

A. The sun's course on the summer solstice in East Lansing, Michigan (≈43° N). Symbols show hourly positions of the sun in the sky from 0500 to 1900 h. The sun moves at a constant rate (15° h⁻¹) along its arc, but the rate of change of the azimuth varies over the day as the sun rises toward the zenith (cross) and then descends again. Shaded sectors show the change in azimuth over two equal (two-hour) periods. The rate of change is 9.5° h⁻¹ in the morning and 37.5° h⁻¹ during the period spanning noon.

B. Alternative method for displaying change in azimuth over the day (the solar ephemeris). Heavy line (*a*) is plot of sun's course shown in A. Also shown are the ephemerides at the equator on the December solstice (*b*), when the azimuth shifts clockwise (left-to-right) across the southern horizon, and on the June solstice (*c*), when the azimuth shifts counter-clockwise across the northern horizon (From Dickinson and Dyer, 1996).

23.5 Communication among honey-bees

Von Frisch constructed a hive with a glass wall through which he could observe the behaviour of the bees inside the hive. He noticed that returning foragers performed dances which attracted the attention of other bees. He identified two types of dance, a round dance (Figure 23.8), and a waggle

Fig. 23.8 The round dance.

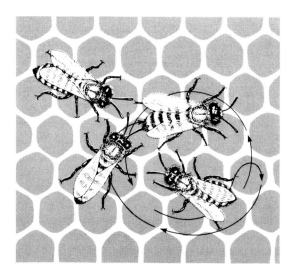

Fig. 23.9 The waggle dance.

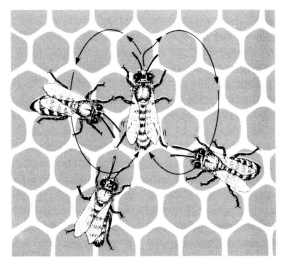

dance (Figure 23.9). At first he thought that the round dance indicated nectar and the waggle dance indicated pollen. Later he discovered that this interpretation was incorrect. He found that foragers returning from food sources differing in distance and direction from the hive performed waggle dances that differed in their details.

The bees perform the dance on vertical sheets of comb in the darkness of the hive. The angle between the axis of the dance and the vertical (Figure 23.10) corresponds to the angle between the direction of the food and the direction of the sun. As the sun moves west the dances rotate counterclockwise. The duration of the waggle portion of the dance corresponds to the distance of the food from the hive (Figure 23.10). The round dance is simply a waggle dance performed to indicate food that is so close that no waggles are necessary. Different geographical races of bees have differing dance dialects. The more primitive tropical honey-bees dance on the horizontal surface

Fig. 23.10 The waggle
dance of the honey-bee. The
angle between the axis of
the dance and the vertical is
the same as the angle
between the food source and
the sun.

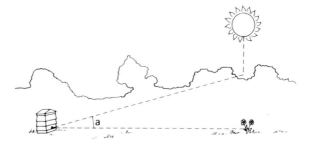

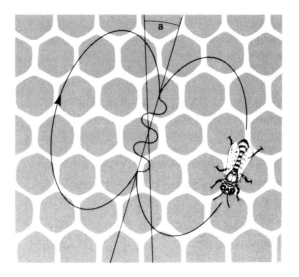

formed by the top of the comb. The axis of the dance points directly in the
direction of the food source. This is also what happens when temperate zone
honey-bees are forced to dance on a horizontal surface.

The returning scout attracts other workers by a display (Figure 23.11)
during which she fans her wings and releases a recruitment pheromone.
However, this is done only if the food source is of worthwhile quality. The
value of the food is judged by the forager in relation to its distance from the
hive, and to the quality of alternative sources of food. The greater the distance of a food source the sweeter it must be to elicit recruitment and dancing. In spring and early summer, when food is generally plentiful, a food
source must be really sweet to elicit recruitment. In late summer and fall,
when food becomes scarce, even low quality food will result in recruitment.
How the other workers obtain information from the waggle dance is not
entirely understood. The returning forager brings back traces of the scent of
the flowers she has visited. The other workers crowd around the dancing bee
and pick up the scent. They quickly learn the food odour and make use of
their memory of it when they arrive near the feeding place. In addition to
odour cues, the bees probably utilize sound cues emitted by the dancing bee
in the darkness of the hive. During the waggle portion of the dance, the bee
emits a sound that has a pulse rate approximately 2.5 times as fast as the rate
of waggling (Wenner, 1962, 1964). This sound, probably in addition to the

Fig. 23.11 The recruitment
display of the honey-bee
(After a photograph by
Gould, 1982).

rate of waggling as felt by the workers crowding around, provides information about the distance of the food source.

It has been suggested by some that the dance is not important in recruiting foragers, and that this is done primarily by means of olfactory cues, as do some other social insects. Much of von Frisch's early experimental data is consistent with this interpretation. However, although honey-bees rely upon odour to locate food under certain circumstances, the more recent experiments of von Frisch and others (Gould, 1976) provide convincing evidence that the dance is the primary system used to communicate the location of food sources. For example, the sensitivity to light of foraging bees can be reduced eight times by painting their ocelli black. The **ocelli** are the three simple eyes on top of the bee's head. If the vertical comb on which the dance takes place is provided with artificial sunlight which the returning ocelli-blackened scouts cannot see, then their dances are orientated with respect to gravity, just as they would normally be. However, the unpainted recruits respond to the artificial sun and this gives the experimenter a means of manipulating the recruits' interpretation of the dance. By altering the position of the artificial sun the recruits can be sent off in a direction determined by the experimenter. This experiment shows that the recruited workers were relying upon the dance to obtain information about the distance and direction of the food.

James Gould (1976) has suggested that honey-bee communication has taken so to long analyse, and has been so controversial, because of its complexity. Not only do the scouts have a sophisticated dance routine, but if deprived of the opportunity to use this type of communication, they can often fall back on the odour system of recruiting foragers. The polarization pattern of the sky can be used instead of direct sunlight. The direction of gravity can be used as a substitute for the sun's direction by bees dancing in the darkness of the hive. The bees' endogenous clock can provide a substitute for the movement of the sun, when the bee is inside the hive, or entrapped by an experimenter. It seems that the bee has an answer for every contingency. As we shall see, this kind of contingency planning is the key to complex behaviour in apparently simple animals.

23.6 The organization of complex behaviour

In temperate climates foraging honey-bees are faced with a variable environment. As the bee leaves the hive on a scouting trip, the sun may be visible, or obscured by clouds. The flowers that were previously rich in nectar may be closed, or dead, or removed by some animal. The odours remembered from the last trip may no longer be present, or may be mixed in with a whole galaxy of new odours.

We can imagine a computer program designed to cope with contingencies such as those that face a foraging honey-bee. As the bee flies out on a scouting trip the program asks a series of questions, each of which is based upon the answer to the previous question. The final outcome is contingent upon the outcome of the previous stage, and so on. This kind of programming can be incorporated into a simple logic network that is specifically designed for a particular task. To some extent bee behaviour appears to be organized in this way.

However, honey-bee behaviour is more complex than a simple IF–THEN flow chart. Bees are capable of adapting to new circumstances by learning. Randolf Menzel investigated how bees learn the colour of the food source by changing the colours of artificial flowers during the approach, landing, feeding and departure phases of a visit. He found that the bee learns about the colour of a flower during the final two seconds before landing on it (Menzel, 1978). The bees learn about the location of landmarks only as they fly away from a flower. If the landmarks are removed during feeding, and replaced when the bee has departed, the bee will not be able to remember them even though they were present when she initially approached the flower. Thus it appears that colour and landmark learning are closely tied to particular aspects of behaviour. Similarly, the bees learn the location of the hive as they depart from it each day. Day-to-day changes in the appearance of the hive, or its surrounding vegetation, do not trouble them. However, if the hive is moved a few feet while bees are out foraging, they have great difficulty in locating it when they return.

Von Frisch showed that bees are able to learn flower odours extremely rapidly, and to distinguish one from among 700 others. The colour, shape, location, and time of opening of flowers are learned progressively less readily. If the scent of a familiar artificial flower is changed, then the bee rapidly learns the new scent, but the colour, shape, etc., which remain unchanged, nevertheless have to be learned all over again (Menzel *et al.*, 1974). In other words, foraging bees appear to learn about the characteristics (scent, shape, colour, location, etc.) of flowers as a package. If the scent is changed experimentally then the whole package has to be relearned. Thus it seems that learning in bees is preprogrammed in the sense that specific types of learning take place in particular situations, which are characterized by the bees' behaviour, such as leaving the hive, approaching a flower, leaving a flower, etc.

To explain honey-bee behaviour we do not have to invoke any special mental powers or cognitive abilities. We are tempted to account for it in terms of sets of procedural rules. This may prove complicated if we are to encompass the animal's total behavioural repertoire. It is always possible, moreover, that some feature of honey-bee behaviour will be discovered that will force us to abandon the procedural mode of explanation. We discuss this issue in Chapter 25.

Many vertebrates are capable of behaviour that appears to be much more complex than that of honey-bees. At the same time, much of vertebrate behaviour is rather stereotyped. It may be that the mental abilities of vertebrates are overestimated, and there is not such a big gap between them and honey-bees. In the chapters that follow we shall be addressing this kind of question in various ways. In considering the more sophisticated aspects of vertebrate behaviour, we will have at the back of our minds the complex behaviour of honey-bees.

Points to remember

- Honey-bees provide many examples of complex behaviour in a seemingly primitive animal.
- Their foraging involves sophisticated feats of navigation, including landmark recognition, the use of the sun as a compass and the use of the polarization of light in the sky and of aspects of the earth's magnetic field.
- In recognizing flowers bees make use of their sense of smell, colour vision and visual-pattern detection.
- In communicating their findings to others, bees can indicate the distance and direction of food sources.
- The evidence suggests that the complex behaviour of bees is the result of the systematic application of simple programming rules.

Further reading

Dyer, F.C. (1994) Spatial cognition and navigation in insects. In *Behavioral Mechanisms in Evolutionary Ecology*. Real, L.A. (ed.) University of Chicago Press, Chicago, IL.

Frisch, K. von (1967) *The Dance Language and Orientation of Bees*, Belknap, Cambridge, MA.

Lindauer, M. (1961) *Communication Among Social Bees*, Harvard University Press, Cambridge, MA.

Animal economics

John Goss-Custard (1977a) studied foraging in the redshank (*Tringa totanus*), a wading bird that hunts for food along the seashore and on mudflats. He found that when these birds are feeding exclusively on polychaete worms (*Nereis diversicolor* and *Nephthys hombergi*), they tend to pass over the smaller worms and to select those over a certain size. Their size preference is influenced by the rate at which they encounter the larger worms but not by their encounter rate with small worms. This finding is consistent with the view that the redshank's foraging strategy is designed to maximize energy profitability; that is, they select those worms that provide the greatest amount of energy per unit of energy expended on foraging. The smaller worms are not so profitable because of the relatively low net energy returns on time spent foraging for them. Taken at face value, these results might suggest that the redshank make decisions about which prey to take on the basis of energy trade-offs. However, Goss-Custard (1977b) also found that when the amphipod crustacean *Corophium* was available in addition to polychaete worms, the birds tend to select *Corophium*, as illustrated in Figure 24.1. He was able to discount the possibility that the habitat typical of *Corophium* was one in which polychaete worms were harder to find because he observed that some birds concentrated on worms while the majority was feeding on *Corophium*. On the

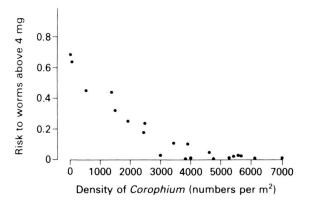

Fig. 24.1 Index of the risk of worms being eaten by redshank, in relation to the local density of *Corophium*. Each point represents the average for a particular study site (After Goss-Custard, 1977b).

Table 24.1 Comparison of the rates at which redshank obtained energy in three sites where they mainly took *Corophium* with the rates they would have achieved by taking worms instead (After Goss-Custard, 1977a)

Site	Rate of ingesting energy (cal min⁻¹)	
	Potential rate from *Nereis* alone	Actual rate from mainly *Corophium*
9	234	88
10	224	70
11	185	93

Table 24.2 The effort expanded in three sites on collecting 1 kcal by birds feeding mainly on *Corophium* as they actually did, and on *Nereis* alone (After Goss-Custard, 1977a)

Site	Distance searched in metres		Number of pecks and probes made		Time spent swallowing prey (s)	
	Corophium	*Nereis*	*Corophium*	*Nereis*	*Corophium*	*Nereis*
9	103	42	470	165	62	48
10	150	44	671	167	121	49
11	106	56	543	198	79	48

basis of previous work, Goss-Custard had formed the hypothesis that redshank achieved a higher rate of net energy intake by feeding on *Corophium* than by taking worms. However, when he came to analyse the energy content of the prey (Table 24.1) the energy costs of obtaining prey (Table 24.2), Goss-Custard found that the birds would have obtained between two and three times more energy per minute by taking worms exclusively than they obtained by feeding on *Corophium*. Clearly, energy was not the only factor relevant to foraging redshank when *Corophium* was available. Presumably, the *Corophium* contain something other than energy that is important to the redshank.

This example illustrates the complexity of foraging behaviour (see also Chapter 9). In order to find out what is happening in this type of situation, we need to understand certain **economic** concepts. These are the subject matter of this chapter.

24.1 Functional aspects of decision-making

There are two basic principles of economic decision-making. The first is that it must be rational, and the second is that it must involve some evaluation of the pros and cons of the situation. The basis of **rational decision-making** is that it should be self-consistent, a property that is usually called **transitivity of choice**. Suppose a person has to choose between options A, B and C. If A is preferred to B, and B to C, then it is rational to expect that A will be preferred to C. We can now write a consistent order of preference, A>B>C. These relationships among A, B and C are said to be transitive. If A>B and B>C, but A is not preferred to C, then the decision to choose C above A is irrational, and the relationships among A, B and C are intransitive.

Economists base their theory upon the concept of the rational economic person, and this implies that all preference relationships are transitive (Edwards, 1954).

Rationality does not necessarily imply the use of reason. People may make some decisions as a result of reasoning, but they may also make rational decisions in a purely automatic way, as if designed or programmed to do so. People also make irrational decisions. In fact, it is impossible to prove that human choices are transitive because to do so would involve repeated choice experiments under identical circumstances. This is not possible because circumstances are never exactly the same twice, if only because the memory of having made one choice changes the circumstances for the next (Edwards, 1961). Economists have to take transitivity of choice as a working hypothesis, an assumption upon which the elementary theory is based. There is, however, good experimental evidence that young children and monkeys do make transitive choices in behavioural tests, but that this ability probably does not involve reasoning (McGonigle and Chalmers, 1986). Thus it may be that rational behaviour is a common feature of animal behaviour.

Transitivity of choice implies that something is maximized in the decision-making process. To see that this must be so, let us consider the following situation. Suppose A, B and C can be evaluated numerically in some way. If A has a higher score than B, then we write A>B. A will be chosen over B by a person using a maximization principle (like choose the option with the larger score). If we know that B is larger than C, we can write B>C. If C is chosen over A, it would appear that C has been allocated a higher score than A, but we know that A>B>C, which implies that C has a lower score than A. Thus, if C is chosen over A, C must be preferred even though it has a lower score than A. A person making choices on this intransitive basis could not be choosing the option with the largest number of points; such a person could not be using a maximization principle. If a person's preferences are transitive, however, then we can deduce that he or she is using a maximizing principle, although the person may not be aware of it.

The name given to the quantity that is maximized in the choice behaviour of the rational economic person is **utility**. I may obtain a certain amount of utility from buying china for my collection, a certain amount from sport and a certain amount from reading books. In spending my time and money on these things, I choose in a way that maximizes the amount of satisfaction or utility that I obtain in return. I am not aware of maximizing utility, but (if I am rational) I appear to behave in a way that maximizes utility. Thus, utility is a **notional** measure of the psychological value of goods, leisure, etc. It is called a notional measure because we do not know whether or not it comes into people's choice behaviour. We only assume that they behave as if utility is maximized.

The equivalent of utility in animal behaviour is **benefit** (negative cost). Just as utility is a notional measure of the value of behaviour, so cost is a notional measure of the change in fitness that is associated with an animal's behaviour and its internal state (see Figure 24.2). We do not know whether or not cost enters directly into the decision-making processes of animals. We

Fig. 24.2 Parallel concepts in economics and ethology: (above) general scheme showing the most important contributors (with the dates of their major works); (below) detailed scheme as applied to the individual person or animal (McFarland and Houston, 1981).

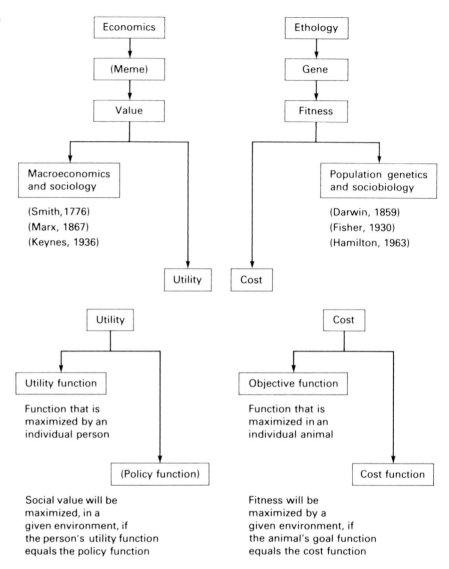

can only test the hypothesis that animals behave in a way that maximizes benefit (minimizes cost). The components of cost include factors associated with the behaviour occurring at a particular time and risks associated with the animal's internal state. Thus, as we see below, the incubating gull incurs physiological cost in keeping the eggs warm, and it incurs costs as a result of its increasing hunger while on the nest. The various costs can be combined to form a **cost function**, which evaluates every aspect of the animal's state and behaviour in terms of its associated cost. In economics, the equivalent (but inverse) function is called a **utility function** (see Figure 24.2). At this point, it may be helpful to analyse a particular example of animal behaviour in economic terms to see how the concept of utility, and other economic concepts, can be useful in the study of animal behaviour.

24.2 The animal as an economic consumer

	Worms	Shrimps
Energy	7	3
Nutrient	7	3

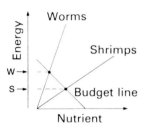

Fig. 24.3 Consequences of eating worms and shrimps in terms of energy and nutrient gained. The budget line is based upon the energy prices of worms and shrimps used in this model (After McFarland and Houston, 1981).

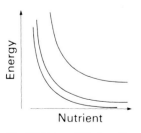

Fig. 24.4 Hypothetical utility functions for nutrient and energy.

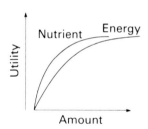

Fig. 24.5 Iso-utility functions for nutrient and energy, based upon Figure 24.4.

We saw above that redshank foraging on mudflats prefer *Corophium* to polychaete worms, even though they obtain much less energy from *Corophium*. We conclude that the *Corophium* must provide something other than energy that is of importance to the redshank. What might an economist have to say about this situation?

To simplify matters, we will refer to the redshank prey as worms and shrimps, and we will assume that each prey provides a certain amount of energy and a certain amount of some unknown nutrient. The energy and nutrient are present in different proportions in the two prey, as shown in Figure 24.3. This means that the consequences of eating two prey are different and can be portrayed as different arrows in a diagram representing the possible consequences of choice behaviour.

The economist would ask next about the relative price of worms and shrimps. The redshank have to expend more energy to obtain shrimps than to obtain the equivalent energy return from worms. The price of shrimps in energy terms is about twice that of worms. Figure 24.3 shows how many worms and shrimps a bird could obtain for a given amount of energy spent on foraging. This is equivalent to the amount of goods A and B that a shopper could obtain for a given amount of money. The amount of energy an animal has at the time of foraging acts as a constraint on what it can purchase. The energy obtained by foraging is not immediately available for use because the food has to be digested, just as the money a person earns by working is not immediately available to spend. The budget line in Figure 24.3 represents the constraints on foraging that are imposed by the bird's available energy. This means that, within a given period of time, the bird cannot obtain more worms and shrimps than the quantities represented inside the triangle formed by the budget line.

The next question is what utility does a redshank derive from energy and nutrient? The utility of such commodities usually obeys a law of diminishing returns; that is, if an animal already has a good supply of energy, a little extra does not add much to the utility. If the animal is short of energy, however, then that same small amount of energy will make a large contribution to utility (see Figure 24.4). The same considerations commonly apply to economics. Thus, I may derive a certain amount of satisfaction or utility from my large china collection. If I add one more piece of china to my collection, then I will increase my satisfaction by a small amount. If I had a smaller collection, however, then adding that same piece would increase my satisfaction by a larger amount.

Figure 24.4 shows hypothetical utility functions for nutrient and energy, and Figure 24.5 shows how these functions can be combined to give a set of **iso-utility curves**. These curves join all points of equal utility, and Figure 24.5 shows that the redshank may obtain the same utility from a large amount of energy combined with a small amount of nutrient as from a small amount of energy combined with a large amount of nutrient. The shape of the iso-utility curve is determined by the shape of the corresponding utility

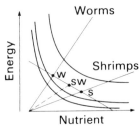

Fig. 24.6 Iso-utility functions from Figure 24.5 superimposed on Figure 24.3. The dotted line indicates the best possible mixture of worms and shrimps, assuming that there is no cost of changing between them (After McFarland and Houston, 1981).

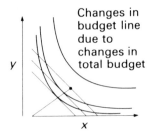

Fig. 24.7 Different budget lines result from differences in the amount of energy initially available to the animal. Note that the optimal preference does not change.

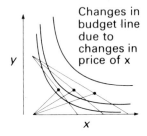

Fig. 24.8 Changes in the budget line due to changes in the price of x. Note the changes in optimal preference.

functions shown in Figure 24.4. In terms of the economic analogy, I might derive the same total utility from purchasing two pieces of china and one tennis ball as from one piece of china and six tennis balls. For this reason, economists often call iso-utility curves **indifference curves**.

Suppose we now combine Figures 24.3 and 24.5, as shown in Figure 24.6. We can now see that certain combinations of worms and shrimps will yield higher utility than others. The least utility is obtained by foraging for worms alone (w), more is obtained from shrimps alone (s), but even more is obtained by a certain combination of worms and shrimps (sw). Two main factors affect the combination that yields the highest utility: the shape of the iso-utility curves (determined by the utility functions) and the slope of the budget line (determined by the relative process of the two commodities).

In consumer economics, if a person has more money to spend, then more goods can be purchased and there is a parallel shift in the budget line, as illustrated in Figure 24.7. If the price of a particular item is lowered, then more can be purchased for the same amount of money, and this also has the effect of making the budget triangle larger (see Figure 24.8). However, a price change for a single item alters the slope of the budget line, which does not occur if the overall budget is increased. Comparison of Figures 24.7 and 24.8 shows that there is little change in the optimal preference when the overall budget is altered but that changes in price cause a considerable shift in optimal preferences.

In likening an animal to an economic consumer, we regard cost or utility as the common currency of decision-making. We distinguish between these concepts and the energy cost or price associated with particular activities or goods. Although energy cost is a common factor in the alternative possible activities in the examples we have discussed so far, this is not always the case. It is a mistake, therefore, to regard energy as a common currency for decision-making even though it may be a common factor in particular cases. Energy will be a factor common to all alternatives only if energy availability acts as a constraint in the circumstances being considered. An economic example may help to make this distinction clear.

Suppose a person enters a supermarket to purchase a variety of goods. As we have seen, utility functions will be associated with each possible item of purchase, and the utility of particular goods will depend upon the shopper's preferences and experience. Factors such as the palatability, visual attractiveness, list of ingredients, etc. will influence the utility associated with food items. The concept of utility enables very different items to be compared. A person may obtain the same utility from a packet of mixed herbs as from a bottle of soda. Utility is thus the currency by which different items are evaluated. However, another currency common to all items is money. This will be important if the shopper has a limited amount to spend because it acts as a constraint upon the amount he or she can purchase. However, although every item in the supermarket has a marked price, money is not always the relevant constraint. A shopper who is in a great hurry may carry more money than can be spent in the time available. Time is thus the constraint that bites, and it may be that the shopper can choose among bottles of soda more quickly than among different packets of mixed herbs. Time taken to choose each item will determine the slope of the time-budget line, and it may turn out that the shopper's choice when pressed for time is different from that of the shopper with limited cash.

The lesson to be learned from this example is that many possible constraints act upon an animal's choice behaviour. These constraints may vary with circumstances, but it is only the constraint that bites at the time that is important. For this reason, time and energy should not be confused with utility or with cost (in terms of fitness). In drawing the analogy between an animal and an economic consumer, cost is equivalent to utility, energy is equivalent to money and time is equivalent to time. The animal can earn energy (money) by foraging (working) and may spend it upon various other activities. Over and above the basic continuous level of metabolic (subsistence) expenditure, the animal can save energy (bank money) by hoarding food or depositing fat, or it can spend it upon various activities including foraging (working). When the price of activities is high, the animal is subject to a tight energy-budget constraint, and when the price is reduced the animal experiences an increase in real income and the budget constraint is relaxed (McFarland and Houston, 1981).

24.3 Time and energy budgets

If we have a limited amount of money to spend on a variety of goods and activities, we often partition it among the different purposes. We usually review our expenditure over a particular period of time such as a day, a week or a year, and we call this a budget. The budgeting process may occur in advance of any expenditure, or it may occur in retrospect. In the one case we allocate particular sums to particular purposes, while in the other we review the expenditure that has already occurred. In either case, the notion of budgeting implies a certain discipline in spending money.

As we have seen, money in consumer economics can be seen as analogous to energy in animal behaviour. It is natural, therefore, to ask whether or not animals have energy budgets. However, we have seen also that energy is only one type of constraint that impinges on animal behaviour. Another important constraint is time, and we can also ask whether time budgets are relevant to animal behaviour. A time or energy budget should not simply be an account of how an animal spent its time or energy. An animal whose use of time and energy was completely chaotic would have no budget. However, we can expect that natural selection will design animals in a way that their available time and energy is put to maximum use. It seems reasonable, therefore, to expect that animals will treat time and energy as valuable resources and will budget accordingly.

Bernd Heinrich (1979) likened the foraging bumblebee (*Bombus*) to a shopper:

'A bee starting to forage in a meadow with many different flowers faces a task not unlike that confronting an illiterate shopper pushing a cart down the aisle of a supermarket. Directly or indirectly, both try to get the most value for their money. Neither knows beforehand the precise contents of the packages on the shelf or in the meadow. But they learn by experience.'

The bees are dependent upon flowers for the energy required to rear the young, but they may have to expend considerable amounts of energy in foraging. Bumblebees are able to exist in cold climates by virtue of their remarkable themoregulatory physiology. They can maintain a high body temperature at a low environmental temperature. This enables them to be active, although it involves a high rate of energy expenditure. They can conserve energy by greatly reducing their activity level and conserving heat. When food resources from flowers are scarce, the bees nevertheless manage to make a profit by foraging slowly. When food is abundant, they raise their body temperature and forage rapidly. Thus, they budget their energy expenditure in accordance with the prevailing circumstances.

A foraging bumblebee spends most of its time travelling. In moving from flower to flower, bees try to keep their flight time and distance to a minimum. They fly between 11 and 20 km h^{-1} and spend only 2–4 minutes inside the nest between foraging trips. Simple calculations (Heinrich, 1979) show that the time budget is more important than the energy budget for a foraging bumblebee. Suppose, for example, that one bee has flowers close to the nest and can forage there continuously, while another bee is foraging 3 km from the nest. If the second bee flies at 15 km h^{-1}, it must spend 24 minutes travelling time per trip. Foraging on fireweed, both bees could collect a honeycropful (about 30 milligrams of sugar) of nectar in about 10 minutes. The commuting bee would thus collect 30 milligrams of sugar in 34 minutes but would expend about 3 milligrams in flight metabolism. Thus, commuting takes up about two-thirds of the time per trip but only one-tenth of the energy. The bee foraging close to the nest would obtain 102 milligrams of sugar in a foraging trip of the same duration. Thus, to make commuting worthwhile, the flowers far away from the nest would have to be 3.4 times more rewarding than those close by.

Bumblebees change their foraging behaviour in response to changes in nectar abundance. The more nectar they find per flower, the more they search other flowers in the vicinity. Heinrich carried out experiments in which he laid screens of bridal veil over some patches of clover and left others unscreened. The bees depleted the nectar in the unscreened clover, but that in the screened clover accumulated. When the screening was removed, the bees could visit both rich and poor areas of clover. In nectar-rich areas they probed about twelve florets per head and made short flights between heads. In areas of low nectar the bees probed only about two florets on each flower head and made longer flights between heads as shown in Figure 24.9. In this way the bumblebees concentrated their foraging in the more profitable areas.

The energy costs of foraging depend partly upon the environmental temperature. When foraging on fireweed the bees stop for only one or two seconds at each flower and remove tiny drops of nectar with a dab of the tongue. On the other types of flower, however, they may remain a number of minutes (see Figure 24.10). While perching on a flower, the bees do not allow their flight mechanism to cool, but they keep it at flight temperature so

Fig. 24.9 Changes in foraging behaviour of bumblebees foraging in two patches of white clover. One (left) was utilized by many bumblebees and had only 0.003 mg sugar per flower top, the other (right) had been screened with bridal veil to allow nectar to accumulate to a level of 0.01 mg sugar per flower. The graphs show that the bees made longer flights between flower heads when nectar supplies were low (After Heinrich, 1979).

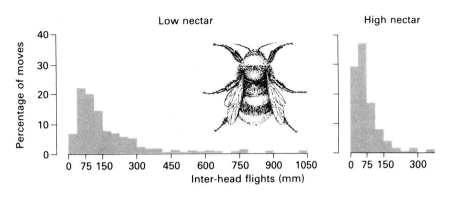

Fig. 24.10 The proportions of time that bumblebees spend in flight and perching are closely related to the kind of flowers visited and not much affected by air temperature. Pale bars indicate results obtained at 20 °C, dark bars show results obtained at 30 °C (After Heinrich, 1979).

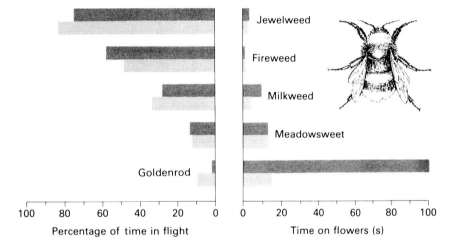

that they are ready to fly without delay. During prolonged perching they maintain the thorax at a temperature of about 32 °C by shivering, but they do not waste energy heating the abdomen, which is not involved in powered flight (see Figure 24.11). At environmental temperatures of about 25 °C, it is not necessary to heat the thorax during periods of inactivity, but below this temperature the bee has to pay progressively more to maintain efficient foraging (see Figure 24.11). Some flower species produce more nectar than others, and it is possible to calculate the relative costs of foraging on different types of flower at different temperatures (see Figure 24.12). Bumblebees forage on the profitable rhododendron flowers over a wide range of temperature, but they do not forage on lambkill and wild cherry at low temperatures because they cannot forage quickly enough to offset the extra energy required for thermoregulation (Heinrich, 1979).

Bumblebee foraging illustrates how time and energy act as constraints upon foraging efficiency. When the bees have to travel some distance to find productive flowers, then time becomes a limiting factor, and it is worthwhile for the bee to expend energy in order to save time. When foraging on

Fig. 24.11 (a) Body temperature of bumblebees at different air temperatures, while foraging from profitable flowers. (b) Calculated foraging costs at different air temperatures for a worker bee weighing 0.2 g, regulating its thoracic temperature at 30 °C, and spending half its time in flight and half perched on flowers (After Heinrich, 1979).

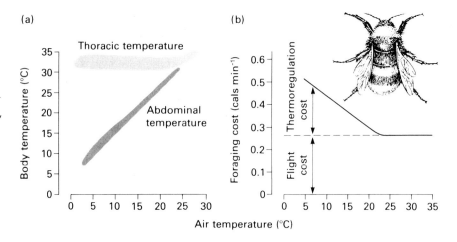

Fig. 24.12 Calculated foraging costs for queen bees (0.5 g) visiting different types of flower (After Heinrich, 1979).

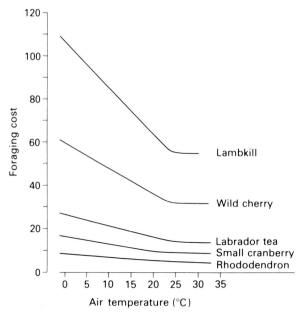

relatively unproductive flowers or when foraging at a low temperature, the bee may take more time in order to save energy.

Interactions between time and energy are important in many aspects of animal behaviour. The animal has to budget for each particular type of activity and for the particular consequences of each. In such circumstances trade-off is inevitable, and has been demonstrated many times in animal behaviour (see Chapter 10). In addition, the animal must budget for the cost of changing from one activity to another. When an animal changes from one activity to another, there may be a cost involved in the process of changing. For example, a pigeon feeding in a cornfield will become thirsty as a result of eating dry seeds. It may have to fly a mile to obtain water. In addition to the physiological cost of the journey, the bird will spend some time simply travelling from the feeding place to the drinking place. During this period it will not receive any of the benefits of feeding or drinking. Moreover, there may be risks involved in the journey, such as exposure to hawks or farmers with guns. Thus, the cost of changing is the decrement of fitness that arises during the period in which the animal is changing from one activity to another and receiving benefits from neither. The cost may involve loss of valuable time, expenditure of energy or risk from predators. It can be shown theoretically that the **cost of changing** should be budgeted as if it were part of the cost of the activity that is about to be performed.

We can distinguish between the **instantaneous cost** that arises within each unit of the relevant period of time and the **cumulative cost** that occurs over the whole of the period under consideration (see Figure 24.13). The important thing is that a change in behaviour is worthwhile if the cumulative cost a short time after changing is less than it would have been if the animal had not changed behaviour at all. I showed by experiment that, when the cost of changing is high, doves change between feeding and drinking less often than they would otherwise do. Stephan Larkin and I obtained evidence that the birds do indeed allocate the cost of changing to the cost of the behaviour that is about to be performed (Larkin and McFarland, 1978). They do not change their behaviour until they have accounted for this cost. Larkin (1981) found that the patterns of feeding and drinking in doves were altered in a predictable manner when the cost of changing from one to the other was increased in terms of time or of energy expenditure required (see McFarland, 1989b).

In addition to minute-to-minute considerations, animals take a more global account of their time and energy budgets. In general, we can expect behaviour that has the prime function of promoting the survival of the individual will take precedence over behaviour that promotes other aspects of fitness, such as territorial mating and parental behaviour. However, species vary considerably. Some aspects of behaviour are essential, but others like thermoregulatory behaviour may be important only when physiological mechanisms cannot cope. Thus, drinking is a daily necessity for some species, but other species can manage without drinking at all.

Each activity has value in terms of fitness, and animals must be designed to allocate priorities to activities in a general way, as well as from minute to

Fig. 24.13 Cost of changing from one activity to another. (a) The instantaneous cost of activity A is higher than that of B, but to change from A to B the animal must do the changing behaviour C, which has an even higher instantaneous cost. (b) At time T the accumulated cost of the transition A–B–C is the same as it would have been if the animal had remained doing A (After Larkin and McFarland, 1978).

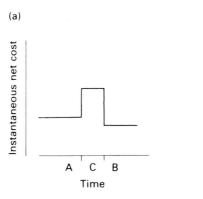

(a)

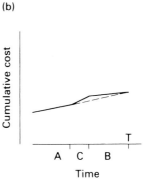

(b)

minute. Alasdair Houston and I approached this problem by considering a measure of the cost to an animal of abstaining from each activity in its natural repertoire (Houston and McFarland, 1980). If an animal did no feeding, for instance, the cost would be high, but if it abstained from grooming the cost might be relatively low. An animal that had a high motivational tendency for both feeding and grooming but that did not have time to do both would sacrifice less, in terms of fitness, if it devoted its available time to feeding.

Suppose an animal fills its typical day with useful activities, as illustrated in Figure 24.14. In an environment that was much the same from day to day, the animal would adjust its activities to the time available. Suppose, however, there is a change in the environment such that it now takes very much longer to obtain the normal amount of food (see Figure 24.15). The animal can respond to the changed circumstances by spending the same amount of time feeding as before and settling for less food. Alternatively, it could insist on the same amount of food as usual, or it could compromise between the two extremes. If the animal spent a long time obtaining the usual amount of food, then there would be less time available for all the other activities in its repertoire. These would have to be squashed into the remaining time. We found that the extent to which an activity resists squashing can be represented by a single parameter, which we called **resilience**. In the case where the animal feeds for the normal amount of time and ends up with a reduced food intake (Figure 24.14), the resilience of feeding is relatively low because feeding has not compressed the other activities even though the animal's hunger is increased. In the case where the animal insists on the normal amount of food (Figure 24.15), the resilience of feeding is relatively high because feeding ousts the other activities from the time available without itself being curtailed in any way.

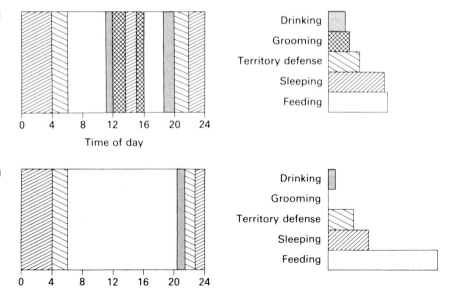

Fig. 24.14 Diagram showing how an animal might spend its time throughout the day.

Fig. 24.15 Diagram showing how the animal in Figure 24.14 might adjust its daily routine when required to spend much more time to obtain the same amount of food.

Behavioural resilience is a measure of the extent to which each activity can be squashed in terms of time by other activities in the animal's repertoire. It also reflects the importance of an activity in a long-term sense. During periods when time is a budget constraint, activities with low resilience will tend to be ignored. Indeed, if an activity completely disappears from an animal's repertoire when time is rationed, we might call it a luxury or leisure activity.

Behavioural resilience has been measured directly in few cases because of the practical difficulties involved. It is necessary to keep a record of the behaviour around the clock for a number of days in a situation where available time can be manipulated. Observations of this kind have been carried out by David Croft (1975) on female canaries (*Serinus canarias*). The aim of the study was to investigate how photoperiod length affects the activity and hormonal balance of the birds. By manipulating photoperiods, Croft effectively was altering the time available for activity because canaries are inactive in the dark. Croft found that the birds spend the same amount of time feeding on short days as on long days but that they fed more efficiently on long days. During long days the birds expend more energy in various activities, so this result is not entirely unexpected. Birds kept on long days spent more daytime inactive and sleeping than birds on short days. Since the birds can sleep at night, it seems most likely that birds kept on long days filled spare time with sleep. This view is supported by the fact that birds that were building a nest spent less time sleeping during the day.

Croft calculated the time available for nest building after time necessary for feeding, grooming and travelling from place to place had been taken into account. He found out that there is adequate time for nest building on long days but barely enough time on short days. The interpretation of the observations is complicated, however, by the fact that long days stimulate hormone production so that these birds are more motivated to build a nest than birds kept on short days. The nest-building behaviour of birds kept on long days is more efficient than that of birds on short days, and less time is wasted on unnecessarily repetitive gathering and building movements. Thus, although **resilience** is a concept that is theoretically distinct from that of **motivation** (McFarland and Houston, 1981; McFarland, 1989b), it is difficult to isolate its effects in studies of time budgets.

However, resilience can be measured indirectly by means of **demand functions**. These are used by economists to express the relationship between the price and the consumption of a commodity. For example, when the price of coffee is increased in the supermarket, people continue to buy about the same amount as before, perhaps a little less (see Figure 24.16). As the price of fruit is increased, however, the demand for fruit falls off. When the price of fresh fish is increased, demand declines markedly. Presumably, people are willing to pay more to maintain their normal coffee-drinking habits. Demand for coffee is said to be **inelastic**. If the price of fresh fish increases, however, people tend to buy less and to switch to substitute foods such as meat or canned fish. Demand for fish is said to be **elastic**.

Exactly analogous phenomena occur in animal behaviour. If an animal expends a certain amount of energy in a particular activity, then it usually does less of that activity if the energy requirement is increased (see Figure

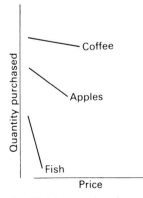

Fig. 24.16 Examples of economic demand curves. Demand for coffee is inelastic, demand for fish is elastic (both axes have log scales and arbitrary origins).

Fig. 24.17 Mean response rates of groups of Siamese fighting fish (*Betta splendens*) swimming through a tunnel for food reward (right), or for the opportunity to display to their mirror image (left). The abscissa shows the number of responses required to obtain a reward (After Hogan *et al.*, 1970).

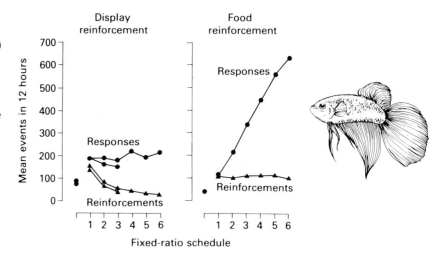

24.17). Numerous studies have shown that demand functions in animals follow the same general pattern as those of humans (Kagel *et al.*, 1980). For example, Steven Lea and Tim Roper (1977) found that demand for food was inelastic in rats required to work (press a lever) to obtain food rewards. As the work required (lever presses per reward) increased, the rats continued to work for about the same amount of food. However, these investigators also found that elasticity of demand for food pellets increased when sucrose was available as a substitute.

The parallel between the demand phenomena of animals and people is a topic of considerable interest (e.g. Allison, 1979, 1983; Lea, 1978; Rachlin, 1980). The elasticity of demand functions gives an indication of the relative importance of the commodities (or activities) on which the person (or animal) spends his or her money (or energy). There is a close relationship between elasticity of demand and resilience (Houston and McFarland, 1980). Thus, demand functions can be used as indirect measures of resilience. If activity A has higher resilience than activity B, then A will tend to show a relatively inelastic demand function and B will show an elastic one (McFarland and Houston, 1981).

24.4 Animal and human economics

Human economics is the study of those activities which involve exchange transactions among people. Such transactions may include reciprocal exchange, barter market, or a price market (Harris, 1985; McFarland, 1989b). An important economic activity is that of choosing how to allocate scarce resources among a variety of alternative uses. As we have seen, some economic principles apply not only to humans but to other animal species. All animals have to allocate scarce resources among competing ends. The scarce means may be energy, nutrients or time. The competing ends may be growth and reproduction, in the long term; or alternative activities, in the case of short-term decision-making.

As we saw in Chapter 9 the !Kung San are hunter–gatherers, who work to obtain calories which are shared among all members of the group, plus any visitors. The consumption of each person can therefore be assumed to be the same, in relation to their requirements. On this basis, it is possible to construct a model of the economy of the !Kung San (McFarland, 1989b). The San do not live a hand-to-mouth existence. They may eat a little of what they obtain in the field, but most is transported to the camp and shared out. Most of what the foragers bring back cannot be eaten right away, but must be processed. Mongongo nuts must be cracked and roasted. Meat is cooked, or cut into strips and hung up to dry. Just as in the case of animal foraging (see Chapter 9), the handling time should be taken into account. In terms of hours per week, the work load of men and women is as follows: the average man works a 44.5 hour week in total, earning just under 500 kcal h^{-1} (2 MJ h^{-1}). Women earn 631 kcal h^{-1} (2.64 MJ h^{-1}) working a 40.1 hour week. We can use these figures to calculate the notional wage earned by hunters and gatherers. The BMR (basal metabolic rate) for men is 1400 kcal per day (5850 kJ), while the working requirement is 2250 kcal per day (9405 kJ), which is 1.60 as a multiple of BMR. The MBMR (multiple of the BMR) is a convenient unit of comparison of individuals of different bodysize. The energy requirement for a working woman is 1.59 MBMR, so the calorie requirement for working men and women is the same, when taken as the MBMR.

We can use these requirements to set up a preliminary model of the labour supply situation. In human economics, a worker divides the time available into time spent working and time spent at leisure. Every hour of work deprives the worker of one leisure hour. On this plot of daily hours of work against total income, we can represent **wage rates** as lines fanning out from the origin, as shown in Figure 24.18. The higher the wage rate, the greater the slope of the line. Assuming that a worker is free to choose the amount of work done each day, we can expect that there will be a trade-off between work and leisure, resulting in a set of **indifference curves** tangent to the wage rate lines. All points on a particular indifference curve have equal utility for the worker. The optimum number of working hours, at a particular wage rate, is given by the point on the line which attains the

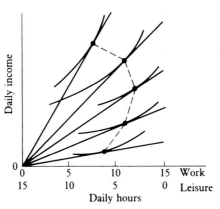

Fig. 24.18 A backward-bending labour supply curve (dotted line) (From McFarland, 1989).

greatest utility; that is the point at which the highest possible indifference curve is reached. By joining the optimum point on each wage rate line (by a dotted line) we attain the **labour supply curve** of economic theory. Such curves have been obtained for people, using money as a measure of income, but they can also be obtained for animals, using energy as a measure of income (e.g. Allison, 1983). Thus we are treating money and energy as equivalent, and taking foraging (earning energy) as synonymous with work (earning money).

There is quite a lot that we can say *a priori* about the labour supply situation in animals. First, there must be a sufficient income to sustain the basal metabolic rate. So we can assume that an animal would be prepared to work nearly 24 hours a day at the lowest sustainable wage rate, otherwise it will die. There may be some animals willing to sacrifice their lives for some activity other than foraging, but these must operate above the minimal wage rate in order to have energy available for these other activities. Animals are able to go for long periods without food only when they have considerable unearned income to draw upon. Second, we can expect that there will always be a maximum income, earned or otherwise, beyond which the animal is simply not interested. We cannot expect to find animal millionaires who accumulate energy simply for the sake of it. The maximum income is that beyond which the animal derives no utility from further income. Third, we can assume that our animal will forage less when given free food, and this means that the labour supply curve must move downwards from left to right, as illustrated in Figure 24.19.

The San, like animals, are subject to similar metabolic constraints on their labour supply curve, and if we use a calorie scale of income, we obtain a picture something like that illustrated in Figure 24.20. We know (from Lee, 1979) that 60 per cent of the total calorie income comes from gathering wild plants. The average daily income is 2355 kcal (9.84 MJ), so the average daily income from gathering is some 1400 kcal (5.85 MJ), shown as wage rate G in Figure 24.20. The model indicates that it should take six hours per day to produce this income. A man-day of hunting brings in 7230 kcal (30.2 MJ) on

Fig. 24.19 Labour supply curve obtained when some non-labour income is provided in a natural environment (From McFarland, 1989).

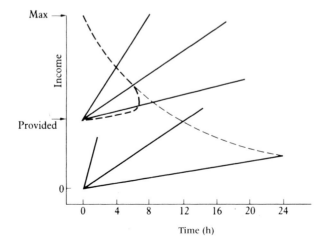

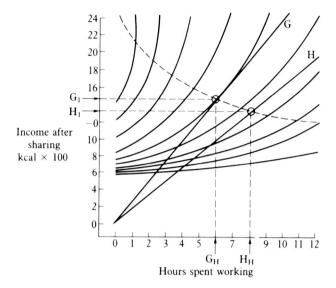

Fig. 24.20 A simple economic model for San foraging. G is the wage rate for gathering, H is that for hunting (From McFarland, 1989).

average, compared with 12 000 kcal (50.16 MJ) for a person-day of gathering. So the hunting wage rate (H) is only 60 per cent of the gathering wage rate. On this basis, the model predicts an eight hour day for hunting, with an income of about 1250 kcal (5.23 MJ) after sharing. In fact, the women usually work a six hour day and the men an eight hour day, when foraging (Lee, 1979; McFarland, 1989b).

These earnings are purely notional, in the sense that their value is dependent upon an exchange process. If there are many people in the camp, then the rate of exchange is low. If there are few, then the exchange rate is high. What a hunter or forager brings back from a particular day's work is of no direct benefit, because it is pooled with the produce from other foragers and shared out to all members of the group. It is as if the forager received a wage that is determined by the success of the company as a whole. In evolutionary terms, as we saw in Chapter 9, the San seem to be practising reciprocal altruism. Among humans in general, an altruistic act may take many forms, including baby minding, loan of tools, or a political favour. In advanced human societies it may be possible to quantify such favours in terms of money, but the San have no money. It seems likely that some forms of economic altruism existed before money was invented, and that money had its origins in exchange of favours (McFarland, 1989b).

Points to remember

- Animals can be assumed to make rational decisions in the sense that their choices are consistent and transitive.

- There are analogies between human consumer economics and the costs and benefits incurred by animals. Money is analogous to energy, and utility is analogous to benefit (increment in fitness). Time and energy budgets can be analysed on this basis.

Further reading

Cuthill, I.C. and Houston, A.I. (1997) Managing time and energy. In Krebs, J.R. and Davies, N.B. (eds) *Behavioural Ecology,* 4th edn. Blackwell, Oxford.

Heinrich, B. (1979) *Bumblebee Economics.* Harvard University Press, Cambridge, MA.

McFarland, D.J. and Houston, A. (1981) *Quantitative Ethology: The State Space Approach.* Pitman, London.

Stephens, D.W. and Krebs, J.R. (1986) *Foraging Theory.* Princeton University Press, Princeton, NJ.

Animal robotics

Since 1990 there has been considerable growth in research on animal-like robots (usually called animats), either real or simulated on a computer. This research shows that roboticists have much to learn from animal behaviour, and vice versa.

In this chapter we look at some aspects of animal robotics, and relate these to problems of animal behaviour. First, we consider animals as automata, and show how very simple robots can collectively exhibit complex behaviour akin to that of ants and termites. Second, we consider how mobile robots might be self-refuelling, and see what animal-like qualities are required for energy autonomy. Third, we look at robot learning and compare it with some aspects of animal learning. Finally, we look at cognitive architectures for robots, and see how this field can help us to understand issues of animal cognition.

25.1 Animals as automata

Originally, an automaton was a machine in the form of a doll or animal, which operated by clockwork to perform particular operations in a life-like manner (Cohen, 1966; McCorduck, 1979). An example is the traditional Japanese *Karakuri* doll-like automaton (Figure 25.1), which shuffles forward, bows, and proffers a cup of tea on a tray.

An automaton is a machine whose behaviour is entirely determined by its state and by outside forces, as in any physical system. Some automata act unconditionally, as in the *Karakuri* automaton. This robot has no sensors. It has a fixed behavioural routine that depends entirely upon its current state. Others are capable of acting conditionally. That is, when in one state they behave in one way and when in another state they behave in a different way. The change in state may result from having accomplished a particular activity. For example, a car-washing machine starts off in a particular state and consequently starts a particular behavioural routine, such as brushing the car from front to back. Having finished the routine the machine arrives in a different state, and starts a new routine. Depending upon the type of car,

Fig. 25.1 Japanese *Karakuri* robot from the eighteenth century (Japanese Embassy).

the machine is able to vary its behaviour to some extent, each variation being triggered by a change of state.

An important feature of automata is that they are situated. **Situated actions** are actions taken in the context of particular, concrete circumstances (Suchman, 1987). The concrete circumstances partly determine the action to be taken, and largely determine the consequences of the action. For example, in an automatic car, decisions about changing gear are entirely situated, in the sense that they are determined by the current circumstances relating to the engine revs, speed, etc. In a car with a manual gear shift the equivalent decisions are only partly situated.

Primitive robots, such as those used on assembly lines, behave in a largely unsituated manner. Like the *Karakuri* robot they complete a behavioural routine once it is started, and are insensitive to changing circumstances. Such robots are said to have a 'preprogrammed architecture'. We can find evidence of this type of behaviour in animals (see Box 25.1).

At the other extreme, we can imagine an animal that was purely reactive in the sense that all its behaviour could be accounted for in terms of its per-

Box 25.1 Stereotyped nesting activities of a digger wasp

In his study of the digger wasp, *Ammophila campestris*, Gerard Baerends (1941) found that the female, when about to lay an egg, digs a hole, kills or paralyses a moth caterpillar, carries it to the hole, deposits an egg on the caterpillar and stows it away in the hole. The female wasp then repeats this procedure with the second and subsequent eggs. Meanwhile, the first egg has hatched, and the larva has begun to consume the caterpillar. The wasp now returns to the first hole and provisions it with more caterpillars. She then may start another hole, or she may provision the second hole, depending upon the circumstances. In this way the female wasp may maintain up to five nests simultaneously, as shown in the figure.

Baerends found that the wasps inspect all the holes each morning before leaving for the hunting grounds. By robbing a hole he could make the wasp bring more food than usual, and by adding caterpillars he could induce her to bring less food than usual. However, he could manipulate the wasp in this way only if he made changes to the nest before the wasp's first visit of the day. Changes made later in the day had no effect. The female wasp seems to operate by simple rules. There is a standard routine for laying an egg, which involves digging a hole and providing a caterpillar. There is a standard early morning inspection routine that usually determines which nest will be provisioned during the day. There is a standard stopping routine by which the wasp closes up the nest when sufficient caterpillars have been supplied. Although the wasp is capable of assessing the extent of provisions when she visits a nest, she does not always use this ability. Moreover, each routine, once started, is followed to its conclusion. Thus, a wasp will go on and on provisioning a nest if the caterpillars are removed systematically each time they are supplied. This example shows that complex behaviour can be programmed on the basis of a set of rigid rules. The wasp behaves in an automaton-like way, although it may have some routines for extricating itself from difficulties, like removing an obstacle from the burrow.

Diagram of the nesting activities of an individual female *Ammophila* wasp (After Baerends, 1941).

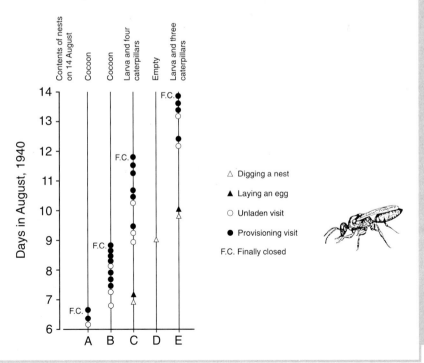

ception of the outside world. A robot example is shown in Box 25.2. This type of robot has a purely reactive architecture. The robot's behaviour is thoroughly situated because it is determined entirely by the robot's place in the world – its situation – in contradistinction to preprogrammed robots which simply ignore the world, and to motivated robots whose response to the world is partly determined by their internal state (see below).

As an example of an animal as a situated automaton, let us look at navigation in ants. Of course, it is difficult for us to be certain that these animals are simply automata, but the research that has been done by Wehner and his co-workers strongly suggests this. It appears that the *Cataglyphis* ants of the Sahara Desert navigate by path integration. They continually measure all distances covered and all angles turned, and integrate these linear and angular

---

Box 25.2   A situated automaton in the form of a simple mobile robot

This robot has a single ultrasound sensor at the front, which provides information about the distance of an object directly in front of the robot. This is the only sensory information available to the robot. Yet the robot is able to move around at high speed without hitting objects. How is this done?

1. By making the speed of the robot proportional to the distance to an obstacle, the robot slows down as it approaches the object.

2. By making the angle of the front steering wheels proportional to the distance to an obstacle, the robot turns more sharply the nearer it gets to an object.

3. If the robot is turning to avoid an object, but the distance is getting smaller, then the robot must be turning the wrong way. So by changing the direction of steering when the distance is getting smaller, but not when it is getting larger, the robot turns away from the obstacle.

This robot was designed by David McFarland and made by Marinus Maris (see Pfeiffer and Scheier, 1994). It is able to avoid objects at high speed by following the three simple rules described above. The robot is purely reactive in the sense that its behaviour is determined solely by reaction to external stimuli. The robot's place in the world at any instant (its situation) determines its behaviour.

components into a continually updated vector that points towards home. Wehner *et al.* (1996) show that the ants accomplish this by automatic sensory processing similar to that described in Box 25.2.

Cataglyphis uses a compass to monitor the angular components of its movements. This compass utilizes the pattern of polarized light in the sky (Figure 25.2), as do some bees (see Chapter 23). The photoreceptors in a particular region of the ant's (and bee's) eye are sensitive to directional oscillation in this pattern. The pattern of polarization changes with the elevation of the sun above the horizon. The ant is equipped with a neural template which resembles the polarization pattern when the sun is at the horizon, but differs from it for all other elevations (Wehner, 1994). The best possible match between the template and the external pattern is achieved when the insect is aligned with the solar (or anti-solar) meridian. As the animal rotates about its vertical body axis the match decreases systematically. Thus the best match gives the zero point on the compass, and this deteriorates as the ant selects other compass directions.

Because there is a discrepancy between the internal template and the external pattern, mismatches occur whenever only parts of the external pattern are available. If an animal experiencing the entire sky pattern is suddenly presented with a small patch of sky, then navigational errors will arise.

Fig. 25.2 (a) Two-dimensional representation of the pattern of polarization in the sky for two elevations of the sun (black disc): 25° (left) and 60° (right). The directions of polarized light are represented by the orientation of the black bars. The sizes of the black bars indicate the degree (percentage) of polarization. (b) The ant's internal representation of the sky as derived from behavioural experiments. The open bars indicate where in the sky the ant assumes a particular polarization plane (E-vector) to be. This template is used for all elevations of the sun. (c) Outward and homeward paths of the ant *Cataglyphis* (see also Fig. 9.3) (From Wehner, 1997).

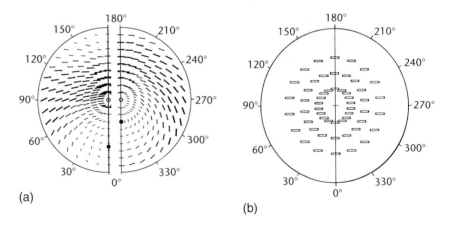

(a)

(b)

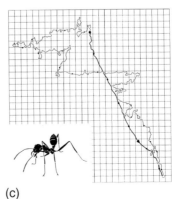

(c)

However, such errors do not occur when the animal is continuously presented with the same patch of sky. Thus the animal utilizes a temporary, if arbitrary, zero point for its compass, and integrates its angular movements in relation to this. Because the animal forages for short periods of time (measurable in tens of minutes), it does not matter that the basic setting of the compass is not always accurate, since it is changes in relation to the (temporary) reference point that are important.

The foraging ant has simply to integrate, within one foraging trip, the angular deviations that it makes and the associated distance travelled since it left home. This information is continually integrated to provide an angular value which, when taken in relation to the current arbitrary compass setting, points directly to home. Thus the animal has an automatic device that points homewards throughout its wanderings. By taking in information from the environment, and transforming it automatically into other information, the animal is behaving as a situated automaton, similar to that described in Box 25.2.

The situatedness of animals is well illustrated by the **stigmergic** behaviour of ants and termites. The principle of stigmergy was first recognized and named by the French biologist Grasse (1959). It is essentially the production of behaviour that is a consequence of the effects produced in the local environment by previous behaviour.

For example, when termites start to build a nest they modify their local environment by making little mud balls, each of which is impregnated by pheromone. Initially, the termites deposit their mud balls at random. The probability of depositing one increases as the sensed concentration of pheromone increases. After the first few random placements the other termites tend to deposit their mud balls in the same place, so that small columns are formed. The pheromone from neighbouring columns causes the tops of columns to lean towards neighbouring columns. Eventually the tops meet, forming arches, the basic building units of the nest. As other stigmergic processes come into play, involving water vapour, carbon dioxide concentrations, and the presence of the queen, the whole complex nest structure is produced. This may include the royal cell, brood nurseries, air conditioning, larders, and communication and foraging tunnels.

Stigmergic behaviour has also been demonstrated in simple mobile robots. As an example, let us consider a simple type of robot made by Beckers *et al.* (1994). They built battery-powered robots on a 21 × 17 cm platform. A 12 V motor-powered wheel was positioned at the mid-point of each long side, with a castor wheel at the mid-point of one of the shorter sides; this allows the robot to move forwards or backwards in a straight or curved trajectory, and to turn on the spot. Each robot carries a 17 cm wide aluminium forward-facing, C-shaped horizontal scoop with which it can push objects (Figure 25.3a). The objects used are circular pucks, 4 cm in diameter and 2.5 cm in height. The robots are equipped with two IR (infra-red) sensors for obstacle avoidance, and a microswitch which is activated by the scoop when a certain number of pucks are pushed. For the experiments reported here, this number is set to three.

The robots have only three behaviours, and only one is active at any time.

(a)

(b)

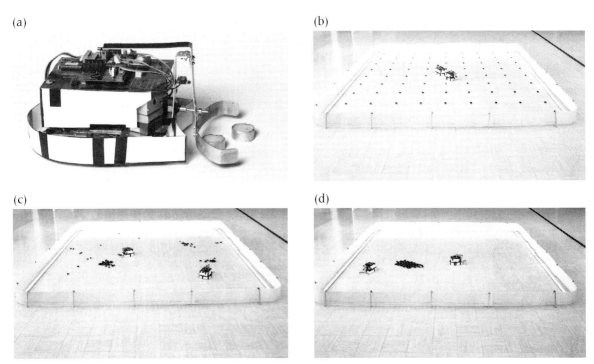

(c)

(d)

Fig. 25.3 Collective behaviour in robots. (a) Here we have an example of a very simple robot. Notice that each robot carries an aluminium forward-facing C-shaped horizontal scoop with which it can push objects. The objects used are circular pucks. Each robot is equipped with two active infra-red sensors for obstacle avoidance, and a microswitch which is activated by the scoop when a certain number of pucks are pushed. For the experiments reported here, this number is set to three. (b) Arena with robots and pucks in the starting position. (c) Clusters formed at about the halfway point. (d) Final cluster. (*Photographs: Owen Holland*).

When no sensor is activated, a robot executes the default behaviour of moving in a straight line until an obstacle is detected by the IR sensors, or until the microswitch is activated (pucks are not detected by the IR sensors). On detecting an obstacle, the robot executes the obstacle avoidance behaviour of turning on the spot away from the obstacle and through a random angle; the default behaviour then takes over again, and the robot moves in a straight line in the new direction. If the robot is pushing pucks when it encounters the obstacle, the pucks will be retained by the scoop throughout the turn. When the scoop pushes three or more pucks, the microswitch is activated; this triggers the puck-dropping behaviour, which consists of backing up by reversing both motors for one second (releasing the pucks from the scoop), and then executing a turn through a random angle, after which the robot returns to its default behaviour and moves forwards in a straight line. The obstacle avoidance behaviour has priority over the puck-dropping behaviour.

If we look at the results of the experiments reported by Beckers *et al.* (1994) we see apparently purposive behaviour. The experiments are conducted in an arena, in which 80 pucks are set out in a grid pattern (Figure 25.3b). At the outset a number of robots (which may be 1, 2, 3, 4, or 5) are placed in the middle of the arena, and the experiment starts when these are switched on by the experimenter. The results are similar irrespective of the number of robots used in the experiment.

From a qualitative point of view, each experiment has three more or less distinct phases, regardless of the number of robots involved. At the start, the arena contains only single pucks. In the first phase, a robot typically moves forwards collecting pucks into the scoop one at a time. When three have been gathered, the robot drops them, leaving them as a cluster of three, and moves off in another direction. Within a short time, most pucks are in small clusters which cannot be pushed around (Figure 25.3c). In the second phase, the robot removes one or two pucks from clusters by striking the clusters at an angle with the scoop. The pucks removed in this way are added to other clusters when the robot collides with them. Some clusters grow rapidly in this phase. After a time, there will be a small number of relatively large clusters. The third and most protracted phase consists of the occasional removal of a puck or two from one of the large clusters, and the addition of these pucks to one of the clusters, often to the one they were taken from in the first place. The process eventually results in the formation of a single cluster (Figure 25.3d). In every case, however many robots are used, the experiment begins with the pucks set out in a grid pattern and ends with them in a single cluster. To the naive onlooker, the behaviour of the robots seems purposive. The robots work on the grid pattern, transforming it into a single cluster of pucks, an obvious and easily identifiable end point.

25.2 Energy autonomy

Robots require energy, and we usually think of this as being supplied by their human owners. Robots will have to be energy self-sufficient in the future, to carry out tasks that are useful to people, such as work in areas of high radioactivity, planetary exploration, or deep-sea monitoring. This will make them more like animals.

Energy autonomy concerns the ability of the robot to obtain its own energy. The first person to demonstrate that this could be done in robots was Grey Walter (see Box 25.3). Like animals, robots may have differing degrees of energy autonomy (Table 25.1). Robots that periodically visit a recharging station are of particular interest to ethologists, because they are similar to foraging animals.

To be truly self-sufficient, an agent must exhibit both behavioural stability and market viability. Behavioural stability implies that the agent does not succumb to irrecoverable debt of any vital resource. The **vital resources** are those that enable the agent to carry out its design tasks, and may include energy, time, tools, etc. **Irrecoverable debt** is not simply a question of running out of a vital resource, but may include debts, the repayment of which engenders other debts. Behavioural instability may occur if the agent runs out of a vital resource, or if the servicing of debts takes so much time and/or energy that the agent is not able to carry out its design tasks. **Market viability** amounts to pleasing the robot's employer (see below). The employer

Box 25.3 Grey Walter and the first self-charging robot

Grey Walter (1950) was the first to demonstrate that a machine could be an 'imitation of life' in the sense that it could obtain its own energy through its own behaviour. Owen Holland (1997) has resurrected the original robot, made a replica, and shown that it is the precursor of modern behaviour-based robots.

Grey Walter's robot was a simple electro-mechanical device on wheels. It had a single photo-receptive 'eye' which continually scanned the environment, enabling the robot to exhibit both positive and negative phototaxis (see Chapter 14). The robot is attracted to a light source of moderate intensity, but avoids bright light, so it tends to hover near a light source without approaching too close. As the battery runs down the balance between positive and negative phototaxis shifts,

and the robot is increasingly attracted to a bright light. The robot has a 'hutch' with a bright light inside, and it can enter the hutch and recharge its battery.

When the robot shell hits against an obstacle a switch is activated and this triggers reflex obstacle-avoidance behaviour which overrides the response to light. Thus, if the path towards a light source is obstructed, the robot is often able to circumvent the obstacle and continue its progress. By fixing a lamp onto the front of the robot, Grey Walter was able to demonstrate quite complex responses to mirrors and to other similar robots. Overall, he was able to demonstrate that a range of activities could be achieved by interaction between two basic behaviour-based mechanisms: phototaxis and obstacle avoidance.

(Photograph: Owen Holland).

will be satisfied if the agent is behaviourally stable, provided that the agent is also able to perform the tasks that it is designed to perform, and provided also that the running costs are acceptable.

Self-sufficient agents must have a degree of **autonomy**, because they must have the freedom to decide for themselves when to refuel, when to perform certain activities, etc. (see McFarland and Bosser, 1993; McFarland, 1995a).

Table 25.1 Energy autonomy of animals and robots

Animals unable to provide for themselves and robots recharged by man	Energy self-sufficient animals Self-recharging robots		Agents dependent only on the sun for energy
	Energy source supplied by man	Energy obtained from the natural environment	
PARASITES	DOMESTIC ANIMALS	WILD ANIMALS	PLANTS
LAB ROBOTS TOYS	OFFICE ROBOTS	FORAGING ROBOTS	SOLAR ROBOTS
Reproduction dependent on host	Reproduction dependent on own behaviour		Reproduction dependent on energy gained
Success dependent on host	Success dependent on satisfying owner through own behaviour		Success dependent on energy gained

The degree of autonomy usually equates with the number of **basic resources** that the agent has to manage. Simple self-sufficient agents are selfish in the sense that they manage their own resources regardless of other agents that may be operating in the ecosystem. Such agents usually have to **trade off** between refuelling activities and activities designed to please the employer (call this **working**).

There are two basic resources that must be provided by the robot's environment if the robot is to be self-sufficient and economically viable. These are energy (E), which the robot must be able to obtain in some way, and M, which can be obtained by working. M can stand for memory of amount of work done, merit points for work done, or money, or market viability. Each time the robot does a certain amount of work (i.e. fulfils part of the task that is useful to the owner) it earns a unit of M.

When we consider a single self-sufficient robot, it is evident that it should perform a **basic cycle** of activities to maintain viability. The robot goes through a cycle of work, find fuel, refuel. When **working**, M is gained and E lost. At some point the robot breaks off work and goes to **find fuel**. This also leads to a reduction in E, but what about M?

The answer to this question depends upon the attitude of the owner of the robot. M represents the utility of the robot's work from the point of view of the owner. There are three basic possibilities, as outlined in Figure 25.4.

If the owner is primarily interested in robots that spend as much time as possible doing useful work, irrespective of energy expenditure, then the owner will not gain utility from the time that is spent not working. This means that the robot should be designed so that M declines during unproductive time, as shown in Figure 25.4(a).

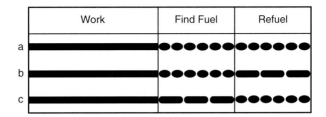

Fig. 25.4 In all three cases the work done by the robot provides positive utility (solid line) to the owner. In (a) the owner attaches negative utility (dotted line) to all time spent not working. In (b) the owner attaches negative utility to unproductive energy loss and is neutral (dashed line) towards time spent refuelling. In (c) the owner is neutral towards the time spent finding fuel, but attaches negative utility to providing the robot with fuel (From McFarland and Spier, 1997).

If the owner is concerned about energy expenditure on activities that are not productive, then M should decline during that period of the basic cycle, as shown in Figure 25.4(b). If, on the other hand, the owner is concerned to minimize energy expenditure across the board, then it makes sense for the robot to pay for its fuel. In Figure 25.4(c) M is earned during **work** and spent during **refuel**. In other words, M is like money which the robot earns by working and spends on fuel. These possibilities can be summarized as basic cycles, as shown in Figure 25.5

McFarland and Spier (1997) discuss a particular example, set up as an experiment at the AI Laboratory of the VUB, Brussels, under the direction of Luc Steels. They devised a robot environment in which there is a single system battery that has a constant unalterable rate of energy inflow (McFarland, 1994; Steels, 1994). In the absence of any robots there is a constant rate of outflow of energy due to a number of lamps that are housed in boxes in the robot arena. In the absence of any robots, the system is in balance, the rate of outflow of energy equalling the rate of inflow.

Once we introduce robots that visit the recharging station, then the average rate of outflow of energy is increased because energy is drained from the system battery when robot batteries are recharged. To keep the system in balance, the robots have to turn off the lamps in the arena. This they can do by

Fig. 25.5 Three different types of basic cycle, represented in the EM plane: (a) M declines throughout all unproductive time, i.e. when the robot is not working. (b) M declines only when there is unproductive energy loss, i.e. when the robot is searching for a source of fuel. (c) M declines when the robot is refuelling, i.e. the robot is effectively paying for its fuel with M (From McFarland and Spier, 1997).

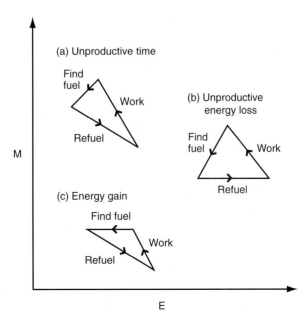

knocking on the boxes in which the lamps are housed. The robots thus have two main tasks:

1. To recharge their on-board batteries from time to time.

2. To reduce energy waste in the ecosystem.

They accomplish the first task by seeking the recharging station, and they accomplish the second task by seeking out lamp boxes and extinguishing the lamps, an activity called **tending**.

The lamps are, in effect, parasites that feed upon the energy in the ecosystem. Even when a lamp has been extinguished by a robot, it gradually regrows to its former energy-dissipating magnitude. So the robots must keep eradicating the parasites to maintain sufficient energy in the ecosystem for recharging their own batteries.

Here we have a miniature ecosystem, in which the work that a robot does indirectly contributes to the resources that it needs for survival. The robots log their own work progress. In other words, the robot remembers the amount of progress that it has made in carrying out the task. In practice, the robot gains one unit of M for each knock that it delivers during tending. When the robot arrives at the recharging station, it receives one unit of E (energy) for each unit of M that it gives up. (This **exchange rate** is under experimental control.) Each robot can decide for itself how much to spend at the recharging station, but if it has no M it cannot obtain any E.

The viability of the robot in this situation depends upon a number of factors. The rate at which a robot can earn M depends upon factors such as the spacing between boxes, the obstacles in the arena, and the M content of the boxes. These are under experimental control. The robot has to find the recharging station at the relevant time. When it is within a certain range it can home in on the station by phototaxis.

The stability of the basic cycle depends upon two main types of factor: the nature of the environment, and the decisions made by the robot. Thus the system can become unstable because the environment is just too difficult for the robot to cope with. An animal equivalent might be an environment in which food was very scarce. The system could also become unstable simply because the robot made bad decisions, such as ignoring an opportunity to recharge (i.e. when near the recharging station). Thus behavioural stability and decision-making are closely connected. Examples of stable and unstable basic cycles, obtained from experiments at the VUB, are shown in Figures 25.6 and 25.7.

The self-sufficient robot must have some degree of autonomy because it must have the freedom to do that behaviour that is in its own vital interests. For example, a robot that is able to obtain its own energy by visiting some kind of recharging station must be able to decide for itself when to discontinue its current behaviour and commence the appetitive behaviour that leads to recharging. Ideally, the robot would base this decision partly upon information about its own on-board energy level, and partly upon other considerations, such as its current motivation for work (e.g. foraging). The robot could start looking for fuel as soon as its battery level reached a certain

Fig. 25.6 A stable basic cycle taken from an experiment with a real robot. The energy and memory values are recorded by the robot and downloaded after the experiment. Vertical jumps have been inserted between cycles for clarity (From McFarland and Spier, 1997).

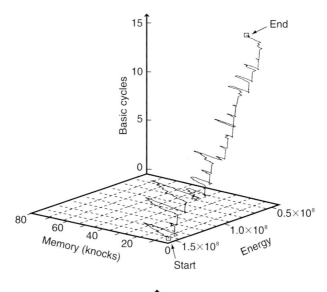

Fig. 25.7 An unstable basic cycle taken from an experiment with a real robot. Data and conventions as in Figure 25.6 (From McFarland and Spier, 1997).

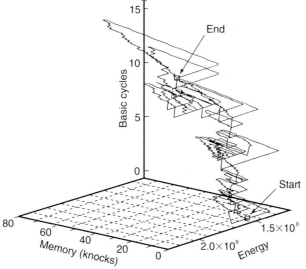

threshold, but this is a sub-optimal design. Like an animal, the robot ought to be able to trade off the motivational alternatives. The ability of a robot to do this is an aspect of its **motivational autonomy** (see McFarland, 1995a).

25.3 Robot learning

We have seen (Chapter 19) that animals can learn to associate two events if the relationship between them conforms to what we normally call a causal relationship. Thus, animals can learn that one event (the cause) predicts another event (the effect) or that one event predicts that another event (non-effect) will not occur. They can also learn that certain stimuli predict no consequences in a given situation or that a class of stimuli (including the animal's own behaviour) is causally irrelevant. The conditions under which these types of associative learning occur are those that we would expect on the hypothesis that animals are designed to acquire knowledge about the causal relationships in their environment. Thus, the animal must be able to distinguish potential causes from contextual cues, and for this to occur there must be some surprising occurrence that draws the animal's attention to particular events, or the events must be (innately) relevant to particular consequences. If these conditions are not fulfilled, contextual cues may overshadow potential causal events, or learning may be blocked by prior association with a now irrelevant cue. Thus, the conditions under which associative learning occurs are consistent with our commonsense views about the nature of causality.

In his book *The Organization of Behaviour*, Donald Hebb (1949) introduced various hypotheses about the neural substrate of learning and memory, including what is now known as the Hebb synapse. According to Hebb (1949, p. 62):

'When an axon of cell A is near enough to excite cell B or repeatedly or persistently takes part in firing it, some growth process or metabolic change takes place in one or both cells such that A's efficiency, as one of the cells firing B, is increased.'

This statement can be translated into a precise quantitative expression, and incorporated into a computer algorithm to provide a **Hebbian rule**. Such a rule can be used to form associations between one stimulus and another. Dynamic associations enable the network to learn to predict that one stimulus pattern will be followed at a later time by another, as in classical conditioning. A simple neural model of the classical-conditioning process is shown in Figure 25.8. There is a strong unmodifiable synapse from US to R, which

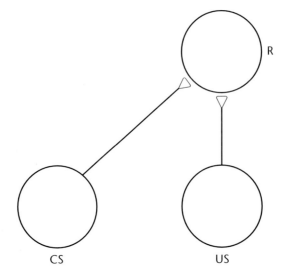

Fig. 25.8 Hebbian classical conditioning. The unconditional stimulus (US) elicits a response in the postsynaptic cell (R). Temporal coincidence of the response and the conditional stimulus (CS) strengthens the synapse between CS and R.

ensures that the US automatically evokes the response. There is a modifiable synapse from CS to R, which in the naive animal is initially weak.

Modifications of the Hebbian rule can be used to produce many of the conditioning phenomena known in animals (Tesauro, 1986; Klopf, 1982; Sutton and Barto, 1981). Klopf (1989) proposes a 'drive-reinforcement' model of neuronal function, which is implemented within a single complex neuron. Halperin (1991) outlines a mini-network with similar capabilities. Both these models make some attempt to incorporate the motivational element of reinforcement, and are able to reproduce many of the aspects of conditioning discovered by experimental psychologists.

Neural networks consist of a large number of inter-connected artificial 'neurons'. These may be simple signal-transmitting devices, or more complex attempts to mimic real neurons. The performance of a neural network depends upon:

■ the properties of the individual neurons

■ the architecture which specifies which neurons are connected to which, and what type the connections are (e.g. excitatory or inhibitory)

■ the rules governing strength changes in those connections that are variable

Connectionism is neurally inspired modelling (Arbib, 1987), using the above properties of neural nets. It should be distinguished from neural modelling, which is an attempt to model the neural activity of real brains. Connectionist models are of many types (see Arbib, 1987, for a review), and some are capable of learning.

The three main types of learning employed in connectionism are self-organizing, supervised learning and reinforcement learning. Self-organizing systems change their properties on the basis of a single uniform learning rule, no matter what the task (Arbib, 1987). Examples include Hopfield nets (Hopfield, 1982; Hopfield and Tank, 1986) and Boltzmann machines (Hinton *et al.*, 1984). In supervised learning, the system is 'trained' to behave in the desired manner by the judicious administration of reinforcement at appropriate stages in the learning process. Examples are the 'perceptrons' of Rosenblatt (1962), and Minsky and Papert (1969). In reinforcement learning, the agent continually receives sensory feedback from the consequences of its own behaviour and, in addition, receives a (usually scalar) reinforcement signal. Thus reinforcement learning is learning by trial and error from performance feedback. The reinforcer evaluates the consequences of the behaviour generated by the agent, but does not indicate 'correct' behaviour. Sutton (1991) reviews the major steps in the development of reinforcement learning over the previous decade (for a brief review see McFarland and Bosser, 1993, chapter 11).

Theories of learning sometimes turn out to have unexpected effects, so we should not be surprised to discover emergent properties in simple learning systems designed for robots. An example is provided in the interesting study by Verschure *et al.* (1991) of distributed adaptive control. Starting with a neural model of a basic classical-conditioning mechanism, these authors developed a strongly situated system for robot control. When simulated, this system exhibited emergent anticipatory behaviour.

The system is equipped with a Hebb-type learning mechanism which allows it to combine sensory inputs (CSs) with reflex responses (URs) which are triggered by a set of USs. The US–UR relationship is pre-tuned (in the authors' terminology it has a **predefined value system**, see below), and the agent is motivated to keep moving until the target is touched by the front-end, or snout.

This basic design is translated into a number of neural fields, as illustrated in Figure 25.9. Information from the environment is provided by a range finder, which projects sensory input to the CS field. This field is connected to two US fields by modifiable weights (initially set at 0), shown as large synapses in Figure 25.9. The US fields have opposite signs. One (–ve) receives information from a collision detector, and the other (+ve) from a goal detector, which locates the direction of the target.

Both US fields are connected to a UR field in which the motor responses are stored. In essence, these responses are: turn left, turn right, retract, and advance. Substantial activity in the –ve US field gives rise to avoidance behaviour. Following a collision, the robot will retract and turn through a predefined angle opposite to the direction of the collision (in the simulations a standard avoidance angle of 9° was used). The +ve US field is capable of locating the target, categorizing its position into an inner area (within 5° of the straight-ahead) and an outer area (beyond 5°). Location of the target within the inner area causes a 1° turn towards it, while location in the outer area causes a 9° turn. Thus the robot turns slowly if already nearly headed towards the target, and quickly if something is in its outer field of view. In this way the robot is able to approach objects.

Fig. 25.9 Connections between the neural fields of the neutral net controlling the behaviour of a simulated mobile robot (After Verschure *et al.*, 1991).

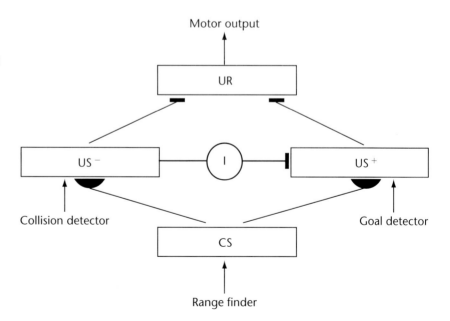

The UR field consists of a number of **command neurons** (Kupfermann and Weiss, 1978) which code specified motor responses. Whenever a specific command neuron is activated a specific motor response is automatically executed. The connections between the US fields and UR field are prewired and not modifiable. They are indicated by small synapses in Figure 25.9. In addition, there is a specific inhibitory relationship between the −ve and +ve US fields. Activity in the −ve US field inhibits the output of the +ve US field. This prevents the robot from bumping into things if it is currently involved in approach behaviour. In other words, approach behaviour can be temporarily overruled by avoidance behaviour.

The model robot was tested using the ASSIM simulation program (Krose and Dondorp, 1989). With this simulator the robot architecture can be detailed, with all connections and sensor transfer functions specified mathematically. The mobile robot can be positioned in a predefined environment and its behaviour studied. This was done in two stages. First the architecture relating to avoidance behaviour was tested on its own (the +ve US field and the inhibitory relation being omitted). Second, simulations with the complete architecture were performed.

The arena for the first experiment was a rectangular space with five identical obstacles located at the centre and near the four corners. The first 500 steps are shown in Figure 25.10(a). After colliding with the central obstacle, the robot turns left, and makes a number of collisions with other obstacles. Figure 25.10(b) shows the robot trajectory between steps 900 and 1000. The robot has successfully learned to avoid the obstacles and no collisions are now evident. The relative roles of the collision detector and the range finder over the first 1000 trials are illustrated in Figure 25.11. During the first 100 steps 29 avoidance movements are initiated as a result of physical collision and eight by the range finder. Thereafter, the initial avoidance reflex (UR) resulting from physical collision (US) is supplanted by the conditional reflex (CR) initiated through learned association with data from the range finder (CS). Thus the rise in avoidance movements initiated by the range finder (Figure 25.11) is effectively a learning curve.

Fig. 25.10 Outline of paths traced by simulated mobile robot (a) during first 500 trials (note collisions with square obstacles), and (b) during trials 900–1000 (note lack of collisions) of a learning session. (After Verschure *et al.*, 1991.)

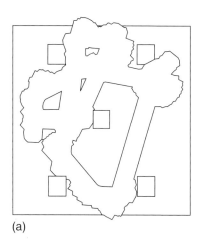

(a)

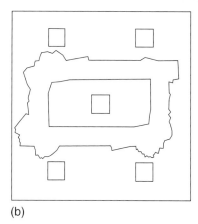

(b)

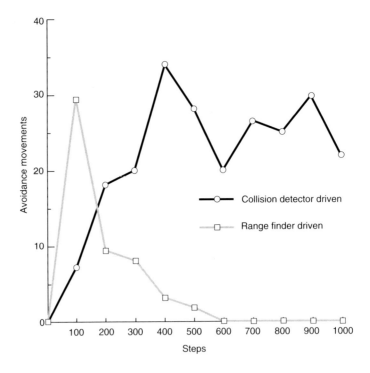

Fig. 25.11 Relative roles of collision detector and range finder during 1000 learning trials. (After Verschure *et al.*, 1991).

The arena for the second experiment was a rectangular space containing a number of obstacles and a doorway at one end. Beyond the doorway is a target which is attractive to the robot within the range shown by the circle in Figure 25.12. If the robot approaches the target it stops moving as it enters the inner circle. Figure 25.12 shows superimposed paths taken by the robot over 20 runs. During the first run the robot collided 23 times and took 139

Fig. 25.12 Outline of paths traced by simulated mobile robot learning to reach a target (inner circle) (After Verschure *et al.*, 1991).

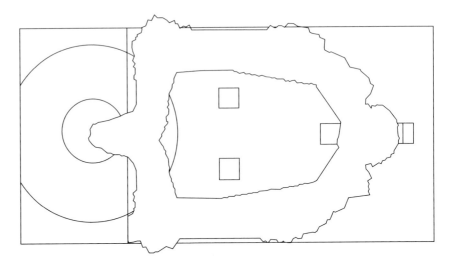

steps to reach the target. No collisions took place after the eleventh run, showing that the robot had successfully learned to avoid obstacles and reach the target.

This robot develops associations between specific sensory inputs and responses, which are driven by learning criteria provided by a predefined value system (see below). For example, during the trajectory illustrated in Figure 25.10(a), every collision establishes a sense–act reflex which is the basis for conditioning, in that it provides the associative mechanisms with a criterion (reinforcement) that guides the coupling of more sophisticated sensors (the range finders) to the relevant environmental features (the proximity of objects). In other words, learning to anticipate collisions takes place because certain range-finder data becomes associated with the consequences of a collision. As the robot gains experience over time it slowly shifts from collision-driven behaviour to range-finder-driven behaviour (see Figure 25.11) and thus starts anticipating obstacles. This anticipatory behaviour is emergent, because there is no component of the system designed to do anticipation *per se*.

It is interesting to note that the early animal learning theorists also understood this property of simple conditioning mechanisms. Thus Hull (1931) proposed a number of intermediate mechanisms derived from the basic laws of conditioning. In particular, in repeating a sequence of activities leading to a goal, as in a rat running through a maze, there are always stimuli that have become conditioned to the goal as a result of previous trials. In Hull's system these give rise to fractional antedating goal responses which are reinforced because they are followed by rewarding consequences. Subsequently, as explained by Mackintosh (1974), such theories became incorporated into incentive theory (incentive being anticipation of reinforcement). Thus an animal psychologist should not be surprised to find emergent anticipatory responses in robot behaviour, but should be impressed to see it demonstrated in such a simple system.

25.4 Cognitive architectures

As we saw in Chapter 19, there is an empirical issue in animal behaviour research as to whether or not animals possess explicit representations and are able to manipulate such explicit knowledge. An animal that does not have such representations can only have procedural knowledge. It can learn, but is has no cognitive ability.

The ability to manipulate explicit representations is necessary for cognition. Indeed, some scientists go further, and insist that cognition requires symbol manipulation. Where the explicit knowledge is a proxy for an object, property or event, it is usually called a **sign**. There is a straightforward one-to-one relationship between a sign and its referent (Hendriks-Jansen, 1996). **Symbols** are not straightforward proxies for objects. When explicit knowledge takes the form of a symbol, it leads the agent to conceive of the object. What a symbol signifies is an act of conception, which is reactivated anew on each occurrence. As Pylyshyn (1984, p. 15) points out: 'organisms can respond selectively to properties of the environment that are not specifiable physically, such properties as being beautiful . . .'.

Within the fields of classical artificial intelligence and robotics it is taken for granted that if you judge a behaviour to be intelligent you are committed to see the behaviour as resulting from rational, mental, cognitive, symbolic processing. The view of classical artificial intelligence is that cognition involves two things:

1. There must be internal (mental) representations of the world (including aspects of the agents themselves).

2. There have to be operations, or computations over these representations to yield new beliefs, such as beliefs about necessary means to accomplish some goal.

The representations are articulated in some sort of symbol system and, together, they make up the knowledge base from which an agent is supposed to reason and to decide what to do next. Within the field of artificial intelligence, intelligent behaviour is seen in terms of sense–think–act cycles. The relevant internal processing is conceived in terms of symbol manipulation in which a program delineates a series of instructions that effectively specify the information processing to be carried out by the machine. The aim is to write for every kind of interesting intelligent behaviour a program that, when implemented on a machine, enables the machine to exhibit the relevant intelligent behaviour.

This view is very different from the behaviour-based view, which seeks to apply Morgan's canon (see Chapter 1) by insisting on empirical evidence for mechanisms that are supposed to control behaviour. Behaviour-based robotics is much more biologically based. First, the notion of intelligence is very different from that of artificial intelligence (McFarland and Bosser, 1993). Second, the philosophical outlook is very different (Hendriks-Jansen, 1996). Third, classical artificial intelligence is widely considered not to have lived up to its promises, and the behaviour-based approach has been more successful in building robots (Brooks, 1986, 1989, 1991). However, this does not mean that we should abandon the possibility of cognitive processes. It may well be that some animals do have genuine cognition, and it may well be that cognitive processes can be useful in robots. Indeed, the robot is an ideal tool with which to investigate these matters, and there are two basic ways in which this can be done:

1. Robots can be used to investigate whether it is possible to employ cognitive processes to produce the behaviour that is thought to be due to cognition in animals. We look at an example of this in the next chapter.

2. Robots can be used to investigate whether aspects of animal behaviour that are thought to be due to cognition can be performed by robots that are known (by the designer) to have no cognitive ability.

In Chapter 19 we saw that there is evidence that rats can integrate separately formed associations, though whether a cognitive explanation is required remains a matter of controversy. Instrumental learning experiments currently provide the most convincing evidence of true cognition in rats (Heyes and Dickinson, 1990; Heyes, 1993). Specifically, the issue is whether an instrumental action (such as a rat pressing a bar to obtain a food reward) is mediated by **knowledge** of the contingency between the action and its outcome, whether this knowledge is conceived by expectation or belief (Dickinson and Balleine, 1994). Evidence that this is the case comes from experiments involving outcome-devaluation (e.g. Adams and Dickinson, 1981; Colwill and Rescorla, 1985), incentive learning (e.g. Balleine and Dickinson, 1991; Balleine, 1992), and the irrelevant incentive effect (e.g. Dickinson, 1986; Dickinson and Balleine, 1990). These, and numerous other experiments, strongly suggest that the control of instrumental action in rats is mediated by two processes. The first involves classical learned association between contextual stimuli and the outcome (e.g. reward) presented during instrumental training. The second is mediated by explicit knowledge of the contingency between the action and its outcome, and controls the value assigned to this outcome.

A typical outcome-devaluation experiment involves the experimental design illustrated in Table 25.2. Two groups of rats are trained to press a lever to obtain one type of reinforcer (food pellets or sucrose solution), and two other groups are trained to pull a chain to obtain the other type of reinforcer. After training the rats are removed to another apparatus, in another room, where they are given a treatment. They are allowed free access to a reinforcer, either when satiated (in some experiments) or when injected with lithium chloride (LiCl) which makes them sick (a procedure called food aversion conditioning, and described in Chapter 18). Typical experiments are those of Adams and Dickinson (1981) and Colwill and Rescorla (1985). Both found that the rats' response associated with the 'devalued' reinforcer was significantly attenuated (Figure 25.13). Colwill and Rescorla conclude that the 'structure underlying instrumental performance clearly contains detailed information about the nature of the reinforcer'. Dickinson (1985, 1989) takes the view that the results of outcome-devaluation experiments cannot be explained in purely procedural terms, but 'must be encoded in a proposition-like form so that it can be operated on by a practical inference process to generate the instrumental performance'. This is clearly a cognitive view involving manipulation of explicit (proposition-like) knowledge.

Table 25.2 Design of outcome-devaluation experiment

Half of a four group balanced experimental design (this one following Adams and Dickinson, 1981) can be summarized as:		
Lever press → Reinforcer 1 Chain pull → Reinforcer 2	Reinforcer 1 → LiCl	Lever press v Chain pull?
Lever press → Reinforcer 1 Chain pull → Reinforcer 2	Reinforcer 2 → LiCl	Lever press v Chain pull?
with the other two groups being those that receive the opposite instrumental contingencies.		

Fig. 25.13 Data extracted from Colwill and Rescorla (1985) showing outcome devaluation in rats. (a) Outcome devaluation following LiCl poisoning. (b) Outcome devaluation following reinforcer-specific satiety. Both (a) and (b) show that the behaviour associated with the devalued reinforcer is performed less than the alternative (From Spier, 1997).

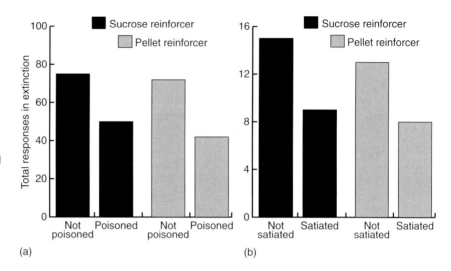

Clearly, in the case of animal work, it is extremely difficult to determine whether a cognitive explanation is necessary to account for a particular behavioural phenomenon. As we see in Chapter 26, for many apparently cognitive phenomena a non-cognitive explanation can be suggested. An advantage of the robotics approach is that there can be no argument about whether a mechanism involves cognition, since the designer can determine this.

Spier and McFarland (1998) set out to test whether the outcome-devaluation effect could be demonstrated by a robot using purely procedural learning algorithms. Using the apparatus described in Box 25.4, they programmed the animat with a purely procedural control algorithm, in effect setting up a simulated automaton. They then incorporated learning rules, maintaining the purely procedural status of the robot. They then set up an instrumental learning experiment, involving both lever press and chain pull, and showed that the agent showed learning curves typical of rats (see Box 25.4).

After 'training' the agent was subjected to reinforcer-devaluation experiments, using both satiation and aversive types of devaluation. The results demonstrated that the typical outcome-devaluation effect was occurring in this robot (see Figure 25.14). Here we have a situation in which a simulated robot, programmed with purely procedural algorithms containing no symbolic or explicit representations, is behaving in the same way as a rat in a typical outcome-devaluation experiment. This study shows that a purely procedural account of the outcome-devaluation effect is possible, and that a cognitive account is not necessary to explain the observed behaviour. Of course, this does not mean that the rat is not in fact using a cognitive method of solving the problem, but it does nullify the main argument supporting the cognitive position. Here we have an example of a robot being used as a tool to investigate a long-standing problem of animal behaviour.

Box 25.4 A simulated Skinner box

A computer simulation of a Skinner box used to test theories of animal learning. The animat (circle and radius) can obtain food from the food hopper (FH) by pressing a lever (LP), or pulling a chain (CP), but not by responding to an irrelevant cue (IC) (From Spier, 1997).

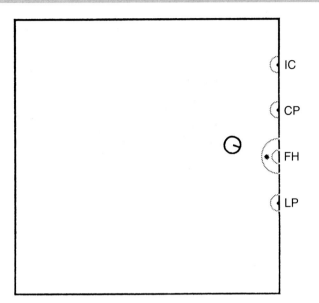

The naive animat learns to lever press and chain pull by trial and error, showing learning curves similar to those of rats (After Spier, 1997).

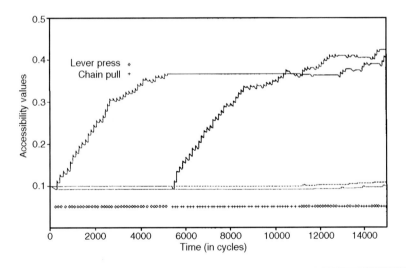

Fig. 25.14 (a) Responses during extinction of activities LP, CP, and IC following reinforcer devaluation in a simulated robot. 'A devalued' means that the response of lever pressing (LP) was devalued by 'poisoning'. 'B devalued' means that chain pull (CP) was devalued. (b) Responses during extinction of activities following reinforcer devaluation by satiation. 'A hungry' means that LP was devalued by satiation. 'B hungry' means that CP was devalued by satiation. Compare with the results for rats shown in Figure 25.13 (From Spier, 1997).

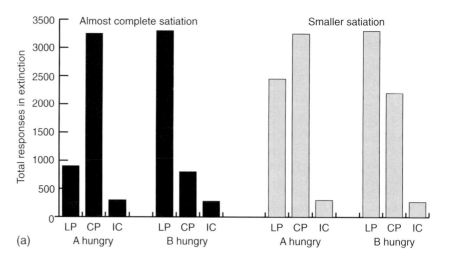

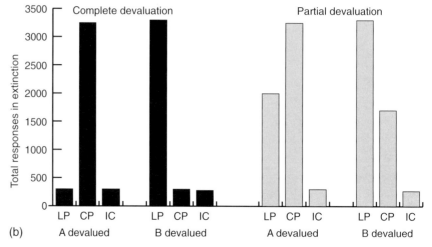

Points to remember

- The point of animal robotics is that instead of attempting to control the behaviour of the robot directly, and make it do what we want it to do, we design the robot to be autonomous, to want to do what we want it to do. This means that the robot should have a degree of self-sufficiency and autonomy. Thus animal robotics treats robots and animals as similar.

- An automaton is a machine whose behaviour is entirely determined by its state and by outside forces, as in any physical system. An important feature of automata is that they are situated.

- Situated actions are actions taken in the context of particular, concrete circumstances.

- The self-sufficient robot must have some degree of autonomy because it must have the freedom to do that behaviour which is in its own vital interests.

- Robots can be used to investigate both the possibility of employing symbolic processes to produce the behaviour that is thought to be due to cognition in animals, and whether aspects of animal behaviour that are thought to be due to cognition can be performed by robots that are known (by the designer) to have no cognitive ability.

Further reading

Benhamou, S. (1996) No evidence for cognitive mapping in rats. *Anim. Behav.* **52**, 201–212.

Benhamou, S. (1997) Path integration by swimming rats. *Anim. Behav.* **54**, 321–328.

Hendriks-Jansen, H. (1996) *Catching Ourselves in the Act.* MIT Press, Cambridge, MA.

McFarland, D. and Boesser, T. (1993) *Intelligent Behaviour in Animals and Robots.* MIT Press, Cambridge, MA.

The mentality of animals

Profile of Edward Tolman

(Photograph: Brendan McGonigle).

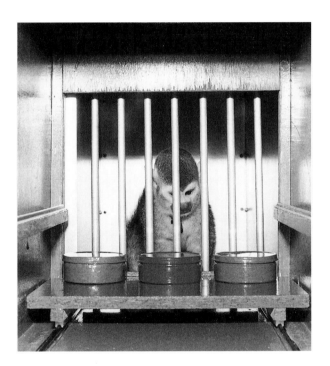

In this final group of three chapters we consider the controversial question of whether there is any affinity between the mental life of humans and other animals. Chapter 26 looks at language and mental representation, including questions such as whether apes are capable of language and whether animals have intentions. Chapter 27 looks at intelligence in animals, at tool use (sometimes taken as a sign of intelligence) and at cultural evolution. In Chapter 28 we discuss self-awareness and emotion in animals. We explore questions about consciousness and suffering in relation to animal welfare. Finally, we discuss the problem of animal suffering from an evolutionary perspective.

Edward Tolman (1886–1959)

Edward Tolman, an American psychologist, was a thorn in the flesh of his behaviourist contemporaries. His thinking was, in many respects, ahead of its time. He can be said to be the father of the modern cognitive approach to animal behaviour.

Unlike other cognitive theorists of his time, such as George Romanes or Wolfgang Kohler, Tolman's view was not mentalistic. His system was purposive, but not anthropomorphic. He believed that animals behaved in a purposive manner, but did not assume they had mental images of their goals.

Tolman regarded himself as a behaviourist, but he espoused a molar rather than a molecular behaviourism. In the molar view, behavioural acts have distinctive properties of their own, which can be described independently of the particular physiological processes responsible for the behaviour. Molecular behaviourism, on the other hand, is reductionist in that it seeks to account for behaviour in terms of the underlying physics and physiology. Tolman's major publication was *Purposive Behavior in Animals and Men* (1932), but he sub-sequently published many papers criticizing contemporary views, both by argument and experiment, and refining his own cognitive perspective.

Tolman's (1932) cognitive view of Pavlovian conditioning has much in common with modern thinking (Rescorla, 1978). For instance, he believed that the animal learns about the reinforcer, and not simply because of the reinforcer. He attacked the stimulus–response theory prevalent in his time and was a pioneer of the view that the conditional stimulus is a sign that some other event is to follow. Tolman also pioneered the idea of cognitive maps. His view was that animals acquire items of knowledge, or cognitions, which are organized so they can be utilized when needed. Challenging the prevalent view that animals 'learn by doing', Tolman demonstrated by experiment that animals could learn general features of a room or maze without performing the relevant behaviour. Tolman's evidence suggests that the animal acquires a cognitive map indicating how the relevant causal or spatial features of the environment relate to each other.

Tolman's theories were both hard-headed and sophisticated, and they were supported by experiments that provided a challenge other theories had difficulty in meeting. The criticism that Tolman's theories left the animal 'buried in thought' (Guthrie, 1952) without predicting its behaviour can be countered by the observation that the molecular behaviourist view leaves the animal 'lost in action'.

Language and mental representation

Do animals have a mental life similar to that of humans, or are they mindless automata? For the past hundred years this question has troubled animal psychologists and ethologists. During this period the prevailing opinion has swung between the two extremes. There has been a vast increase in our knowledge of animal behaviour and physiology, but the more we know, the more difficult the problem seems to become. In this chapter, and subsequent chapters, we draw together some lines of argument that have developed within different branches of animal psychology, and we attempt to arrive at a position from which the student can begin to form his or her own opinions about the mental life of animals.

There are many aspects of human behaviour which seem to set us apart from other animals. It used to be thought that only humans could make and use tools, but we now know that many other species have this ability (see Chapter 27). As our knowledge of animal behaviour has improved, so the difference between humans and other animals has appeared to diminish. Some human abilities, however, are difficult to substantiate in other species. One of these is language.

Language seems to us to be a uniquely human feature. Perhaps it is the only feature that distinguishes us from other animals. Whether or not this is true, we should not allow our anthropomorphic sentiments to cloud our judgement about the possibility of language in other animals. Unfortunately, there has long been a tendency to define language in a way that ensures it could only occur in humans. The most blatant of these attempts assert that language requires consciousness that only humans possess or that language depends upon speech of which only humans are capable. This is not an acceptable scientific procedure because it introduces an insurmountable bias into the investigation.

Unfortunately, language is not easy to define in objective terms because it has many necessary attributes. For example, we can agree that language is an aspect of communication but that not all communication is language. As we have seen (Chapter 22) humans make considerable use of non-verbal communication which is similar in principle to that of other animals. The question is, does human language embody principles that are not found in other animals?

Language

Human language usually involves speech, but it does not always require speech, as in the case of Morse code. Language is symbolic, but some aspects of communication in bees appear to be symbolic. Language is learned during a specifically sensitive period of development, but so is the song of some birds. Language can convey information about situations that are not immediate but distant in space and time. Some animal alarm calls have these attributes. Some aspects of language, like its grammatical rules, seem to separate it from other aspects of animal behaviour, but even this is controversial. In exploring the occurrence of language in the animal kingdom, we have to tread carefully.

A number of the characteristic features of human language can be found in other animals. For example, the signals employed in human language are arbitrary in the sense that they do not physically resemble the features of the world they represent. This abstract quality is found also in the communicative behaviour of honey-bees (*Apis mellifera*), the study of which was pioneered by Karl von Frisch (see Chapter 23).

The honey-bee dance is symbolic in a number of respects. Thus the rate of the waggle dance is indicative of the distance of the food source from the hive. The precise relationship between the rate of dancing and the distance is a matter of local convention. Different geographic races seem to have different dialects. Thus, one waggle indicates about 75 metres to a German honey-bee, about 25 metres to an Italian bee and only 5 metres to an Egyptian bee. Provided all the bees in the colony agree on the convention, it does not matter what precise values are used. Some researchers argue (e.g. Hinde, 1974) that since the symbols (direction and speed of dance) are physically related to the direction of the food source, they cannot be arbitrary. However, in any symbolic system that represents a range of values, there is bound to be some correspondence between the range of symbols and the reality. Another complication is that, although aspects of bee communication may seem symbolic to us, they may not be represented in the bee in a symbolic form. This is an aspect that we discuss below.

Other ethologists (e.g. Gould and Gould, 1982) regard the honey-bee dance as an example of arbitrary convention, arguing, for example that the bees could take north as a reference point instead of the sun. The honey-bee dance refers to situations that are remote from the communicating animal. This feature is widely considered to be an important property of human language. The dance can refer not only to food sources as remote in space (as far as 10 kilometres away) but also those that may have been visited some hours previously. During the intervening period, the forager bee keeps a mental track of the sun's movement and corrects the dance accordingly (see Chapter 23).

Another feature of human language is that it is an open system into which new messages can be incorporated. The bee dance can refer to new sources of food, but this seems to be a rather restricted example of openness. However, the dance can be used also to direct bees to water, to propolis (a type of tree sap used to caulk the hive) and to possible new hive sites at swarming time (Gould, 1981).

Of course, some features of human language like its acoustic quality, are not found in bee language. Some of these features may be seen in other aspects of animal communication, like bird song. Undoubtedly, human language is more sophisticated and complex than that of animals. But does this mean that there is a qualitative difference between human and animal communication, or is it merely a matter of degree? The question remains controversial.

If animals are capable of language, we might expect apes to be nearest to humans in this respect. The vocalizations and facial expressions of apes are subtle and complex, and it is possible that they converse among themselves, using a language we do not understand. There have been various attempts to discover whether or not apes are capable of language as we know it. The first of these were attempts to train chimpanzees to copy human speech. After several years of training, an orangutan was able to produce the words papa and cup. After prolonged training, the chimpanzee Viki managed the words **mamma**, **papa**, **cup** and **up** (Hayes and Nissen, 1971). In both cases the words were enunciated poorly, and it became apparent that apes simply do not have the vocal apparatus necessary to reproduce human speech sounds. In the chimpanzee and in the human fetus, the larynx is positioned high in the vocal tract, whereas in humans it is in a low position, as illustrated in Figure 26.1. This arrangement makes it possible for humans to alter the shape of the pharyngeal cavity with the tongue and so produce a wide range of modulated sounds. Chimpanzees and other apes simply are not capable of producing these sounds (Jordan, 1971; Lieberman, 1975).

Even though apes cannot speak, there remain other important questions

Fig. 26.1 Head and neck of adult human (a) and adult chimpanzee (b) (After Lieberman, 1975).

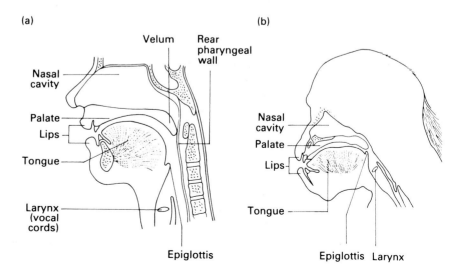

to be answered. Do they communicate among themselves in a language-like manner? Are they capable of understanding human speech sounds? Can they be trained to use human language through a medium other than speech?

Chimpanzees have a repertoire of about thirteen sounds, and they can provide gradations between them. They use the sounds both for communication at a distance and at close quarters. They can recognize the voices of individuals they know, and they continually make use of sounds to keep contact with each other in undergrowth or among other obstacles to visual contact. From the intention movements, facial expressions and scents and sounds produced by a chimpanzee, another member of the group usually can tell its identity, position, motivational state and likely behaviour. These are not, however, attributes of true language, so we must look for evidence of symbolic communication about the outside world.

A number of species have been reported to give alarm calls that differ according to the type of danger. Adult vervet monkeys (*Cercopithecus aethiops*) give different alarm calls when they sight a python, a leopard or a martial eagle. When they hear the alarm, the other monkeys take action appropriate to the predator that has been sighted. If it is a snake they look down, and if it is an eagle they look up. If they hear the leopard alarm, then they run into the trees (Seyfarth *et al.*, 1980). This phenomenon is called **functional reference**. It involves signals that have production specificity, discrete structure and context independence. The signals contain enough information about the circumstances for conspecifics to respond appropriately. The phenomenon has been discovered in chickens, and other species (Evans and Marler, 1995), and it appears that many animals have the ability to modulate their signalling behaviour, inhibiting calls when potential receivers are absent, or socially inappropriate, and increasing signal rate in other conditions. Thus signal production in animals is not simply a reflex response to sign stimuli, but may be flexibly dependent upon social context.

Emil Menzel (1974, 1979) conducted experiments with chimpanzees to see if they could convey information about the location of food. He set up a series of tests for six chimpanzees that were kept in an area of field. Accompanied by one of the group, he hid food in the field and then released all six chimpanzees and allowed them to search for the food. Usually, the group ran enthusiastically directly to the food and found it very quickly. However, the chimp that had seen the food hidden did not necessarily lead the group. When he hid a snake instead of food, the chimps approached it cautiously with evident signs of fear. In one experiment, Menzel showed one cache of food to one chimpanzee and another cache to a different chimp. When all the chimps were released, they would usually make for the more desirable of the two sources of food. Thus, there might be more food in one location than in the other, or there might be fruit in one location and less desirable vegetables in the other.

It seems that chimpanzees can make deductions about their environment that are based upon the behaviour of a companion. The chimp who knows the location of the food indicates by his actions and emotions the desirability and direction of the goal. There is no direct communication about the location of the food, but the other chimpanzees are intelligent enough to draw their own conclusions (Menzel and Johnson, 1976). Some evidence indicates that chimps can learn to point to objects in the laboratory (Terrace, 1979; Woodruff and Premack, 1979) and that they can learn to use human pointing as a clue as to the whereabouts of food (Menzel, 1979). However, they do not seem to use pointing or other indications of direction in communicating among themselves. Modulation of food calling, in relation to social context, also occurs in chickens (Evans and Marler, 1995), and it seems that this type of communication flexibility is not confined to primates.

26.2 Teaching apes to converse

Although it seems that chimpanzees normally do not communicate about objects remote in time or space, it may be that they could be taught to do so. If we knew the extent to which apes were capable of handling various features of language, we perhaps could gain some understanding of our own abilities.

The Kelloggs reared a chimpanzee called Gua in their home (Kellogg and Kellogg, 1933). They claimed that she learned to understand 95 words and phrases in about eight months, about the same as their son Donald, who was three months older. Gua was tested by being given a card with four pictures on it. Another chimpanzee, called Ally, was also reared in a human house and gained some understanding of speech. She was taught the gestures that correspond to various words in Ameslan (American Sign Language). She could make the correct sign when she heard the word spoken (Fouts *et al.*, 1976). Other experiments, with a gorilla (Patterson, 1978) and a dog (Warden and Warner, 1928), suggest that these animals are able to associate sounds and visual cues.

Some experimenters have tried to investigate the extent to which primates have voluntary control over the sounds they make. In one experiment, rhesus monkeys were required to bark when a green light came on and to produce a coo to a red light. The monkeys learned to produce the correct sound to obtain food reward (Sutton, 1979). An orangutan has been trained to make three different sounds to obtain food, drink or contact with the keeper (Laidler, 1978), and a chimpanzee has been trained to bark to induce a human to play (Randolph and Brooks, 1967). As a control, the chimpanzee was also trained to initiate play by touching its human companion when she was facing, and barking when she had her back to the chimp. These experiments suggest that apes and monkeys may have some limited voluntary control of the sounds they produce. Without special training, they do not seem to mimic the sounds they hear, even when they are living with a human family (Kellogg, 1968). The fact that they readily imitate human actions, however, suggests that sound is not a medium that apes readily learn to use for communication beyond their normal limited repertoire (Passingham, 1982).

Once it had become clear that speech was not a necessary aspect of language, and that the ability to produce or respond to sounds was not necessarily relevant to the question of whether or not animals were capable of language, then the way was open to investigate language by manipulating visual symbols (Gardner and Gardner, 1969; Premack, 1970). The Gardners started with a single chimpanzee, called Washoe, to whom they taught Ameslan. In this language, words are represented by gestures of the hand and arm. Washoe was trained from the age of eleven months, and by the age of five years she had mastered 132 signs (Gardner and Gardner, 1975). Washoe spontaneously learned to combine signs into strings of two to five words. Among her first combinations were 'come open' and 'gimme sweet' (Gardner and Gardner, 1971). The Gardners have since trained two chimpanzees from birth, and these progressed more quickly than Washoe. Others have been taught by Roger Fouts (1975), and a gorilla has been taught to use hand signs and to respond to spoken English commands (Patterson, 1978). Herbert Terrace (1979) and co-workers trained a chimpanzee called Nim Chimpsky to use Ameslan. This was a thorough study in which a transcript was kept of every sign Nim ever made and every combination he produced.

David Premack (1976, 1978) taught a chimpanzee called Sarah to read and write. He used coloured plastic shapes to represent words. The shapes in no way resembled the things to which they referred (Figure 26.2). The shapes were presented on a vertical magnetic board, and Sarah could answer questions by placing appropriate shapes on the board. Sarah mastered 120 plastic symbols, although she was not pressed to attain a large vocabulary (Premack, 1976). She could carry out commands and answer questions using several symbols in combination. Other chimpanzees also have been trained in this method by Premack and his colleagues.

Another method was employed by Duane Rumbaugh (1977). Using an artificial grammar called Yerkish (von Glasersfeld, 1977), a chimpanzee called Lana learned to operate a computer keyboard that displayed word symbols on a screen. The computer was programmed to recognize grammatical and

Fig. 26.2 Example of a question asked of the chimpanzee Sarah by means of plastic shapes. Sarah could respond by choosing one of the alternatives below (After Premack, 1976).

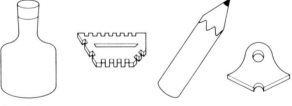

Question
'Is A (the bottle) the same as B (the pencil)?'

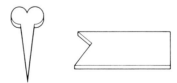

Alternatives:
'Yes' or 'no'

incorrect usage of the symbols and to reward Lana accordingly. The advantage of this approach is that Lana was able to converse with the computer at any time of day instead of having to attend formal sessions. Other chimpanzees have also been trained to converse with each other by the computer-based method (Savage-Rumbaugh *et al.*, 1978, 1980).

In assessing the results of these experiments, it is necessary to bear in mind the possibility that the chimps might cheat. They might make use of involuntary hints given by their trainers, or they might simply learn a succession of tricks, like a circus animal.

The Gardners (1978) tested Washoe under conditions in which the trainer did not know the answer to the question. Washoe was required to name the object shown on a slide by making the appropriate sign to a nearby person who could not see the slide. A second person could see Washoe's gestures without being seen by Washoe and without being able to see the slides. Washoe was asked to name 32 items, each of which was shown to her four times. She gave the corrrect answer to 92 out of 128 questions. Similar tests have been conducted on some of the other chimps used in these studies (Patterson, 1979; Premack, 1976; Rumbaugh, 1977).

It is possible that the chimps learn what to do when they perceive a certain cue, just as a circus animal learns what to do upon receipt of cues from its trainer. In order to discover whether or not the chimps understand the meaning of the signs and symbols they manipulate, it is necessary to conduct an experiment in which the chimp is required to name the object in a situation different from the one in which it learned to name it. Various tests have been conducted with this point of view (e.g. Gardner and Gardner, 1978; Savage-Rumbaugh *et al.*, 1980), and the results indicate that chimps can genuinely name objects. Sometimes, moreover, chimps do this spontaneously. Thus, Nim would give the sign for dog upon seeing a dog or a picture of a dog or upon hearing the bark of a dog (Terrace, 1979).

The evidence suggests that chimpanzees can come to understand the meaning of words in the sense that they can name things genuinely. However, in many other aspects of human language, their abilities are not so evident. Some of these are primarily of interest in assessing the cognitive abilities of chimpanzees. Of particular interest is the question of whether or not the chimp can incorporate new messages into its repertoire. We considered this issue to be of some importance in assessing the bee dance (see Chapter 23).

It appears that chimpanzees do sometimes coin novel phrases. Thus, Washoe is reported to have invented the word 'candy drink' for watermelon and to have called a swan a 'water bird'. Such instances are difficult to interpret, however, because of the possibility that apparently novel usage is the result of simple generalization. For example, Washoe was introduced to the sign of a flower by being offered a real flower. She learned the sign but applied it not only to flowers but also to tobacco and cooking smells. Apparently, Washoe associated the sign with the smell of the flower and generalized to other odours (Gardner and Gardner, 1969).

Another problem is that the chimps sometimes make novel word combinations that appear to make no sense. Nim's favourite food was banana, and he often combined this word with other words such as drink, tickle or toothbrush. It is possible that 'banana toothbrush' is a request for a banana and for a toothbrush to clean the teeth after eating the banana. However, this seems unlikely because a banana and a toothbrush would not be on view at the same time, and Nim never asked for objects that he could not see (Ristau and Robbins, 1982). Perhaps the bizarre combinations of words are an example of word play, similar to that found in children. Washoe was observed to sign to herself when playing alone, much as children talk to themselves.

Thus, we can say the attempts to teach chimpanzees and other apes various types of human language have had limited success. The apes seem to be able to attain a standard equivalent to that of a young child. The difference between ape and child may simply be one of intelligence, but it is also possible that humans have an innate language acquisition device, as first suggested by Noam Chomsky (1972). In any event, the experiments with apes certainly have uncovered abilities that were not suspected previously, and they have given us considerable insight into apes' cognitive abilities.

Many of the fundamentals of langue are not specific to *homo sapiens* (Savage-Rumbaugh and Brakke, 1996; Evans and Marler, 1995), but the question of whether the difference between man and other animals is one of capacity (e.g. memory) or is a genuine structural difference remains controversial.

26.3 Language and cognition

We have seen that chimpanzees, and other apes, can be taught to converse with humans by using sign language or by reading and writing with plastic shapes or symbols presented by a computer. The evidence suggests that chimpanzees can learn the meaning of words in the sense that they can name things. They can master a vocabulary of more than a hundred words, and it appears that they sometimes coin novel phrases.

A question of considerable interest is whether or not the various ape language projects reveal **cognitive** abilities. In Chapter 19 we discussed the distinction between knowing how and knowing that. An ape may know how to ask for a reward in the sense that it can learn to make the appropriate gesture. However, this is not the same as knowing that making a particular gesture will obtain a reward for it. To know that is to understand a relationship beyond the mere coupling of stimulus and response. In humans, knowing how may cover complex skills such as fast typing or golf, where the performer can achieve a good result without understanding, or being able to describe, the relationship between the objective of the behaviour and its performance. In other cases, people clearly do know about the processes that led them to a particular objective and can describe them. The question of whether or not the results of ape language experiments allow us to make this distinction is discussed by Ristau and Robbins (1981, 1982).

Some of the results of the Lana project (Rumbaugh and Gill, 1977) are relevant here. Lana could use phrases made up of plastic symbols to request specific rewards. She might write: 'Please machine give banana.' This can be described readily as knowing how to obtain a banana. However, attempts were also made to teach Lana to name two items, banana and M&M candy. She was presented with either a banana or an M&M on a tray and asked (via the experimenter's computer keyboard): '?What name-of this'. The correct response (via a keyboard) was: 'banana name-of this'. Correct responses were rewarded. Lana required 1600 trials to learn this task, even though she had requested these items previously, hundreds of times, by using the standard phrase: 'Please machine give banana'. This suggests that Lana did not know the meaning of the original symbols that she had manipulated previously to request an M&M or banana. However, it is worth noting that Lana originally was not required to know the meaning and that requesting items in a routine knowing-how way could have become a habit that was difficult to break away from. This interpretation is supported by the fact that subsequent training, involving naming other items, results in much more rapid success. The initial difficulties in labelling may have been the result of an inability to distinguish the requirements of the new situation from that in which the banana and M&M symbols were used originally.

In experiments with Sarah, Premack (1976) taught the concept 'name-of' in the following way. The plastic symbol for apple was placed in front of a real apple with a gap between them. Sarah was required to place a new plastic chip, intended to mean 'name-of,' into this gap. She thus apparently created the sentence '"Apple" name-of object apple.' 'Not name-of' was formed by gluing the normal negative symbol onto the plastic symbol for 'name-of'. When the symbol for 'apple' was placed in front of a banana with a gap between, Sarah was required to select the correct symbol to fill the gap (Figure 26.3). In this case, the choice of 'not name-of' would earn Sarah a reward. She was able to use these labels correctly in both naming tests and sequences with other plastic symbols.

The ability to learn that abstract shapes are symbolic of objects in the real world suggests a kind of 'knowing that' that is similar to explicit or declarative representation. However, it is difficult to devise a demonstration

Fig. 26.3 Procedure used to teach Sarah the concept 'name-of'. (a) The 'name-of' symbol is placed between the symbol for apple and a real apple. (b) The 'name-of' symbol is placed between the banana symbol and a real banana. (c) The 'name-of' symbol preceded by a 'not' symbol, is placed between the banana symbol and an apple (After Premack and Premack, 1972).

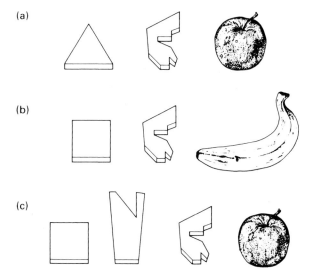

(a)

(b)

(c)

that is divorced from all procedural possibilities. Perhaps Sarah was merely learning that the two shapes 'name-of' and 'not name-of' were responses that should be made when two things (i.e. the apple and its lexicon) were equivalent. Much of the controversy about animal language hinges around the nature of mental representation, and we now look at this question.

26.4 Mental representation

In Chapter 19 the differences between procedural, explicit and declarative representations were discussed. Briefly, a **declarative representation** is a representation of knowledge about something. It is knowledge **that** such and such is the case, and can be declared to be the case. Because animals cannot use language to declare things, the animal equivalent of a declarative representation is an **explicit representation**. Knowledge **how** to perform some behaviour involves **procedural representations**, a set of instructions relating to some procedure. In a declarative or explicit system, knowledge is represented in a form that corresponds to a statement or proposition describing a relationship among events in the world. This form of representation does not commit the agent to use the information in any particular way. In a procedural system, however, the form of representation directly reflects the use to which the knowledge will be put.

If a human subject is told that A is bigger than B, and B is bigger than C, and then asked if A is bigger than C, we would expect a normal adult human to be able to infer that C is smaller than A, from the information provided. Such a problem is called a **transitive inference problem**. To us it seems logical that if A>B>C, then A>C, and a person capable of logical thinking should be able to attain the correct answer to this type of question. It is usually taken for granted that the ability to solve such problems involves manipulation of declarative representations.

Many transitivity experiments have been carried out on human adults and children, but McGonigle and Chalmers (1986) point to two main problems with the majority of such experiments. The first is that people learn an evaluative ordering more readily from better to worse than from worse to better. Consequently, when asking a question of the type 'if x is smaller than y, etc.', it is difficult not to introduce an incongruity between the form of the question and the subject's directional bias. The second problem concerns 'mental distance'. This concept implies that the more remote the items to be compared 'in the mind's eye', the faster decisions are made. Such results have been well established by Trabasso and Riley (1975), but it is not clear what these phenomena imply for the nature of mental representation.

McGonigle and Chalmers (1986) suggest that these phenomena may be due to prelogical structures, in which case they may be present in subjects that are incapable of using conventional logical procedures. In order to circumvent these methodological problems, they suggest two types of experimental strategy. First, to deal with the congruity issue, they suggest an alternative paradigm known as 'internal psychophysics' (Moyer, 1973). This requires subjects to decide as rapidly as possible the serial relationships between objects from memory alone. For example, subjects might be presented with a pair of names of animals such as 'hen' or 'elephant' and asked to denote the larger by pressing a switch below a panel bearing a printed name, or a picture (scaled to a standard size). 'As the knowledge representation is assumed to be established prior to the task, the problem of congruity endemic to the logical task can be eliminated. Now there is no basis for ambiguity. The degree of mapping achieved has to be between the question and the representation, and not between the question, the informing statement and the representation' (McGonigle and Chalmers, 1986).

Second, to investigate the question of prelogical structures, McGonigle and Chalmers (1986) conducted experiments with younger subjects, 'whose failures in logical tasks have been documented (e.g. Inhelder and Piaget, 1964) and monkeys (not well-known for their logical skills)' (McGonigle and Chalmers, 1986, p. 146). They carried out experiments in which six- and nine-year-old children were required to compare the sizes of familiar animals, presented either as written names (lexical mode), or as pictures (pictorial mode) of a standard size, as described above. They measured the time taken to compare symbols, and they also required the children to verify statements of size relationship in the conventional manner (e.g. is a cow smaller than a cat?). Their results with the conventional methods show the 'symbolic distance effect' obtained by Trabasso and Riley (1975). That is, the time taken to compare stimuli varies inversely with the distance between the stimuli along the dimension being judged. Along the size dimension, therefore, the time taken to compare the relative sizes of 'cat' and 'whale' is less than taken for 'cat' versus 'fox'. This is the type of result usually obtained with adults (Moyer, 1973; Paivio, 1975).

Similar results were obtained when six-year-old children were presented with pictures scaled to equal size, as illustrated in Figure 26.4. Overall,

Fig. 26.4 Symbolic distance effect produced by six-year-old children comparing sizes of an insect and a mouse (After McGonigle and Chalmers, 1986).

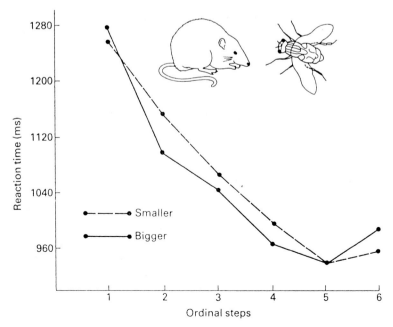

Fig. 26.5 Squirrel monkey (a) and child (b) performing similar transitivity tasks. (*Photograph: Brendan McGonigle*).

(a)

(b)

McGonigle and Chalmers (1984) found that children as young as six show a significant symbolic distance effect in both pictorial and lexical modes when the simple comparative question (bigger or smaller?) is used in the test. They also found marked categorical asymmetry, particularly in the lexical mode. Not only was the time taken to judge an item as 'big' faster than that to judge one as 'small', but even for items judged as 'small' it was faster to deny that they were 'big' than to affirm that they were 'small'.

McGonigle and Chalmers (1986) report a series of experiments on squirrel monkeys, designed to test their abilities in transitive inference problems, in a situation similar to that used for testing children (Figure 26.5). In one experiment five monkeys were required to learn a series of conditional size discriminations such that within a series of size objects (ABCDE) they had to choose the larger or largest one of a pair or triad if, say, the objects were black; if white, they had to choose the smaller or smallest one (McGonigle and Chalmers, 1980). They found that there was a significant and consistent effect of direction of processing, such that decisions following the 'instruction' to find the bigger were made more quickly than those following the signal to find the smaller. Through practice, the animals became progressively faster, yet the absolute difference between the 'instruction' conditions remained invariant. Figure 26.6 summarizes some of the results of this experiment, and compares them with the results of similar experiments on children.

In another experiment, based on a modification of a five-term series problem given to very young children by Bryant and Trabasso (1971), they tested monkeys on transitive inference tasks in which the animals were trained to solve a series of four discrimination problems. Each monkey was confronted with a pair of differently coloured containers that varied in weight (A>B). When B was chosen reliably over A, the monkey moved to the next problem

Fig. 26.6 Categorical and contrastive effects produced by (a) monkeys and (b) children (After McGonigle and Chalmers, 1986).

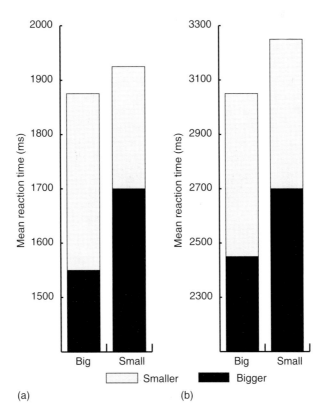

(B>C, where C must be chosen), and so on until the entire series was performed correctly. Only two weight values were used throughout the series, so no specific weight could be uniquely identified with the stimuli B, C or D. When the monkeys had learned to achieve a high level of performance on all four training pairs, regardless of presentation order, transitivity tests were given. In these, novel pairings were presented, representing all ten possibilities from the five-term series. The results showed impeccable transitivity, indistinguishable from the results obtained with six-year-old children (Bryant and Trabasso, 1971). Analysis of decision times revealed a significant distance effect in which the decision times for non-adjacent comparisons were significantly shorter than those for solving the training pairs, as shown in Figure 26.7.

On all major points of comparison, McGonigle and Chalmers (1986) found that the monkeys were identical in performance to young humans. Similar profiles in six-year-old children, using both non-verbal and verbal forms of the same task, have also been reported (McGonigle and Chalmers, 1984). So neither the nature of the task within a species, nor comparison of performance between species, seems to affect the conclusion that the symbolic distance effect (usually taken as evidence of cognitive processing in human adults and older children), and asymmetry in the direction of encoding (a characteristic feature of human transitive inference) occur in subjects unable to perform formal logical tasks. McGonigle and Chalmers (1986) come to the reasonable conclusion that the ability to order items transitively is a prelogical phenomenon.

While it is clear that monkeys do not solve transitive inference problems by manipulating declarative representations, it seems likely that the serial ordering involved in some of the more complex problems could be represented explicitly. In other words, some animals may be capable of non-verbal thinking (McGonigle, 1987). Evidence that this may well be the case comes from experiments in which children and monkeys are given complex serial search problems (McGonigle and Chalmers, 1998). Some of these problems enable the subject to develop their own stimulus classification and

Fig. 26.7 Symbolic distance effect in monkeys: (a) all comparisons, (b) non-end-anchored comparison (After McGonigle and Chalmers, 1986).

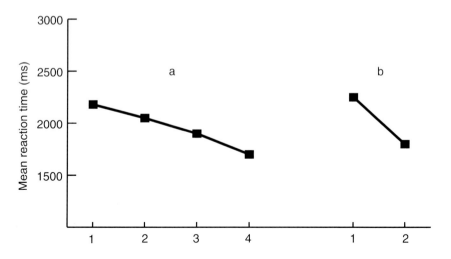

Box 26.1 Object sorting by the Edinburgh R2 robot

This robot is set the task of finding five objects and transporting them to its workbench. Then it must arrange them in a line in a particular order, as shown in the photograph. To achieve all this the robot has to have particular **competences**, such as that required for navigation, transporting an object, object discrimination, seriation, etc. As is the case with animals, the **state** (see Chapters 15 and 25) of the robot determines the **task gram-** mar, enabling the robot to interpret information in a way that is appropriate to the task in hand. Within this general framework, questions can be asked about the way this information must be represented to carry out the different kinds of task. Especially pertinent is the question of whether symbol manipulation is a necessary part of sorting behaviour (see McGonigle, 1999, for further details).

search strategy. The evidence suggests that the subjects develop cognitive strategies that economize on time, effort and memory load. Such an ability implies a certain type of **cognitive architecture** in which there is a degree of both modularity and flexibility. Such ideas can be tested out on robots (see Chapter 25). Thus a robot can be set tasks similar to those given to monkeys and children, as shown in Box 26.1.

26.5 Intentional behaviour

Human behaviour can be said to be intentional when it involves some representation of a goal that is instrumental in guiding behaviour. Thus, if I have a mental picture of the desirable arrangement of books on my shelf, and if this mental representation guides my behaviour in placing books on the shelf, then I can be said to place them intentionally. If, however, I place books on the shelf haphazardly, or on the basis of simple habit rules, then my arrangement of the books may not be intentional. For my behaviour to be intentional, my mental representation of the book arrangement does not have to be a conscious one. Although consciousness and intentionality are sometimes linked, it is better to treat them separately (Dennett, 1978).

Ethologists have long recognized intention movements (see Chapter 21) in animals as indications of what the animal was about to do. Both human observers and members of the animal's species can predict the future behaviour of an animal from its intention movements, and it may seem silly to assume that the animal cannot anticipate the next steps in its own behaviour (Griffin, 1976). It is more likely, however, that intention movements are merely the initial stages of behaviour patterns that are terminated prematurely, either because the animal is in a motivational conflict or because its attention is diverted to other possible aspects of behaviour. Indeed, it is difficult to imagine how such incipient behavioural fragments could be avoided in an animal with a complex repertoire of activities. Although some so-called intention movements may have become ritualized during the course of evolution, this does not mean that they are intentional in the ordinary sense of the word.

Another aspect of animal behaviour that has the appearance of intentionality is the injury-feigning distraction display of certain birds. When an incubating bird, like a sandpiper (*Ereunetes mauri*), is disturbed by a ground predator, it may leave the nest and act as though injured, trailing an apparently broken wing and luring the predator away from the nest. When the predator has been led a safe distance from the nest, the bird suddenly regains its normal behaviour and flies away (Brown, 1962; Skutch, 1976). While most ethologists are content to account for this type of behaviour in terms of ritualized display, some (e.g. Griffin, 1981, p. 135) wish to keep open the possibility that the birds are behaving intentionally. Griffin notes that the possibility of animals having mental states or intentional behaviour is often most vehemently denied in cases were the evidence is weakest. The problem is to know what does count as evidence. In Griffin's view, evidence is most likely to come from studies of animal communication, particularly that of primates.

Do the various studies of chimpanzee communication throw any light on this question? George Woodruff and David Premack (1979) explored the chimpanzee's ability in intentional communication by setting up a situation in which a human and a chimp could cooperate or compete in obtaining food. They communicated by means of non-verbal signals about the location of hidden food. When the human was cooperative and gave all the food that was found to the chimpanzee, the chimpanzee could successfully send and receive behavioural signals about the location of the food. When the human and chimp were in competition with each other and the human kept the food that was found, the chimpanzee learned to mislead the human by withholding information and by discounting misleading behavioural cues given by the human. The chimpanzee's behaviour suggests that chimps can infer purposes, or intentions, from human behaviour and that they can have some knowledge of the human's perception of their own behaviour.

Donald Griffin (1981) comments on cases (Ruppel, 1969) in which a mother Arctic fox was competing for food with several of her nearly full-grown young. In order to reach the food first, a young fox might, for example, urinate in its mother's face. After several such tricks, the mother gave false alarm calls and seized the food for herself when the young ran off. Griffin notes that it is difficult to interpret such behaviour without postulating short-term intentions and plans on the parts of both mother and young. However, we must ask if these animals are really inferring motives in others or simply learning effective means of obtaining food in various situations.

In the case of primates, the situation is even more complicated, on account of their sophisticated social behaviour (see Chapter 10). To unravel the complications, we need to focus on communication. A purely procedural form of communication would provide other animals with fairly reliable information about the state of the sender, and about the behaviour that it is likely to follow. If the communication is honest then the sender is open to exploitation by conspecifics. Thus if an incubating gull gives an alarm call and flies off the nest, a neighbour can seize the opportunity to steal its eggs. If the procedural communication is dishonest, then the sender will eventually be discriminated against. Thus the animal that persistently cries 'wolf' to gain an advantage, runs the risk of having its communication devalued or discounted (see also Chapter 21). Procedural communication has a function, but it has no 'meaning' in the sense that it is part of a symbolic system (see Chapter 25). The advantage of a symbolic communication is that it need give no clear indication of what behaviour is likely to follow. This makes it difficult for rivals and enemies to take advantage of the situation. The question of whether there is any hard evidence that non-human primates are capable to symbolic communication is a matter of current controversy.

In recent years there has developed a considerable interest in the topic of deception (Mitchell and Thompson, 1986; Trivers, 1985). This follows from evolutionary arguments to the effect that communication is not simply the transmission of information, but can be expected to be designed (by natural selection) to manipulate other individuals (Dawkins and Krebs, 1978). However, it is evident that a large proportion of the deception practised by animals is entirely procedural, as in the numerous instances of mimicry (Trivers, 1985). Some of the literature seems very confused and equivocal on this point (see Kummer *et al.*, 1990 for a critical review), and much of the evidence is anecdotal. For example, de Waal (1982) in his study of chimpanzee social behaviour notes that:

'Yeroen hurts his hand during a fight with Nikkie . . . Yeroen walks past the sitting Nikkie from a point in front of him to a point behind him and the whole time Yeroen is in Nikkie's field of vision he hobbles pitifully, but once he has passed Nikkie his behaviour changes and he walks normally again. For nearly a week Yeroen's movement is affected in this way whenever he knows Nikkie can see him.'

Some laboratory studies suggest that chimpanzees can make use of cognitive deception. In one study the development of a new behaviour, pointing, was used advantageously to indicate the correct or incorrect container. Chimpanzees rarely point in laboratory or field situations, though they do understand pointing by humans. A second observation concerns a chimpanzee Sadie, who pointed to an empty container when asked which contained food. When the human lifted the container and discovered no food, 'Sadie's head snaps in the direction of the other container, which she knew contained the food' (Premack and Woodruff, 1978).

It is possible to imagine both procedural and declarative explanations of these data. How these might be distinguished remains an unresolved question (Heyes, 1993, 1994). Despite claims to the contrary (e.g. Byrne and Whiten, 1988; Whiten and Byrne, 1986) convincing evidence of **mental state attribution** remains elusive. Indeed, even some studies of children are difficult to interpret. In one study, children aged two–three participated in a game in which a treasure could be hidden in one of several cups by a puppet that left inky tracks. When children were asked to hide the treasure so that an adult who had left the room would not find it, they used a sponge to wipe away tell-tale tracks, or made false trails to empty cups (Chandler *et al*, 1989). This result seems to show that the children were capable of attributing ignorance, or false belief, to another person, and of acting to encourage such false belief. Subsequent studies show that children under four, who erase tracks and leave misleading trails, do not behave in a way that is consistent with this interpretation. When asked where the adult would believe the treasure to be hidden, they pointed to the cup that actually contained the treasure (Sodian, *et al.*, 1991). It appears that children become capable of intentional, rather than functional, deception after the age of four (Heyes, 1994).

It appears at times that the argument between behaviourist and cognitive explanations of behaviour is never ending (see, e.g., the correspondence on Premack and Woodruff's paper in *Behavioural and Brain Sciences*, 1978). For any set of behavioural observations, it seems, each side can come up with an alternative explanation. However, Daniel Dennett (1978) notes that in considering the behaviour of a complex system, we can take a number of different stances that are not necessarily contradictory. One is the design stance. If one knows exactly how a system is designed, one can predict its designed response to any particular situation. In ordinary language, we use this stance in predicting what will happen when we manipulate an object with a known function; thus, 'strike the match and it will light'. In biology, we use the design stance in making predictions based upon the theory of natural selection. In fact, we have two stances at our disposal: the **maximize-inclusive-fitness** stance, and the **selfish gene-eye** view advocated by Dawkins (1976, 1982) (see Chapter 6).

Dennett (1978) also advocates an **intentional stance** to the behaviour of complex systems. This stance assumes that the system under investigation is an intentional system and that it possesses certain information and beliefs and is directed by certain goals. In advocating this stance, Dennett is not attempting to refute behavioural or physiological explanations but to offer a higher level of explanation for the behaviour of systems that are so complex that other stances become unmanageable:

'An intentional system is a system whose behaviour can be (at least some-times) explained and predicted by relying on ascriptions to the system of beliefs and desires (and other intentionally characterised features) – what I will call intentions here, meaning to include hopes, fears, intentions, perceptions, expectations, etc. There may, in every case be other ways of predicting and explaining the behaviour of an intentional system – for instance, mechanistic or physical ways – but the intentional stance may be the handiest or most effective or in any case a successful stance to adopt, which suffices for the object to be an intentional system.' (Dennett, 1978)

Dennett (1983) maintains that ethologists and others studying animal behav-iour from a cognitive viewpoint are in need of a descriptive language and method that is open minded and capable of empirical verification. He pro-poses that intentional system theory can fulfil this role. He quotes the fol-lowing example from Seyfarth *et al.* (1980):

'Vervet monkeys give different alarm calls to different predators. Recordings of the alarms played back when predators were absent caused the monkeys to run into the trees for leopard alarms, look up for eagle alarms, and look down for snake alarms. Adults call primarily to leopards, martial eagles, and pythons, but infants give leopard alarms to various mammals, eagle alarms to many birds, and snake alarms to various snakelike objects. Predator classifi-cation improves with age and experience.

In adopting an intentional stance toward vervet monkeys, we regard the animal as an intentional system and attribute beliefs, desires and rationality to it, according to the type or order of intentional system that is appropriate. A zero-order (or nonintentional) account of a particular monkey, called Tom, who gives a leopard alarm call in the presence of another vervet, might be as follows: Tom is prone to three types of anxieties – leopard anxiety, eagle anxiety and snake anxiety – each producing a characteristic vocalization that is produced automatically, without taking account of its effect upon other vervet monkeys.'

A first-order intentional account might maintain that Tom wants to cause another monkey, Sam, to run into the trees. Tom uses a particular vocaliz-ation to stimulate this response in Sam. A second-order intentional account goes a step further in maintaining that Tom wants Sam to believe that there is a leopard in the vicinity and that he should run into the trees. A third-order account might say that Tom wants Sam to believe that Tom wants Sam to run into the trees.

Dennett maintains that the question of which order of intentionality is appropriate is an empirical question. For example, suppose a lone male vervet monkey, travelling among groups and out of hearing of other vervets, silently seeks refuge in the trees upon seeing a leopard. We then might be inclined to dismiss the zero-order account. Vervet monkeys can recognize the individual voices of other members of the group, so if Tom's leopard alarm call was broadcast from a tape recorder in a situation where Sam could see Tom, then we might wish to abandon the third-order intentional expla-nation. If we observed that Sam did take to the trees under such circumstances, this behaviour would not be rational, and rationality is an assumed property of an intentional system (Dennett, 1983).

The assumption that intentional systems are rational enables us to construct what Dennett (1983) calls the Sherlock Holmes method:

'In *A Scandal in Bohemia*, Sherlock Holmes' opponent has hidden a very important photograph in a room and Holmes wants to find out where it is. Holmes has Watson throw a smoke bomb into the roof and yell 'fire' when Holmes' opponent is in the next room, while Holmes watches. Then, as one would expect, the opponent runs into the room and takes the photograph from where it is hidden. Not everyone would have devised such an ingenious plan for manipulating an opponent's behaviour, but once the conditions are described, it seems very easy to predict the opponent's actions.' (Cherniak, 1981, p. 161)

The Sherlock Holmes method is not foolproof, however, as shown by the following story. As a graduate student, I used to experiment at making pigeons frustrated, by presenting food on some occasions and not others or by presenting food that the bird could not obtain. One particular pigeon used to behave rather aggressively toward me, and I formed the impression that he did not like my treatment of him. One day I inadvertently performed a Sherlock Holmes experiment and allowed the pigeon to escape from his cage. He immediately marched over to the tangle of electrical wires that controlled the apparatus and started to pull them apart with his beak. Feeling rather shaken and guilty, I quit the room to find coffee. Upon returning I realized, from his behaviour, that the pigeon regarded the wires as nest material and that his aggression toward me was typical of the early stages of courtship.

Nevertheless, such an approach might work in certain circumstances. For example, in the Premack and Woodruff (1978) experiments, Sadie points to the box containing food when the cooperative human enters the room and to the box containing no food when the uncooperative human enters. How could we test the theory that Sadie intended to deceive the uncooperative human? Dennett (1983) suggests that:

'We introduce all the chimps in an entirely different context to transparent plastic boxes; they should come to know that since they – and anyone else – can see through them, anyone can see, and hence come to know, what is in them. Then on a one-trial, novel behaviour test, we can introduce a clear plastic box and an opaque box one day, and place the food in the clear plastic box. The competitive trainer then enters, and lets Sadie see him looking right at the plastic box. If Sadie still points to the opaque box, she reveals, sadly, that she really doesn't have a grasp of the sophisticated ideas involved in deception. Of course this experiment is still imperfectly designed. For one thing, Sadie might point to the opaque box out of despair, seeing no better option. To improve the experiment, an option should be introduced that would appear better to her only if the first option was hopeless, as in this case. Moreover, shouldn't Sadie be puzzled by the competitive trainer's curious behaviour? Shouldn't it bother her that the competitive trainer, on finding no food where she points, just sits on the corner and 'sulks' instead of checking out the other box? Shouldn't she be puzzled to discover that her trick keeps working? She should wonder: can the competitive trainer be that stupid? Further, better designed experiments with Sadie – and other creatures – are called for.'

Points to remember

■ Non-verbal communication occurs as part of everyday human behaviour and has some features in common with animal communication.

■ Language should not be defined as a uniquely human activity because there are many features of animal communication that are language-like.

■ Apes cannot learn to speak, but they can learn to communicate with humans using symbols to represent words.

■ According to Chomsky (1972), the possession of human language is associated with a specific type of mental organization, not simply with a higher degree of intelligence.

■ The question of whether ape language involves cognition hinges upon distinctions such as 'knowing how' and 'knowing that'.

■ Although mental images are thought to be important in human behaviour, it seems that they are not necessary for the control of similar behaviour in animals.

Further reading

Gardner, B.T. and Gardner, R.A. (1969) Teaching sign language to a chimpanzee. *Science* **165**, 664–72.

Griffin, D.R. (1982) *Animal Mind–Human Mind*. Springer-Verlag, Berlin.

Passingham, R.E. (1982) *The Human Primate*. Freeman, New York.

Rumbaugh, D.M. (1977) *Language Learning by a Chimpanzee*. Academic Press, New York.

Terrace, H.S. (1979) *Nim*. Eyre Methuen, London.

Intelligence, tool use and culture

Darwin believed in the evolutionary continuity of mental capabilities, and he opposed the widely held view that animals were merely automatons, far inferior to humans. In his *Descent of Man* (1871) Darwin argued that 'animals possess some power of reasoning' and that 'the difference in mind between man and higher animals, great as it is, certainly is one of degree and not kind'. The intelligence of animals was exaggerated by Darwin's disciple George Romanes, whose publication of *Animal Intelligence* (1882) was the first attempt at a scientific analysis of animal intelligence. Much of Romanes' evidence was anecdotal, and his book was full of stories told by respectable members of Victorian society.

Romanes defined intelligence as the capacity to adjust behaviour in accordance with changing conditions. His uncritical assessment of the abilities of animals provoked a revolt by Conway Lloyd Morgan (1894) and the subsequent behaviourists, who attempted to pin down animal intelligence in terms of specific abilities. They were critical of the anecdotal approach and set up strict criteria for attributing mental capacities to animals. Morgan inspired Edward Thorndike to investigate trial-and-error learning (see Chapter 17), and he thus indirectly founded the behaviourist school of animal psychology. In his book *An Introduction to Comparative Psychology* (1894), Morgan suggested that higher faculties evolved from lower ones, and he proposed a psychological scale of mental abilities.

Although the idea of a ladder-like evolutionary scale of abilities had a considerable influence upon animal psychology, it is not an acceptable view today. Studies of brain structure (e.g. Hodos, 1982) and of the abilities of various species (e.g. Macphail, 1982) make it abundantly clear that different species in different ecological circumstances exhibit a wide variety of types of intelligence. This makes intelligence difficult to define, but it emphasizes the importance of studying animal intelligence from a functional point of view as well as investigating the mechanisms involved.

Another viewpoint, which comes from animal robotics (see Chapter 25), is that intelligence is not really a matter of what mechanisms an animal possesses, but rather a question of whether the animal can produce the best response to the problem at hand. Let us take a simple example. Figure 27.1(a) shows a prototype for a brick-laying robot. The problem is that when the

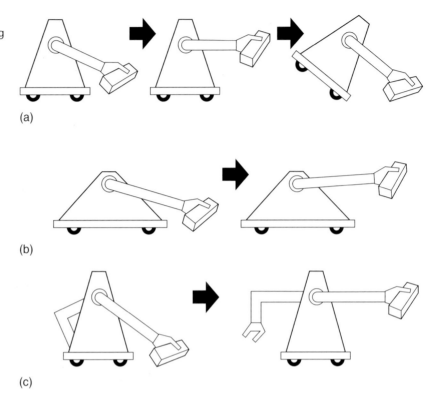

Fig. 27.1 Design problems associated with a brick-laying robot. (a) The robot is unstable once a brick is grasped. (b) Stability is achieved by altering the body shape at the expense of reduced height and increased weight. (c) Stability is achieved by the intelligent behaviour of using the free arm as a counter weight (From McFarland and Bosser, 1993).

(a)

(b)

(c)

robot picks up a brick it falls over due to the weight of the brick. A possible solution that does not involve intelligent behaviour is shown in Figure 27.1(b). Altering the shape of the robot, particularly broadening its base, alters the centre of gravity, so that the robot no longer falls over when it picks up a brick. This is not a good solution because the altered shape implies decreased height and possible increased weight, both undesirable traits in a bricklayer because they lead to lack of reach and increased fuel requirements. The robot is no longer an economic proposition. Figure 27.1(c) shows an alternative solution. The robot is equipped with a counterweight (either a special limb, or another arm capable of holding another brick), which can be raised at the appropriate time to prevent the robot falling over when it lifts up a brick.

It may be objected that the intelligence, or lack of it, applied to this problem is a property of the designer rather than of the robot. We have to be careful here. It is intelligent behaviour of robot Mark C to raise the counterbalance when it picks up a brick. The question of whether the robot thinks up this solution itself, or does it automatically, is not really relevant. Robot Mark A, which has no counterbalance, cannot perform this intelligent behaviour, no matter how much thinking power it has. Similarly, robot Mark B, however big its on-board computer, cannot overcome the fact that it is overweight. So intelligent behaviour is not simply a matter of cognition, but a product of the behavioural capacity and the environmental circumstances.

In this chapter we look at some of the ways in which scientists have tackled the problem of assessing the intelligence of animals. We look in some detail at tool use as an example of behaviour most people would regard as indicating intelligence, and we discuss cultural transmission as a means by which animals acquire new forms of behaviour.

27.1 Comparative aspects of intelligence

There are two main ways of assessing the intelligence of animals. One is to make a behavioural assessment and the other is to study the brain. In the past, both these approaches have been dominated by the idea that there is a linear progression from lower, unintelligent animals with simple brains to higher, intelligent animals with complex brains. A survey of the animal kingdom as a whole tends to confirm this impression (see Chapter 11), but when we look closely at specific cases, we find many apparent anomalies. These are not exceptions to an overall rule but are due to the fact that evolution does not progress in a linear manner. It diverges among a multiplicity of routes, each involving adaptation to a different set of circumstances. This means that animals may exhibit considerable complexity in some respects but not others and that different species may reach equivalent degrees of complexity along different evolutionary routes.

In comparing the brains of animals of different species, we can expect to find a relationship between the relative size of a particular structure and the degree of sophistication of the behaviour the structure controls. The more an animal makes use of a particular aspect of behaviour in adapting to its environment, the larger the number of neurons and interconnections will be present in the appropriate parts of the brain. This is easy to see when comparing specialized parts of the brain like those concerned with different sensory processes (see Chapter 12). It is less easy to interpret when we come to look at the more general-purpose parts because they may have enlarged as a result of different selective pressures in different species (Jerison, 1973).

Many traditional ideas about the evolution of the vertebrate brain have been challenged. Thus, it has been claimed that, contrary to popular belief, there is no progressive increase in relative brain size in the sequence fish, reptile, bird, mammal or in the relative size of the forebrain in the sequence lamprey, shark, bony fish, amphibian, reptile, bird, mammal (Jerison, 1973). Indeed, some sharks have forebrains equivalent to those of mammals in relative size (Northcutt, 1981). It was long thought that the telencephalon of sharks and bony fish is dominated by the olfactory sense, but it is now claimed that the representation of the olfactory sense in this region is no greater in non-mammalian vertebrates than in mammals (Hodos, 1982). The idea that an undifferentiated forebrain is characteristic of lower vertebrates has also been challenged (Hodos, 1982).

In reviewing our understanding of animal intelligence in the light of modern knowledge of neuroanatomy, William Hodos (1982) comes to the conclusion that:

'If we are to find signs of intelligence in the animal kingdom and relate them to developments in neural structures, we must abandon the unilinear, hierarchial models that have dominated both searches. We must accept a more general definition of intelligence than one closely tied to human needs and values. We must accept the fact that divergence and nonlinearities characterize evolutionary history, and we must not expect to find smooth progressions from one major taxon to another. Finally, we must not allow ourselves to be biased by our knowledge of the mammalian central nervous system in our search for neural correlates of intelligence in other vertebrate classes. Without such changes in our thinking, we would appear to have little hope of progressing any further than we have in our attempt to understand the relationships between the human mind and the animal mind and their respective neural substrates.'

We now turn to the question of how to assess animal intelligence behaviourally. Since intelligence tests for people were introduced by Albert Binet in 1905, considerable progress has been made in improving and refining them. This has been possible largely because different tests can be evaluated by checking on the subsequent educational progress of individuals. Modern I.Q. tests are reasonably accurate in predicting how well a person will progress in intellectual achievement. However, difficulties remain, especially in attempting to compare the general intelligence of people from differing cultural backgrounds. The difficulties in assessing the intelligence of animals are much more severe because there is no way of checking on the validity of a test and because animals of different species differ greatly in their performance in particular respects.

Until recently, attempts to assess animal intelligence concentrated upon abilities that normally would be taken as a sign of intelligence in humans. A modern I.Q. test includes various subtests designed to assess a person's memory, arithmetic and reasoning power, language ability and ability to form concepts. Pigeons appear to have a prodigious ability to form concepts such as water, tree, human beings, etc. (Herrnstein, 1985). Are we to take this as a sign of great intelligence? In discussing language abilities in animals, we came to the conclusion that human abilities, in this respect, far exceed those of any other animal, however well trained. Does this show that humans have greater general intelligence or merely that human intelligence is highly specialized in the use of language?

In comparing the intelligence of different species, it is difficult to devise a test that is not biased in one way or another. Many of the early tests of animal problem-solving ability were unreliable (Warren, 1973). Sometimes the same test, with the same species, gave different results according to the type of apparatus employed. Sometimes the same test using the same apparatus gives surprisingly different results, as we see in Box 27.1.

Various attempts have been made to discover whether or not animals can master problems that require the learning of a general rule. Animals can be trained to choose from an array one item that matches a sample. Primates learn to solve this type of problem quickly, but pigeons require a large number of trials. Harry Harlow (1949) devised a test to measure the ability of animals to follow rules and to make valid inferences. Instead of testing

Box 27.1   Mental chronometry in humans and pigeons

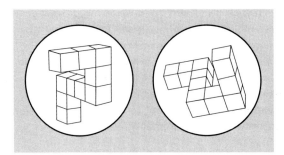

(a) Examples of a pair of shapes used in mental chronometry. This is an identical pair differing by 80° rotation (After Shepard and Metzler, 1971).

Mental chronometry uses the time required to solve a spatial problem as an index of the processes involved (Posner, 1978). For example, a classic study (Shepard and Metzler, 1971) requires human subjects to view pairs of drawings of three-dimensional objects. On each trial the subject has to indicate whether two objects are the same in shape or are mirror images. As well as being different in shape, the objects can also differ in orientation, as shown in Figure (a). It is found that the time required to indicate whether or not the two objects are the same shape increases, in a regular manner, with the angular difference between the pairs of drawings presented to the subject, as in Figure (b). The usual conclusion is that the subjects mentally rotated an internal representation of one object to make it line up with the other before they compared the shape of the two representations.

Although the precise nature of the representation used in this type of task is a matter of controversy (Cooper, 1982; Kosslyn, 1981), the most straightforward interpretation is that some form of mental imagery is involved and that the processes of rotation and comparison are carried out in series. The fact that reaction time is a function of presentation angle is taken to show that mental rotation takes time (about 30 milliseconds for every 20°) (Cooper and Shepard, 1973).

(b) Time taken to decide whether pairs of similar shapes are the same, plotted as a function of the orientation of one shape relative to the other (After Shepard and Metzler, 1971).

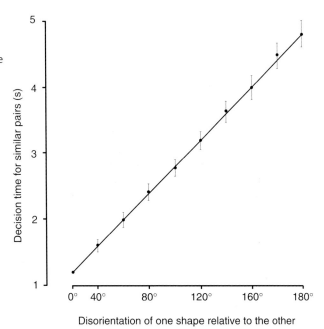

Box 27.1 *(cont)*

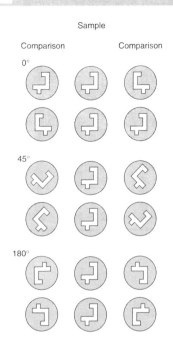

Sample

Comparison   Comparison

0°

45°

180°

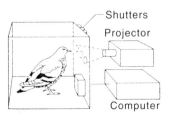

Shutters

Projector

Computer

(c) Experimental apparatus used to display symbols to pigeons. The birds were trained to indicate which of the two comparison symbols the sample symbol most resembled. The comparison symbols were presented at 0, 45, and 180° rotation (After Hollard and Delius, 1983).

Valerie Hollard and Juan Delius (1983) trained pigeons in a Skinner box to discriminate between shapes and their mirror images presented in various orientations, as shown in Figure (c). They then measured the pigeons' reaction times in tests for rotational invariance. When this part of the study was complete, the chamber was disassembled, and the test panel, containing the lights and keys, was presented to human subjects in a series of similar tests. In this way a direct comparison of the performance of pigeons and humans could be made on the basis of the same stimulus patterns.

The result showed that pigeons and humans were capable of similar accuracy, as judged by the errors made. However, whereas the human reaction time increased with the angular disparity between the sample and comparison forms, as shown by previous studies, the pigeon reaction time remained unaffected by the angular rotation, as shown in Figure (d). It appears that pigeons are able to solve this type of problem more efficiently than humans, presumably through some parallel form of processing. This result not only has implications for the assessment of intelligence in animals but also raises questions about the validity of studies of mental representations.

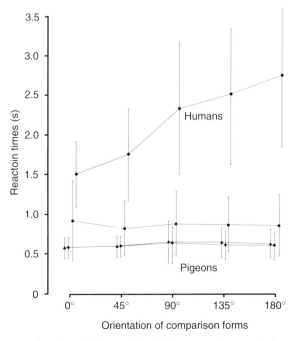

(d) Mean reaction times as a function of the rotation of the comparison symbols. Data from nine pigeons and 22 humans. Note that the human reaction time increases with the angle rotation, but the pigeon reaction time does not (After Hollard and Delius, 1983).

Fig. 27.2 Series of discrimination problems used to test learning sets.
(a) Simple discrimination (the arrow indicates the correct choice). (b) Reversal task (the animal has to reverse its originally correct choice). (c) Conditional task (one object is correct when both are grey, the other when both are white). (d) Matching task (the animal has to match the sample as left on tray). (e) Oddity task (the animal has to choose the odd man out) (After Passingham, 1981).

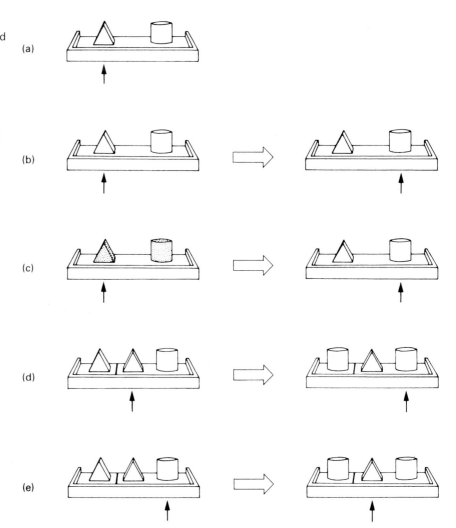

monkeys on a single visual discrimination like that shown in Figure 27.2, Harlow presented them with a series of tests, all of which require the same rule. Thus, the animal might be given a series of discrimination problems of the type shown in the figure. Although the objects are changed from problem to problem the rule is that the food reward is to be found always under the same object, trial after trial, irrespective of the position of the object. If the animal improves over a series of such problems, it is said to have acquired a **learning set**.

As shown in Figure 27.2, various types of problem can be used to test an animal's ability to learn that a general rule is common to a set of problems and that the correct choice is governed by a single principle. Despite the criticism that the ability of different species to master learning sets depends largely upon the way the tests are set up (Hodos, 1970), there do seem to be genuine differences among species (see Figure 27.3) once these criticisms have been taken into account (Passingham, 1981). When these animals are

Fig. 27.3 Visual discrimination learning set in mammals. The percentage correct on the second trial of each problem is plotted against the number of problems given (After Passingham, 1981).

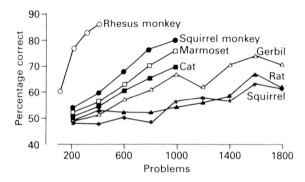

ranked in terms of their rate of improvement over a series of problems, their rank can be predicted from an index of brain development (Riddell, 1979; Passingham, 1982). This index is an estimate of the number of nerve cells in the brain that are additional to those needed for the control of bodily functions (Jerison, 1973). Thus, it seems that tests of animal intelligence can be devised that are similar to those used in human intelligence and that differentiate members of different species.

The claim that such tests are genuine measures of intelligence is strengthened by the finding that performance in such tests is correlated with an index of brain size. Similar results are obtained with the other types of test illustrated in Figure 27.2. Thus, rhesus monkeys and chimpanzees, but not cats, improve more rapidly on a series of object discrimination problems if they have prior experience of **reversal** problems (in which the correct solution is periodically reversed) (Warren, 1974). Common principles apply to the two types of problem which these primates are able to make use of, while cats do not seem to have this ability. Similar differences between cats and monkeys occur with series of **oddity** problems (in which the animal has to select the odd symbol from a group) (Warren, 1965). Critics of these types of test argue that the test inevitably is presented in a way that makes it easy for animals of some species but not for others (MacPhail, 1982). Even if the demonstrated differences are taken at face value, they represent only one aspect of intelligence, and it is not surprising that monkeys and apes perform well on tests derived from human I.Q. tests because all are primates.

Recent studies of pigeons, carried out by Juan Delius and his co-workers at the University of Bochum, indicate some of the problems in conducting tests of this type. There are three ways in which an animal could master a match-to-sample, or an oddity-from-sample, task:

■ it may learn a series of separate sample-specific stimulus–response connections

■ it may learn a set of discriminations based upon the different overall stimulus configurations

■ it may learn a single generalized concept-like rule that is independent of the type of stimulus to which it is applied

There has long been controversy as to the performance of pigeons on such tasks. Reviewing the situation at the time, Carter and Werner (1978) concluded that pigeons did not learn identity/oddity concepts, but learned discrete stimulus–response rules. Zentall and Hogan (1978) came to the opposite conclusion. More recent reviewers, Macphail (1982) and Mackintosh (1983), came to conflicting opinions.

Part of the problem is that some studies of identity/oddity tasks in pigeons employed tasks that invited the pigeons to use more simple solutions than those required for concept learning. Delius and his co-workers set out to establish whether pigeons could learn concept-like rules and apply them in a novel situation, but they took care to familiarize the pigeons with all the stimuli so that their strong novelty aversion would not interfere with their performance during the crucial tests. This is a precaution not always taken in previous studies. Moreover, the Delius group reasoned that concepts were likely to be learned on the basis of examples, so they provided some pigeons with many examples, and others with few examples, during training.

Using an apparatus and procedure similar to that described for pigeons in Box 27.1 and presenting similar stimuli, they showed that pigeons can learn to select stimuli on the basis of oddity, and can transfer this ability to stimuli not used during training. Of the three possible explanations outlined above, only that involving a general concept-like rule can account for these results. Moreover, the pigeons trained with many examples performed better than those trained with few examples of the oddity task. Most previous studies of identity/oddity in the pigeon have used relatively few training exemplars, and this may account for their meagre evidence for the concept hypothesis. There can now be little doubt that an abstract concept of oddity (and by implication identity) can control the behaviour of pigeons in suitable situations. Whether the pigeons have to learn to use the concept, or whether they already have it and learn to use it in a particular situation, remains an open question.

27.2 Tool use in animals

The ability to use tools has long been regarded as an aspect of intelligence, and the ability to make tools once was regarded as a factor that set humans apart from other animals. Now that we know much more about tool using in animals, the issue is not so clear cut, although the manufacture of tools is regarded widely as an important influence upon human evolution.

Tool use can be defined as the use of an external object as a functional extension of the body to attain an immediate goal (van Lawick-Goodall, 1970). This definition excludes some cases of manipulation of objects by animals and includes others. For example, we saw that crows will take whelks up into the air and drop them on a rock to break them open (Chapter 8). Other birds have similar behaviour. Thus song thrushes (*Turdus philomelos*) hold snails in their beak and smash them against a rock anvil, while ravens (*Corvus corax*) and some vultures (*Gypaetus barbatus*) drop bones in order to crack them open to feed on the marrow. The use of a stone anvil to break open food items, however, does not count as tool using because the anvil is not an extension of the animal's body.

Fig. 27.4 A vulture about to hurl a stone at an ostrich egg (*Photograph: J.M. Pearson, Biofotos*).

The Egyptian vulture (*Neophron percnopterus*) is known to break ostrich eggs by throwing them against a stone anvil. This does not count as tool using. However, the vultures also may carry a stone into the air and drop it onto an ostrich nest or pick up a stone in the beak and throw it at an egg, as shown in Figure 27.4. These uses of a stone do count as tool use because the stone can be regarded as an extension of the vulture's body.

An animal that scratches or rubs itself against a tree is not using the tree as a tool, but an elephant or horse that picks up a stick to scratch itself is using the stick as an extension of its body for a short-term purpose. However, a bird that carries twigs to build a nest is using the twig as material and not as an extension of its body. A nest normally is not regarded as a tool for raising the young because it achieves a long-term rather than a short-term objective.

The Galapagos woodpecker finch (*Cactospiza pallida*) probes for insects in crevices in the bark of trees by holding a cactus spine in its beak, as shown in Figure 27.5. This clearly counts as tool using by the earlier definition, but should it count as a sign of intelligence? From the functional point of view, the use of a cactus spine to probe for food is an intelligent solution to a particular problem. A human who hit upon this solution to the problem would normally be regarded as showing signs of intelligence. Those who wish to judge intelligence purely on the basis of appropriate responses to circumstances have to allow that the finch is behaving intelligently.

Fig. 27.5 A Galapagos finch using a stick to probe for prey (*Photograph: Alan Root, Survival Anglia Ltd, Oxford Scientific Films Ltd*).

If the behaviour of the woodpecker finch were largely innate, then we probably would not want to count its probing behaviour as intelligent. Observations of a juvenile woodpecker finch that had been taken from the nest as a fledgling (Eibl-Eibesfeldt, 1967) showed that the bird manipulated twigs from an early age. If the hungry bird was presented with an insect in a hole, it would drop the twig and try to obtain the insect with its beak. Gradually, the bird began to probe for insects with a twig, and it seems likely that learning plays some part in the development of the behaviour. Does this make us more inclined to regard the behaviour as an indication of intelligence? We have to be very careful here. Even if it is true that learning plays a part in the development of the behaviour, it seems likely that woodpecker finches are predisposed genetically to learn this particular type of manipulation in much the same way that some birds are predisposed to learn particular types of song. Conversely, the probing behaviour is the functional equivalent of intelligent behaviour, and we must resist the temptation to say that the bird is not really intelligent just because it is a bird and not a mammal. We must be careful to use the same criteria in judging the bird's behaviour as we would in judging similar behaviour in a chimpanzee.

Chimpanzees in the wild have been observed to make use of sticks, twigs and grass stems to probe for food items. Grass stems may be used to probe for termites, as illustrated in Figure 27.6. These are chosen with care and may be modified to make them more suitable for their purpose. For example, if the end becomes bent, the chimps may bite it off (van Lawick-Goodall, 1970). Juvenile chimps may manipulate grass stems during play, but they do not use them for probing for food until they are about three years old. Even then they are often clumsy and may select tools that are not appropriate for the task. The skill required in probing for termites does not appear to be learned easily. Wild chimpanzees have also been observed to use sticks to obtain honey from bees' nests and to dig up plants with edible roots. They may use leaves as a sponge to obtain drinking water from a hole in a tree or to clean various parts of the body. Although tool use has been studied most intensively in wild chimpanzees, it has also been observed in other wild primates. Thus, baboons may use stones to squash scorpions and twigs to probe for insects (Kortlandt and Kooij, 1963).

The ability to use tools in the natural situation probably develops in the individual through a mixture of imitative and instrumental learning. In these respects tool using in primates is difficult to separate from the development of probing behaviour in the woodpecker finch. Some biologists, while

Fig. 27.6 A chimpanzee using a twig to probe for termites (*Photograph: Peter Davey; by courtesy of Bruce Coleman Ltd*).

admitting that tool using is not in itself a sign of intelligence, argue that it sets the stage for truly intelligent behaviour that involves innovation.

That innovations do occur among chimpanzees probing for termites is suggested by comparison of the methods used by different chimpanzee populations. Chimpanzees at Gombe, in East Africa, use twigs without previously peeling off the bark. They may use each end of the twig to probe in turn. Chimps at Okorobiko, in Central Africa, usually peel the bark off a twig before using it as a probe, and they use only one end. Chimpanzees from Mount Assirik in Senegal, West Africa, do not use twigs as a probe but instead use relatively large sticks to make holes in the termite mound through which they pick out the termites by hand (McGrew *et al.*, 1979). These differences suggest that a certain amount of variation exists within a population that may lead to innovations that suit local circumstances. The technique of fishing for termites is learned by imitation and is passed through the population by cultural tradition.

That some aspects of behaviour that are typical of a certain population are maintained by cultural means is well documented (Bonner, 1980; Mundinger, 1980). In a few cases, inventions have been observed and their spread through the population recorded. A celebrated case concerns the Japanese macaques (*Macaca fuscata*) of Koshima Island (Kawamura, 1963). To bring the monkeys into the open where they could be observed more readily, scientists supplemented their diet by scattering sweet potatoes on the beach. A 16-month-old female, called Imo, was observed to wash the sand off her potatoes in a stream. She continued this practice on a regular basis and soon was imitated by other monkeys, particularly those of her own age. Within ten years, the habit had been acquired by the majority of the population, with the exception of adults more than twelve years old and infants of less than a year. Two years later Imo invented another food-cleaning procedure. Scientists had been scattering grain on the beach, and the monkeys picked grains up one at a time. Imo gathered handfuls of mixed sand and grain and threw them into the sea. The sand sank and the grain was scooped easily from the water surface. The new procedure spread through the population in a manner similar to that of potato washing. The new behaviour was adopted first by monkeys of Imo's own age. Mothers learned from juveniles and adult males were the last to catch on.

Was Imo an especially intelligent monkey? On the one hand, we might argue that she learned in a manner similar to the so-called insight learning of Kohler's apes, discussed in Chapter 19. Imo made her discoveries by chance and learned to exploit them. She did not necessarily have any special insight into the situation. On the other hand, many human inventions came about in a similar way. If the invention leads to a genuine improvement in the monkeys' circumstances, it counts as a form of adaptive behaviour arrived at through the efforts of a single individual. As judged by the results, Imo appears to be highly intelligent, but if we require that intelligence involves certain mechanisms like reasoning, then we need to know more about Imo's thought processes before we make a judgement.

27.3 | Cultural aspects of behaviour

Evolution occurs as a result of natural selection, and inheritance of acquired characteristics is not normally possible. However much an individual animal adapts to its environment, whether by learning or by physiological adaptation, the acquired adaptations cannot be passed to the offspring by genetic means. So much is widely accepted among biologists. However, as we have seen, information can be passed from parent to offspring by imitation and by imprinting. In general, the passage of information from one generation to the next by non-genetic means is known as **cultural exchange**.

We have seen (Chapter 20) that sensitive periods of learning occur in the early life of many animals. During such periods they often learn from their parents. For example, the white crowned sparrow will remember its parents' song provided it hears it during the sensitive period between ten and 50 days old. Individuals prevented from hearing the song of their own species during this period never produce a proper white crowned sparrow song during later life. Whereas the juvenile white crowned sparrow will not learn songs that are much different from the song of its own species, other birds, such as the bullfinch, will learn the song of a completely different species. Thus a juvenile bullfinch fostered by a canary will adopt canary song. The tendency to copy the song of the parents leads to regional variations even among species that will learn only songs similar to that of their own species. In the region of San Francisco, for example, populations of white crowned sparrows separated by only a few miles have distinct dialects (Marler and Tamura, 1964).

Animal dialects represent an elementary form of tradition. Juvenile white crowned sparrows are inevitably exposed to the song dialect characteristic of the locality in which they are born, because their sensitive period of learning occurs before they are mobile. Other forms of traditional behaviour include the migration routes of some mammals and birds. Geese, ducks and swans migrate in flocks composed of mixed juveniles and adults. The juveniles learn the route that is characteristic of their population, stopping at traditional rest places, and breeding and over-wintering localities (Schmidt-Koenig, 1979). There are many other examples. Reindeer also show fidelity to traditional migration routes and calving grounds. Migratory salmon hatch in freshwater streams and migrate to the sea. The juveniles become imprinted on the odour of their native stream and return to the same stream as adults to spawn in the traditional places. Some game trails of deer and other mammals are known to have been used for centuries. In parts of Britain and Germany, where roads have been built across the traditional migratory routes of toads, conservationists stand guard during the breeding season to prevent the toads from being run over by motorists.

Simple forms of traditional behaviour do not require any special learning abilities, or any special teaching. They arise as an inevitable consequence of the circumstances in which the young are raised and of the tendency of juvenile animals to become imprinted upon their habitat, their parents and their peers. In some animals, however, more complex forms of learning are involved. We have seen how tool use in primates can spread through a population, probably as a result of imitative learning.

Imitation is not necessarily a sign of high intelligence, however (Heyes, 1994). Animals may copy each other as a result of simple social facilitation. Many animals eat more when fed in groups than when fed alone. This has been demonstrated experimentally in chickens, puppies, fish, and opossums. If a chicken is allowed to eat until completely satiated, and is then introduced to others that are still feeding, the satiated bird will resume eating. Domestic chicks also have a tendency to peck when others peck, even if there is no food available. Chicks tend to peck at the same type of food-like particle as the mother hen. The tendency to concentrate on the type of food being eaten by others has also been demonstrated in sparrows and chaffinches (Davis, 1973).

Avoidance of noxious food can also be socially facilitated. Attempts to kill large flocks of the common crow in the United States, by providing poisoned bait, were not successful because the majority avoided the bait after a few individuals had been poisoned. Similarly, the reaction of just one rat to novel food may be sufficient to determine the reactions of other rats in the group. If the one rat eats the food then others will join in, but if the rat sniffs the food and rejects it, the others will reject it. Sometimes, the pioneer rat urinates on the bait, thus warning others to reject it (Rozin, 1976).

Social attraction to, and avoidance of, food often result from a tendency to investigate places where other members of the species have been observed, rather than direct copying of the behaviour of others. This is probably the case with the tradition of stealing milk by blue tits in England (Figure 27.7). For many years milk has been delivered to houses in England and left on the doorstep early in the morning. Blue tits, and sometimes great tits, peck through the foil tops of the bottles and help themselves to the rich cream that floats on the top of the milk. Hinde and Fischer (1951) reported that this practice first appeared in particular localities and gradually spread, suggesting that the birds were learning from each other.

Another interesting example of traditional feeding behaviour is seen in Mike Norton-Griffiths' study of young oystercatchers (Norton-Griffiths, 1967, 1969). This is a shorebird that feeds on mussels, a bivalve mollusc with a hard shell. The adults use one of two methods to obtain the flesh. They may hammer very hard with their beaks at the weakest point of the shell, or they may insert their bill into the open siphon when the mussel is under water and cut the adductor muscle which holds the two halves of the shell together. Adult birds are either hammerers or stabbers, and the two techniques are not used by the same bird. Moreover, both members of a mated pair use the same method.

By switching the eggs of hammerers and stabbers, Norton-Griffiths showed that the young learn the method of opening mussels from their parents. The appropriate movements for both hammering and stabbing are seen in all birds at a young age, but they do not seem to practise these movements. It is possible that both methods are to a certain extent innate. How the juvenile oystercatchers acquire the necessary skills is not fully understood. They stay with their parents for 18–26 weeks, whereas the young of oystercatchers that

Fig. 27.7 A blue tit opening
a milk bottle to obtain food
(*Photograph: John Markham;
by courtesy of Bruce Coleman
Ltd*).

feed mainly on other prey stay with their parents for only six–seven weeks. Those born into the mussel-feeding tradition have to be fed by their parents while they serve their apprenticeship. Perhaps they learn by observation, or perhaps the necessary coordination requires a long time to develop.

From these examples it is obvious that culture and tradition do not require great intelligence on the part of individuals. Although highly developed culture and highly developed intelligence are both primarily human traits, they are not necessarily causally related. There are many cultural practices found in primitive human societies that are biologically adaptive, but not obviously rational. For example, most of the traditional methods of cooking maize (corn) by the indigenous people of the New World involve some kind of alkali treatment. Often the maize is boiled for about 40 minutes in water containing wood ash, lye, or dissolved lime. It is then eaten directly, or converted into dough, tortillas, etc. The alkali treatment is practised for purely traditional reasons, although it is said by some to make the food more palatable. However, the alkali treatment has important nutritional consequences. The local maize has low levels of available lysine, a nutritionally essential amino-acid. Most of the lysine in maize occurs as part of an indigestible protein called glutelin. The alkali treatment breaks up the glutelin, releases the lysine and greatly improves the nutritional quality of the food. Some kind of alkali treatment is found in all indigenous cultures that rely on maize, and it is probable that natural selection, acting through malnutrition, has eliminated those people who did not follow the traditional practice.

Thus we see that although culture can short-circuit biological heredity, and lead to very rapid evolution, the products of cultural evolution are still subject to natural selection. Although some learning ability is essential for cultural change, no great intelligence is necessary, and those people who follow traditional practices are often unable to give a rational explanation of their behaviour.

Points to remember

- It is difficult to compare intelligence among different species because many have special skills and particular disabilities. Should intelligence be judged by the type of problem solved or by the way it is solved?

- The ability to use tools is widespread in the animal kingdom, and some species can even make their own tools. Should this ability be regarded as an indication of intelligence?

- Culture and tradition, involving the transfer of information from one generation to the next by non-genetic means, occurs in many species. It may result from imprinting or from imitation.

Further reading

Bonner, J.T. (1980) *The Evolution of Culture in Animals*. Princeton University Press, Princeton, NJ.

Eibl-Eibesfeldt, I. (1989) *Human Ethology*. Aldine de Gruyter, New York.

Galef, B.G. Jr (1995) Why behaviour patterns that animals learn socially are locally adaptive. *Anim. Behav.* **49**, 1325–1334.

Macphail, E.M. (1982) *Brain and Intelligence in Vertebrates*. Clarendon Press, Oxford.

Animal awareness and emotion

28.1   Self-awareness in animals

28.2   Physiological aspects of emotion

28.3   Consciousness and suffering

28.4   The evolutionary perspective

Emotion is an important part of human experience. At the commonsense level we can all agree what we mean by emotion, and we usually agree on the attributes of the different emotions. At the scientific level, however, emotion poses considerable problems.

Emotion has subjective, physiological and behavioural manifestations that are difficult to reconcile with each other. At the subjective level, emotion is an essentially private experience. There is no way we can know what the emotional experiences of another person are like. We tend to assume that they are the same as our own experiences, but we have no logical way of verifying this. When it comes to the subjective emotional experience of animals, we are in even more difficulty. We tend to assume that animals similar to ourselves, like primates, have similar emotional experiences and that dissimilar animals, like insects, probably have rather different experiences, if any. However, this is a commonsense and not a scientific view. In scientific terms, we cannot assume that animals have particular subjective feelings any more than we are entitled logically to make such assumptions about other people.

In physiological terms, emotional states in humans are typically accompanied by **autonomic** changes, but these are not a reliable guide to particular emotional states. In animals especially, while physiological indexes such as increased heart rate or changes in hormonal balance do provide evidence of emotional arousal – e.g. fear, aggression or of a sexual nature – the emotions are poorly differentiated. In other words, most animals (at least vertebrates) react to stressors in roughly the same way (Toates, 1995) (see Chapter 15), whether their emotional response is one of fear, of aggression or of a sexual nature. Thus, although some insight can be gained from physiological investigations, interpretation of the physiological responses is difficult.

Charles Darwin (1872) stressed the communicative aspect of emotion. As we saw in Chapter 21, he postulated that facial expressions and other behavioural signs of emotion had evolved from protective responses and other utilitarian aspects of behaviour. Although Darwin's evolutionary thesis, as developed and expanded by the early ethologists, is accepted widely, his ideas about emotion were naive. Darwin and his disciple George Romanes did not hesitate in labelling animal emotions in human terms. Thus, the expression of a dog that had done wrong was taken to indicate shame (Darwin, 1872), fish experienced jealousy and parrots had a sense of pride in their utterances (Romanes, 1882). This anthropomorphic approach led to a revolt among psychologists like Conway Lloyd Morgan (1894), who advocated an approach devoid of speculation about the private thoughts and feelings of animals.

In this chapter, we explore the inner life of animals from the subjective, physiological and behavioural viewpoints and discuss the implications of this research for the welfare of animals and their treatment by human beings.

28.1 Self-awareness in animals

The behaviourist attitude that the private mental experiences of animals cannot be a subject of scientific investigations was dominant for the first three-quarters of the twentieth century. During this period various scientists like Edward Tolman (1932), dissented, but they did not have much influence (see Griffin, 1976, for a review). The behaviourist position seems unassailable on logical grounds, but it can be circumvented in various ways. One argument is that, although we cannot prove that animals have subjective experiences, it may be true nevertheless. What difference would it make if it were true? Another approach is to argue that it is unlikely, on evolutionary grounds, that there is a marked discontinuity between humans and other animals in this respect.

Donald Griffin (1976), who was one of the first to mount a concerted attack on the behaviourist position, uses both these arguments. In his opinion, the study of animal communication is most likely to provide evidence that 'they have mental experiences and communicate with conscious intent'. However, the study of animal language in more recent years has not fulfilled its early promise. The interpretation of the behaviour of chimpanzees trained in some aspect of human language remains controversial, and there is doubt that it will ever reveal much about the private experiences of these animals (Ristau and Robbins, 1982; Terrace, 1979). There have been various attempts to investigate self-awareness in animals by other means, and it is to these that we now turn.

Are animals aware of themselves in the sense that they know what posture they are adopting and what action they are taking? Sensory information from the joints and muscles is available to the brain, so it seems that animals ought to be aware of their behaviour. In an experiment designed to test this, rats were trained to press one of four levers, depending upon which of four activities the animal was engaged in when a buzzer sounded (Beninger *et al.,* 1974). If a signal occurred when the rat was grooming, for example, then it would require to press the grooming lever to obtain a food reward. The rats learned to press a different lever depending on whether they were grooming, walking, rearing up or remaining still when the buzzer sounded. The results of similar studies (Morgan and Nicholas, 1979) show that rats can base operant behaviour on signals emanating from their behaviour as well as from the external environment. In a sense, the rats must be aware of their actions, but this does not necessarily mean that they are conscious of them; they may be aware of their actions in the same way that they are aware of external stimuli.

Many animals respond to a mirror as if they were seeing another member of their species. However, there is some evidence that chimpanzees and orang-utans can recognize themselves in a mirror. Young wild-born chimpanzees will use mirrors to groom parts of the body that they cannot otherwise see. Gallup (1977, 1979) painted small patches of red dye on the eyebrow and opposite ear of chimps while they were under a light anaesthetic. When they awoke from the anaesthetic, he confirmed that they did not touch these parts of their body more than normal. Then he provided the chimps with a mirror. The chimps examined their reflections and repeatedly touched their own dyed eyebrows and ears. As we see below, similar results have been obtained with human infants.

Does the ability to respond to parts of one's body seen in a mirror indicate self-awareness? After twenty years of research, this remains a controversial question (see Parker *et al.*, 1994, for reviews). The question is not far removed from a wider question. Does the ability to imitate the actions of others indicate self-awareness? Chimpanzees are extraordinarily proficient at imitating each other and human beings. Although true imitation has to be separated carefully from other forms of social learning (Davis, 1973; Heyes, 1994)(see Chapter 27), there is some doubt about its occurrence among the primates. For example, the chimpanzee Viki was fostered by the Hayes family. They demonstrated a series of 70 acts that Viki was encouraged to copy. Many of them she had never seen before, and she copied ten of these the first time she saw them. Viki learned to produce 55 acts in response to the appropriate demonstration (see Figure 28.1). She also learned fairly complicated household tasks such as washing dishes and dusting furniture

Fig. 28.1 Viki imitating a photograph of herself (Drawn from a photograph).

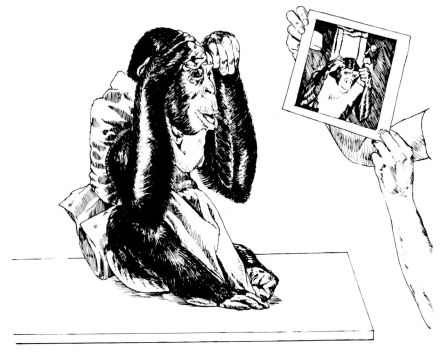

(Hayes and Hayes, 1952). Many of these actions were imitated spontaneously, without prompting, although some (e.g. Heyes, 1994) doubt that true imitation involving mental state attribution is demonstrated by these studies, because of the extensive training involved.

Although the ability to imitate is sometimes taken as a sign of intelligence, its occurrence in very young human babies and in a wide variety of non-mammalian species must cast some doubt on this view. The study of bird song shows that some form of imitation occurs in the song learning of many species and that some species are especially proficient at imitating sounds. Parrots and mynah birds can reproduce human sounds with extraordinary fidelity (Nottebohm, 1976).

To be able to imitate, an animal must perceive the external auditory or visual example and match it with a set of motor instructions of its own. For example, a baby that imitates an adult sticking out its tongue must somehow associate the sight of the tongue with its own motor instructions for sticking out its tongue. The baby does not have to be aware of the fact that it has a tongue, it merely has to connect a particular perception with a particular set of motor commands. How this is done is a mystery, but the question of whether imitation necessarily involves self-awareness is debatable.

Part of the problem is that we need to be clear about what we mean by self-awareness. As Griffin (1982) notes, many philosophers distinguish between awareness and consciousness. Awareness is a form of perception, while consciousness involves a special kind of self-awareness, which is not simple awareness of parts of one's body or of processes occurring within the brain. Consciousness, in this view, involves a propositional awareness that it is I who am feeling or thinking, I am the animal aware of the circumstances. We have reviewed some of the evidence that animals are aware in the perceptual sense. However, the ability to report on one's actions, to imitate the actions of others or to recognize one's image in a mirror need not necessarily require consciousness as defined here.

A dissociation between conscious and unconscious perception can occur in people with brain damage. Some people with damage to certain areas of the brain used in visual processing report that they are partially blind. They cannot name objects presented to them in certain parts of the visual field. They claim that they cannot see such objects, yet when asked to point they often can do so accurately (Weiskrantz, 1980). One patient could guess accurately whether he was being shown a horizontal or diagonal line, yet he was unaware of seeing anything (Weiskrantz et al., 1974). This phenomenon, called blind sight, is due to damage to parts of the brain responsible for visual awareness, which leaves intact other parts of the brain involved in visual processing. These other parts enable patients to make correct judgements even though they are not aware of what they see.

28.2 Physiological aspects of emotion

Emotional states are accompanied by an increase in the activity of the autonomic nervous system. This is a division of the vertebrate nervous system that serves the internal organs such as the blood vessels, heart, intestines, lungs and certain glands. The **autonomic** nervous system is controlled by the brain and provides two types of innervations that have antagonistic effects upon internal organs. The **sympathetic** nerve pathways become active under conditions of stress or exertion and have an emergency function. They have the effect of increasing the blood supply to the muscles, brain, heart and lungs, of increasing the heart rate and of reducing the blood supply to the intestine and peripheral parts of the body. It is also intimately related to the hypothalamic–pituitary–adrenocortical axis (see Chapter 15). The **parasympathetic** nervous system is anatomically distinct from the sympathetic, and it serves a recuperative function, restoring the blood supply to normal and counteracting other effects of sympathetic activity. Emotional arousal is accompanied by sympathetic activity. In humans this includes increased heart rate, sweating and changes in peripheral blood circulation so that the face becomes pale or flushed.

The psychologist William James proposed that the subjective feelings of emotion are generated by the sensory receptors that are involved in the emotional reaction. For example, a frightening stimulus would elicit certain behavioural and physiological changes, and these would give rise to the subjective experience of fear. As James (1890) put it, when we meet a bear we run. We do not run because we are afraid; we are afraid because we run. Carl Lange, a contemporary of James, proposed a similar explanation of emotional experience, and this view is now known as the James–Lange theory.

Walter Cannon (1927) pointed out that the sympathetic arousal was much the same whatever emotion was being experienced. He showed that cats exhibit emotional behaviour, such as hissing and spitting, even after all autonomic sensory feedback has been eliminated. Cannon concluded that autonomic activity is not necessary for emotional experience independently of its effects on autonomic activity. When human subjects are injected with the hormone adrenaline, a wide range of sympathetic activity occurs, including sweating, increased heartbeat rate, etc. However, the subjects do not report emotional experiences. Some report the physical symptoms and others say that they feel a cold emotion, as if they were angry or afraid (Landis and Hunt, 1932).

Autonomic arousal may provide a basis for emotional experience, but it provides no differentiation among the various emotional situations. However, the situation, as perceived by the subject, may lead to an interpretation of the autonomic arousal in terms of the appropriate emotional experience. This is the hypothesis proposed by Stanley Schachter and J. Singer (1962). They injected human subjects with adrenaline, pretending it was a vitamin supplement. Some subjects were told that the injection would cause an increase in heart rate, flushing, etc. (i.e. the real effects of adrenaline), while others were told it might have side effects such as itching or numbness (i.e. they were misinformed). All subjects then were asked to sit in a waiting room prior to a vision test. In this waiting room they were observed by a hidden experimenter and were accompanied by a confederate of the experimenter. In one test the confederate behaved in a sullen and angry manner, while in another test he was frivolous and playful. When the subjects were asked to rate their emotional experience following their stay in the waiting room, marked differences were found between the two test conditions. Those who had been misinformed about the injection felt happier than the correctly informed subjects when in the presence of the playful confederate. They felt angrier and more resentful than the controls when in the presence of the sullen confederate. Although it has been the subject of some criticism (e.g. Plutchik and Ax, 1970), this experiment does suggest that both our perceived physiological state and our assessment of the external situation contribute to our emotional feelings.

In evaluating the subjective experiences of animals from a physiological viewpoint, it is possible to make direct comparisons with humans, but none of these is entirely satisfactory. For example, do animals experience pain? When subjected to stimuli that would be painful to humans, many animals have similar stress responses (see Chapter 15). Pain-killing drugs, which alleviate physiological and behavioural reactions to pain in humans, produce corresponding effects on standard behavioural indexes of pain in animals. However, the interpretation of such findings rests largely on the behavioural index used, and there are a number of criteria that could be used (Bateson, 1991). Thus, a simple withdrawal reflex probably would not be a good indication of pain (as subjectively experienced) because such reflexes are widespread in the animal kingdom and occur in very primitive animals. Even the criterion of crying out in pain is difficult to evaluate. While a dog or monkey would scream with pain if severely injured, an antelope torn to pieces by a predator remains relatively silent. Screams may serve to elicit help from other members of the group, but they may also endanger those that are attracted by the calls.

Comparative aspects of brain anatomy are sometimes used to assess the likelihood that animals have subjective experiences similar to those of humans. Usually, a fairly simple criterion such as the size of the brain or the cortex is used as an index in an assumed progression from simple to more intelligent animals (see also Bateson, 1991). However, thorough anatomical studies reveal that some brain structures show marked differences when different animal groups are compared. Hodos (1982) suggests that some of these specialized areas may contribute to a type of mental activity that normally is associated with the cortex in humans. Moreover, in assessing the intelligence of animals in relation to their brain structure, it is often difficult to provide a fair test that is devoid of anthropomorphic assumptions (Macphail, 1982).

28.3   Consciousness and suffering

The question of consciousness in animals is fraught with difficulties. The spectrum of scientific opinion is very wide. There are those who believe that consciousness does not occur in animals, and there are those who maintain that it occurs in most animals. There are those who believe that consciousness is not a suitable subject for scientific study, and there are those who think it is a neglected topic. The situation is confounded by the difficulty of arriving at an acceptable definition of consciousness.

Donald Griffin (1976) defines consciousness as the presence of mental images and their use by the animal to regulate its behaviour. This is not far removed from the definition given in the *Oxford English Dictionary*. To be conscious is to be 'aware of what one is doing or intending to do, having a purpose and intention in one's actions' (see Griffin, 1982). According to Griffin (1976), 'an intention involves mental images of future events in which the intender pictures himself as a participant and makes a choice as to which image he will try to bring to reality'. Although Griffin and others (e.g. Thorpe, 1974) see intention and consciousness as part and parcel of the same phenomenon, this is not a universal view. As we saw in Chapter 26, it is possible to imagine intentional behaviour that does not involve consciousness.

Consciousness is thought by many researchers to involve more than simple perceptual awareness. For example, Humphrey (1978) sees consciousness as self-knowledge used to predict the behaviour of other individuals, and Hubbard (1975) suggests that it involves knowledge of oneself as distinct from other selves. Such knowledge could be used as a basis for communication, but this does not mean that consciousness necessarily involves language. We can agree with Passingham (1982) that 'spoken language has revolutionized thought. The use of language for thought vastly amplifies the level of intelligence that can be achieved. Animals think, but people can think in a totally new way, with a completely different code'. Undoubtedly, the advent of language has altered the way we think about ourselves. It is hard for us to imagine consciousness without language. However, this does not entitle us to assume that animals that have no language, or only primitive language, do not have consciousness. We saw previously that some sort of self-awareness can be demonstrated in animals with no language equivalent to ours. Therefore, we should not equate language and consciousness.

Do animals have to be conscious to suffer? At the commonsense level, we are inclined to suppose that they do. When we are unconscious we do not suffer pain or mental anguish because parts of our brain are deactivated. However, we do not know whether these parts are involved only in consciousness or in consciousness plus other aspects of brain activity. Thus, we cannot use the fact that we do not experience pain when unconscious to infer that consciousness and suffering go hand in hand. It may be that whatever makes us unconscious also stops pain but that the two are not causally connected.

We have no conception of what the conscious experiences of animals might involve, if they exist. Therefore, we can draw no conclusions about the relationship between consciousness and suffering in animals. In our ignorance, it would be wrong to assume that suffering in animals is confined to those that are intelligent, that use language or that show evidence of conscious experience.

There are both dangers and virtues in using ourselves as models of what animals might feel (M. Dawkins, 1980). The scientific basis for the analogy between the mental experiences of ourselves and other animals is weak. It would not be good scientific practice to come to conclusions about the mental experiences of animals on the basis of such evidence. In contrast, we do come to conclusions about the mental experiences of other people on the basis of analogy with ourselves. When we see another person suffering or hear their cries of pain, we do not ignore them because we cannot prove that their mental experiences are the same as ours. We give them the benefit of the doubt and go to their aid. Perhaps we should also give members of other species the benefit of the doubt.

28.4 The evolutionary perspective

It is natural for us to offer help to other people in distress. Although there may be some differences among cultures, our sympathy probably has an innate basis. Even when we are being cruel to another person, we recognize that we are causing them to suffer. We automatically assume that their mental experiences will be similar to those that we would experience in the same situation. It would appear that we are designed by natural selection to assume that other people have similar mental experiences to ourselves. Why?

Some researchers have argued (e.g. Humphrey, 1979; Crook, 1980) that the evolution of close-knit societies made recognition of others advantageous. By recognition of others we mean some awareness of others as beings with feelings similar to our own, a phenomenon often called **empathy**. This recognition might facilitate an ability to respond to the individual attributes of others and perhaps to the development of a language capable of expressing this sympathy. Griffin (1981) suggests that this may have occurred in other species and 'that animals that are consciously aware of their sociobiological goals can achieve them more effectively than would otherwise be the case' (p. 147). Certainly, the social and political life of some primates seems complicated enough to warrant such a development (de Waal, 1982). The problem is that we can always invent a plausible adaptive advantage for an observed or supposed trait, and such speculation does not lead very far.

Some recent experiments involving human infants throw some light on this problem. Doris Bischof-Köhler (1990) investigated the onset of empathy in infants. Her reasoning was as follows. The affective preconditions of empathy, such as the ability to discriminate emotions, and responsiveness to emotional contagiousness, develop during the first year. The cognitive requirements for empathy, such as self-objectification and mental imagery, emerge during the second year. Since these two abilities are necessary for recognition of one's own mirror image, she formed the hypothesis that empathy should spontaneously occur in every child as soon as he or she is able to recognize his or her own mirror image.

Mirror recognition was tested by the 'rouge test' method, similar to that employed by Gallup (1977, 1979; see also Parker *et al.*, 1994, and Chapter 27) on chimpanzees, and known to work with infants before age two (Amsterdam, 1972). Empathy was evaluated by determining whether one infant would go to the aid of another in distress. When presented with a situation in which another child demonstrates distress, the empathic child shows emotional concern and compassion, whereas children who have not yet reached the empathy stage either remain indifferent, or respond to the emotional atmosphere by seeking comfort for themselves. The studies were con-

Fig. 28.2 Stills from a video showing a child going to the assistance of another, and pointing to a spot on the face seen in the mirror (*Photograph: Doris Bischof-Köhler*).

ducted on a group of infants in a kindergarten, and consisted of an empathy experiment and a test of mirror-recognition. These were conducted separately by two different investigators, who were kept strictly uninformed about each other's results until the evaluation was completed.

The results showed that, between 16 and 24 months of age, there was a transition from non-recognition to mirror-recognition, and a simultaneous transition from non-empathizers to empathizers, as shown in Figure 28.2. Moreover, these transitions occurred at the same age in a given child (Figure 28.3). These results are encouraging in that they demonstrate that objective tests of hypotheses concerning self-awareness are possible, even if it is not yet possible to test hypotheses concerning the function of such attributes.

Another approach is to concentrate on the possible evolutionary origins of more widespread and simple attributes. Why, for example, might animals evolve to feel pain? If we imagine a population of animals equipped with simple avoidance behaviour devoid of pain or suffering, we can postulate a mutant that had some primitive awareness of pain. Could such a mutant successfully invade the population? Obviously, for the new trait to have a selective advantage, it would have to make some difference to the animal's behaviour. If the mental experience were entirely private and ineffectual, it

Fig. 28.3 Empathy and self-recognition. Children that show self-recognition tend to show empathy towards other children in distress (From Bischof-Köhler, 1990).

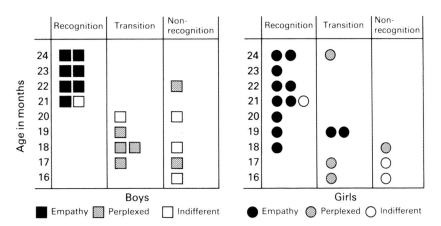

is hard to see how natural selection could act upon it. Perhaps the trait would improve the animal's ability to communicate with its kin, or perhaps it somehow could enable it to learn more effectively about the dangers of the environment. We do not know, but we can at least improve the cutting edge of our speculations by conforming to the recognized procedures of evolutionary biology. Could the proposed new trait successfully invade the population? McFarland (1989b) has argued that the concept of suffering could usefully apply to an aversive, declarative (see Chapters 19 and 25) state of an animal, but not to a state that is part of a routine behavioural procedure. In other words, to suffer an animal would have to know that it was suffering. Thus some cognitive ability is a precondition for suffering.

In evaluating animal suffering from an evolutionary viewpoint, we should give the animal the benefit of the doubt and assume that it suffers in situations that it normally would take steps to avoid. If animals do suffer, they presumably do so in response to circumstances that are functionally disadvantageous. In general, we can expect animals to be designed to choose situations that lead to increased fitness and to avoid those likely to result in decreased fitness. Perhaps, therefore, we can use the choice behaviour of animals as a guide to their welfare.

In 1880 Herbert Spencer suggested that the subjective experience of pleasure and pain evolved to help animals to choose suitable habitats and living conditions. While it is well known that wild animals have marked habitat preferences (Lack, 1937; Hilden, 1965; Partridge, 1978), this does not mean that subjective feelings are necessary for them to exercise a choice. However, we can expect that animal habitat preferences are related to their welfare in the evolutionary sense. In the case of domestic or laboratory breeds of animals, the connection between preference and fitness is not so clear cut. Nevertheless, the choices made by such animals probably offer a reasonable rule-of-thumb guide to their welfare (M. Dawkins, 1980).

An illustration of the value of this approach comes from attempts to specify the conditions under which battery hens should be kept. In the United Kingdom, the government set up a committee to inquire into the welfare of animals kept under intensive livestock husbandry systems. In its report, the committee recommended that the floors of battery cages should not be made of fine-gauge hexagonal wire because they thought that this was uncomfortable for the hens to stand on (Brambell, 1965). However, when battery hens were given preference tests between this type of floor and a floor made of the heavy rectangular metal mesh recommended by the government committee, the researchers found that the birds preferred the hexagonal wire floors (Hughes and Black, 1973). Photographs taken from below the floors showed that their feet gained more support from the fine mesh.

Extensive investigations of habitat preference in domestic fowl have been carried out by Marian Dawkins. In one study she offered hens a choice between a battery cage and an outside run in the garden. When the birds arrived at a T junction in a choice corridor, they could see the outside run on one side and the battery house environment on the other. The hens were free to enter both environments, and they were confined for five minutes in the one they chose before being tested again. Hens that just previously had been living in an outside run chose the outside run from the very first choice test. Hens that had been living in battery cages initially chose the battery cage environment (Figure 28.4). However, after repeated choice tests they preferred the outside run. Thus, even a short recent experience of the outside run was sufficient to alter their preferences (M. Dawkins, 1976, 1977).

Choice tests can be criticized on the grounds that the preferences of animals reflect a number of different influences. Thus, habitat preferences may be influenced by genetic factors, imprinting, familiarity and recent experience. Genetic differences can be investigated by testing different genetic strains of domestic animals (M. Dawkins, 1980). An animal's early experience may have long-term effects upon its habitat preferences but these can be controlled in investigations on domestic animals.

In conducting choice tests, care must be taken to ensure that the animals are familiar with the alternatives. Many animals avoid novel or unfamiliar

Fig. 28.4 A hen from one of Marian Dawkins' experiments. Hens which have been kept in a battery for a number of weeks may initially select the familiar battery cage in preference tests between battery and outdoor cages (*Photograph: Tony Allen*).

situations, and it is obviously possible that they might prefer a particular alternative once they become familiar with it. Farmers sometimes say that their animals must like the conditions in which they are kept because they return there after being freed. This is similar to the argument that some people prefer to be in prison because they find it difficult to adjust to life outside after a prolonged sentence. The problem has to do with the interaction between familiarity and recent experience. The initial choice of an animal may be its most recently familiar surroundings, but this preference can wear off quickly, as with the battery hens discussed earlier.

Preference tests have been criticized on the grounds that they do not discriminate between the animal's short-term and long-term preferences. Ian Duncan (1977, 1978) showed that hens will chose to enter a trap nest to lay their eggs, even though this results in confinement for several hours without food. The preference for the nests can be so strong that the hens will enter them day after day. However, such phenomena must be judged in relation to the alternatives available to the animal. How would an animal budget its time if it were able to choose among different daily routines? In the wild we might expect animals to arrange their daily routines in an optimal manner (see Chapter 17), but adjustment between one daily routine and another might be slow. Animals have both short-term and long-term mechanisms of adjustment to environmental changes, and these affect behaviour in complex ways (McFarland, 1989b). Marian Dawkins (1983, 1990) has suggested that measures of behavioural resilience (see Chapter 24) might go some way to resolving these difficulties. It can be argued that preferences exhibited under pressure are more likely to be reliable than those that result from idle whim. Domestic animals are not normally subject to time and energy constraints in the way that wild animals are, so perhaps we should provide this kind of pressure when conducting preference tests. Economists know that the apparent preferences of consumers may change radically with changes in prices. The equivalent demand function in animals (see Chapter 24) would give us a more coherent picture of animal preferences.

If the evidence were available, then we might be able to argue on evolutionary grounds that departures from the natural way of life that the animal was willing to pay to restore would be a good index of welfare. In what other way could suffering have evolved?

Points to remember

■ Animals may be aware of their own body-state and behaviour, but this does not necessarily involve consciousness.

■ Physiological manifestations of emotion occur in many animals, but this does not mean that animals experience emotion in the way that humans do.

■ In judging whether animals are suffering, we have to make assumptions about their mental state which are not testable scientifically.

■ Evolutionary considerations lead to the suggestion that animals usually know what is best for them and make choices on this basis.

Further reading

Dawkins, M.S. (1980) *Animal Suffering*. Chapman and Hall, London.

Dols, M., Kasanmoentalib, S., Lijmbach, S., Rivas, E. and van den Bos, R. (eds) (1997) *Animal Consciousness and Animal Ethics*. Van Gorcum, Assen, Netherlands.

Gallup, G.G., Jr. (1979) Self-awareness in primates. *American Scientist* **67**, 417–421.

Griffin, D.R. (1976, 1981) *The Question of Animal Awareness*. The Rockefeller University Press, New York.

Parker, S.T., Mitchell, R.W. and Boccia, M.L. (eds) (1994) *Self-awareness in Animals and Humans*. Cambridge University Press, Cambridge.

References

Able, K.P. (1982) The effects of overcast skies on the orientation of free-flying nocturnal migrants. In Papi, F. and Wallraff, H.G. (eds) *Avian Navigation*, pp. 38–49, Springer-Verlag, Berlin.

Adams, C.D. and Dickinson, A. (1981) Instrumental responding following reinforcer devaluation. *Quart. J. Exp. Psychol.* **33B**, 109–112.

Adler, H.E. (1971) Orientation: sensory basis. *Ann. N.Y. Acad. Sci.* **188**, 1–408.

Adolph, E.F. (1972) Some general concepts of physiological adaptations. In Yousef, M.K., Horvath, S.M. and Bullard, R.W. (eds) *Physiological Adaptations*, 1–7. Academic Press, New York.

Alcock, J. (1984) *Animal Behaviour, An Evolutionary Approach,* 3rd edn. Sinauer Associates Inc., Sunderland, MA.

Allison, J. (1979) Demand economics and experimental psychology. *Behav. Sci.* **24**, 403–415.

Allison, J. (1983) Behavioral substitutes and complements. In Melgren, R.L. (ed) *Animal Cognition and Behavior*. North-Holland, Amsterdam.

Amoore, J.E. (1963) Stereochemical theory of olfaction. *Nature* **198**, 271–272.

Amsterdam, B.K. (1972) Mirror self-image reactions before age two. *Develop. Psychol.*, **5**, 297–305.

Andersson, M. (1986) Evolution of condition dependent sex ornaments and mating preferences: sexual selection based on viability differences. *Evolution* **40**, 804–816.

Andrew, R.J. (1956a) Some remarks on behaviour in conflict situations, with special reference to Emberiza spp. *Brit. J. Anim. Behav.* **4**, 41–45.

Andrew, R.J. (1956b) Normal and irrelevant toilet behaviour in Emberiza spp. *Brit. J. Anim. Behav.* **4**, 85–91.

Andrew, R.J. (1956c) Intention movements of flight in certain passerines, and their use in systematics. *Behaviour* **10**, 179–204.

Andrew, R.J. (1963) The origins and evolution of the calls and facial expressions of the primates. *Behaviour* **20**, 1–109.

Anthoney, T.R. (1968) The ontogeny of greeting, grooming and sexual motor patterns in captive baboons. *Behaviour* **31**, 358–372.

Arbib, M.A. (1987) *Brains, Machines and Mathematics,* 2nd edn. Springer-Verlag, New York.

Archer, J. (1988) *The Behavioural Biology of Aggression*. Cambridge University Press, Cambridge.

Argyle, M. (1972) Non-verbal communication in human social interaction. In R.A. Hinde (ed.), *Non-Verbal Communication*. Cambridge University Press, Cambridge.

Aschoff, J. (1960) Exogenous and endogenous components in circadian rhythms. *Cold Spring Harbor Symp. Quant. Biol.* **25**, 11–28.

Aschoff, J. (1965) Circadian rhythms in man. *Science* **148**, 1427–1432.

Assem, J. van den (1967) Territory in the three-spined stickleback, *Gasterosteus aculeatus. Behaviour* Suppl. **16**, 1–164.

Baerends, G.P. (1941) Fortpflanzungsverhalten und Orientierung der Grabwespe Ammophilia campestris Jur. *Tijdschr. Ent.* **84**, 68–275.

Baerends, G.P. (1950) Specialisations in organs and movements with a releasing function. *Symp. Soc. Exp. Biol.* **IV**, 337–360.

Baerends, G.P. (1975) An evaluation of the conflict hypothesis as an explanatory principle for the evolution of displays. In Baerends, G., Beer, C. and Manning, A. (eds)
Function and Evolution in Behaviour. Clarendon Press, Oxford.

Baerends, G.P., Brouwer, R. and Waterbolk, H. Tj. (1955) Ethological studies on *Lebistes
reticulatus* (Peters): I. An analysis of the male courtship pattern. *Behaviour* **8**,
249–334.

Baerends, G.P. and Drent, R.H. (1970) The herring gull and its egg. Part I. *Behaviour*
Suppl. XVIII, 265–310.

Baerends, G.P. and Drent, R.H. (1982) The herring gull and its egg. Part II. *Behaviour*
82, 1–416.

Baerends, G.P. and Kruijt, J.P. (1973) Stimulus selection. In Hinde, R.A. and Stevenson-
Hinde, J. (eds) *Constraints on Learning: Limitations and Predispositions*, pp. 23–50.
Academic Press, London.

Bakeman, R. and Brownlee, J.R. (1982) Social rules governing object conflicts in tod-
dlers and preschoolers. In Rubib, K.H. and Ross, H.S. (eds) *Peer Relations and Social
Skills in Childhood*, pp. 99–111. Springer, New York.

Baker, A.G. and Mackintosh, N.J. (1977) Excitatory and inhibitory conditioning following uncorrelated presentations of the CS and US. *Anim. Learning and Behav.* **5**,
315–319.

Baker, R.R. (1981) Man and other vertebrates: a common perspective to migration and
navigation. In Aidley, D.J. (ed.) *Animal Migration.* Cambridge University Press,
Cambridge.

Baker, R.R. and Bellis, M. (1995) *Human Sperm Competition.* Chapman and Hall,
London.

Baldwin, E. (1948) *An Introduction to Comparative Biochemistry*, 3rd edn. Cambridge
University Press, Cambridge.

Balleine, B. (1992) Instrumental performance following a shift in primary motivation
depends upon incentive learning. *J. Exp. Psychol. Anim. Behav. Processes* **18**,
236–250.

Balleine, B. and Dickinson, A. (1991) Instrumental performance following reinforcer
devaluation depends upon incentive learning. *Quart. J. Exp. Psychol.* **43b**, 279–
296.

Barnard, C.J. (1984) (ed) *Producers and Scroungers.* Croom Helm, London; Chapman &
Hall, New York.

Barnard, C.J. and Stevens, H. (1981) Prey size selection in lapwings in lapwing/gull
associations. *Behaviour* **77**, 1–22.

Barrington, E.J.W. (1968) *The Chemical Basis of Physiological Regulation.* Scott, Foresman
and Company, IL.

Bastock, M. (1956) A gene mutation which changes a behaviour pattern. *Evolution* **10**,
421–439.

Bastock, M., Morris, D. and Moynihan, M. (1953) Some comments on conflict and
thwarting in animals. *Behaviour* **6**, 66–84.

Bateman, A.J. (1948) Intra-sexual selection in *Drosophila. Heredity* **2**, 349–368.

Bateson, P.P.G. (1964) An effect of imprinting on the perceptual development of
domestic chicks. *Nature* **202**, 421–422.

Bateson, P.P.G. (1966) The characteristics and context of imprinting. *Biol. Rev.* **41**,
177–220.

Bateson, P.P.G. (1973) Internal influences on early learning in birds. In Hinde, R.A. and
Stevenson-Hinde, J. (eds) *Constraints on Learning*, pp. 101–116. Academic Press,
London.

Bateson, P.P.G. (1979) How do sensitive periods arise and what are they for? *Anim. Behav.* **27**, 470–486.

Bateson, P.P.G. (1980) Optimal outbreeding and the development of sexual preferences in Japanese quail. *Z. Tierpsychol.* **53**, 231–244.

Bateson, P.P.G. (1990) Is imprinting a special case? *Phil. Trans. R. Soc. Lond. B* **203**, 125–131.

Bateson, P.P.G. (1991) Assessment of pain in animals. *Anim. Behav.* **42**, 827–839.

Bateson, P.P.G., Lotwick, W. and Scott, D.K. (1980) Similarities between the faces of parents and offspring in Bewick's swans and the differences between mates. *J. Zool. Lond.* **191**, 61–74.

Bateson, P.P.G. and Reese, E.P. (1969) The reinforcing properties of conspicuous stimuli in the imprinting situation. *Anim. Behav.* **17**, 692–699.

Bateson, P.P.G. and Wainwright, A.A.P. (1972) The effects of prior exposure to light on the imprinting process in domestic chicks. *Behaviour* **42**, 279–290.

Beckers, R., Holland, O.E. and Deneubourg, J.L. (1994) From local actions to global tasks: stigmergy and collective robotics. In Brooks, R.A. and Maes, P. (eds) *Artificial Life IV*, pp. 181–189. MIT Press, Cambridge, MA.

Beer, C.G. (1970) Individual recognition of voice in the social behaviour of birds. *Adv. Study Behav.* **3**, 27–74.

Bekesy, G. von (1952) Mechanics of hearing. *Nature* **169**, 241–242.

Bekesy, G. von (1960) *Experiments on Hearing.* McGraw-Hill, New York.

Bekhterev, V.M. (1913) La psychologie objective. Alcan, Paris.

Bell, W.J. (1991) *Searching Behaviour.* Chapman and Hall, London.

Beninger, R.J., Kendall, S.B. and Vanderwof, C.H. (1974) The ability of rats to discriminate their own behaviours. *Can. J. Psychol.* **28**, 79–91.

Bentley, D.R. (1971) Genetic control of an insect network. *Science* **174**, 1139–1141.

Bentley, D.R. and Hoy, R.R. (1972) Genetic control of the neuronal network generating cricket (*Teleogryllus gryllus*) song patterns. *Anim. Behav.* **20**, 478–492.

Benvenuti, S. and Wallraff, H.G. (1985) Pigeon navigation: site simulation by means of atmospheric odours. *J. Comp. Physiol.* **156**, 737–746.

Bernstein, J.I. (1970) Anatomy and physiology of the central nervous system. In Hoar, W.S. and Randall, D.J. (eds) *Fish Physiology*, vol. 4, pp. 1–90. Academic Press, New York.

Berthold, P. (1973) Relationship between migratory restlessness and migration distance in six sylvia species. *Zool. Jb. Syst.* **115**, 594–599.

Berthold, P. (1974) Circannual rhythms in birds with different migratory habits. In Pengelley, E.T. (ed.) *Circannual Clocks*, Academic Press, New York.

Berthold, P. (1984a) The endogenous control of bird migration: a survey of experimental evidence. *Bird Study* **31**, 19–27.

Berthold, P. (1984b) The control of partial migration in birds: a review. *The Ring* **10**, 253–265.

Bertram, B.C.R. (1980) Vigilance and group size in ostriches. *Anim. Behav.* **28**, 278–286.

Biebach, H. (1983) Genetic determination of partial migration in the European robin (*Erithacus rubecula*). *Auk* **100**, 601–606.

Biebach, H. (1985) Sahara stopover in migratory flycatchers: fat and food affect the time program. *Experientia* **41**, 695–697.

Biebach, H., Freidrich, W. and Heine, G. (1986) Interaction of bodymass, fat, foraging and stopover period in trans-Sahara migrating passerine birds. *Oecologia* **69**, 370–379.

Biederman, G.B., D'Amato, M.R. and Keller, D.M. (1964) Facilitation of discriminated avoidance learning by dissociation of CS and manipulandum. *Psychon. Sci.* **1**, 229–230.

Bischof, H.-J. (1983) Imprinting and cortical plasticity: a comparative review. *Neurosci. Biobehav. Rev.* **7**, 213–225.

Bischof, N. (1975) Comparative ethology of incest avoidance. In Fox, R. (ed.) *Biosocial Anthropology*, pp. 37–67. Malaby Press, New York.

Bischof-Köhler, D. (1990) The development of empathy in infants. In Lamb, M.E. and Keller, H. (eds) *Infant development: perspectives from German-speaking countries*. Lawrence Erlbaum, Hillsdale.

Black, R. (1971) Hatching success in the three-spined stickleback (*Gasterosteus aculeatus*) in relation to changes in behaviour during the parental phase. *Anim. Behav.* **19**, 532–541.

Blakemore, R.P. (1975) 'Magnetic bacteria'. *Science* **190**, 377–379.

Bleek, D. F. (1930) *Rock Paintings in South Africa*. Methuen, London.

Blest, A.D. (1957) The evolution of protective displays in the Saturnioidea and Sphingidae (Lepidoptera). *Behaviour* **11**, 257–309.

Blest, A.D. (1961) The concept of ritualization. In Thorpe, W.H. and Zangwill, O.L. (eds) *Current Problems in Animal Behaviour*. Cambridge University Press, London.

Bligh, J. (1976) Reproduction. In Bligh, J., Cloudsley-Thompson, J.L. and Macdonald, A.G. *Environmental Physiology of Animals*. Blackwell, Oxford.

Blough, D.S. (1955) Method for tracing dark adaptation in the pigeon. *Science* **121**, 703–704.

Blurton-Jones, N.G. (1968) Observations and experiments on the causation of threat displays of the great tit (*Parus major*). *Anim. Behav. Monogr.* **1**, 2.

Blurton-Jones, N.G. (1972) *Ethological Studies of Child Behaviour*. Cambridge University Press, New York.

Blurton-Jones, N. (1975) Ethology, anthropology and childhood. In Fox, R. (ed.) *Biosocial Anthropology*. pp. 69–92. Malaby Press, London.

Blurton-Jones, N.G. and Sibly, R.M. (1978) Testing adaptiveness of culturally determined behaviour: do bushman women maximize their reproductive success by spacing births widely and foraging seldom? In Reynolds, V. and Blurton-Jones, N. (eds) *Human Behaviour and Adaptation*, Taylor and Francis, London.

Bodmer, W.F. and Cavalli-Sforza, L.L. (1976) *Genetics, Evolution, and Man*. W.H. Freeman, San Francisco, CA.

Boeuf, Le, B.J. (1974) Male–male competition and reproductive success in elephant seals. *Amer. Zool.* **14**, 163–176.

Bohner, J. (1983) Song learning in the zebra finch (*Taeniopygia guttata*): selectivity in the choice of a tutor and accuracy of song copies. *Anim. Behav.* **31**, 231–237.

Bolles, R.C. (1967) *Theory of Motivation*. Harper and Row, New York.

Bolles, R.C. (1970) Species-specific defense reactions and avoidance learning. *Psychol. Rev.* **77**, 32–48.

Bonner, J.T. (1980) *The Evolution of Culture in Animals*. Princeton University Press, Princeton, NJ.

Bookman, M.A. (1978) Sensitivity of the homing pigeon to an earth-strength magnetic field. In Schmidt-Koenig, K. and Keeton, W.T. (eds) *Animal Migration, Navigation and Homing*, pp. 127–134. Proceedings in Life Sciences. Springer-Verlag, New York.

Booth, D.A. (1978) (ed.) *Hunger Models. Computable Theory of Feeding Control*. Academic Press, London.

Brambell, F.W.R. (1965) (Chairman) *Report of the Technical Committee to Enquire into the Welfare of Animals kept under Intensive Livestock Husbandry Systems*. Cmnd 2836. H.M.S.O., London.

Braun, H.W. and Geiselhart, R. (1959) Age differences in the acquisition and extinction of the conditioned eyelid response. *J. Exp. Psychol.* **57**, 386–388.

Bray, O.E., Kenelly, J.J. and Guarino, J.L. (1975) Fertility of eggs produced on territories of vasectomized red-winged blackbirds. *Wilson Bull.* **87**, 187–195.

Breland, K. and Breland, M. (1961) The misbehaviour of organisms. *Am. Psychol.* **16**, 661–664.

Brockmann, H.J. and Barnard, C.J. (1979) Kleptoparasitism in birds. *Anim. Behav.* **27**, 487–514.

Brockmann, H.J. and Dawkins, R. (1979) Joint nesting in a digger wasp as an evolutionarily stable preadaptation to social life. *Behaviour* **71**, 203–245.

Brockmann, H.J., Grafen, A. and Dawkins, R. (1979) Evolutionarily stable nesting strategy in a digger wasp. *J. Theor. Biol.* **77**, 473–496.

Brogden, W.J. (1939) Sensory pre-conditioning. *J. Exp. Psychol.* **25**, 323–332.

Brogden, W.J., Lipman, E.A. and Cullen, E. (1938) The role of incentive in conditioning and extinction. *Am. J. Psychol.* **51**, 109–117.

Brooks, R. (1986) A robust layered control system for a mobile robot. *IEEE J. Robotics and Automation*, RA-2, April, 14–23.

Brooks, R. (1989) A robot that walks; emergent behaviour from a carefully evolved network. *Neural Computation* **1**, 253–262.

Brooks, R. (1991) Challenges for complete creature architectures. In Meyer, J. and Wilson, S. (eds) *From Animals to Animats*. MIT Press, Cambridge, MA.

Brower, J.V. (1958) Experimental studies of mimicry in some North American butterflies. I. The Monarch, *Danaus plexippus*, and Viceroy, *Limenitis archippus archippus*. *Evolution* **12**, 32–47.

Brower, L.P. (1969) Ecological chemistry. *Scientific American* 220 (Feb.), 22–29. Offprint 1133.

Brown, C.R. (1988) Enhanced foraging efficiency through information centers: a benefit of coloniality in cliff swallows. *Ecology* **69**, 602–613.

Brown, C.R. and Brown, M.B. (1986) Ectoparasitism as a cost of coloniality in cliff swallows (*Hirundo pyrrhonata*). *Ecology* **67**, 1206–1218.

Brown, J.L. (1975) *The Evolution of Behaviour*. W.W. Norton and Co., New York.

Brown, P.L. and Jenkins, H.M. (1968) Auto-shaping of the pigeon's key-peck. *J. Exp. Anim. Behav.* **11**, 1–8.

Brown, R.G.B. (1962) The aggressive and distraction behaviour of the western sandpiper *Ereunetes mauri*. *Ibis* **104**, 1–12.

Bryant, P.E. and Trabasso, T. (1971) Transitive inferences and memory in young children. *Nature* **232**, 456–458.

Budgell, P. (1970a) Modulation of drinking by ambient temparature changes. *Anim. Behav.* **18**, 753–757.

Budgell, P. (1970b) The effect of changes in ambient temperature on water intake and evaporative water loss. *Psychon. Sci.* **20**, 275–276.

Bullock, T.H. (1977) *Introduction to Nervous Systems*. W.H. Freeman, San Francisco, CA.

Bunning, E. (1967) *The Physiological Clock*. Springer-Verlag, New York.

Burchfield, S.R., Woods, S.C. and Elich, M.S. (1980) Pituitary adrenocortical response to chronic intermittent stress. *Physiol. Behav.* **24**, 297–302.

Burghardt, G.M. (1970) Chemical perception in reptiles. In Johnson, J.W., Moulton, D.G. and Turk, A. (eds) *Communication by Chemical Signals*, Appleton-Century-Crofts, New York.

Burghardt, G.M. (1975) Chemical preference ploymorphism in newborn garter snakes, *Thamnopphis sirtalis. Behaviour* **52**, 202–225.

Burrows, M. and Horridge, G.A. (1974) The organisation of inputs to motoneurons of the locust metathoracic leg. *Phil. Trans. R. Soc. B.* **269**, 49–94.

Buss, D.M. (1989) Sex differences in human mate preferences: evolutionary hypotheses tested in 37 cultures. *Behavioural and Brain Sciences* **12**, 1–49.

Bykov, K.M. (1957) *The Cerebral Cortex and the Internal Organs*. Chemical Publishing Co., New York.

Byrne, R.W. and Whiten, A. (1988) *Machiavellian Intelligence*. Clarendon Press, Oxford.

Cade, T.J. and Dybas, J.A. (1962) Water economy of the budgerigar. *Auk* **79**, 345–364.

Campbell, C. (1969) Development of specific preferences in thiamine-deficient rats: evidence against mediation by aftertastes. Unpublished master's thesis, Library of the University of Illinois at Chicago Circle.

Cannon, W.B. (1927) The James–Lange theory of emotions: a critical examination and an alternative theory. *Am. J. of Psychol.* **39**, 106–124.

Cannon, W.B. (1932) *The Wisdom of the Body*. Kegan Paul, London.

Capretta, P.J. (1961) An experimental modification of food preferences in chickens. *J. Comp. Physiol. Psychol.* **54**, 238–242.

Carter, D.E. and Werner, T.J. (1978) Complex learning and information processing by pigeons: a critical analysis. *J. Exp. Anal. Behav.* **29**, 565–601.

Caspari, E. (1972) Sexual selection in human evolution. In Campbell, B. (ed) *Sexual Selection and the Descent of Man*. Heinemann, London.

Catchpole, C.K. and Slater, P.J.B. (1995) *Bird Song*. Cambridge University Press, Cambridge.

Cavalli-Sforza, L.L. (1969) Genetic drift in an Italian population. *Scientific American*, August 1967.

Cavalli-Sforza, L.L. and Bodmer, W.F. (1971) *The Genetics of Human Populations*. W.H. Freeman, San Francisco, CA.

Cavalli-Sforza, L.L. and Feldman, M.W. (1974) Cultural versus biological inheritance: phenotypic transmission from parent to children (a theory of the effect of parental phenotypes on children's phenotype). *Am. J. of Human Genetics* **25**, 618–637.

Chagnon, N.A. (1968) *Yanomamo: The Fierce People*. Holt Reinhart and Winston, New York.

Chagnon, N.A., Neel, J.V., Weitkamp, L., Gershowitz, H. and Ayres, M. (1970) The influence of cultural factors on the demography and pattern of gene flow from the Makiritare to the Yanomama Indians. *Am. J. Phys. Anthrop.* **32**, 339–350.

Chance, M.R.A. (1960) Köhler's chimpanzees – how did they perform? *Man* **60**, 130–135.

Chandler, M.J., Fritz, A.S. and Hala, S. (1989). Small-scale deceit: deception as a marker of 2-, 3-, and 4-year olds' theories of mind. *Child Development* **60**, 1263–1277.

Chappell, J. (1997) An analysis of clock-shift experiments: is scatter increased and deflection reduced in clock-shifted homing pigeons? *J. Exp. Biol.* **200**, 2269–2277.

Chappell, J. and Guilford, T. (1997) The orientational salience of visual cues to the homing pigeon. *Anim. Behav.* **53**, 287–296.

Cherry, C. (1957) *On Human Communication*. MIT Press, Cambridge, MA.

Cherniak, C. (1981) Minimal rationality. *Mind* **90**, 161–183.

Chitty, D. (1954) *Control of Rats and Mice. Vol. 2. Rats*. Oxford University Press, Oxford.

Chomsky, N. (1972) *Language and Mind*. Harcourt Brace and Jovanovich, New York.

Chrousos, G.P., Loriaux, D.L. and Gold, P.W. (1988) The concept of stress and its historical development. In Chrousos, G.P., Loriaux, D.L. and Gold, P.W. (eds) *Mechanisms of Physical and Emotional Stress*, pp. 3–7. Plenum Press, New York.

Church, R.M. (1963) The varied effects of punishment of behaviour. *Psychol. Rev.* **70**, 369–402.

Church, R.M. (1969) Response suppression. In Campbell, B.A. and Church, R.M. (eds) *Punishment and Aversive Behaviour*, pp. 111–156. Appleton-Century-Crofts, New York.

Church, R.M. (1978) The internal clock. In Hulse, S.H., Fowler, H. and Honig, W.K. (eds) *Cognitive Processes in Animal Behaviour*. Lawrence Erlbaum, Hillsdale, N.J.

Cloudsley-Thompson, J.L. (1980) *Biological Clocks*. Weidenfeld and Nicolson, London.

Clutton-Brock, T.H. and Albon, S.D. (1979) The roaring of red deer and the evolution of honest advertisement. *Behaviour* **69**, 145–170.

Clutton-Brock, T.H. and Harvey, P.H. (1977) Primate ecology and social organisation. *J. Zool. Lond.* **183**, 1–39.

Clutton-Brock, T.H. and Harvey, P.H. (1979) Comparison and adaptation. *Proc. R. Soc. Lond. B.* **205**, 547–565.

Cohen, J. (1966) *Human Robots in Myth and Science*. Allen and Unwin, London.

Colwill, R.C. and Rescorla, R.A. (1985) Postconditioning devaluation of a reinforcer affects instrumental responding. *J. of Exp. Psychol.: Anim. Behav. Processes* **11**, 120–132.

Coon, C.S. (1962) *The Origin of Races*. Alfred A. Knopf, New York.

Cooper, L.A. (1982) Internal representation. In Griffin, D.R. (ed.) *Animal Mind – Human Mind*. Springer-Verlag, New York.

Cooper, L.A. and Shepard, R.N. (1973) The time required to prepare for a rotated stimulus. *Mem. Cognit.* **1**, 246–250.

Cooper, R.M. and Zubek, J.P. (1958) Effects of enriched and restricted early environments on the learning ability of bright and dull rats. *Can. J. Psychol.* **12**, 159–164.

Coppinger, R.P. (1969) The effect of experience and novelty on avian feeding behaviour with reference to the evolution of warning coloration in butterflies. Part I: reactions of wild caught adult Blue Jays to novel insects. *Behaviour* **35**, 45–60.

Coppinger, R.P. (1970) The effect of experience and novelty on avian feeding behavior with reference to the evolution of warning coloration in butterflies. II. Reactions of naive birds to novel insects. *Am. Nat.* **104**, 323–335.

Coss, R.G. and Owings, D.H. (1978) Snake-directed behavior by snake naive and experienced California ground squirrels in a simulated burrow. *Z. Tierpsychol.* **48**, 421–435.

Cox, C.R. and Le Boeuf, B.J. (1977) Female incitation of male competition: a mechanism of mate selection. *Am. Nat.* **111**, 317–335.

Croft, D.B. (1975) *The Effect of Photoperiod Length on the Diurnal Activity of Canaries*. Unpublished Ph.D. thesis, University of Cambridge.

Crook, J.H. (1964) The evolution of social organization and visual communication in the weaverbirds (Ploceidae). *Behaviour* Suppl. **10**, 1–178.

Crook, J.H. (1966) Gelada baboon herd structure and movement. A comparative report. *Symp. Zool. Soc. Lond.* **18**, 237.

Crook, J.H. (1972) Sexual selection, dimorphism, and social organization in the primates. In Campbell, B. (ed.) *Sexual Selection and the Descent of Man 1871–1971*, pp. 231–281. Aldine, Chicago.

Crook, J.H. (1980) *The Evolution of Human Consciousness*. Oxford University Press, New York.

Crook, J.H. and Gartlan, J.S. (1966) Evolution of primate societies. *Nature* **210**, 1200–1203.

Croze, H.J. (1970) Searching image in carrion crows. *Z. Tierpsychol. Beiheft.* **5**, 86.

Cullen, E. (1957) Adaptations in the kittiwake to cliff-nesting. *J. Comp. Psychol* **99**, 275–302.

Curio, E. (1975) The functional organization of anti-predator behaviour in the pied flycatcher: a study of avian visual perception. *Anim. Behav.* **23**, 1–115.

Daan, S. (1981) Adaptive daily strategies in behavior. In Ashoff, J. (ed.) *Handbook of Behavioral Neurobiology, Vol. 4: Biological Rhythms*. Plenum Press, New York.

Daanje, A. (1950) On locomotory movements in birds and the intention movements derived from them. *Behaviour* **3**, 48–98.

Dallman, M.F., Akana, S.F., Cascio, C.S., Darlington, D.N., Jacobsen, L. and Levin, N. (1987) Regulation of ACTH secretion: variations on a theme of B. In Clark, J.H. (ed.) *Recent Progress in Hormone Research, Vol. 43.* Academic Press, Orlando.

Daly, M. and Wilson, M. (1978) *Sex, Evolution and Behaviour.* Duxbury Press, North Scituate, MA.

Dampier, W. (1929) *History of Science.* Cambridge University Press, John Murray, London.

Darwin, C. (1841) *The Zoology of the Voyage of H.M.S. 'Beagle', Pt. III. Birds, by J. Gould. Notice of their habits and ranges, by Charles Darwin*, pp. 98–106.

Darwin, C. (1859) *On the Origin of Species by Natural Selection or the Preservation of Favoured Races in the Struggle for Life.* John Murray, London.

Darwin, C. (1868) *The Variation of Animals and Plants under Domestication.* John Murray, London.

Darwin, C. (1871) *The Descent of Man and Selection in Relation to Sex.* John Murray, London.

Darwin, C. (1872) *The Expression of the Emotions in Man and Animals.* John Murray, London.

Davenport, J. (1985) *Environmental Stress and Behavioural Adaptation.* Croom Helm, London.

Davies, N.B. (1978) Ecological questions about territorial behaviour. In Krebs J.R. and Davies, N.B. (eds) *Behavioural Ecology.* Blackwell Scientific Publications, Oxford.

Davies, N.B. and Brooke, M. de L. (1988) Cuckoos versus reed warblers: adaptations and counteradaptations. *Anim. Behav.* 36, 262–284.

Davies, N.B. and Brooke, M. de L. (1989) An experimental study of coevolution between the cuckoo *Cuculus canorus* and its hosts. I. Host egg discrimination. *J. Anim. Ecol.* **58**, 207–224.

Davies, N.B. and Halliday, T.R. (1977) Optimal mate selection in the toad, *Bufo bufo. Nature* **269**, 56–58.

Davies, N.B. and Houston, A.I. (1981) Owners and satellites: the economics of territory defence in the pied wagtail, *Motacilla alba. J. Animal Ecol.* **50**, 157–180.

Davies, N.B. and Houston, A.I. (1984) Territory economics. In Krebs, J.R. and Davies, N.B. (eds) *Behavioural Ecology*, 2nd edn. Blackwell, Oxford.

Davis, J.M. (1973) Imitation: a review and critique. In Bateson, P.P.G. and Klopfer, P.H. (eds) *Perspectives in Ethology.* Plenum Press, New York.

Davis, J.M. (1975) Socially induced flight reactions in pigeons. *Anim. Behav.* **23**, 597–601.

Dawkins, M. (1971a) Perceptual changes in chicks: another look at the 'search image' concept. *Anim. Behav.* **19**, 566–574.

Dawkins, M. (1971b) Shifts of 'attention' in chicks during feeding. *Anim. Behav.* **19**, 575–582.

Dawkins, M. (1976) Towards an objective method of assessing welfare in domestic fowl. *App. Anim. Ethol.* **2**, 245.

Dawkins, M. (1977) Do hens suffer in battery cages? Environmental preference and welfare. *Anim. Behav.* **25**, 1034.

Dawkins, M.S. (1980) *Animal Suffering.* Chapman and Hall, London.

Dawkins, M.S. (1983) Battery hens name their price: consumer demand theory and the measurement of ethological 'needs'. *Anim. Behav.* **31**, 1195–1205.

Dawkins, M.S. (1990) From and animal's point of view: motivation, fitness and animal welfare. *Behav. Brain Sci.* **13**, 1–61.

Dawkins R. (1976) *The Selfish Gene.* Oxford University Press, Oxford.

Dawkins, R. (1978) Replicator selection and the extended phenotype. *Z. Tierpsychol.* **47**, 61–76.

Dawkins, R. (1979) Twelve misunderstandings of kin selection. *Z. Tierpsychol.* **51**, 184–200.

Dawkins, R. (1980) Good strategy or evolutionarily stable strategy? In Barlow G.W. and Silverberg, J. (eds) *Sociobiology: Beyond Nature/Nurture?*, pp. 331–367. Westview Press, Boulder, CO.

Dawkins, R. (1982) *The Extended Phenotype.* W.H. Freeman, Oxford.

Dawkins, R. and Brockmann, H.J. (1980) Do digger wasps commit the Concorde fallacy? *Anim. Behav.* **28**, 892–896.

Dawkins, R. and Carlisle, T.R. (1976) Parental investment, mate desertion and a fallacy. *Nature* **262**, 131–133.

Dawkins, R. and Krebs, J.R. (1978) Animal signals: information or manipulation? In Krebs, J.R. and Davies, N.B. (eds) *Behavioural Ecology*, pp. 282–309. Blackwell Scientific Publications, Oxford.

De Boer, S.F., Slangen, J.L. and Van der Gugten, J. (1988) Adaptation of plasma catecholamine and corticosterone responses to short-term repeated noise stress in rats. *Physiol. and Behav.* **44**, 237–280.

Delius, J.D. (1986) Komplexe Wahrehmungsleistungen bei Tauben. *Spektrum der Wissenschaft* **4**, 46–58.

Delius, J.D. and Emmerton, J. (1978) Sensory mechanisms related to homing in pigeons. In Schmidt-Koenig, K. and Keeton, W.T. (eds) *Animal Migration. Navigation and Homing*, pp. 35–41. Proceedings in Life Sciences. Springer-Verlag, Berlin.

Delius, J.D. and Nowak, B. (1982) Visual symmetry recognition by pigeons. *Psychol. Res.* **44**, 199–212.

Delius, J.D., Perchard, R.J. and Emmerton, J. (1976) Polarized light discrimination by pigeons and an electroretinographic correlate. *J. Comp. Physiol. Psychol.* **90**, 560–571.

Delius, J.D. and Vollrath, F.W. (1973) Rotation compensation reflexes independent of the labyrinth: neurosensory correlates in pigeons. *J. Comp. Physiol. Psychol.* **83**, 123–134.

Denenberg, V.H. (1962) An attempt to isolate critical periods of development in the rat. *J. Comp. Physiol. Psychol.* **55**, 813–815.

Dennett, D.C. (1978) *Brainstorms.* Bradford Books, VT, USA; Harvester Press, UK, 1979.

Dennett D.C. (1983) Intentional systems in cognitive ethology: the 'Panglossian paradigm' defended. *Behav. Brain Sci.* **6**, 343–390.

Denton, E.J. and Warren, F.J. (1957) The photosensitive pigments in the retinae of deep-sea fish. *J. Mar. Biol. Assoc. UK* **36**, 651–662.

De Ruiter, J.R. (1986) The influence of group size on predator scanning and foraging behaviour of wedgecapped capuchin monkeys (*Cebus olivaceus*). *Behaviour* **98**, 240–258.

Dewsbury, D.A. (1982) Dominance rank, copulatory behavior, and differential reproduction. *Quart. Rev. Biol.* **57**, 135–159.

Dewsbury, D.A. (1984) Sperm competition in muroid rodents. In Smith, R.L. (ed.) *Sperm Competition and the Evolution of Animal Mating Systems.* Academic Press, New York.

Dickinson, A. (1980) *Contemporary Animal Learning Theory.* Cambridge University Press, Cambridge.

Dickinson, A. (1985) Actions and habits: the development of behavioural autonomy. *Phil. Trans. R. Soc. Lond. B* **308**, 67–78.

Dickinson, A. (1986) Re-examination of the role of the instrumental contingency in the sodium-appetite irrelevant incentive effect. *Quart. J. of Exp. Psychol.* **38B**, 161–172.

Dickinson, A. (1989) Expectancy theory in animal conditioning. In Klein, S. and

Mower, R. (eds) *Contemporary Learning Theories: Pavlovian Conditioning and the Status of Traditional Learning Theory*. LEA, New York.

Dickinson, A. and Balleine, B. (1990) Motivational control of instrumental performance following a shift from hunger to thirst. *Quart. J. of Exp. Psychol.* **42B**, 413–431.

Dickinson, A. and Balleine, B. (1994) Motivational control of goal-directed action. *Anim. Learning Behav.* **22**, 1–18.

Dickerson, J. and Dyer, F. (1996) How insects learn about the sun's course: alternative modelling approaches. In Maes, P., Mafaric, M.J., Meyer, J., Pollock J. and Wilson. S. (eds) *Animals to Animats, Vol. 4.* MIT Press, Cambridge, MA.

Dittus, W.P.J. (1977) The social regulation of population density and age–sex distribution in the toque monkey. *Behaviour* **63**, 281–322.

Dobbing, J. (1976) Vulnerable period in brain growth and somatic growth. In Roberts, D.F. and Thompson, A.M. (eds) *The Biology of Human Growth*. Taylor and Francis, London.

Dobzhansky, T. (1964) *The Heredity and the Nature of Man*. Harcourt, Brace and World, New York.

Domjan, M. and Wilson, N.E. (1972) Contribution of ingestive behaviours to taste-aversion learning in the rat. *J. Comp. Physiol. Psychol.* **80**, 403–412.

Dorst, J. and Dandelot, P. (1970) *A Field Guide to the Large Mammals of Africa*. Houghton Miflin, Boston, MA.

Doty, R.W. (1968) Neural organisation of deglutition. In Heidel, W. (ed.) *Handbook of Physiology – Alimentary canal, IV*. American Physiological Society, Washington.

Dowling, J.E. and Boycott, B.B. (1966) Organisation of the primate retina: electron microscopy. *Proc. Roy. Soc. Lond. B.* **166**, 80–111.

Drent, R.H. (1970) Functional aspects of incubation in the herring gull. In Baerends, G.P. and Drent, R.H. (eds) *The Herring Gull and its Egg, Part I, Behaviour* **82**, 1–132.

Duke-Elder, S. (1958) *System of Ophthalmology I. The Eye in Evolution*. Henry Kimpton, London.

Dunbar, R.I.M. (1988) *Primate Social Systems*. Croom Helm, London.

Duncan, I.J.H. (1977) Behavioural wisdom lost. *Appl. Anim. Ethol.* **3**, 193.

Duncan, I.J.H. (1978) The interpretation of preference tests in animal behaviour. *Appl. Ethol.* **4**, 197.

Duncan, P. and Vigne, N. (1979) The effect of group size in horses on the rate of attacks by blood-sucking flies. *Anim. Behav.* **27**, 623–625.

Dyer, F.C. (1985) Nocturnal orientation by the Asian honey bee, *Apis dorsata*. *Anim. Behav.* **33**, 769–774.

Dyson-Hudson, R. and Smith, E.A. (1978) Human territorality: an ecological reassessment. *Am. Anthropol.* **80**, 21–41.

Edmunds, M. (1974) *Defence in Animals*. Longman, Harlow.

Edwards, W. (1954) The theory of decision making. *Psychol. Bull.* **51** (4), 380–417.

Edwards, W. (1961) Behavioural decision theory. *Ann. Rev. Psychol.* **12**, 473–498.

Ehrman, L. and Parsons, P.A. (1976) *The Genetics of Behavior*. Sinauer Associates, Sunderland, MA.

Eibl-Eibesfeldt, I. (1967) Concepts of ethology and their significance in the study of human behaviour. In Stevenson, H.W. (ed.) *Early Behaviour: Comparative and Developmental Approaches*. Wiley, New York.

Eibl-Eibesfeldt, I. (1970) *Ethology. The Biology of Behavior*. Holt, Rinehart and Winston, New York.

Eibl-Eibesfeldt, I. (1989) *Human Ethology*. De Gruyer, New York.

Eiseman, C.H., Jorgensen, W.K., Kemt, D.J., Rice, M.J., Gibb, B.W., Webb, P.D. and Zalucki, M.P. (1984) Do insects feel pain? A biological view. *Experientia*, **40**, 164–167.

Ekman, P. (1971) Universals and cultural differences in facial expressions of emotion. *Nebraska Symposium on Motivation*, 204–284.

Ekman, P. and Friesen, W.V. (1969) The repertoire of nonverbal behavior: categories, origins, usage and coding. *Semiotica* **1**, 49–98.

Ellis, H. (1906) *Sexual Selection in Man*. F.A. Davis, Philadelphia, PA.

Emlen, S.T. (1967) Migratory orientation in the Indigo Bunting, *Passerina cyanea*. Part 1: The evidence for use of celestial cues. *Auk* **84**, 309–342. Part II: Mechanisms of celestial orientation. *Auk* **84**, 463–489.

Emlen, S.T. (1972) The ontogenetic development of orientation capabilities. In Galler, S.R., Schmidt-Koenig, K., Jacobs, G.J. and Belleville, R.E. (eds) *Animal Orientation and Navigation*. pp. 191–210. NASA SP-262. US Govt. Printing Office, Washington, D.C.

Emlen, S.T. (1978) The evolution of cooperative breeding in birds. In Krebs J.R. and Davies, N.B. (eds) *Behavioural Ecology, an Evolutionary Approach*. Blackwell Scientific Publications, Oxford.

Emlen, S.T. (1997) Predicting family dynamics in social vertebrates. In Krebs J.R. and Davies, N.B. (eds) *Behavioural Ecology*, 4th edn. Blackwell, Oxford.

Emlen, S.T. and Emlen, J.T. (1966) A technique for recording migratory orientation of captive birds. *Auk* **83**, 361–367.

Estes, W.K. (1944) An experimental study of punishment. *Psychol. Monogr.* **57** (263), 98, 109–112.

Etienne, A.S. (1969) Analyse der schlagauslosenden Bewegungsparameter einer punkt-formigen Beuteattrappe bei der Aeschnalarve. *Z. vergle. Physiol.* **64**, 71–110.

Evans, C.S. and Marler, P. (1995) Language and animal communication: parallels and contrasts. In Roitblat, H.L. and Meyer, J. (eds) *Comparative Approaches to Cognitive Science*. MIT Press, Cambridge, MA.

Evans, S.M. (1966) Non-associative avoidance learning in Nereid polychaetes. *Anim. Behav.* **14**, 102–106.

Evarts, E.V.J. (1968) Motorneuron firing in pyramidal tract correlated with movement. *Neurophsyiol* **31**, 14–27.

Ewert, J.P. (1980) *Neuroethology*. Springer, Heidelberg.

Ewert, J.P. (1987) Neuroethology of releasing mechanisms: prey-catching in toads. *Behav. Brain Sci.* **10**, 337–405.

Ewert, J.P. and Burghagen, H. (1979) Ontogenetic aspects of visual 'size-constancy' phenomena in the midwife toad *Alytes obstetricians* (Laur). *Brain Behav. Evolution* **16**, 99–112.

Ewert, J.P. and Hock, F. (1972) Movement sensitive neurons in the toad's retina. *Expl. Brain Res.* **16**, 41–59.

Ewert, J.P. and Traud, R. (1979) Releasing stimuli for antipredator behaviour in the common toad *Bufo bufo* (L). *Behaviour* **68**, 170–180.

Falconer, D.S. (1960) *Introduction to Quantitative Genetics*. Oliver and Boyd, Edinburgh.

Fernald, A. (1985) Four-month-old infants prefer to listen to motherese. *Infant Behav. Develop.* **8**, 181–195.

Ferson, L. von and Delius, J.D. (1989) Long term retention of many visual patterns by pigeons. *Ethology* (in press).

Fincke, O.M. (1982) Lifetime mating success in a natural population of the damsel fly, *Enallagma hageni* (Walsh) (Odonata: Coenagrionidae). *Behav. Ecol. Sociobio.* **10**, 293–302.

Fisher, R.A. (1918) The correlation between relatives on the supposition of Mendelian inheritance. *Trans. Roy. Soc. Edinb.* **52**, 399–433.

Fisher, R.A. (1930) *The Genetical Theory of Natural Selection*. Oxford University Press, Oxford. (2nd edn, 1958).

Fitzsimons, J.T. (1971) The physiology of thirst: a review of the extraneural aspects of the mechanisms of drinking. In Stellar, E. and Sprague, J.M. (eds) *Progress in Physiological Psychology, Vol. 4*. pp. 119–201. Academic Press, New York.

Fitzsimons, J.T. (1976) The physiological basis of thirst. *Kidney Int.* **10**, 3–11.

Fitzsimons, J.T. and Le Magnen, J. (1969) Eating as a regulatory control of drinking in the rat. *J. Comp. Physiol. Psychol.* **67**, 273–283.

Follett, B.K. (1973) Circadian rhythms and photoperiodic time measurement in birds. *J. Reprod. Fert.*, Suppl. **19**, 5–18.

Ford, E.B. and Huxley, J.S. (1927) Mendelian genes and rates of development in *Gammarus chevreuxi. Brit. J. Exp. Biol.* **5**, 112–133.

Foree, D.D. and LoLordo, V.M. (1973) Attention in the pigeon: the differential effects of food-getting vs. shock-avoidance procedures. *J. Comp. Phsyiol. Psychol.* **85**, 551–558.

Foster, W.A. and Treherne, J.E. (1981) Evidence for the dilution effect in the selfish herd from fish predation on a marine insect. *Nature,* **295** 466–467.

Fouts, R.S. (1975) Capacities for language in great apes. In Tuttle, R.H. (ed.) *Socioecology and Psychology of Primates.* pp. 371–390. Mouton, The Hague.

Fouts, R.S., Chown, B. and Goodwin, L. (1976) Transfer of signed responses in American Sign Language from vocal English stimuli to physical object stimuli by a chimpanzee (*Pan troglodytes*). *Learn. and Motiv.* **7**, 458–475.

Fox, R. (1972) Alliance and constraint: sexual selection in the evolution of human kinship systems. In Campbell B. (ed.) *Sexual Selection and the Descent of Man 1871–1971.* Heinemann, London.

Fraenkel, G.S. and Gunn, D.L. (1940) *The Orientation of Animals.* Clarendon Press, Oxford (Dover Books, 1961).

Fretwell, S.D. (1972) *Populations in a Seasonal Environment.* Princeton University Press, Princeton, NJ.

Freud, S. (1915) Instincts and their vicissitudes. *Coll. Papers Vol. IV.* Hogarth Press, London.

Frisch, K. von (1967) *The Dance Language and Orientation of Bees.* (Translatated by L.E. Chadwick) Harvard University Press, Cambridge MA.

Frisch, K. von and Lindauer, M. (1954) Himmel und Erde in Konkurrenz bei der Orientierung der Bienen. *Naturwissenschaften* **41**, 245–253.

Fry, F.E.J. and Hochachka, P.W. (1970). Fish. In Whittow, G.C. (ed.) *Comparative Physiology of Thermoregulation, Vol. 1,* pp. 79–134. Academic Press, New York.

Fuller, J.L. and Thompson, W.R. (1960) *Behaviour Genetics.* Wiley, New York.

Futuyma, D.J. (1979) *Evolutionary Biology,* pp. 314–316. Sinauer Associates, Sunderland, MA.

Futuyama, D.J. (1986) *Evolutionary Biology,* 2nd edn. Sinauer Associates, Sunderland, MA.

Gallagher, J.E. (1977) Sexual imprinting: a sensitive period in Japanese quail (*Coturnix coturnix japonica*). *J. Comp. Physiol. Psychol.* **91**, 72–78.

Gallup, G.G., Jr. (1977) Self-recognition in primates. A comparative approach to the bidirectional properties of consciousness. *Am. Psychol.* **32**, 329–338.

Gallup, G.G., Jr. (1979) Self-awareness in primates. *Am. Sci.* **67**, 417–421.

Galton, F. (1869) *Herditary Genius: An Inquiry into its Laws and Consequences.* Macmillan, London.

Galton, F. (1883) *Inquiries into Human Faculty and its Development.* Macmillan, London.

Gandolfi, G., Mainardi, D., Rossi, A.C. (1968) La reazione di paura e lo svantaggio individuale dei pesci allarmisti (experimenti con modelli). *Zoologia* **102**, 8–14.

Garcia, J., Ervin, F.R., Yorke, C.H. and Koelling, R.A. (1967) Conditioning with delayed vitamin injection. *Science* **155**, 716–718.

Garcia, J., Kimeldorf, D.J. and Hunt, E.L. (1961) The use of ionizing radiation as a motivating stimulus. *Psychol. Rev.* **68**, 383–385.

Garcia, J., Kimmeldorf, D.J. and Koelling, R.A. (1955) Conditioned aversion to saccharin resulting from exposure to gamma radiation. *Science* **122**, 157–158.

Garcia, J. and Koelling, R.A. (1966) Relation of cue to consequence in avoidance learning. *Psychon. Sci.* **4**, 123–124.

Garcia, J., Kovner, R. and Green, K.F. (1970). Cue properties vs. palatability of flavours in avoidance learning. *Psychon. Sci.* **20**, 313–314.

Garcia, J., McGowan, B.K., Ervin, F.R. and Koelling, R.A. (1968) Cues: their effectiveness as a function of the reinforcer. *Science* **160**, 794–795.

Gardner, B.T. and Gardner, R.A. (1969) Teaching sign language to a chimpanzee. *Science* **165**, 664–672.

Gardner, B.T. and Gardner, R.A. (1971) Two-way communication with an infant chimpanzee. In Schrier, A.M. and Stollnitz, F. (eds) *Behaviour of Non-Human Primates. Vol. IV.* Academic Press, New York.

Gardner, B.T. and Gardner, R.A. (1975) Evidence for sentence constituents in the early utterances of child and chimpanzee. *J. Exp. Psychol.* **104**, 244–267.

Gardner, R.A. and Gardner, B.T. (1978) Comparative psychology and language acquisition. *Ann. N.Y. Acad. Sci.* **309**, 37–76.

Gass, C.L., Angehr, E. and Centa, J. (1976) Regulation of the food supply by feeding territoriality in the rufous hummingbird. *Can. J. Zool.*, **54**, 2046–2054.

Gaston, A.J. (1976) Factors affecting the evolution of group territories of babblers (*Turdoides*) and long-tailed tits. Unpublished dissertation. Oxford University.

Geist, V. (1971) *Mountain Sheep: a Study in Behavior and Evolution.* University of Chicago Press, Chicago, IL.

Gibson, E.J. and Walk, R.D. (1956) The effect of prolonged exposure to visually presented patterns on learning to discriminate them. *J. Comp. Physiol. Psychol.* **49**, 239–242.

Gibson, E.J., Walk, R.D. and Tighe, T.J. (1959) Enhancement and deprivation of visual stimulation during rearing as factors in visual discrimination learning. *J. Comp. Physiol. Psychol.* **52**, 74–81.

Gill, F.B. and Wolf, L.L. (1975) Economics of feeding territorality in the golden-winged sunbird. *Ecology* **56**, 333–345.

Gill, F.B. and Wolf, L.L. (1977) Non-random foraging by sunbirds in a patchy environment. *Ecology* **58**, 1284–1296.

Glasersfeld, E. von (1977) Linguistic communication: theory and definition. In Rumbaugh, D. (ed.) *Language Learning by a Chimpanzee,* pp. 55–71. Academic Press, New York.

Goldsmith, T.H. (1972) The natural history of invertebrate visual pigments. In Dartnall, H.J.A. (ed.) *Handbook of Sensory Physiology,* pp. 685–719. Springer-Verlag, Berlin.

Goodhart, C.B. (1964) A biological view of toplessness. *New Scientist* **23**, 588–560.

Gormezano, I. (1966) Classical conditioning. In Sidowski, J.B. (ed.) *Experimental Methods and Instrumentation in Psychology.* McGraw-Hill, New York.

Goss-Custard, J.D. (1970) Feeding dispersion in some overwintering wading birds. In Crook J.H. (ed.) *Social Behavior in Birds and Mammals,* pp. 3–34. Academic Press, London.

Goss-Custard, J.D. (1976) Variation in the dispersion of redshank (*Tringa totanus*) on their winter feeding grounds. *Ibis* **118**, 257–263.

Goss-Custard, J.D. (1977a) Optimal foraging and size-selection of worms by redshank *Tringa totanus. Anim. Behav.* **25**, 10–29.

Goss-Custard, J.D. (1977b) Predator responses and prey mortality in the redshank

Tringa totanus (L.) and a preferred prey *Corophium volutator* (Pallas). *J. Anim. Ecol.* **46**, 21–36.

Gottlieb, G. (1961) The following-response and imprinting in wild and domestic duck-lings of the same species (*Anas platyrhynchos*). *Behaviour* **18**, 205–228.

Gottlieb, G. (1963) A naturalistic study of imprinting in wood ducklings (*Aix sponsa*). *J. Comp. Physiol. Psychol.* **56**, 86–91.

Gottlieb, G. (1970) Ontogenesis of sensory function in birds and mammals. In Tobach, E. (ed.) *Biopsychology of Development*. Academic Press, New York.

Gottlieb, G. (1971) *Development of Species Identification in Birds*. University of Chicago Press, Chicago, IL.

Gould, J.L. (1976) The dance language controversy. *Quart. Rev. Biol.* **57**, 211–244.

Gould, J.L. (1980) The case for magnetic-field sensitivity in birds and bees (such as it is). *American Scientist* **68**, 256–267.

Gould, J.L. (1981) Language. In McFarland, D.J. (ed.) *The Oxford Companion to Animal Behaviour*. Oxford University Press, Oxford.

Gould, J.L. (1982) *Ethology*. W.W. Norton, New York.

Gould, J.L. and Gould, C.G. (1982) The insect mind: physics or metaphysics? In Griffin, D.R. (ed.) *Animal Mind – Human Mind*. Dahlem Konferenzen. Springer-Verlag, Berlin.

Gould, S.J. (1978) *Ever Since Darwin*. Burnett, London.

Gouzoules, H., Gouzoules, S. and Fedigan, L. (1982) Behavioral dominance and repro-ductive success in female Japanese monkeys (*M. fuscata*). *Anim. Behav.* **30**, 1138–1151.

Gouzoules, S. and Gouzoules, H. (1987) Kinship. In Smuts, B.B., Cheney, D.L., Seyfarth, R.M., Wrangham R.W. and Struhsaker, T.T. (eds) *Primate Societies*. University of Chicago Press, Chicago, IL.

Grafen, A. (1982) How not to measure inclusive fitness. *Nature* **298**, 425–426.

Grafen, A. (1984) Natural selection, kin selection and group selection. In Krebs, J.R. and Davies, N.B. (eds) *Behavioural Ecology,* 2nd edn. Blackwell, Oxford.

Grasse, P.P. (1959) La reconstruction du nid et les coordinations inter-individuelles chez *Bellicositermes natalensis* et *Cucitermes* sp. La theorie de la stigmergie: Essai d'inter-pretation des termites constructeurs. *Ins. Soc.* **6**, 41–83.

Gray, J. (1950) The role of peripheral sense organs during locomotion in the verte-brates. *Sym. Soc. Exp. Biol.* **4**, 112–126.

Green, S. and Marler, P. (1979) The analysis of animal communication. In Marler, P. and Vanderbergh, J.G. (eds) *Handbook of Behavioural Neurobiology. Vol. 3*, pp. 73–158. Plenum Press, New York.

Grether, W.F. (1938) Pseudo-conditioning without paired stimulation encountered in attempted backward conditioning. *J. Comp. Psychol.* **25**, 91–96.

Griffin, D.R. (1958) *Listening in the Dark*. Yale University Press, New Haven, CT.

Griffin, D.R. (1976) *The Question of Animal Awareness*. Rockefeller University Press, New York.

Griffin, D.R. (1981) *The Question of Animal Awareness*, 2nd edn. Rockefeller University Press, New York.

Griffin, D.R. (1982) *Animal Mind – Human Mind*. Springer-Verlag, Berlin.

Grinnell, J. (1917) The niche relationships of the California thrasher. *Auk* **21**, 364–382.

Grogham, P.C. (1976) Ionic and osmotic regulation of aquatic animals. In Bligh, J., Cloudsley-Thompson J.L. and Macdonald, A.G. (eds) *Environmental Physiology of Animals,* Blackwell Scientific Publications, Oxford.

Grohmann, J. (1939) Modifikation oder Funktionsreifung? Ein Beitrag zur Klarung der wechselseitigen Beziehungen zwischen Instinkthandlung und Erfahrung. *Z. Tierpsychol.* **2**, 132–144.

Grossen, N.E. and Kelley, M.J. (1972) Species-specific behaviour and acquisition of avoidance behaviour in rats. *J. Comp. Physiol. Psychol.* **81**, 307–310.

Guilford, T. and Dawkins, M.S. (1987) Search images not proven: a reappraisal of recent evidence. *Anim. Behav.* **35**, 1838–1845.

Guilford, T. and Dawkins, M. (1991) Receiver psychology and the evolution of animal signals. *Anim. Behav.* **42**, 1–14.

Guiton, P. (1959) Socialisation and imprinting in Brown Leghorn chicks. *Anim. Behav.* **7**, 26–34.

Gumma, N.R., South, F.E. and Allen, J.N. (1967) Temperature preference in golden hamsters. *Anim. Behav.* **15**, 534–537.

Gurney, M.E. and Konishi, M. (1980) Hormone-induced sexual differentiation of brain and behaviour in zebra finches. *Science* **208**, 1380–1383.

Guthrie, D.M. (1980) *Neuroethology: An Introduction.* Blackwell Scientific Publications, Oxford.

Guthrie, E.R. (1952) *The Psychology of Learning.* Harper, New York.

Gwinner, E. (1971) A comparative study of circannual rhythms in warblers. In Menaker, M. (ed.) *Biochronometry.* National Academy of Science, Washington, D.C.

Gwinner, E. (1972) Endogenous timing factors in birds migration. In Galler, S.R., Schmidt-Koenig, K., Jacobs, G.J., and Belleville, R.E., (eds) *Animal Orientation and Navigation.* NASA, Washington, D.C.

Gwinner, E. (1986a) Circannual rhythms in the control of avian migrations. *Adv. Study of Behav.* **16**, 191–228.

Gwinner, E. (1986b) Internal rhythms in bird migration. *Scientific American* **254**, 84–92.

Gwinner, E. and Wiltschko, W. (1978) Endogenously controlled changes in migratory direction of the garden warbler, *Sylvia borin. J. Comp. Physiol.* **125**, 267–273.

Hailman, J.P. (1965) Cliff-nesting adaptations of the Galapagos swallow-tailed gull. *Wilson Bull.* **77**, 346–362.

Haldane, J.B.S. (1955) Population genetics. *New Biol.* **18**, 34–51.

Halliday, T.R. (1974) The sexual behaviour of the smooth newt, *Triturus vulgaris* (Urodela, Salamandridae). *J. Herpetol.* **8**, 277–292.

Halliday, T.R. (1976) The libidinous newt. An analysis of variations in the sexual behaviour of the smooth newt, *Triturus vulgaris. Anim. Behav.* **24**, 398–414.

Halliday, T.R. (1977a) The effects of experimental manipulation of breathing behaviour on the sexual behaviour of the smooth newt, *Triturus vulgaris. Anim. Behav.* **25**, 39–45.

Halliday, T.R. (1977b) The courtship of European newts: an evolutionary perspective. In Taylor, D.H. and Guttman, S.I. (eds) *The Reproductive Biology of Amphibians.* Plenum, New York.

Halliday, T.R. (1978) Sexual selection and mate choice. In Krebs, J.R. and Davies, N.B. (eds) *Behavioural Ecology: An Evolutionary Approach.* Blackwell Scientific Publications, Oxford.

Halliday, T.R. (1980) *Sexual Strategy.* Oxford University Press, Oxford.

Halperin, J.R.P. (1991) Machine motivation. In Meyer, J. and Wilson, S. (eds) *From Animals to Animats.* MIT Press, Cambridge, MA.

Ham, R.G. and Veomett, M.J. (1980) *Mechanisms of Development.* C.V. Mosby, St Louis, MO.

Hamilton, W.D. (1964) The genetical theory of social behaviour (I and II). *J. Theoret. Biol.* **7**, 1–16; 17–32.

Hamilton, W.D. (1979) Wingless and fighting males in fig wasps and other insects. In Blum, M.S. and Blum, N.A. (eds) *Sexual Selection and Reproductive Competition in Insects.* Academic Press, London.

Hamilton, W.D. and Zuk, M. (1982) Heritable true fitness and bright birds: a role for parasites? *Science* **218**, 384–387.

Hansen, P. (1979) Vocal learning: its role in adapting sound structures to long distance propagation and a hypothesis on its evolution. *Anim. Behav.* **27**, 1270–1271.

Hanson, J.D., Larson, M.E. and Snowdon, C.T. (1976) The effects of control over high intensity noise on plasma cortisol levels in rhesus monkeys. *Behav. Biol.* **16**, 333–340.

Harcourt, A.H. (1979) Social relationships between adult male and female gorillas in the wild. *Anim. Behav.* **27**, 325–342.

Harlow, H.F. (1949) The formation of learning sets. *Psychol. Rev.* **56**, 51–65.

Harris, G.W., Michael, R.P. and Scott, P.P. (1958) Neurological site of action of stilboestrol in eliciting sexual behaviour. In *Ciba Foundation Symposium on the Neurological Basis of Behaviour*. Churchill, London.

Harris, L.J., Clay, J., Hargreaves, F. and Ward, A. (1933) Appetite and choice of diet. The ability of the vitamin B deficient rat to discriminate between diets containing and lacking the vitamin. *Proc. Roy. Soc. Lond. B* **113**, 161–190.

Harris, M. (1985) *Culture, People, Nature*. 4th edn. Harper and Row, New York.

Harvey, P.H. and Harcourt, A.H. (1984) Sperm competition, testes size, and breeding systems in primates. In Smith, R.L. (ed.) *Sperm Competition and the Evolution of Animal Mating Systems*. Academic Press, New York.

Harvey, P.H., Martin, R.D. and Clutton-Brock, T.H. (1987) Life histories in comparative perspective. In Smuts, B.B., Cheney, D.L., Seyfarth, R.M., Wrangham R.W. and Struhsaker, T.T. (eds) *Primate Societies*. University of Chicago Press, Chicago.

Harvey, P. and Pagel, M.D. (1991) *The Comparative Method in Evolutionary Biology*. Oxford University Press, Oxford.

Haskell, P.T. (1961) *Insect Sounds*. Witherby, London.

Hasler, A.D. (1960) Homing orientation in migrating fishes. *Ergebnisse der Biologie* **23**, 94–115.

Hasler, A.D. and Schwassman, H.O. (1960) Sun orientation of fish at different latitudes. *Cold Spring Harbor Symp. Quant. Biol.* **25**, 411–429.

Hayes, K.J. and Hayes, C. (1952) Imitation in a home-raised chimpanzee. *J. Comp. Physiol. Psychol.* **45**, 450–459.

Hayes, K.J. and Nissen, C.H. (1971) Higher mental functions of a home-raised chimpanzee. In Schrier, A.M. and Stollnitz, F. (eds) *Behaviour of Nonhuman Primates, Vol. 4*. Academic Press, New York.

Hebb, D.O. (1949) *The Organization of Behavior*. Wiley, New York.

Hebb, D.O. (1953) Heredity and environment in mammalian behaviour. *Brit. J. Anim. Behav.* **1**, 43–47.

Hegner, R.E., Emlen, S.T. and Demazo, N.J. (1982) Spatial organisation at the white-fronted-bee-eater. *Nature Lond.* **298**, 264–266.

Heiligenberg, W. (1976) The interaction of stimulus patterns controlling aggressiveness in the cichlid fish *Haplochromis burtoni*. *Anim. Behav.* **24**, 452–458.

Heiligenberg, W., Kramer, U. and Schulz, V. (1972) The angular orientation of the black eye-bar in *Haplochromis burtoni* (Cichlidae: Pisces) and its relevance to aggressivity. *Z. vergl. Physiol.* **76**, 168–176.

Heinrich, B. (1979) *Bumblebee Economics*. Harvard University Press, Cambridge, MA.

Heinroth, O. (1911) *Beitrage zur Biologie, insbesondere Psychologie und Ethologie der Anatiden*. Verhandlungen des internationalen Ornithologenkongresses, Berlin.

Helmholtz, H. von (1867) *Handbuch der Physiologischen Optik*. Voss, Leipzig.

Hendriks-Jansen, H. (1996) *Catching Ourselves in the Act*. MIT Press, Cambridge, MA.

Henton, W.W., Smith, J.C. and Tucker, D. (1966) Odour discrimination in pigeons. *Science* **153**, 1138–1139.

Herrnstein, R.J. (1985) Riddles of natural categorisation. *Phil. Trans. Roy. Soc. Lond. B.* **308**, 129–144.

Herrnstein, R.J., Loveland, D.H. and Cable, C. (1976) Natural concepts in pigeons. *J. Exp. Psychol., Anim. Behav. Processes* **2**, 285–302.

Hess, E.H. (1959a) Imprinting. *Science* **130**, 133–141.

Hess, E.H. (1959b) The conditions limiting critical age of imprinting. *J. Comp. Physiol. Psychol.* **52**, 515–518.

Heyes, C.M. (1993) Imitation, culture, and cognition. *Anim. Behav.* **46**, 999–1010.

Heyes, C.M. (1994) Social cognition in primates. In Mackintosh, N.J. (ed.) *Animal Learning and Cognition*. Academic Press, London.

Heyes, C. and Dickinson, A. (1990) The intentionality of animal action. *Mind and Language* **5**, 87–104.

Hilden, O. (1965) Habitat selection in birds. *Ann. Zool. Fenn.* **2**, 53–75.

Hill, K.G., Loftus-Hills, J.J. and Gartside, D.F. (1972) Premating isolation between the Australian field crickets, *Teleogryllus commodus* and *T. oceanicus. Aust. J. Zool.* **20**, 153–163.

Hinde, R.A. (1952) The behaviour of the great tit (*Parus major*) and some other related species. *Behaviour*, suppl. 2.

Hinde, R.A. (1959) Unitary drives. *Anim. Behav.* **7**, 130–141.

Hinde, R.A. (1960) Energy models of motivation. *Sym. Soc. Exp. Biol.* **14**, 199–213.

Hinde, R.A. (1966) *Animal Behaviour*. McGraw-Hill, New York.

Hinde, R.A. (1970) *Animal Behaviour*, 2nd edn. McGraw-Hill, New York.

Hinde, R.A. (1974) *Biological Bases of Human Social Behaviour*. McGraw-Hill, New York.

Hinde, R.A. (1976) Interactions, relationships and social structure. *Man* **11**, 1–17.

Hinde, R.A. (1979) *Towards Understanding Relationships*. Academic Press, London.

Hinde, R.A. (1987) In Smuts, B.B., Seyfarth, R.M., Wrangham, R.W. and Struhsaker, T.T. (eds) *Private Societies*. University of Chicago Press, Chicago, IL.

Hinde, R.A. and Fisher, J. (1951) Further observations on the opening of milk bottles by birds. *Brit. Birds* **44**, 393–396.

Hinton, G.E., Sejnowski, T.J. and Ackley, D.H. (1984) A learning Boltzman machine. *Cognitive Sci.* **9**, 147–169.

Hirsch, J. (1963) Behaviour genetics and individuality understood. *Science* **142**, 1436–1442.

Hirsch, J. (1967) (ed.) *Behaviour–Genetic Analysis*. McGraw-Hill, New York.

Hirsch, J. and Erlenmeyer-Kimling, L. (1962) Individual differences in behaviour and their genetic basis. In Bliss, E.L. (ed.) *Roots of Behaviour*. Harper and Row, New York.

Hodgkin, A.L. and Huxley, A.F. (1945) *J. Physiol.* **104**, 176–195.

Hodos, W. (1970) Evolutionary interpretation of neural and behavioural studies of living vertebrates. In Schmitt, F.O. (ed.) *The Neuro-Sciences; Second Study Program*, pp. 26–39. Rockefeller University Press, New York.

Hodos, W. (1982) Some perspectives on the evolution of intelligence and the brain. In Griffin, D.R. (ed.) *Animal Mind – Human Mind*. Springer-Verlag, Berlin.

Hodos, W. and Campbell, C.B.G. (1969) Scala Naturae: Why there is no theory in comparative psychology. *Psychol. Rev.* **76**, 337–350.

Hoelzer, G.A. (1989) The good parent process of sexual selection. *Anim. Behav.* **38**, 1067–1078.

Hoffman, H.S., Searle, J.L., Toffrey, S. and Kozma, F., Jr. (1966) Behavioural control by an imprinted stimulus. *J. Exp. Anal. Behav.* **9**, 177–189.

Hoffman, K. (1954) Versuche zu der im Richtungsfinden der Vogel enthaltenen Zeitschatzung. *Z. Tierpsychol.* **11**, 453–475.

Hogan, J.A., Kkeist, S. and Hutchings, C.S.L. (1970) Display and food as reinforcers in the Siamese fighting fish (*Betta splendens*). *J. Comp. Physiol. Psychol.* **70**, 351–357.

Holland, O. (1997) Grey Walter: the pioneer of real artificial life. In Langton C. and Shimohara, K. (eds) *Artificial Life V*. MIT Press, Cambridge, MA.

Holland, P.C. (1977) Conditioned stimulus as a determinant of the form of the Pavolvian conditioned response. *J. Exp. Psychol. Anim. Behav. Processes* **3**, 77–104.

Holland, P.C. and Straub, J.J. (1979) Differential effects of two ways of devaluing the unconditioned stimulus after Pavlovian appetitive conditioning. *J. Exp. Psychol. Anim. Behav. Processes* **5**, 65–78.

Hollard, V.D. and Delius, J.D. (1983) Rotational invariance in visual pattern recognition by pigeons and humans. *Science* **218**, 804–806.

Holst, E. von (1939) Entwurf eines Systems der Lokomotorischen Periodenbildungen bei Fischen. *Z. Verg. Physiol.* **26**, 481–528.

Holst, E. von (1954) Relations between the central nervous system and the peripheral organs. *Br. J. Anim. Behav.* **2**, 89–94.

Holst, E. von (1973) *The Behavioural Physiology of Animals and Man*. Methuen, London.

Holst, E. von and Mittelstaedt, H. (1950) Das Reafferenzprinzip. *Naturwissenschaften* **37**, 464–476.

Hooff, J.A.R.A.M. van (1972) A comparative approach to the phylogeny of laughter and smiling. In Hinde, R.A. (ed.) *Non-Verbal Communication*, pp. 209–238. Cambridge University Press, Cambridge.

Hooff, J.A.R.A.M. van (1976) The comparison of facial expression in man and higher primates. In von Cranach, M. (ed.) *Methods of Inference from Animal to Human Behaviour*, pp. 165–196. Mouton, The Hague.

Hoogland, R., Morris, D. and Tinbergen, N. (1957) The spines of sticklebacks (*Gasterosteus* and *Pygosteus*) as means of defence against predators (*Perca* and *Essox*). *Behaviour* **10**, 205–236.

Hopfield, J. 1982. Neural networks and physical systems with emergent collective computational properties. *Proc. Nat. Acad. Sci.* **79**, 2554–2558.

Hopfield, J., and Tank, D.W. (1986) Computing with neural circuits: A model. *Science* **233**, 625–632.

Hord, R.M. (1982) *Digital Image Processing of Remotely Sensed Data*. Academic Press, New York.

Horn, H.S. (1978) Optimal tactics of reproduction and life history. In Krebs, J.R. and Davies, N.B. (eds) *Behavioural Ecology: An Evolutionary Approach*. Blackwell Scientific Publications, Oxford.

Houston, A.I. and McFarland, D.J. (1976) On the measurement of motivational variables. *Anim. Behav.* **24**, 459–475.

Houston, A.I. and McFarland, D.J. (1980) Behavioural resilience and its relation to demand functions. In Staddon, J.E.R. (ed.) *Limits to Action: The Allocation of Individual Behaviour*, pp. 177–203. Academic Press, New York.

Howard, I.P. (1982) *Human Visual Orientation*. Wiley, New York.

Howard, I.P. and Templeton, W.B. (1966) *Human Spatial Orientation*. Wiley, London.

Howard, R.D. (1978) The evolution of mating strategies in bullfrogs, *Rana catesbiana*. *Evolution* **32**, 850–71.

Howard, R.R. and Brodie, E.D. (1971) Experimental study of mimicry in salamanders involving *Notophthalmus viridescens viridescens* and *Pseudotriton ruber schencki*. *Nature* **233**, 277.

Hoy, R.R. (1974) Genetic control of acoustic behavior in crickets. *Am. Zool.* **14**, 1067.

Hoy, R.R. and Paul, R.L. (1973) Genetic control of song specificity in crickets. *Science* **180**, 82–83.

Hubbard, J.I. (1975) *The Biological Bases of Mental Activity*. Addison-Wesley, Reading, MA.

Hubel, D.H. and Wisel, T.N. (1965) Receptive fields and functional architecture in two nonstriate visual areas (18 and 19) of the cat. *J. Neuro-physiol.* **28**, 229–289.

Hudson, B.B. (1950) One trial learning in the domestic rat. *Genetic Psychology Monographs* **41**, 99–145.

Hughes, A. (1977) The topography of vision in mammals. In Crescitelli, F. (ed.) *Handbook of Sensory Physiology VII/5*, pp. 613–756. Springer-Verlag, Berlin.

Hughes, B.O. and Black, A.J. (1973) The preference of domestic hens for different types of battery cage floor. *Br. Poultry Sci.* **14**, 615.

Hull, C.L. (1931) Goal attraction and directing ideas conceived as habit phenomena. *Psychol. Rev.* **38**, 487–506.

Humphrey, N.K. (1978) Nature's psychologists. *New Scientist* **78**, 900.

Humphrey, N. (1979) Nature's Psychologists. In Josephson, B. and Ramachandra, B.S. (eds) *Consciousness and the Physical World*. Pergamon: New York. Reprinted in Sunderland, E. and Smith, M.T. (1980) *The Exercise of Intelligence*. Garland STPM Press, New York.

Hunter, M.L. and Krebs, J.R. (1979) Geographical variation in the song of the great tit (*Parus major*) in relation to ecological factors. *J. Anim. Ecol.* **48**, 759–785.

Hutchinson, G.E. (1957) Concluding remarks. *Cold Spring Harbor Symp. Quant. Biol.* **22**, 415–427.

Huxley, J.S. (1914) The courtship habits of the great crested grebe (*Podiceps cristatus*); with an addition to the theory of sexual selection. *Proc. Zool. Soc. Lond.* **2**, 491–562.

Immelmann, K. (1969) Uber den Einfluss fruhkindlicher Erjahrungen auf die geschiechtliche Objektfixierung bei Estrildiden. *Z. Tierpyschol.* **26**, 677–691.

Immelmann, K. (1972) Sexual and other long-term aspects of imprinting in birds and other species. In Lehrman, D.S., Hinde, R.A. and Shaw, E. (eds) *Advances in the Study of Behaviour, Vol. 4*. Academic Press, New York.

Immelmann, K. (1984) The natural history of bird learning. In Marler, P. and Terrace, H.S. (eds) *The Biology of Learning*, Dahlem Konferenzen 1984. Springer-Verlag, Berlin.

Immelmann, K. (1985) Sexual imprinting in Zebra finches – mechanisms and biological significance. *Acta XVIII Congressus Internationalis Ornithologicus*. Moscow, 1982, Nauka, 1985.

Inhelder, B. and Piaget, J. (1964) *The Early Growth of Logic in the Child*. Routledge & Kegan Paul, London.

Irons, W. (1979) Cultural and biological success. In Chagnon, N. and Irons, W. (eds) *Evolutionary Biology and Human Social Behavior*. Duxbury Press, North Scituate, MA.

James, W. (1890) *The Principles of Psychology*. Holt, New York.

Jarman, P.J. (1974) The social organisation of antelope in relation to their ecology. *Behaviour* **48**, 215–267.

Jenkins, H.M. and Harrison, R.H. (1958) Auditory generalisation in the pigeon. Air research and development command. TN 58–443, Astia document 158248.

Jennings, T. and Evans, S.M. (1980) Influence of position in the flock and flock size on vigilance in the starling *Sturnus vulgaris*. *Anim. Behav.* **30**, 634–635.

Jerison, H.J. (1973) *Evolution of the Brain and Intelligence*. Academic Press, New York.

Johnstone, R. (1997) The evolution of animal signals. In Krebs, J.R. and Davies, N.B. (eds) *Behavioural Ecology*. 4th edn. Blackwell, Oxford.

Jones, F.R.H. (1955) Photo-kinesis in the ammocoete larva of the brook lamprey. *J. Exp. Biol.* **32**, 492–503.

Jordan, J. (1971) Studies of the structure of the organ of voice and vocalization in the chimpanzee. I. *Folia morphol.* **30**, 97–126. III. *Folia morphol.* **30**, 322–340.

Kaas, J.H., Guillery, R.W. and Allman, J.M. (1972) Some principles of organisation in the dorsal lateral geniculate nucleus. *Brain Behav. Evol.* **6**, 253–299.

Kacelnik, A. (1979) The foraging efficiency of great tits (*Parus major*) in relation to light intensity. *Anim. Behav.* **27**, 237–41.

Kacelnik, A. and Krebs, J.R. (1983) The dawn chorus of the great tit (*Parus major*): proximate and ultimate causes. *Behaviour* **82**, 287–309.

Kagel, J.H., Battalio, R.C., Green, L. and Rachlin, H. (1980) Consumer demand theory applied to choice behaviour of rats. In Staddon, J.E.R. (ed.) *Limits to Action*, Academic Press, New York.

Kaissling, K.E. and Priesner, E. (1970) Die Riechschwelle des Seidenspinners. *Naturwissenschaften* **57**, 23–28.

Kalat, J.W. and Rozin, P. (1971) Role of interference in taste-aversion learning. *J. comp. Physiol.* **77**, 53–58

Kamin, L.J. (1969) Predictability, surprise, attention and conditioning. In Campbell, B.A. and Church, R.M. (eds) *Punishment and Aversive Behaviour*. Appleton-Century-Crofts, New York.

Kaplan, J.R. (1978) Fight interference and altruism in rhesus monkeys. *Am. J. Phys. Anthropol.* **47**, 241–249.

Kavaliers, M. (1989) Evolutionary aspects of the neuromodulation of nociceptive behaviours. *Am. Zool.* **29**, 1345–1353.

Kawamura, S. (1963) The process of sub-culture propagation among Japanese macaques. In Southwick, C.H. (ed.) *Primate Social Behaviour*. pp. 82–90. Van Nostrand, New York.

Keeton, W.T. (1971) Magnets interfere with pigeon homing. *Proc. Nat. Acad. Sci. USA* **68**, 102–106.

Keeton, W.T. (1972) *Biological Science*. 2nd edn. Norton, New York.

Kellogg, W.N. (1968) Chimpanzees in experimental homes. *Psychol. Rec.* **18**, 489–498.

Kellogg, W.N. and Kellogg, I.A. (1933) *The Ape and the Child*. McGraw-Hill, New York.

Kendon, A. (1967) Some functions of gaze direction in social interaction. *Acta psychol.* **26**, 1–47.

Kennedy, D. (1976) Neural elements in relation to network function. In Fentress, J.C. (ed.) *Simpler Networks and Behavior*. Sinauer, Sunderland, MA.

Kennedy, J.S. (1945) Classification and nomenclature of animal behaviour. *Nature* **156**, 754.

Kettlewell, H.B.D. (1955) Selection experiments on industrial melanism in the Lepidoptera. *Heredity* **9**, 323–342.

Kettlewell, H.B.D. (1956) Further selection experiments on industrial melanism in the Lepidoptera. *Nature* **175**, 934.

Kettlewell, H.B.D. (1973) *The Evolution of Melanism*. Oxford University Press, Oxford.

Kling, J.W. and Riggs, L.A. (1971) *Experimental Psychology*. Methuen, London.

Klopf, A.H. (1982) *The Hedonistic Neuron: a Theory of Memory, Learning, and Intelligence*. Hemisphere, New York.

Klopf, A.H. (1989) Classical conditioning phenomena predicted by a drive-reinforcement model of neuronal function. In Byrne, J.H. and Berry, W.O. (eds) *Neural Models of Plasticity*. Academic Press, San Diego, CA.

Klopfer, P.H. and Gamble, J. (1966) Maternal 'imprinting' in goats: the role of chemical senses. *Z. Tierpsychol.* **23**, 588–592.

Knudsen, H. (1981) The hearing of the barn owl. *Scientific American* **245** (Dec), 83–91.

Kodric-Brown, A. and Brown, J.H. (1978) Influence of economics, interspecific competition, and sexual dimorphism on territoriality of migrant rufous hummingbirds. *Ecology* **59**, 285–296.

Koenig, W. (1990) Opportunity of parentage and nest destruction in polygynandrous acorn woodpeckers: an experimental study. *Behav. Ecol.* **1**, 55–61.

Kohler, W. (1925) *The Mentality of Apes.* Harcourt Brace, New York.

Konishi, M. (1965) The role of auditory feedback in the control of vocalisation in the white-crowned sparrow. *Z. Tierpsychol.* **22**, 770–778.

Korringa, P. (1947) Relations between the moon and periodicity in the breeding of marine animals. *Ecol. Monogr.* **17**, 347–381.

Kortlandt, A. (1940) Eine Ubersicht der angeborenen Verhaltensweisen des Mitteleuropaischen Kormorans (*Phalacrocorax carbo sinensis*). *Arch. Neerl. Zool.* **14**, 401–442.

Kortlandt, A. and Kooij, M. (1963) Protohominid behaviour in primates (preliminary communication). *Symp. Zool. Soc. Lond.* **10**, 61–88.

Kosslyn, M. (1981) The medium and the message in mental imagery; a theory. *Psychol. Rev.* **88**, 46–66.

Kramer, G. (1951) Eine neue Methode zur Erforschung der Zugorientierung und die bisher damit erzielten Ergebnisse. *Proc. X. Intern. Ornithol. Congr. Uppsala* 1950, 269–280.

Krebs, J.R. (1971) Territory and breeding density in the great tit, *Parus major* L. *Ecology* **52**, 2–22.

Krebs, J.R. (1977) The significance of song repertoires: the Beau Geste hypothesis. *Anim. Behav.* **25**, 475–478.

Krebs, J.R. and Davies, N.B. (1981) *An Introduction to Behavioural Ecology.* Blackwell Scientific Publications, Oxford.

Krebs, J.R. and Davies, N.B. (1993) *An Introduction to Behavioural Ecology*, 3rd edn. Blackwell, Oxford.

Kreithen, M.L. (1978) Sensory mechanisms for animal orientation – can any new ones be discovered? In Schmidt-Koenig, K. and Keeton, W.T. (eds) *Animal Migration, Navigation and Homing*, pp. 25–34. Proceedings in Life Sciences. Springer-Verlag, Berlin.

Kreithen, M.L. and Keeton, W.T. (1974) Detection of changes in atmospheric pressure by the homing pigeon, *Columba livia. J. Comp. Physiol.* **89**, 73–82.

Krieckhaus, E.E. (1970) 'Innate recognition' aids rats in sodium regulation. *J. Comp. Physiol. Psychol.* **73**, 117–122.

Krose, B.J.A. and Dondorp, E. (1989) A sensor simulation system for mobile robots. In Kanade, T., Groen, F.C.A. and Herzberger, L.O. (eds) *Intelligent Autonomous Systems 2.* Elsevier, Amsterdam.

Kruijt, J.P. (1964) Ontogeny of social behaviour in Burmese red junglefowl (*Gallus gallus spadiceus* Bonnaterre). *Behaviour* Suppl. **12**.

Kruuk, H. (1964) Predators and anti-predator behaviour of the black-headed gull (*Larus ridibundus* L.) *Behaviour* Suppl. **11**, 1–130.

Kruuk, H. (1972) *The Spotted Hyena. A Study of Predation and Social Behaviour.* The University of Chicago Press, Chicago, IL.

Kummer, H. (1968) *Social Organization of Hamadryas Baboons.* University of Chicago Press, Chicago, IL.

Kummer, H. (1978) On the value of social relationships to non-human premates: a heuristic scheme. *Social Science Information* **17**, 687–705.

Kummer, H., Dasser, V. and Moyningen-Muene, P. (1990) Exploring primate social cognition: some critical remarks. *Behaviour* **112**, 84–98.

Kuo, Z.Y. (1932) Ontogeny of embryonic behaviour in Aves: IV The influence of embryonic movements upon the behavior after hatching. *J. Comp. Psychol.* **14**, 109–122.

Kuo, Z.Y. (1967) *The Dynamics of Behaviour Development.* Random House, New York.

Kupfermann, I. and Weiss, K.R. (1978) The command neuron concept. *Behav. Brain Sci.* **1**, 3–39.

Lack, D. (1937) The psychological factor in bird distribution. *British Birds* **31**, 130.

Lack, D. (1943) *The Life of the Robin*. Cambridge University Press, Cambridge.

Lack, D. (1954) *The Natural Regulation of Animal Numbers*. Oxford University Press, Oxford.

Lack, D. (1966) *Population Studies of Birds*. Clarendon Press, Oxford.

Lack, D. (1968) *Ecological Adaptations for Breeding in Birds*. Methuen, London.

Laidler, K. (1978) Language in the orang-utan. In Lock, A. (ed.) *Action, Gesture and Symbol*, pp. 133–155. Academic Press, New York.

Laming, P.R. (1981) *Brain Mechanisms of Behaviour in Lower Vertebrates*. Cambridge University Press, Cambridge.

Landis, C. and Hunt, W.A. (1932) Adrenalin and emotion. *Psychol. Rev.* **39**, 467–485.

Landsberg, J.W. (1976) Posthatch age and developmental age as a baseline for determination of the sensitive period for imprinting. *J. Comp. Physiol. Psychol.* **90**, 47–52.

Landsberg, J.W. (1981) Hormones and filial imprinting. *Proceedings of the 7th International Ornithological Congress, Berlin 1978*, 837–841.

Lanz, P. and McFarland, D. (1995) On representation, goals and cognition. *Int. Studies Philos. of Sci.* **9**, 121–133.

Larkin, S. (1981) Time and Energy in Decision-making. Unpublished D.Phil. thesis. University of Oxford.

Larkin, S. and McFarland, D.J. (1978) The cost of changing from one activity to another. *Anim. Behav.* **26**, 1237–1246.

Lawick-Goodall, J. van (1970) Tool-using in primates and other vertebrates. In Lehrman, D.S., Hinde, R.A. and Shaw, E. (eds) *Advances in The Study of Behavior*. Academic Press, New York.

Lea, S.E.G. (1978) The psychology and economics of demand. *Psychol. Bull.* **85**, 441–466.

Lea, S.E.G. and Roper, T.J. (1977) Demand for food on fixed-ratio schedules as a function of the quality of concurrently available reinforcement. *J. Exp. Analysis Behav.* **27**, 371–380.

Leach, E. (1972) The influence of cultural context on non-verbal communication in man. In Hinde, R.A. (ed.) *Non-Verbal Communication*. Cambridge University Press, Cambridge.

Lee, R.B. (1972) The !Kung Bushmen of Botswana. In Bicchieri, M.G. (ed.) *Hunters and Gatherers Today*. Holt Rinehart Winston, New York.

Lee, R.B. (1979) *The !Kung San. Men, Women and Work in a Foraging Society*. Cambridge University Press, Cambridge.

Lee, R.B. and DeVore, I. (eds) *Man the Hunter*. Aldine, Chicago, IL.

Lehrman, D.S. (1953) A critique of Konrad Lorenz's theory of instinctive behaviour. *Quart. Rev. Biol.* **28**, 337–363.

Lehrman, D.S. (1955) The physiological basis of parental feeding behaviour in the ring dove (*Streptopelia risoria*). *Behaviour* **7**, 241–286.

Lehrman, D.S. (1961) Gonadal hormones and parental behaviour in birds and infrahuman mammals. In Young, W.C. (ed.) *Sex and Internal Secretion*. Williams and Wilkins, Baltimore, MD.

Lehrman, D.S. (1970) Semantic and conceptual issues in the nature–nurture problem. In Aronson, L.R., Tobach, E., Lehrman, D.S. and Rosenblatt, J.S. (eds) *Development and Evolution of Behaviour*. Freeman, San Francisco, CA.

Lendrem, D. (1986) *Modelling in Behavioural Ecology*. Croom Helm, Portland, OR.

Leong, C.Y. (1969) The quantitative effect of releasers on the attack readiness of the fish *Haplochromis burtoni* (Cichlidae: Pisces). *Z. Vergl. Physiol.* **65**, 29–50.

Leroy, Y. (1964) Transmission du parametre frequence dans le signal acoustique des hybrides F1 et P X P1, de deux Grillons: *Teleogryllus commodus* Walker et *F. oceanicus* Le Guillon (Orthopteres, ensiferes). *C.R. Acad Sci.* **259**, 892–895.

Lettvin, J.W., Maturana, H.R., McCulloch, W.S. and Pitts, W.H. (1959) What the frogs eye tells the frogs brain. *Proc. I.R.E.* **47**, 1940–1951.

Lewis, R.A. (1975) Social influences on marital choice. In Dragastin, S.E. and Elder, G.H. (eds) *Adolescence in the Life Cycle*, pp. 211–225. Wiley, New York.

Lieberman, P. (1975) *On the Origins of Language.* MacMillan, New York.

Lima, S., Valone, T.J. and Caraco, T. (1985) Foraging-efficiency-predation-risk tradeoff in the grey squirrel. *Anim. Behav.* **33**, 155–165.

Lindauer, M. (1957) Sonnenorientierung der Bienan unter der Aequatorsonne und zur Nachtzeit. *Naturwiss.* **44**, 1–6.

Lindauer, M. (1959) Angeborene und erlerne Komponenten in der Sonnenorientierung der Bienen. *Z. Vergl. Physiol.* **42**, 43–62.

Lindauer, M. (1960) Time-compensated sun orientation in bees. *Cold Spring Harbor Symp. Quant. Biol.* **25**, 371–377.

Lindauer, M. (1961) *Communication Among Social Bees.* Harvard University Press, Cambridge, MA.

Lindauer, M. (1963) Kompassorientierung. *Ergeb. Biol.* **26**, 158–181.

Linker, E., Moore, M.E. and Galanter, E. (1964) Taste thresholds, detection models, and disparate results. *J. Exp. Psychol.* **67**, 59–66.

Loftus-Hills, J.J. and Littlejohn, M.J. (1992) Reinforcement and reproductive character displacement in *Gastrophryne carolinensis* and *G. olivacea* (Anura; Microhylidae): a reexamination. *Evolution* **46**, 896–906.

Logan, F.A. and Boice, R. (1969) Aggressive behaviours of paired rodents in an avoidance context. *Behaviour* **34**, 161–183.

Lombardi, C.M. and Delius, J.D. (1988) Size invariance of pattern recognition in pigeons. In Commons, M.L., Kosslyn, S.M. and Herrnstein, R.J. (eds) *Pattern Recognition and Concepts in Animals, People and Machines.* Lawrence Erlbaum Associates, Hillsdale, NJ.

Lorenz, K. (1932) Betrachtungen uber das Erkennen der arteigenen Triebhandlungen der Vogel. *J. Ornithol* **80**, 50–98.

Lorenz, K. (1935) Der Kumpan in der Umwelt des Vogels. *J. Ornithol.* **83**, 137–213.

Lorenz, K. (1937) Uber die Bildung des Instinktbegriffes. *Naturwiss.* **25**, 289–300, 307–318, 324–331.

Lorenz, K. (1939) Vergleichende Verhaltensforschung. *Zoo. Anz. Suppl. Bd.* **12**, 69–102.

Lorenz, K. (1943) Die angeborenen Formen moglicher Erfahrung. *Z. Tierpsychol.* **5**, 235–409.

Lorenz, K. (1950) The comparative method in studying innate behaviour patterns. *Sym. Soc. Exp. Biol.* **4**, 221–268.

Lorenz, K. (1965) *Evolution and Modification of Behaviour.* University of Chicago Press, Chicago, IL.

Loschiavo, S.R. (1968) Effect of oviposition on egg production and longevity in *Trogoderma parabile* (Coleoptera: Dermestidae). *Canad. Ent.* **100**, 86–89.

Louw, G.N. and Holm, E. (1972) Physiological, morphological and behavioural adaptations of the ultrapsammophilous Namib Desert lizard *Aporosaura anchietae* (Bocage). *Madogua* **1**, 67–85.

Lovejoy, C.O. (1981) The origin of man. *Science* **211**, 341–350.

Lush, J.L. (1940) Intra-sire correlations or regressions of offspring on dam as a method of estimating heritability of characteristics. *Thirty-third Annual Proceedings of the American Society of Animal Production*, 293–301.

Lythgoe, J.N. (1979) *The Ecology of Vision*. Clarendon Press, Oxford.

MacArthur, R.H. and Wilson, E.O. (1967) *The Theory of Island Biogeography*. Princeton University Press, Princeton, NJ.

Macdonald, D.W. (1983) Ecology of carnivore social behaviour. *Nature* **301**, 379–384.

Mackay, D.M. (1972) Formal analysis of communicative processes. In Hinde, R.A. (ed.) *Non-verbal Communication*. Cambridge University Press, Cambridge.

Mackintosh, N.J. (1973) Stimulus selection: learning to ignore stimuli that predict no change in reinforcement. In Hinde, R.A. and Stevenson-Hinde, J. (eds) *Constraints on Learning*. pp. 75–96. Academic Press, London.

Mackintosh, N.J. (1974) *The Psychology of Animal Learning*. Academic Press, London.

Mackintosh, N.J. (1976) Overshadowing and stimulus intensity. *Anim. Learning Behav.* **4**, 186–192.

Mackintosh, N.J. (1983) *Conditioning and Associative Learning*. Clarendon Press, Oxford.

Macphail, E.M. (1982) *Brain and Intelligence in Vertebrates*. Clarendon Press, Oxford.

MacRoberts, M.H. and MacRoberts, B.R. (1976) Social organization and behavior of the acorn woodpecker in central coastal California. *Ornithol. Monogr.* **21**, 1–115.

Maier, S.F. and Seligman, M.E.P. (1976) Learned helplessness. Theory and evidence. *J. Exp. Psycho. General* **105**, 3–46.

Major, P.F. (1978) Predator–prey interactions in two schooling fishes. *Caranx ignobilis* and *Stolephorus purpureus. Anim. Behav.* **26**, 760–777.

Makkink, G.F. (1936) An attempt at an ethogram of the European avocet (*Recurvirostra avosetta* L.) with ethological and psychological remarks. *Ardea* **25**, 1–60.

Malott, R.W. and Siddall, J.W. (1972) Acquisition of the people concept in pigeons. *Psychol. Reports* **31**, 3–13.

Manning, A. (1956) The effect of honey-guides. *Behaviour* **9**, 114–139.

Marler, P. (1970) A comparative approach to vocal learning: song development in white-crowned sparrows. *J. Comp. Physiol. Psychol.* (Suppl.), **71**, 1–25.

Marler, P. and Mundinger, P. (1971) Vocal learning in birds. In Moltz, H. (ed.) *The Ontogeny of Vertebrate Behavior*. Academic Press, New York.

Marler, P. and Nelson, D. (1992) Neuroselection and big learning in birds: species universals in culturally transmitted behaviour. *Sernis. Neurosci.* **4**, 415–423.

Marler, P. and Tamura, M. (1964) Culturally transmitted patterns of vocal behavior in sparrows. *Science* **146**, 1483–1486.

Marshall, A.J. (1970) Environmental factors other than light involved in the control of sexual cycles in birds and mammals. In Benoit, J. and Assenmacher, I. (eds) *La Photoregulation de la Reproduction chez les Oiseaux et les Mammiferes*. C.N.R.S. Editeur, Paris.

Marten, K. and Marler, P. (1977) Sound transmission and its significance for animal vocalisation. I. Temperate habitats. *Behav. Ecol. Sociobiol.* **2**, 271–290.

Marten, K., Quine, D. and Marler, P. (1977) Sound transmission and its significance for animal vocalisation. II. Tropical forest habitats. *Behav. Ecol. Sociobiol.* **2**, 291–302.

Martin, G.R. (1977) Absolute visual threshold and scotopic spectral sensitivity in the tawny owl, *Strix aluco. Nature* **268**, 636–638.

Martin, R.D. (1981) Relative brain size and basal metabolic rate in terrestrial vertebrates. *Nature* **293**, 56–60.

Mast, S.O. (1911) *Light and the Behavior of Organisms*. Wiley, London.

Masterson, F.A. (1970) Is termination of a warning signal an effective reward for a rat. *J. Comp. Phsyiol. Psychol.* **72**, 471–475.

Matthews, G.V.T. (1955) *Bird Navigation*. Cambridge University Press, Cambridge.

Matthews, G.V.T. (1968) *Bird Navigation,* 2nd edn. Cambridge University Press, Cambridge.

Mayer, J. (1955) Regulation of energy intake and the body weight: the glucostatic theory and lipostatic hypothesis. *Annals of the New York Academy of Sciences* **63**, 15–43.

Mayes, J. and Thomas, D.W. (1967) Regulation of food intke and obesity. *Science* **156**, 328–337.

Maynard Smith, J. (1958) Sexual selection. In Barnett, S.A. (ed.) *A Century of Darwin,* pp. 231–244. Harvard University Press, Cambridge, MA.

Maynard Smith, J. (1964) Group selection and kin selection. *Nature* **201**, 1145–1147.

Maynard Smith, J. (1976) Sexual selection and the handicap principle. *J. Theor. Biol.* **57**, 239–242.

Maynard Smith, J. (1978a) *The Evolution of Sex.* Cambridge University Press, Cambridge.

Maynard Smith, J. (1978b) The ecology of sex. In Krebs, J.R. and Davies, N.B. (eds), *Behavioural Ecology, an Evolutionary Approach.* Blackwell Scientific Publications, Oxford.

Maynard Smith, J. (1982) *Evolution and the Theory of Games.* Cambridge University Press, Cambridge.

McClearn, G.E. and DeFries, J.C. (1973) *Introduction to Behavioral Genetics.* W.H. Freeman, San Francisco, CA.

McCorduck, N. (1979) *Machines Who Think.* W.H. Freeman, San Francisco, CA.

McDonald, D.L. (1972) Some aspects of the use of visual cues in directional training of homing pigeons. In Galler, S.R., Schmidt-Koenig, K., Jacobs, G.J. and Belleville, R.E. (eds) *Animal Orientation and Navigation,* pp. 293–304. NASA SP-262, US Govt Printing Office, Washington, DC.

McDonald, D.L. (1973) The role of shadows in directional training and homing of pigeons, *Columba livia. J. Exp. Zool.* **183**, 267–280.

McDougall, W. (1908) *An Introduction to Social Psychology.* Methuen, London.

McFarland, D.J. (1965) Hunger, thirst and displacement pecking in the Barbary Dove. *Anim. Behav.* **13**, 292–300.

McFarland, D.J. (1971) *Feedback Mechanisms in Animal Behaviour.* Academic Press, London.

McFarland, D.J. (1973) Stimulus relevance and homeostasis. In Hinde, R.A. and Stevenson-Hinde, J. (eds) *Constraints on Learning.* Academic Press, London.

McFarland, D.J. (1974) Time-sharing as a behavioral phenomenon. In Lehrman, D.S., Rosemblatt, J.S., Hinde, R.A. and Shaw, E. (eds) *Advances in The Study of Behavior.* Academic Press, New York.

McFarland, D.J. (1976) Form and function in the temporal organisation of behaviour. In Bateson, P.P.G. and Hinde, R.A. (eds) *Growing Points in Ethology,* pp. 55–93. Cambridge University Press.

McFarland, D. (1989a) The teleological imperative. In Montefiore, A. and Noble, D. (eds) *Goals, No Goals, and Own Goals.* Unwin Hyman, London.

McFarland, D. (1989b) *Problems of Animal Behaviour.* Longman, London.

McFarland, D. (1994) Towards cooperative robots. In Cliff, D., Husbands, P., Meyer, J. and Wilson D. (eds) *From Animals to Animats, 3.* MIT Press, Cambridge, MA.

McFarland, D. (1995a) Autonomy and self-sufficiency in robots. In Steels, L. and Brooks, R. (eds) *The Artificial Life Route to Artificial Intelligence. Building Situated Embodied Agents.* Lawrence Erlbaum, New Haven, CT.

McFarland, D. (1995b) Opportunity versus goals in the control of behaviour in robots, animals and people. In Roitblatt, H. and Meyer, J. (eds) *Comparative Approaches to Cognitive Science.* MIT Press, Cambridge, MA.

McFarland, D. and Bosser, T. (1993) *Intelligent Behaviour in Animals and Robots*. MIT Press, Cambridge, MA.

McFarland, D.J. and Budgell, P. (1970) The thermoregulatory role of feather movements in the Barbary dove (*Streptopelia risoria*). *Physiol. Behav.* **5**, 763–771.

McFarland, D.J. and Houston, A. (1981) *Quantitative Ethology. The State Space Approach*. Pitman, London.

McFarland, D.J. and Sibly, R.M. (1972) 'Unitary drives' revisited. *Anim. Behav.* **20**, 548–563.

McFarland, D. and Spier, E. (1997) Basic cycles, utility and opportunism in self-sufficient robots. *Robotics and Autonomous Systems* **20**, 179–190.

McFarland, D.J. and Wright, P. (1969) Water conservation by inhibition of food intake. *Physiol. Behav.* **4**, 95–99.

McGonigle, B.O. (1987) Non-verbal thinking by animals. *Nature* **325**, 110–112.

McGonigle, B.O. (1999) In Holland, O. and McFarland, D. (eds) *Artificial Ethology*. Oxford University Press (in press).

McGonigle, B.O. and Chalmers, M. (1980) On the genesis of relational terms: a comparative study of monkeys and human children. *Antropologia Contemporanea* **3**, 236.

McGonigle, B.O. and Chalmers, M. (1984) The selective impact of question form and input mode on the symbolic distance effect in children. *J. Exp. Child Psychol.* **37**, 525–554.

McGonigle, B. and Chalmers, M. (1986) Representations and strategies during inference. In Myers, T. *et al.* (eds) *Reasoning and Discourse Processes*. Academic Press, London.

McGonigle, B. and Chalmers, M. (1998) Rationality as optimised cognitive self-regulation. In Oaksford, M. and Chater, N. (eds) *Rational Models of Cognition*. Oxford University Press, Oxford.

McGraw, M.B. (1945) *The Neuromuscular Maturation of the Human Infant*. Columbia University Press, New York.

McGrew, W.C. (1975) Patterns of plant food sharing by wild chimpanzees. In Kawai, M, Kondo, S. and Ehara, A. (eds) *Contemporary Primatology: Proceedings of the Fifth Congress of the International Primatological Society*. S. Karger, Basel.

McGrew, W.C., Tutin, C.E.G. and Baldwin, P.J. (1979) Chimpanzees, tools and termites: cross-cultural comparison of Senegal, Tanzania and Rio Muni. *Man* **14**, 185–214.

McLeese, D.W. (1956) Effects of temperaure, salinity and oxygen on the survival of the American Lobster. *J. Fish. Res. Bd. Canada* **13**, 247–272.

Meddis, R. (1975) On the function of sleep. *Anim. Behav.* **23**, 676–691.

Menzel, E.W. (1974) A group of young chimpanzees in a one-acre field. In Schrier, A.M., and Stollnitz, F. (eds) *Behaviour of Nonhuman Primates*. Academic Press, New York.

Menzel, E.W. (1978) Cognitive mapping in chimpanzees. In Hulse, S.H., Fowler H. and Honig, W.K. (eds) *Cognitive Processes in Animal Behavior*, pp. 375–422. Erlbaum, Hillsdale, NJ.

Menzel, E.W. (1979) Communication of object-locations in a group of young chimpanzees. In Hamburg, D.A. and McGown, E.R. (eds) *The Great Apes*, pp. 359–371. Benjamin/Cummings, Menlo Park, CA.

Menzel, E.W. and Johnson, M.K. (1976) Communication and congitive organization in human and other animals. *Ann. NY Acad. Sci.* **2800**, 131–142.

Menzel, R., Erber, J. and Masuhr, T. (1974) Learning and memory in the honey bee. In Barton Browne, L. (ed.) *Experimental Analysis of Insect Behaviour*. Springer-Verlag, New York.

Merkel, F.W. and Wiltschko, W. (1965) Magnetismus und Richtungsfinden zugunruhiger Rotkelchen (*Erithacus rubecula*). *Vogelwarte* **23**, 71–77.

Messenger, J.B. (1968) The visual attack of the cuttlefish *Sepia officinalis. Anim. Behav.* **16**, 342–369.

Meyer, D.B. (1977) The avian eye and its adaptation. In Crescitelli, F. (ed.) *Handbook of Sensory Physiology. Vol III,* pp. 549–611. Springer-Verlag, Berlin.

Meyerriecks, A.J. (1960) Comparative breeding behavior of four species of North American herons. *Nuttal Ornithological Club Publication* **2**, 1–158.

Milinski, M. (1979) An evolutionarily stable feeding strategy in sticklebacks. *Z. Tierpsychol.* **51**, 36–40.

Miller, N.E. (1948) Studies of fear as an acquirable drive. *J. Exp. Psychol.* **38**, 89–101.

Miller, R.S. (1967) Pattern and process in competiton. *Adv. Ecol. Res.* **4**, 1–74

Minsky, M.L. and Papert, S. (1969) *Perceptrons: An Essay on Computational Geometry.* MIT Press, Cambridge, MA.

Mitchell, R.W. and Thompson, N.S. (eds) (1986) *Deception: Perspectives on human and non-human deceit.* State University of New York Press, New York.

Mittelstaedt, M. (1964) Basic control patterns of orientational homeostasis. *Symp. Soc. Exp. Biol.* **18**, 365–385.

Moore, B.R. (1973) The role of directed Pavlovian reactions in simple instrumental learning in the pigeon. In Hinde, R.A. and Stevenson-Hinde J. (eds) *Constraints on Learning.* Academic Press, London.

Moore, F.R. (1980) Solar cues in migratory orientation of the Savannah sparrow (*Passerculus sandwichensis*). *Anim. Behav.* **28**, 684–704.

Moore, M.J. and Capretta, P.J. (1968) Changes in colored or flavored food preferences in chicks as a function of shock. *Psychon. Sci.* **12**, 195–196.

Morgan, C.L. (1894) *Introduction to Comparative Psychology.* Scott, London.

Morgan, C.L. (1900) *Animal Behaviour.* Scott, London.

Morgan, M. and Nicholas, D.J. (1979) Discrimination between reinforced action patterns in the rat. *Learning and Motivation* **10**, 1–22.

Morris, D. (1967) *The Naked Ape.* Cape, London.

Morse, D.H. (1971) The insectivorous bird as an adaptive strategy. *Ann. Rev. Ecol. Syst.* **2**, 177–200.

Morton, E.S. (1975) Ecological sources of selection on avian sounds. *American Naturalist* **109**, 17–34.

Moyer, R.S. (1973) Comparing objects in memory: evidence suggesting an internal psychophysics. *Perception and Psychophysics* **13**, 180–184.

Moynihan, M. (1955) Some aspects of reproductive behaviour in the black-headed gull (*Larus ridibundus ridibundus* L.) and related species. *Behaviour* Suppl. 4, 1–201.

Munck, A., Guyre, P.M. and Holbrook, N.J. (1984) Physiological functions of glucocorticoids in stress and their relation to pharmacological actions. *Endocrine Reviews* **5**, 25–44.

Mundinger, P.C. (1980) Animal cultures and a general theory of cultural evolution. *Ethol. Sociobiol.* **1**, 183–223.

Muntz, W.R.A. (1974) Comparative aspects in behavioural studies of vertebrate vision.

Munz, F.W. (1958) Photosensitive pigments from the retinae of certain deep sea fishes. *J. Physiol.* **140**, 220–5.

Nachman, M. (1970) Learned taste and temperature aversions due to lithium chloride sickness after temporal delays. *J. Comp. Physiol. Psychol.* **73**, 22–30.

Narius, P.M. and Capranica, R.R. (1976) Sexual differences in the auditory system of the tree frog *Eleutherodactylus coqui, Science* **192**, 378–380.

Nethersole Thompson, C. and Nethersole Thompson, D. (1942) Egg-shell disposal by birds. *British Birds* **35**, 162–169, 190–200, 214–224, 241–250.

Nicol, J.A.C. (1965) Migration of chorioidal tapetal pigment in the spur dog, *Squalus acanthias. J. Mar. Biol. Ass. UK* **45**, 405–427.

Niebuhr, V. (1981) An investigation of courtship feeding in Herring Gulls, *Larus Argentatus. Ibis*, **123**, 218–223.

Nisbet, I.C.T. (1973) Courtship feeding, egg-size and breeding success in Common Terns. *Nature* **241**, 141–142.

Nisbet, I.C.T. (1977) Courtship feeding and clutch size in Common Terns *Sterna hirundo*. In Stonehouse, B. and Perrins, C.M. (eds) *Evolutionary Ecology*. Macmillan, London.

Northcutt, R.G. (1981) Evolution of the telencephalon in non-mammals. *Ann. Rev. Neurosci.* **4**, 301–350.

Norton-Griffiths, M.N. (1967) Some ecological aspects of the feeding behaviour of the oystercatcher *Haematopus ostralegus* on the edible mussel *Mytilus edulis. Ibis* **109**, 412–424.

Norton-Griffiths, M.N. (1969) The organisation, control and development of parental feeding in the oystercatcher (*Haematopus ostralegus*). *Behaviour* **34**, 55–114.

Nottebohm, F. (1976) Vocal tract and brain: a search for evolutionary bottlenecks. *Ann. NY Acad. Sci.* **280**, 643–649.

O'Brien, W.J., Slade, N.A. and Vinyard, G.L. (1976) Apparent size as the determinant of prey selection by Bluegill sunfish (*Lepomis machrochirus*). Ecology **57**, 1304–1311.

Oppenheim, R.W. (1974) The ontogeny of behaviour in the chick embryo. In Lehrman, D.S., Hinde, R.A., Shaw, E. and Rosenblatt, J.S. (eds) *Advances in the Study of Behavior, Vol. 5* pp. 133–172. Academic Press, New York.

Orians, G.H. and Pearson, N.E. (1979) On the theory of central place foraging. In Horn, D.J., Mitchel R. and Stair G.R. (eds) *Analysis of Ecological Systems*, pp. 155–177. Ohio State University Press, Columbus, OH.

Orians, G.H. and Wilson, M.F. (1964) Interspecific territories of birds. *Ecology* **45**, 736–745.

Owings, D.H. and Coss, R.G. (1977) Snake mobbing by California ground squirrels: adaptive variation and ontogeny. *Behaviour* **62**, 50–69.

Packer, C. (1977) Reciprocal altruism in *Papio anubis. Nature* **265**, 441–443.

Paivio, A. (1975) Perceptual comparisons through the mind's eye. *Memory and Cognition* **3**, 635–647.

Palmer, J.D. (1973) Biological clocks of the tidal zone. *Scientific American* **229**, 70–79.

Papi, F. (1960) Orientation by night: the moon. *Cold Spring Harbor Symp. Quant. Biol.* **25**, 475–480.

Papi, F. (1976) The olfactory navigation system of the homing pigeon. *Verh. Dr. Zool. Ges.* **69**, 184–205.

Papi, F. (1982) Olfaction and homing in pigeons: ten years of experiments. In Papi, F. and Wallraff, H.G. (eds) *Avian Navigation*, pp. 149–159. Springer-Verlag, Berlin.

Parker, G.A. (1978) Searching for mates. In Krebs, J.R. and Davies, N.B. (eds) *Behavioural Ecology: an Evolutionary Approach*, pp. 214–44. Blackwell Scientific Publications, Oxford.

Parker, G.A. (1979) Sexual selection and sexual conflict. In Blum, M.S. and Blum, N.A. (eds) *Sexual Selection and Reproductive Competition in Insects*, pp. 123–166. Academic Press, New York.

Parker, G.A., Baker, R.R. and Smith, V.G.F. (1972) The origin and evolution of gamete dimorphism and the male-female phenomenon. *J. Theor. Biol.* **36**, 529–553.

Parker, S.T., Mitchell, R.W. and Boccia, M.L. (eds) (1994) *Self-awareness in Animals and Humans*. Cambridge University Press, Cambridge.

Parsons, P.A. (1967) *The Genetic Analysis of Behaviour*. Methuen, London.

Partridge, L. (1978) Habitat selection. In Krebs, J.R. and Davies, N.B. (eds) *Behavioural Ecology: an Evolutionary Approach*. Blackwell Scientific Publications, Oxford.

Passingham, R.E. (1975) Changes in the size and organization of the brain in man and his ancestors. *Brain Behav. Evol.* **11**, 73–90.

Passingham, R.E. (1981) Primate specializations in brain and intelligence. *Symp. Zool. Soc. Lond.* **46**, 361–388.

Passingham, R.E. (1982) *The Human Primate*. W.H. Freeman, Oxford.

Patterson, F.G. (1978) The gestures of a gorilla: language acquisition in another pongid. *Brain and Lang.* **5**, 72–97.

Patterson, F.G. (1979) Linguistic capabilities of a young lowland gorilla. Unpublished dissertation. Stanford University.

Patterson, I.J. (1965) Timing and spacing of broods in the Black-headed Gull *Larus ridibundus*. *Ibis* **107**, 433–459.

Pavlov, I.P. (1927) *Conditioned Reflexes: an Investigation of the Physiological Activity of the Cerebral Cortex*. Oxford University Press, London.

Payne, T.L. (1974) Pheromone perception. In Birch, M.C. (ed.) *Pheromones*. North Holland, Amsterdam and London.

Pearson, K.G. and Iles, J.F. (1970) Central programming and reflex control of walking in the cockroach. *J. Exp. Biol.* **56**, 173–193.

Pengelley, E.T. (1974) *Circannual Clocks – Annual Biological Rhythms*. Academic Press, New York.

Pengelley, E.T. and Asmundson, S.J. (1974) Circannual rhythmicity in hibernating mammals. In Pengelley, E.T. (ed.) *Circannual Clocks – Annual Biological Rhythms*. Academic Press, New York.

Pennycuick, C.J. (1960) The physical basis of astronavigation in birds: theoretical considerations. *J. Exp. Biol.* **37**, 573–593.

Perdeck, A.C. (1958) Two types of orientation in migrating starlings *Sturnus vulgaris* L., and chaffinches, *Fringilla coelebs* L., as revealed by displacement experiments. *Ardea* **46**, 1–37.

Perdeck, A.C. (1967) Orientation of starlings after displacement to Spain. *Ardea* **55**, 194–202.

Perrins, C.M. (1965) Population fluctuation and clutch-size in the great tit (*Parus major* L.) *J. Anim. Ecol.* **34**, 601–647.

Peters, R.S. (1958) *The Concept of Motivation*. Routledge and Kegan Paul, London.

Pfeiffer, R. and Scheier, C. (1994) From perception to action: the right direction? In *From Perception to Action Conference*, Lausanne, 1994. IEEE Computer Society Press, Los Alamos, NM.

Pilbeam, D. (1972) *The Ascent of Man*. Macmillan, New York.

Pinel, J.P., Treit, D., Ladak, F. and MacLennan, A.J. (1980) Conditioned defensive burying in rats free to escape. *Anim. Learning Behav.* **8**, 447–451.

Pinel, J.P.J. and Wilkie, D.M. (1983) Conditioned defensive burying: a biological and cognitive approach to avoidance learning. In Mellgren, R.L. (ed.) *Animal Cognition and Behavior*, pp. 285–318. North Holland, Amsterdam.

Pitcher, T.J., Partridge, B.L. and Wardle, C.S. (1976) A blind fish can school. *Science* **194**, 963–965.

Plutchik, R. and Ax, A.F. (1970) A critique of determinants of emotional state by Schachter and Singer (1962). In Arnold, M.B. (ed.) *Feelings and Emotions: The Loyola Symposium*. Academic Press, New York.

Pomiankowski, A. (1987) Sexual selection: the handicap principle does work – sometimes. *Proc. Roy. Soc. Lond.* **231**, 123–145.

Posner, M.I. (1978) *Chronometric Explorations of Mind*. Erlbaum, Hillsdale, NJ.

Powers, M.E. (1984) Habitat quality and the distribution of algae-grazing catfish in a Panamanian stream. *J. Anim. Ecol.* **53**, 357–374.

Premack, A.J. and Premack, D. (1972) Teaching language to an ape. *Scientific. American*, **227**, (Oct) 92–99.

Premack, D. (1970) A functional analysis of language. *J. Exp. Anal. Behav.* **14**, 107–125.

Premack, D. (1976) *Intelligence in Ape and Man*. Erlbaum, Hillsdale, NJ.

Premack, D. (1978) Chimpanzee theory of mind: Part II. The evidence for symbols in chimpanzee. *Behav. Brain Sci.* **1**, 625–629.

Premack, D. and Woodruff, G. (1978) Does the chimpanzee have a theory of mind? *Behav. Brain Sci.* **1**, 515–526.

Prosser, C.L. (1973) *Comparative Animal Physiology*, 3rd edn. W.B. Saunders, Philadelphia, PA.

Prout, T. (1971) The relation between fitness components and population prediction in *Drosophila*. I: The estimation of fitness components. *Genetics* **68**, 127–149.

Prove, E. (1983) Hormonal correlates of behavioural development in male Zebra Finches. In Balthazart, J., Prove, E. and Gilles, R. (eds) *Hormones and Behaviour in Higher Vertebrates*. Springer-Verlag, Berlin.

Prove, E. and Immelmann, K. (1982) Behavioral and hormonal responses of male Zebra Finches to anti-androgens. *Horm. Behav.* **16**, 121–131.

Provine, R.R. (1981) Wing-flapping development in chickens made flightless by feather mutations. *Developmental Psychobiology*, **14**, 481–486.

Pulliam, H.R. and Caraco, T. (1984) Living in groups: is there an optimal group size? In Krebs, J.R. and Davies N.B. (eds) *Behavioural Ecology*, 2nd edn. Blackwell, Oxford.

Pusey, A. and Packer, C. (1997) The ecology of relationships. In Krebs, J.R. and Davies, N.B. (eds) *Behavioural Ecology*, 4th edn. Blackwell Scientific Publications, Oxford.

Pylyshyn, Z. (1984) *Computation and Cognition. Toward a Foundation for Cognitive Science*. MIT Press, Cambridge, MA. A Bradford Book.

Rachlin, H. (1980) Economics and behavioral psychology. In Staddon, J.E.R. (ed.) *Limits to Action*. Academic Press, New York.

Randolph, M.C. and Brooks, B.A. (1967) Conditioning of a vocal response in a chimpanzee through social reinforcement. *Folia primat.* **5**, 70–79.

Ratner, A., Yelvington, D.B. and Rosenthal, M. (1989). Prolactin and corticosterone response to repeated footshock stress in male rats. *Psychoendocrinology* **14**, 393–396.

Reinberg, A. (1974) Aspects of circannual rhythms in man. In Pengelley, E.T. (ed.) *Circannual Clocks*. Academic Press, New York.

Reitboeck, H.J. and Altmann, J. (1984) A model for size- and rotation-invariant pattern processing in the visual system. *Biol. Cyb* **5**, 113–121.

Rescorla, R.A. (1971) Variations in the effectiveness of reinforcement and non-reinforcement following prior inhibitory conditioning. *Learning and Motivation* **2**, 113–123.

Rescorla, R.A. (1978) Some implications of a cognitive perspective on Pavolvian conditioning. In Hulse, S.H., Fowler, H. and Honig, W.K. (eds) *Cognitive Processes in Animal Behaviour*. Erlbaum, Hillsdale, New Jersey.

Rescorla, R.A. and Cunningham, C.L. (1979) Spatial contiguity facilitates Pavlovian second-order conditioning. *J. Exp. Psychol. Anim. Behav. Processes* **5**, 152–161.

Rescorla, R.A. and Furrow, D.R. (1977) Stimulus similarity as a determinant of Pavlovian conditioning. *J. Exp. Psychol. Anim. Behav. Processes* **3**, 203–215.

Revusky, S.H. (1967) Hunger level during food consumption: effects on subsequent preference. *Psychon. Sci.* **7**, 109–110.

Revusky, S.H. (1971) The role of interference in association over delay. In Honig, W.K. and James, P.H.R. (eds) *Animal Memory*, pp. 55–213. Academic Press, New York.

Revusky, S.H. (1977) Learning as a general process with an emphasis on data from feeding experiments. In Milgram, N.W., Krames, L. and Alloway, T.M. (eds) *Food Aversion Learning*. Plenum Press, New York.

Revusky, S. and Garcia, J. (1970) Learned associations over long delays. In Bower G.H. (ed.) *Psychology of Learning and Motivation*, Vol. 4, pp. 1–83. Academic Press, New York.

Reyer, H.U. (1980) Flexible helper structure as an ecological adaptation in the pied kingfisher, *Ceryle rudis*. *Behav. Ecol. Sociobiol.* **6**, 219–227.

Reyer, H.U. (1984) Investment and relatedness. A cost/benefit analysis of breeding and helping in the pied kingfisher (*Ceryle rudis*). *Anim. Behav.* **32**, 1163–1178.

Reyer, H.U. (1986) Breeder-helper interactions in the pied kingfisher reflect the costs and benefits of cooperative breeding. *Behaviour* **96**, 277–303.

Reyer, H.U., Dittami, J.P. and Hall, M.R. (1986) Avian helpers at the nest: are they psychologically castrated? *Ethology* **71**, 216–228.

Richelle, M. and Lejeune, H. (1980) *Time in Animal Behaviour*. Pergamon Press, Oxford.

Richter, C.P. (1943) Total self-regulatory functions in animals and human beings. *Harvey Lect.* **38**, 63–103.

Richter, C.P. (1955) Self-regulatory. functions during gestation and lactation. *Trans. Conf. Gestation*, Princeton, NJ.

Richter, C.P., Holt, L.E. and Barelare, B. Jr. (1937) Vitamin B1 craving in rats. *Science* **86**, 354–355.

Riddell, W.I. (1979) Cerebral indices and behavioral differences. In Hahn, M.E., Jensen, C. and Dudek, B.C. (eds) *Development and Evolution of Brain Size*, pp. 89–109. Academic Press, New York.

Ridley, M. and Rechten, C. (1981) Female sticklebacks prefer to spawn with males whose nests contain eggs. *Behaviour* **76**, 1–2.

Ristau, C.A. and Robbins D. (1981) Language in the great apes: a critical review. In Rosenblatt, J., Hinde, R.A., Beer, C. and Busnel, M.C. (eds) *Advances in the Study of Behaviour*. Vol. 12. Academic Press, New York.

Ristau, C.A. and Robbins D. (1982) Cognitive aspects of ape language experiments. In Griffin, D.R. (ed.) *Animal Mind – Human Mind*. Springer-Verlag, Berlin.

Roberts, S. (1981) Isolation of an internal clock. *J. Exp. Psychol. Anim. Behav. Processes* **7**, 242–268.

Robinson, M.H. (1970) Insect anti-predator adaptations and the behaviour of predatory primates. *Congr. Latin. Zool.* **II**, 811–836.

Rodgers, W. and Rozin, P. (1966) Novel food preferences in thiamine-deficient rats. *J. Comp. Physiol. Psychol.* **61**, 1–4.

Roeder, K.D. (1963) *Nerve Cells and Insect Behaviour*. Harvard University Press, Cambridge, MA.

Roeder, K.D. (1970) Episodes in insect brains. *American Scientist* **58**, 378–389.

Rohwer, S. and Rohwer, F.C. (1978) Status signalling in Harris sparrows: experimental deceptions achieved. *Anim. Behav.* **26**, 1012–1022.

Rolls, B.J. and Rolls, E.T. (1982) *Thirst*. Cambridge University Press, Cambridge.

Romanes, G.J. (1882) *Animal Intelligence*. Kegan, Paul, Trench, London.

Romer, A.S. (1958) *Vertebrate Paleontology*. Chicago University Press, Chicago, IL.

Root, R.B. (1967) The niche exploitation pattern of the blue-gray gnatcatcher. *Ecol. Monogr.* **37**, 317–350.

Rosenblatt, F. (1962) *Principles of Neurodynamics*. Spartan.

Rothenbuhler, N. (1964) Behavior genetics of nest cleaning in honey bees. 4. Responses of F1 and backcross generations to disease-killed brood. *Am. Zool.* **4**, 111–123.

Rowell, C.H.F. and Horn, G. (1968) Dishabituation and arousal in the response of single nerve cells in an insect brain. *J. Exp. Biol.* **49**, 171–183.

Rowell, T.E. (1966) Forest living baboons in Uganda. *J. Zool.* **149**, 344.

Rowley, I. (1965) White-winged choughs. *Austral. Natur. Hist.* **15**, 81–85.

Rozin, P. (1967) Specific aversions as a component of specific hungers. *J. Comp. Physiol. Psychol.* **64**. 237–242.

Rozin, P. (1968) Specific aversions and neophobia as a consequence of vitamin deficiency and/or poisoning in half-wild and domestic rats. *J. Comp. Physiol. Psychol* **66**, 82–88.

Rozin, P. (1969) Adaptive food sampling in vitamin deficient rats. *J. Comp. Physiol. Psychol.* **69**, 126–132.

Rozin, P. (1976) The evolution of intelligence and access to the cognitive unconscious. *Progress in Psychobiol. Physiol. Psychol.* **6**, 245–276.

Rozin, P. and Kalat, J. (1971) Specific hungers and poison avoidance as adaptive specializations of learning. *Psychol. Rev.* **78**, 459–486.

Rozin, P. and Kalat, J. (1972) Learning as a situation-specific adaptation. In Seligman, M.E.P. and Hager, J. (eds) *Biological Boundaries of Learning.* pp. 66–97. Appleton, New York.

Rozin, P. and Mayer, J. (1961) Thermal reinforcement and thermoregulatory behavior in the goldfish. *Science* **134**, 942–943.

Rumbaugh, D.M. (1977) *Language Learning by a Chimpanzee.* Academic Press, New York.

Rumbaugh, D.M. and Gill, T.V. (1977) Lana's acquisition of language skills. In Rumbaugh, D.M. (ed.) *Language Learning by a Chimpanzee*, pp. 165–192. Academic Press, New York.

Ruppel, G. (1969) Eine 'Luge' als gerichtete Mitteilung beim Eisfuchs (*Alopex lagopus* L.) *Z. Tierpsychol.* **26**, 371–374.

Rusak, B. (1981) Vertebrate behavioral rhythms. In Aschoff, J. (ed.) *Handbook of Behavioural Neurobiology, Vol. 4, Biological Rhythms.* Plenum Press, New York.

Russell, and Woodburne, (1965) *Essentials of Human Anatomy.* Oxford University Press, Oxford.

Rutledge, J.T. (1974) Circannual rhythm of reproduction in male European Starlings (*Sturnus vulgaris*). In Pengelley, E.T. (ed.) *Circannual Clocks.* Academic Press, New York.

Rzoska, J. (1953) Bait shyness, a study in rat behaviour. *Br. J. Anim. Behav.* **1**, 128–35.

Sadoglu, P. (1975) Genetic paths leading to blindness in *Astyanax mexicanus.* In Ali, M.A. *Vision in Fishes* (ed.) pp. 419–26. Plenum Press, New York.

Salzano, F.M., Neel, J.V. and Maybury-Lewis, D. (1967) Further studies on the Xavante Indians. I Demographic data on two additional villages: genetic structure of the tribe. *Am. J. Human Genetics* **19**, 463–489.

Salzer, D.W. and Larkin, G.J. (1990) Impact of courtship feeding on clutch size and third-egg size in glaucous-winged gulls. *Anim. Behav.* **39**, 1149–1162.

Santschi, F. (1911) Observations et remarques critiques sur le mechanisms de l'orientation. *Rev. Suisse Zool.* **19**, 303–338.

Sauer, F. and Sauer, E. (1955) Zur Frage der nachtlichen Zugorientierung von Grasmucken. *Rev. Suisse Zool.* **62**, 250–259.

Saunders, D.S. (1976) *Insect Clocks.* Pergamon Press, Oxford.

Savage-Rumbaugh, E.S. and Brakke, K.E. (1996) Animal language: methodological and interpretive issues. In Bekoff, M. and Jamieson, D. (eds) *Readings in Animal Cognition.* MIT Press, Cambridge, MA.

Savage-Rumbaugh, E.S., Rumbaugh, D.M. and Boysen S. (1978) Symbolic communication between two chimpanzees (*Pan troglodytes*). *Science* **201**, 641–644.

Savage-Rumbaugh, E.S., Rumbaugh, D.M. and Boysen S. (1980) Do apes use language? *American Scientist* **68**, 49–61.

Schachter, S. and Singer, J. (1962) Cognitive, social and physiological determinants of emotional state. *Psychol. Rev.* **69**, 379–399.

Schaller, G.B. (1972) *The Serengeti Lion.* University of Chicago Press, Chicago, IL.

Scharrer, E. (1964) Photo-neuro-endocrine systems: general concepts. *Ann. NY Acad. Sci.* **117**, 13–22.

Scheller, R.H. and Axel, R. (1984) How genes control an innate behaviour. *Scientific American* **250**, 44–52.

Schiller, P. (1952) Innate constituents of complex responses in primates. *Psychol. Rev.* **59**, 177–191.

Schmidt-Koenig, K. (1958) Experimentelle Einflussnahme auf die 24-Stunden- Periodik bei Brieftauben und deren Auswirkungen unter besonderer Berucksichtigung des Heimfindevermogens. *Z. Tierpsychol.* **15**, 301–331.

Schmidt-Koenig, K. (1960) Internal clocks and homing. *Cold Spring Harbor Symp. Quant. Biol.* **25**, 389–393

Schmidt-Koenig, K. (1961) Die Sonne als Kompass im Heim-Orientierungs- system der Brieftauben. *Z. Tierpsychol.* **68**, 221–244.

Schmidt-Koenig, K. (1979) *Avian Orientation and Navigation.* Academic Press, London.

Schmidt-Nielsen, K. (1964) *Desert Animals. Physiological Problems of Heat and Water.* Clarendon Press, Oxford.

Schneider, D. (1969) Insect olfaction: deciphering system for chemical messages. *Science* **163**, 1031–1036.

Schneidermann, N., Fuentes, I. and Gormezano, I (1962) Acquisition and extinction of the classically conditioned eyelid response in the albino rabbit. *Science* **136**, 650–652.

Schneirla, T.C. (1965) Aspects of stimulation and organization in approach/ withdrawal processes underlying vertebrate behavioural development. In Lehrman, D.S., Hinde, R.A. and Shaw, E. (eds) *Advances in the Study of Behavior,* Vol. 1: pp. 1–74. Academic Press, New York.

Schöne, H. (1984) *Spatial Orientation.* Princeton University Press, Princeton, NJ.

Schutz, F. (1965) Sexuelle Pragung bei Anatiden. *Z. Tierpsychol.* **22**, 50–103.

Schutz, F. (1971) Pragung des Sexualverhaltens von Enten und Gansen durch Sozialeindrucke wahrend der Jugendphase. *J. Neuro-visc. Rel.* Suppl. **10**, 339–357.

Schüz, E. (1963) On the northwestern migration divide of the white stork. *Proc. Int. Ornithol. Congr.* **13**, 475–480.

Schüz, E. (1971) *Grundriss der Vogelzugskunde.* Parey Verlag, Berlin and Hamburg.

Schwartz, E. (1974) Lateral-line mechanoreceptors in fishes and amphibians. In A. Fessard (ed.) *Handbook of Sensory Physiology.* Vol. 3, pp. 257–278. Springer-Verlag, New York.

Scott, E.M. and Verney, E.L. (1947) Self-selection of diet. VI. The nature of appetites for B vitamins. *J. Nutr.* **34**, 471–480.

Scott, J.P. and Fuller, J.L. (1965) *Dog Behavior. The Genetic Basis.* University of Chicago Press, Chicago, IL.

Seitz, A. (1940) Die Paarbildung bei einigen Cichliden I. *Z. Tierpsychol.* **4**, 40–84.

Seligman, M.E.P. (1970) On the generality of the laws of learning. *Psychol. Rev.* **77**, 406–418.

Selye, H. (1973) The evolution of the stress concept. *American Scientist* **61**, 692–699.

Senturia, J.B. and Johansson, B.W. (1974) Physiological and biochemical reflections of circannual rhythmicity in the European hedgehog and Man. In Pengelley E.T. (ed.) *Circannual Clocks.* Academic Press, New York.

Sevenster, P. (1968) Motivation and learning in sticklebacks. In Ingle, D. (ed.) *The*

Central Nervous System and Fish Behaviour, pp. 233–245. University of Chicago Press, Chicago, IL.

Sevenster, P. (1973) Incompatibility of response and reward. In Hinde, R.A. and Stevenson-Hinde, J. (eds) *Constraints on Learning. Limitations and Predispositions*. Academic Press, London.

Seyfarth, R.M. (1983) Grooming and social competition in primates. In Hinde, R. (ed.) *Primate Social Relationships: An Integrated Approach*. Blackwell, Oxford.

Seyfarth, R.M. and Cheney, D.L. (1984) Grooming, alliances and reciprocal altruism in vervet monkeys. *Nature* **308**, 541–543.

Seyfarth, R.M., Cheney, D.L. and Marler, P. (1980) Monkey responses to three different alarm calls: evidence of predator classification and semantic communication. *Science* **210**, 801–803.

Shannon, C.E. (1948) Mathematical theory of communication. *Bell Syst. Tech. J.* **27**, 379–423, 623–656.

Sheffield, F.D. (1965) Relation between classical conditioning and instrumental learning. In Prokasy, W.F. (ed.) *Classical Conditioning: A Symposium*. Appleton-Century-Crofts, New York.

Shepard, R.N. and Metzler, J. (1971) Mental rotation of three-dimensional objects. *Science* **171**, 701–703.

Shepher, J. (1971) Self-imposed incest-avoidance and exogamy in second generation Kibbutz adults. Unpublished doctoral dissertation. Rutgers University, New Brunswick, N.J.

Shepher, J. (1983) *Incest – A Biosocial View*. Academic Press, New York/London.

Sheppard, P.M. (1961) Some contributions to population genetics resulting from the study of Lepidoptera. *Adv. Genet.* **10**, 165–216.

Sherman, P.M. (1981) Kinship, demography and Belding's ground squirrel nepolism. *Behav. Ecol. Sociobiol.* **8**, 251–259.

Sherrington, C.S. (1918) Observations on the sensual role of the proprioceptive nerve-supply of the extrinsic ocular muscles. *Brain* **41**, 332–343.

Sherry, D.F., Mrosovsky, N. and Hogan, J.A. (1980) Weight loss and anorexia during incubation in birds. *J. Comp. Physiol. Psychol.* **94**, 89–98.

Shettleworth, S.J. (1972) Constraints on learning. *Adv. Study Behav.* **4**, 1–68.

Shettleworth, S. (1984) Learning and behavioural ecology. In Krebs, J.R. and Davies, N.B. (eds) *Behavioural Ecology*, 2nd edn. Blackwell, pp. 170–194.

Shumake, S.A., Smith, J.C. and Tucker, D. (1969) Olfactory intensity-difference thresholds in the pigeon. *J. Comp. Physiol. Psychol.* **67**, 64–69.

Sibly, R.M. (1983) Optimal group size is unstable. *Anim. Behav.* **31**, 947–948.

Sibly, R.M. and McFarland, D.J. (1974) A state-space approach to motivation. In McFarland, D.J. (ed.) *Motivational Control Systems Analysis*, pp. 213–250. Academic Press, London.

Siegel, R.G. and Honig, W.K. (1970) Pigeon concept formation: successive and simultaneous acquisition. *J. Exp. Anal. Behav.* **13**, 385–390.

Silk, J.B. (1982) Altruism among female *Macaca radiata*: explanations and analysis of patterns of grooming and coalition formation. *Behaviour* **79**, 162–188.

Silk, J.B. (1987) Social behavior in evolutionary perspective. In Smuts, B.B., Cheney, D.L., Seyfarth, R.M., Wrangham, R.W. and Struhsaker, T.T. (ed.) *Primate Societies*. University of Chicago Press, Chicago.

Simmons, J.A. (1971) 'Echolocation in bats: signal processing of echoes for target range. *Science* **171**, 925–928.

Sinclair, A.R.E. (1977) *The African Buffalo*. University of Chicago Press, Chicago, IL.

Singh, D. (1993) Adaptive significance of female physical attractiveness; role of waist-to-hip ratio. *J. Personality and Social Psychol.* **65**, 293–307.

Singh, D. (1994) Waist-to-hip ratio and judgement of attractiveness and healthiness of female figures by male and female physicians. *Int. J. Obesity* **18**, 731–737.

Skinner, B.F. (1937) Two types of conditioned reflex: a reply to Konorski and Miller. *J. Gen. Pyschol.***16**, 272–279.

Skinner, B.F. (1938) *The Behaviour of Organisms*. Appleton-Century-Crofts, New York.

Skinner, B.F. (1953) *Science and Human Behaviour*. Macmillan, New York.

Skinner, B.F. (1975) *About Behaviourism*. Cape, London.

Skutch, A.F. (1969) Life histories of Central American birds. Vol. III Golden naped Woodpecker. *Pacific Coast Avifauna*, **35**, 479–517.

Skutch, A.F. (1976) *Parent Birds and their Young*. University of Texas Press.

Smart, J.L. (1977) Early life malnutrition and late learning ability. A critical analysis. In Oliveria, A. (ed) *Genetics and Intelligence*, pp. 215–235. North Holland, Amsterdam.

Smith, R.L. (ed.) (1984) *Sperm Competition and the Evolution of Animal Mating Systems*. Academic Press, Orlando, FL.

Sodian, B., Taylor, C., Harris, P.L. and Perner, J. (1991) Early deception and the child's theory of mind: false trails and genuine markers. *Child Development*, **62**, 468–483.

Sparks, J. (1982) *The discovery of Animal Behaviour*. Collins Sons & Co Ltd, London.

Spier, E.H. (1997) From reactive behaviour to adaptive behaviour. Unpublished D.Phil. thesis. Oxford University.

Spier, E. and McFarland, D. (1998) SAB 5 paper.

Steels, L. (1994) A case study in the behavior-oriented design of autonomous agents. In Cliff, D., Husbands, P., Meyer J. and Wilson S. (eds) *From Animals to Animats 3*. MIT Press, Cambridge, MA.

Stephens, D.W. and Krebs, J.R. (1986) *Foraging Theory*. Princeton University Press, Princeton, NJ.

Stern, P. (1973) *Principles of Human Genetics*. W.H. Freeman, New York.

Suchman, L. (1987) *Plans and Situated Actions*. Cambridge University Press, New York.

Sutherland, N.S. and Mackintosh, N.J. (1971) *Mechanisms of Animal Discrimination Learning*. Academic Press, New York.

Sutton, D. (1979) Mechanisms underlying vocal control in nonhuman primates. In Steklis, H.D. and Raleigh, M.J. (eds) *Neurobiology of Social Communication*, pp. 45–67. Academic Press, New York.

Sutton, R.S. (1991) Reinforcement learning architectures for animats. In Meyer, J. and Wilson, S. (eds) *From Animals to Animats*. MIT Press, Cambridge, MA.

Sutton, R. and Barto, A.G. (1981) Toward a modern theory of adaptive networks: expectation and prediction. *Psychol. Rev.* **88**, 135–170.

Sweeney, B.M. (1969) *Rhythmic Phenomena in Plants*. Academic Press, New York.

Swenson, R.M. and Vogel, W.H. (1983) Plasma catecholamine and corticosterone as well as brain catecholamine changes during coping in rats exposed to stressful footshock. *Pharmacol. Biochem. and Behav.* **18**, 689–693.

Swets, J.A., Tanner, W.P. and Birdsall, T.G. (1961) Decision processes in perception. *Psychol. Rev.* **68**, 301–340.

Tansley, K. (1965) *Vision in Vertebrates*. Methuen, London.

Terrace, H.S. (1979) *Nim*. Eyre Methuen, London.

Tesauro, G. (1986) Simple neural models of classical conditioning. *Biol. Cybernet.* **55**, 187–200.

Testa, T.J. (1975) Effects of similarity of location and temporal intensity pattern of conditioned and unconditioned stimuli on acquisition of conditioned suppression in rats. *J. Exp. Psychol. Anim. Behav. Processes* **1**, 114–121.

Thoday, J.M. (1953) Components of fitness. *Symp. Soc. Exp. Biol.* **7**, 96–113.

Thompson, D.B.A. and Barnard, C.J. (1984) Prey selection by plovers: optimal foraging in mixed-species groups. *Anim. Behav.* **32**, 534–563.

Thompson, R.F. (1965) *Foundations of Physiological Psychology.* Harper International, New York.

Thorndike, E.L. (1898) Animal intelligence: an experimental study of the associative processes in animals. *Psychol. Rev. Monogr.* Suppl. 2, 8, 1, 16.

Thorndike, E.L. (1911) *Animal Intelligence.* Macmillan, New York.

Thorndike, E.L. (1913) *The Psychology of Learning.* (Educational psychology II.) Teachers College, New York.

Thorndike, E.L. (1932) Reward and punishment in animal learning. *Comp. Psychol. Monogr.* **8**, (39) 26, 27, 47.

Thornhill, R. (1980) Rape in Panorpa scorpion flies and a general rape hypothesis. *Anim. Behav.* **28**, 52–59.

Thorpe, W.H. (1956) *Learning and Instinct in Animals.* Methuen, London.

Thorpe, W.H. (1963) Ethology and the coding problem in germ cell and brain. *Z. Tierpsychol.* **20**, 529–551.

Thorpe, W.H. (1974) *Animal Nature and Human Nature.* Methuen, London.

Thorpe, W.H. (1979) *The Origins and Rise of Ethology.* Heinemann, London.

Tinbergen, L. (1960) The natural control of insects in pinewoods. I. Factors influencing the intensity of predation by song birds. *Arch. Neerl. Zool.* **13**, 265–343.

Tinbergen, N. (1940) Die Ubersprungbewegung. *Z. Tierpsychol.* **4**, 1–10.

Tinbergen, N. (1942) An objective study of the innate behaviour of animals. *Biblioth. Biother.* **1**, 39–98.

Tinbergen, N. (1949) De functie van de rode vlek op de snavel van de zilvermeeuw. *Bijdragen tat de Dierkunde* **28**, 453–465.

Tinbergen, N. (1950) The hierarchical organization of nervous mechanisms underlying instinctive behaviour. *Symp. Soc. Exp. Biol.* **IV**, 305–312

Tinbergen, N. (1951) *The Study of Instinct.* Oxford University Press, Oxford.

Tinbergen, N. (1952) Derived activities: their causation, biological significance, origin and emancipation during evolution. *Quart. Rev. Biol.* **27**, 1–32.

Tinbergen, N. (1953) *The Herring Gull's World.* Collins, London.

Tinbergen, N. (1959) Comparative studies of the behaviour of gulls (Laridae): a progress report. *Behaviour* **15**, 1–70.

Tinbergen, N. (1962) The evolution of animal communication – a critical examination of methods. *Symp. Zool. Soc. Lond.* **8**, 1–6.

Tinbergen, N., Broekhuysen, G.J., Feekes, F., Houghton, J.C.W., Kruuk, H. and Szulc, E. (1962) Egg shell removal by the black-headed gull, *Larus ridibundus* L.: a behaviour component of camouflage. *Behaviour* **19**, 74–118.

Tinbergen, N. and Perdeck, A.C. (1950) On the stimulus situation releasing the begging response in the newly hatched herring gull chick (*Larus a. argentatus* Pont.). *Behaviour* **3**, 1–38.

Tinkle, D.W. (1969) The concept of reproductive effort in its relation to the evolution of life histories of lizards. *Amer. Natur.* **103**, 501–516.

Toates, F.M. (1980) *Animal Behaviour – A Systems Approach.* Wiley, Chichester.

Toates, F. (1995) *Stress.* Wiley, Chichester.

Tolman, E.C. (1932) *Purposive Behavior in Animals and Men.* Appleton-Century, New York. (Reprinted University of California Press, 1949).

Tolman, E.C. (1938) The determiners of behavior at a choice point. *Psychol. Rev.* **45**, 1–41.

Trabasso, T. and Riley, C.A. (1975) On the construction and use of representations involving linear order. In Solso, R.L. (ed.) *Information Processing and Cognition: The Loyola Symposium.* Erlbaum, Hillsdale, NJ.

Treisman, M. (1977) Motion sickness: an evolutionary hypothesis. *Science* **197**, 493–495.

Trivers, R.L. (1971) The evolution of recipocal altruism. *Quart. Rev. Biol.* **46**, 35–57.

Trivers, R.L. (1972) Parental investment and sexual selection. In Campbell, B. (ed.) *Sexual Selection and the Descent of Man.* Aldine, Chicago, IL.

Trivers, R. (1974) Parent-offspring conflict. *Amer. Zool.* **14**, 249–264.

Trivers, R. (1985) *Social Evolution.* Benjamin-Cummings, Menlo Park, CA.

Tryon, R.C. (1942) Individual differences. In Moss, F.A. *Comparative Psychology*, 2nd edn. Prentice-Hall, Englewood Cliffs, NJ.

Uexkull, J. von. (1934) *Streifzuge durch die Umwelten von Tieren und Menschen.* Springer, Berlin. Translated in Schiller, C.H. (ed.) *Instinctive Behaviour.* Methuen, London.

Ullyott, P. (1936) The behaviour of *Dendrocoelum lacteum*: I and II. *J. Exp. Biol.* **13**, 253–264, 265–278.

Ulrich, R.E. and Azrin, N.H. (1962) Reflexive fighting in response to aversive stimulation. *J. Exp. Anal. Behav.* **5**, 511–520.

Van Shaik, C.P., van Noordwijk, M.A., Wasone, M.A. and Sitriono, E. (1983) Party size and early detection in Sumatran forest primates. *Primates* **24**, 211–221.

Van Shaik, C.P. and van Noordwijk, M.A. (1985) The evolutionary effect of the absence of felids on the social organisation of the Simeulue monkey (*Macaca fascicularis fusca*, Miller 1903). *Int. J. Primatol.* **6**, 180–200.

Vehrencamp, S.L. (1983) A model for the evolution of despotic versus egalitarian societies. *Anim. Behav.* **31**, 667–682.

Verschure, P.F.M.J., Krose, B.J.A. and Pfeifer, R. (1991) Distributed adaptive control: the self organisation of structured behavior. *Robotics and Autonomous Systems.*

Vidal, J.M. (1976) L'Empreint chez les animaux. *La Recherche* **63**, 24–35.

Vince, M.A. (1969) Embryonic communiction, respiration and the synchronization of hatching. In Hinde, R. *Bird* (ed.) *Vocalizations*, pp. 233–60. Cambridge University Press, Cambridge.

Waal, F. de (1982) *Chimpanzee Politics.* Jonathan Cape, London.

Wahlsten, D.L. and Cole, M. (1972). Classical and avoidance training of leg flexion in the dog. In Black A.H. and Prokasy W.F. (eds) *Classical Conditioning II: Current Research and Theory.* New York: Appleton-Century-Crofts. pp. 378–408.

Walcott, B. and Walcott, C. (1982) A search for magnetic field receptors in animals. In Papi, F. and Wallraff, H.G. (eds) *Avian Navigation*, pp. 338–343. Springer-Verlag, Berlin.

Walcott, C. (1977) Magnetic fields and the orientation of homing pigeons under sun. *J. Exp. Biol.* **70**, 105–123.

Walcott, C. and Green, R.C. (1974) Orientation of homing pigeons altered by a change in the direction of an applied magnetic field. *Science* **184**, 180–182.

Wallraff, H.D. (1966) Uber die Heimfindeleistungen von Brieftauben nach Haltung in verschiedenartig abgeschirmten Volieren. *Z. Vgl. Physiol.* **52**, 215–259.

Wallraff, H.D. (1970) Weitere Volierenversuche mit Brieftauben: Wahrscheinlicher Einfluss dynamischer Faktoren der Atmosphare auf die Orientierung. *Z. Vgl. Physiol.* **68**, 182–201.

Wallraff, H.D. (1978) Proposed principles of magnetic field perception in birds. *Oikos* **30**, 188–194.

Wallraff, H.D. (1984) Migration and navigation in birds: a present-state survey with some digressions to related fish behaviour. In McCleave *et al.* (eds) *Mechanisms of Migration in Fishes.* Plenum, New York.

Wallraff, H.D., Papi, F., Ioale, P. and Benvenuti, S. (1986) Magnetic fields affect pigeon navigation only while the birds can smell atmospheric odors. *Naturwiss.* **73**, 215.

Wallraff, H.G. (1969) Uber das Orientierungvermogen von Vogeln unter naturlichen und kunstlichen sternmustern. Dressurversuche mit Stockenten. *Ver. Deut. Zool. Ges. Innsbruck* 348–357.

Wallraff, H.G. (1988) Determinants of homing-flight courses in pigeons. *Acta XIX Congressus Internationalis Ornithologici* 330–334.

Wallraff, H.G. (1989) Simulated navigation based on unreliable sources of information (models on pigeon homing. Part 1) and Simulated navigation based on assumed gradients of atmospheric trace gases (Models on pigeon homing. Part 2). *J. Theor. Biol.* **137**, 1–19; **138**, 511–528.

Wallraff, H.G. (1990) Navigation by homing pigeons. *Ethol. Ecol. and Evol.* **2**, 81–115.

Wallraff, H.G. and Foa, A. (1981) Pigeon navigation: charcoal filter removes relevant information from environmental air. *Behav. Ecol. Sociobiol.* **9**, 67–77.

Walter, W.G. (1950) An imitation of life. *Scientific American*, May, 42–45.

Warden, C.J. and Warner L.H. (1928) The sensory capacities and intelligence of dogs, with a report on the ability of the noted dog 'Fellow' to respond to verbal stimuli. *Quart. Rev. Biol.* **3**, 1–28.

Ware, D.M. (1972) Predation by rainbow trout (*Salmo gairdneri*): the influence of hunger, prey density, and prey size. *J. Fisheries Res. Board Canada* **29**, 1193–1201.

Warren, J.M. (1965) Primate learning in comparative perspective. In Schrier, A.M., Harlow, H.F. and Stollnitz, F. (eds) *Behavior of Nonhuman Primates,* 1: 249–281. Academic Press, New York.

Warren, J.M. (1973) Learning in vertebrates. In Dewsbury, D.A. and Rethlingshafer, D.A. (eds) *Comparative Psychology: a Modern Survey*, pp. 471–509. McGraw-Hill, New York.

Warren, J.M. (1974) Possibly unique characteristics of learning by primates. *J. hum. Evol.* **3**, 445–454.

Warriner, C.C., Lemmon, W.B. and Ray, T.S. (1963) Early experience as a variable in mate selection. *Anim. Behav.* **11**, 221–224.

Wasserman, E.A., Franklin, S. and Hearst, E. (1974) Pavlovian appetitive contingencies and approach vs. withdrawal to conditioned stimuli in pigeons. *J. Comp. and Physiol. Pyschol.* **86**, 616–627.

Watson, J.B. (1907) Kinesthetic and organic sensations: their role in the reactions of the white rat to the maze. *Psychol. Monogr.* **8**, 33–49.

Watson, J.B. (1913) Psychology as the behaviorist views it. *Psychol. Rev.* **20**, 158–177.

Watson, J.B. (1914) *Behavior. An Introduction to Comparative Psychology.* Holt, New York.

Watson, J.B. (1916) The place of the conditioned reflex in psychology. *Psychol. Rev.* **23**, 89–116.

Watson, J.B. (1930) *Behaviorism.* Norton, New York.

Weihaupt, J.G. (1964) Geophysical biology. *Bioscience* **14**, 18–24.

Weiskrantz, L. (1980) Varieties of residual experience. *Quart. J. Exp. Psychol.* **32**, 365–386.

Weiskrantz, L., Warrington, E.K., Sanders, M.D. and Marshall, J. (1974) Visual capacity of the hemianopic field following a restricted occipital ablation. *Brain* **97**, 709–728.

Wells, M. (1968) *Lower Animals.* Weidenfeld and Nicholson, London.

Wehner, R. (1982) Himmelsnavigation bei Insecten. Neurophysiologie und Verhalten. *Vierteljahrsschr. Naturforsch. Ges. Zurich* **5**, 1–132.

Wehner, R. (1994) The polarization-vision project: championing organismic biology. In Schildberger, K. and Elsner, N. (eds) *Neural Basis of Behavioural Adaptation.* Fischer, Stuttgart, New York. pp. 103–143.

Wehner, R. (1997) Sensory systems and behaviour. In Krebs, J.R. and Davies, N.B. (eds) *Behavioural Ecology*, 4th edn. Blackwell Scientific Publications, Oxford.

Wehner, R., Michel, B. and Antonsen, P. (1996) Visual navigation in insects: coupling of egocentric and geocentric information. *J. Exp. Biol.* **199**, 129–140.

Wendler, G. (1966) Coordination of walking movements in arthropods. *Symp. Soc. Exp. Biol.* **20**, 229–249.

Wenner, A.M. (1962) Sound production during the waggle dance of the honey bee. *Anim. Behav.* **10**, 79–95.

Wenner, A.M. (1964) Sound communication in honey bees. *Scientific American* **210** (4), 116–124.

Werner, E.E. and Hall, D.J. (1974) Optimal foraging and the size selection of prey by the Bluegill Sunfish (*Lepomis macrochirus*). *Ecology* **55**, 1216–1232.

West, G.C. and Norton, D.W. (1975) Metabolic adaptations in tundra birds. In Vernberg, F.J. (ed.) *Physiological Adaptations to the Environment*. Intext Educational Publ., New York.

Westermark, E. (1891) *The History of Human Marriage*. Macmillan, London.

Whiten, A. (1972) Operant study of sun altitude and pigeon navigation. *Nature* **237**, 405–406.

Whiten, A. and Byrne, R.W. (1986) The St Andrews catalogue of tactical deception in primates. *St Andrews Psychol. Rep.* **10**.

Whitfield, M. (1976) The evolution of the oceans and the atmosphere. In Bligh, J., Cloudsley-Thompson J.L. and Macdonald A.G. (eds) *Environmental Physiology of Animals*. Blackwell Scientific Publications, Oxford.

Whitten, P.L. (1983) Diet and dominance among female vervet monkeys (*Cercopithecus aethiops*). *Am. J. Primatol.* **5**, 139–159.

Wickler, W. (1967) Socio-sexual signals and their intra-specific imitation among primates. In Morris, D. (ed.) *Primate Ethology*. Weidenfeld and Nicholson, London.

Wickler, W. (1968) *Mimicry in Plants and Animals*. Weidenfeld and Nicholson, London.

Wiepkema, P.R. (1971) Positive feedbacks at work during feeding. *Behaviour* **39**, 2–4.

Wiepkema, P.R., Alingh Prins, A.J. and Steffens, A.B. (1972) Gastrointestinal food transport in relation to meal occurrence in rats 1. *Physiology and Behaviour* **9**, 759–763.

Wiessner, P. (1977) Hxaro: a regional system of reciprocity for reducing risk among the !Kung San. Ph.D. Dissertation. University of Michigan, Ann Arbor, University Microfilms.

Wilcoxon, H.C., Dragoin, W.B. and Kral, P.A. (1971) Illness-induced aversions in rat and quail: relative salience of visual and gustatory cues. *Science* **171**, 826–828.

Wiley, H. (1983) The evolution of communication: information and manipulation. In Halliday, T.R. and Slater, P.J.B. (eds) *Animal Behaviour. 2. Communication*. Blackwell Scientific Publications, Oxford.

Wiley, R.H. and Richards, D.G. (1978) Physical constraints on acoustic communication in the atmosphere: implications for the evolution of animal vocalisation. *Behav. Ecol. Sociobiol.* **3**, 69–74.

Wilkinson, G.S. (1984) Reciprocal food sharing in the vampire bat. *Nature, London* **308**, 181–184.

Wilkinson, P.F. and Shank, C.C. (1977) Rutting-fight among musk oxen on Banks Island, Northwest Territories, Canada. *Anim. Behav.* **24**, 756–758.

Williams, D.R. and Williams, H. (1969) Auto-maintenance in the pigeon: sustained pecking despite contingent non-reinforcement. *J. Exp. Anal. Behav.* **12**, 511–520.

Williams, G.C. (1966) *Adaptation and Natural Selection*. Princeton University Press, Princeton, NJ.

Wilson, D.M. (1968) Inherent asymmetry and reflex modulation of the locust flight pattern. *J. Exp. Biol.* **48**, 631–641.

Wilson, E.O. (1975) *Sociobiology. The New Synthesis*. The Belknap Press of Harvard University Press, Cambridge, MA.

Wilson, J.A. (1979) *Principles of Animal Physiology*. 2nd edn. Macmillan, New York.

Wiltschko, R. and Wiltschko, W. (1980) The process of learning sun compass orientation in young homing pigeons. *Naturwiss* **67**, 512–513.

Wiltschko, W. and Wiltschko, R. (1975) The interaction of stars and magnetic field in the orientation system of night migrating birds. *Z. Tierpsychol.* **37**, 337–355; **39**, 265–282.

Wiltschko, W. and Wiltschko, R. (1976) Interrelation of magnetic compass and star orientation in night migrating birds. *J. Comp. Physiol.* **109**A, 91–99.

Wirtz, P. and Wawra, M. (1986) Vigilance and group size in *Homo sapiens. Ethology* **71**, 283–286.

Wolf, A.P. (1966) Childhood association, sexual attraction and the incest taboo: a Chinese case. *Am. Anthropol.* **68**, 883–898.

Wolf, A.P. (1970) Childhood association and sexual attraction. A further test of the Westermarck hypothesis. *Am. Anthropol.* **72**, 503–515.

Woodard, W.T. and Bitterman, M.E. (1973) Pavlovian analysis of avoidance conditioning in the goldfish (*Carassius auratus*). *J. Comp. Physiol.* **82**, 123–129.

Woodhouse, H.C. (1987) Inter- and intragroup aggression illustrated in the rock paintings of South Africa. *S. Afr. J. Ethnol.* **10**, 42–48.

Woodruff, G. and Premack, D. (1979) Intentional communication in the chimpanzee: the development of deception. *Cognition* **7**, 333–362.

Woodworth, R.S. (1918) *Dynamic Psychology*. Columbia University Press, New York.

Wootton, R.J. (1976) *The Biology of the Sticklebacks*. Academic Press, London.

Wrangham, R.W. (1987) Evolution of social structure. In Smuts, B.B., Cheney, D.L., Seyfarth, R.M., Wrangham, R.W. and Struhsaker, T.T. (eds) *Primate Societies*. University of Chicago Press, Chicago, IL.

Wright, P. and McFarland, D.J. (1969) A functional analysis of hypothalamic polydipsia in the Barbary dove (*Streptopelia risoria*). *Physiol. Behav.* **4**, 877–883.

Wright, S. (1921) Systems of mating. *Genetics* **6**, 111–178.

Wurtman, R.J., Axelrod, J. and Kelly, D.E. (1968) *The Pineal*. Academic Press, New York.

Yodlowski, M.L., Kreithen, M.L. and Keeton, W.T. (1977) Detection of atmospheric infrasound by homing pigeons. *Nature* **265**, 725–726.

Young, J.Z. (1960) The failures of discrimination learning following removal of the vertical lobes in Octopus. *Proc. Roy. Soc. B.* **153**, 18–46.

Young, P.T. (1961) *Motivation and Emotion. A Survey of the Determinants of Human and Animal Activity*. Wiley, New York.

Zach, R. (1979) Shell dropping: decision making and optimal foraging in northwestern crows. *Behaviour* **68**, 106–117.

Zahavi, A. (1971) The social behaviour of the white wagtail *Motacilla alba alba* wintering in Israel. *Ibis* **113**, 203–211.

Zahavi, A. (1974) Communal nesting by the Arabian babbler: a case of individual selection. *Ibis* **116**, 84–87.

Zahavi, A. (1975) Mate selection – a selection for a handicap. *J. Theor. Biol.* **53**, 205–214.

Zahavi, A. (1976) Cooperative nesting in Eurasian birds. *Proc. XVI Int. Orn. Congr.* (Canberra, Australia) pp. 685–693.

Zahorik, D.M. and Maier, S.F. (1969) Appetite conditioning with recovery from thiamine deficiency as the unconditioned stimulus. *Psychon. Sci.* **17**, 309–310.

Zahorik, D.M., Maier, S.F. and Pies R.W. (1974) Preferences for tastes paired with recovery from thiamine deficiency in rats. Appetitive conditioning or learned safety? *J. Comp. Physiol. Psychol.* **87**, 1083–1091.

Zentall, T.R. and Hogan, D.E. (1978) Same/different concept learning in the pigeon: the effect of negative instances and prior adaptation to transfer stimuli. *J. Exp. Anal. Behav.* **30**, 177–186.

Zimmerman, J.L. (1971) The territory and its density dependent effect in *Spiza americana. Auk* **88**, 591–612.

Index